CONSCIOUSNESS EXPLAINED

DANIEL C. DENNETT

Illustrated by Paul Weiner

LITTLE, BROWN AND COMPANY

BOSTON NEW YORK TORONTO LONDON

First Paperback Edition

Permissions to use copyrighted material appear on page 492.

Library of Congress Cataloging-in-Publication Data

Dennett, Daniel Clement.
Consciousness explained / Daniel C. Dennett. — 1st ed.
p. cm.
Includes bibliographical references and index.
ISBN 0-316-18065-3 (hc)
ISBN 0-316-18066-1 (pb)
1. Consciousness. 2. Mind and body. I. Title.
B105.C477D45 1991
126 — dc20 91-15614

10 9 8 7 6 5 4
RRD VA

Published simultaneously in Canada by
Little, Brown & Company (Canada) Limited

Printed in the United States of America

For Nick, Marcel, and Ray

ALSO BY DANIEL C. DENNETT

Content and Consciousness

Brainstorms

The Mind's I *(with Douglas R. Hofstadter)*

Elbow Room

The Intentional Stance

CONTENTS

Preface xi

1 Prelude: How Are Hallucinations Possible? 3

1. The Brain in the Vat
2. Pranksters in the Brain
3. A Party Game Called Psychoanalysis
4. Preview

Part I PROBLEMS AND METHODS

2 Explaining Consciousness 21

1. Pandora's Box: Should Consciousness Be Demystified?
2. The Mystery of Consciousness
3. The Attractions of Mind Stuff
4. Why Dualism Is Forlorn
5. The Challenge

3 A Visit to the Phenomenological Garden 43

1. Welcome to the Phenom
2. Our Experience of the External World
3. Our Experience of the Internal World
4. Affect

4 A Method for Phenomenology 66

1. First Person Plural
2. The Third-Person Perspective
3. The Method of Heterophenomenology
4. Fictional Worlds and Heterophenomenological Worlds
5. The Discreet Charm of the Anthropologist
6. Discovering What Someone Is Really Talking About
7. Shakey's Mental Images
8. The Neutrality of Heterophenomenology

Part II AN EMPIRICAL THEORY OF THE MIND

5 Multiple Drafts Versus the Cartesian Theater 101

1. The Point of View of the Observer
2. Introducing the Multiple Drafts Model
3. Orwellian and Stalinesque Revisions
4. The Theater of Consciousness Revisited
5. The Multiple Drafts Model in Action

6 Time and Experience 139

1. Fleeting Moments and Hopping Rabbits
2. How the Brain Represents Time
3. Libet's Case of "Backwards Referral in Time"
4. Libet's Claim of Subjective Delay of Consciousness of Intention
5. A Treat: Grey Walter's Precognitive Carousel
6. Loose Ends

7 The Evolution of Consciousness 171

1. Inside the Black Box of Consciousness
2. Early Days
Scene One: The Birth of Boundaries and Reasons
Scene Two: New and Better Ways of Producing Future
3. Evolution in Brains, and the Baldwin Effect
4. Plasticity in the Human Brain: Setting the Stage
5. The Invention of Good and Bad Habits of Autostimulation
6. The Third Evolutionary Process: Memes and Cultural Evolution
7. The Memes of Consciousness: The Virtual Machine to Be Installed

8 How Words Do Things with Us 227

1. Review: E Pluribus Unum?
2. Bureaucracy versus Pandemonium
3. When Words Want to Get Themselves Said

9 The Architecture of the Human Mind 253

1. Where Are We?
2. Orienting Ourselves with the Thumbnail Sketch
3. And Then What Happens?
4. The Powers of the Joycean Machine
5. But Is This a Theory of Consciousness?

Part III THE PHILOSOPHICAL PROBLEMS OF CONSCIOUSNESS

10 Show and Tell 285

1. Rotating Images in the Mind's Eye
2. Words, Pictures, and Thoughts
3. Reporting and Expressing
4. Zombies, Zimboes, and the User Illusion
5. Problems with Folk Psychology

11 Dismantling the Witness Protection Program 321

1. Review
2. Blindsight: Partial Zombiehood?
3. Hide the Thimble: An Exercise in Consciousness-Raising
4. Prosthetic Vision: What, Aside from Information, Is Still Missing?
5. "Filling In" versus Finding Out
6. Neglect as a Pathological Loss of Epistemic Appetite
7. Virtual Presence
8. Seeing Is Believing: A Dialogue with Otto

12 Qualia Disqualified 369

1. A New Kite String
2. Why Are There Colors?
3. Enjoying Our Experiences
4. A Philosophical Fantasy: Inverted Qualia
5. "Epiphenomenal" Qualia?
6. Getting Back on My Rocker

13 The Reality of Selves 412

1. How Human Beings Spin a Self
2. How Many Selves to a Customer?
3. The Unbearable Lightness of Being

14 Consciousness Imagined 431

1. Imagining a Conscious Robot
2. What It Is Like to Be a Bat
3. Minding and Mattering
4. Consciousness Explained, or Explained Away?

Appendix A (for Philosophers) 457
Appendix B (for Scientists) 464
Bibliography 469
Index 493

PREFACE

My first year in college, I read Descartes's *Meditations* and was hooked on the mind-body problem. Now here was a mystery. How on earth could my thoughts and feelings fit in the same world with the nerve cells and molecules that made up my brain? Now, after thirty years of thinking, talking, and writing about this mystery, I think I've made some progress. I think I can sketch an outline of the solution, a theory of consciousness that gives answers (or shows how to find the answers) to the questions that have been just as baffling to philosophers and scientists as to laypeople. I've had a lot of help. It's been my good fortune to be taught, informally, indefatigably, and imperturbably, by some wonderful thinkers, whom you will meet in these pages. For the story I have to tell is not one of solitary cogitation but of an odyssey through many fields, and the solutions to the puzzles are inextricably woven into a fabric of dialogue and disagreement, where we often learn more from bold mistakes than from cautious equivocation. I'm sure there are still plenty of mistakes in the theory I will offer here, and I hope they are bold ones, for then they will provoke better answers by others.

The ideas in this book have been hammered into shape over many years, but the writing was begun in January 1990 and finished just a year later, thanks to the generosity of several fine institutions and the help of many friends, students, and colleagues. The Zentrum für Interdisziplinäre Forschung in Bielefeld, CREA at the École Polytechnique in Paris, and the Rockefeller Foundation's Villa Serbelloni in Bellagio provided ideal conditions for writing and conferring during the first five months. My home university, Tufts, has supported my work through the Center for Cognitive Studies, and enabled me to present the penultimate draft in the fall of 1990 in a seminar that drew on the faculties and students of Tufts and the other fine schools in the greater Boston area. I also want to thank the Kapor Foundation and the Harkness Foundation for supporting our research at the Center for Cognitive Studies.

Several years ago, Nicholas Humphrey came to work with me at the Center for Cognitive Studies, and he, Ray Jackendoff, Marcel Kinsbourne, and I began meeting regularly to discuss various aspects and problems of consciousness. It would be hard to find four more different approaches to the mind, but our discussions were so fruitful, and so encouraging, that I dedicate this book to these fine friends, with thanks for all they have taught me. Two other longtime colleagues and friends have also played major roles in shaping my thinking, for which I am eternally grateful: Kathleen Akins and Bo Dahlbom.

I also want to thank the ZIF group in Bielefeld, particularly Peter Bieri, Jaegwon Kim, David Rosenthal, Jay Rosenberg, Eckart Scheerer, Bob van Gulick, Hans Flohr, and Lex van der Heiden; the CREA group in Paris, particularly Daniel Andler, Pierre Jacob, Francisco Varela, Dan Sperber, and Deirdre Wilson; and the "princes of consciousness" who joined Nick, Marcel, Ray, and me at the Villa Serbelloni for an intensely productive week in March: Edoardo Bisiach, Bill Calvin, Tony Marcel, and Aaron Sloman. Thanks also to Edoardo and the other participants of the workshop on neglect, in Parma in June. Pim Levelt, Odmar Neumann, Marvin Minsky, Oliver Selfridge, and Nils Nilsson also provided valuable advice on various chapters. I also want to express my gratitude to Nils for providing the photograph of Shakey, and to Paul Bach-y-Rita for his photographs and advice on prosthetic vision devices.

I am grateful for a bounty of constructive criticism to all the participants in the seminar last fall, a class I will never forget: David Hilbert, Krista Lawlor, David Joslin, Cynthia Schossberger, Luc Faucher, Steve Weinstein, Oakes Spalding, Mini Jaikumar, Leah Steinberg, Jane Anderson, Jim Beattie, Evan Thompson, Turhan Canli, Michael Anthony, Martina Roepke, Beth Sangree, Ned Block, Jeff McConnell, Bjorn Ramberg, Phil Holcomb, Steve White, Owen Flanagan, and Andrew Woodfield. Week after week, this gang held my feet to the fire, in the most constructive way. During the final redrafting, Kathleen Akins, Bo Dahlbom, Doug Hofstadter, and Sue Stafford provided many invaluable suggestions. Paul Weiner turned my crude sketches into the excellent figures and diagrams.

Kathryn Wynes and later Anne Van Voorhis have done an extraordinary job of keeping me, and the Center, from flying apart during the last few hectic years, and without their efficiency and foresight this book would still be years from completion. Last and most important: love and thanks to Susan, Peter, Andrea, Marvin, and Brandon, my family.

Tufts University
January 1991

CONSCIOUSNESS EXPLAINED

1

PRELUDE: HOW ARE HALLUCINATIONS POSSIBLE?

1. THE BRAIN IN THE VAT

Suppose evil scientists removed your brain from your body while you slept, and set it up in a life-support system in a vat. Suppose they then set out to trick you into believing that you were not just a brain in a vat, but still up and about, engaging in a normally embodied round of activities in the real world. This old saw, the brain in the vat, is a favorite thought experiment in the toolkit of many philosophers. It is a modern-day version of Descartes's (1641)[1] evil demon, an imagined illusionist bent on tricking Descartes about absolutely everything, including his own existence. But as Descartes observed, even an infinitely powerful evil demon couldn't trick him into thinking he himself existed if he didn't exist: *cogito ergo sum*, "I think, therefore I am." Philosophers today are less concerned with proving one's own existence as a thinking thing (perhaps because they have decided that Descartes settled that matter quite satisfactorily) and more concerned about what, in principle, we may conclude from our experience about our nature, and about the nature of the world in which we (apparently) live. *Might* you be nothing but a brain in a vat? Might you have *always* been just a brain in a vat? If so, could you even conceive of your predicament (let alone confirm it)?

The idea of the brain in the vat is a vivid way of exploring these questions, but I want to put the old saw to another use. I want to use

1. Dates in parentheses refer to works listed in the Bibliography.

it to uncover some curious facts about hallucinations, which in turn will lead us to the beginnings of a theory — an empirical, scientifically respectable theory — of human consciousness. In the standard thought experiment, it is obvious that the scientists would have their hands full providing the nerve stumps from all your senses with just the right stimulations to carry off the trickery, but philosophers have assumed for the sake of argument that however technically difficult the task might be, it is "possible in principle." One should be leery of these possibilities in principle. It is also possible in principle to build a stainless-steel ladder to the moon, and to write out, in alphabetical order, all intelligible English conversations consisting of less than a thousand words. But neither of these are remotely possible in fact and sometimes an *impossibility in fact* is theoretically more interesting than a *possibility in principle*, as we shall see.

Let's take a moment to consider, then, just how daunting the task facing the evil scientists would be. We can imagine them building up to the hard tasks from some easy beginnings. They begin with a conveniently comatose brain, kept alive but lacking all input from the optic nerves, the auditory nerves, the somatosensory nerves, and all the other afferent, or input, paths to the brain. It is sometimes assumed that such a "deafferented" brain would naturally stay in a comatose state forever, needing no morphine to keep it dormant, but there is some empirical evidence to suggest that spontaneous waking might still occur in these dire circumstances. I think we can suppose that were you to awake in such a state, you would find yourself in horrible straits: blind, deaf, completely numb, with no sense of your body's orientation.

Not wanting to horrify you, then, the scientists arrange to wake you up by piping stereo music (suitably encoded as nerve impulses) into your auditory nerves. They also arrange for the signals that would normally come from your vestibular system or inner ear to indicate that you are lying on your back, but otherwise paralyzed, numb, blind. This much should be within the limits of technical virtuosity in the near future — perhaps possible even today. They might then go on to stimulate the tracts that used to innervate your epidermis, providing it with the input that would normally have been produced by a gentle, even warmth over the ventral (belly) surface of your body, and (getting fancier) they might stimulate the dorsal (back) epidermal nerves in a way that simulated the tingly texture of grains of sand pressing into your back. "Great!" you say to yourself: "Here I am, lying on my back on the beach, paralyzed and blind, listening to rather nice music, but probably in danger of sunburn. How did I get here, and how can I call for help?"

But now suppose the scientists, having accomplished all this, tackle the more difficult problem of convincing you that you are not a mere beach potato, but an agent capable of engaging in some form of activity in the world. Starting with little steps, they decide to lift part of the "paralysis" of your phantom body and let you wiggle your right index finger in the sand. They permit the sensory experience of moving your finger to occur, which is accomplished by giving you the kinesthetic feedback associated with the relevant volitional or motor signals in the output or efferent part of your nervous system, but they must also arrange to remove the numbness from your phantom finger, and provide the stimulation for the feeling that the motion of the imaginary sand around your finger would provoke.

Suddenly, they are faced with a problem that will quickly get out of hand, for just how the sand will feel depends on just how you decide to move your finger. The problem of calculating the proper feedback, generating or composing it, and then presenting it to you in real time is going to be computationally intractable on even the fastest computer, and if the evil scientists decide to solve the real-time problem by pre-calculating and "canning" all the possible responses for playback, they will just trade one insoluble problem for another: there are too many possibilities to store. In short, our evil scientists will be swamped by *combinatorial explosion* as soon as they give you any genuine exploratory powers in this imaginary world.[2]

It is a familiar wall these scientists have hit; we see its shadow in the boring stereotypes in every video game. The alternatives open

---

2. The term *combinatorial explosion* comes from computer science, but the phenomenon was recognized long before computers, for instance in the fable of the emperor who agrees to reward the peasant who saved his life one grain of rice on the first square of the checkerboard, two grains on the second, four on the third, and so forth, doubling the amount for each of the sixty-four squares. He ends up owing the wily peasant millions of billions of grains of rice (2^{64}-1 to be exact). Closer to our example is the plight of the French "aleatoric" novelists who set out to write novels in which, after reading chapter 1, the reader flips a coin and then reads chapter 2a or 2b, depending on the outcome, and then reads chapter 3aa, 3ab, 3ba, or 3bb after that, and so on, flipping a coin at the end of every chapter. These novelists soon came to realize that they had better minimize the number of choice points if they wanted to avoid an explosion of fiction that would prevent anyone from carrying the whole "book" home from the bookstore.

for action have to be strictly — and unrealistically — limited to keep the task of the world-representers within feasible bounds. If the scientists can do no better than convince you that you are doomed to a lifetime of playing Donkey Kong, they are evil scientists indeed.

There is a solution of sorts to this technical problem. It is the solution used, for instance, to ease the computational burden in highly realistic flight simulators: use *replicas* of the items in the simulated world. Use a real cockpit and push and pull it with hydraulic lifters, instead of trying to simulate all that input to the seat of the pants of the pilot in training. In short, there is only one way for you to store for ready access that much information about an imaginary world to be explored, and that is to use a *real* (if tiny or artificial or plaster-of-paris) world to store its own information! This is "cheating" if you're the evil demon claiming to have deceived Descartes about the existence of absolutely everything, but it's a way of actually getting the job done with less than infinite resources.

Descartes was wise to endow his imagined evil demon with *infinite* powers of trickery. Although the task is not, strictly speaking, infinite, the amount of information obtainable in short order by an inquisitive human being is staggeringly large. Engineers measure information flow in bits per second, or speak of the *bandwidth* of the channels through which the information flows. Television requires a greater bandwidth than radio, and high-definition television has a still greater bandwidth. High-definition smello-feelo television would have a still greater bandwidth, and *interactive* smello-feelo television would have an astronomical bandwidth, because it constantly branches into thousands of slightly different trajectories through the (imaginary) world. Throw a skeptic a dubious coin, and in a second or two of hefting, scratching, ringing, tasting, and just plain looking at how the sun glints on its surface, the skeptic will consume more bits of information than a Cray supercomputer can organize in a year. Making a *real* but counterfeit coin is child's play; making a *simulated* coin out of nothing but organized nerve stimulations is beyond human technology now and probably forever.[3]

---

3. The development of "Virtual Reality" systems for recreation and research is currently undergoing a boom. The state of the art is impressive: electronically rigged gloves that provide a convincing interface for "manipulating" virtual objects, and head-mounted visual displays that permit you to explore virtual environments of considerable complexity. The limitations of these systems are apparent, however, and they bear out

One conclusion we can draw from this is that we are not brains in vats — in case you were worried. Another conclusion it seems that we can draw from this is that strong hallucinations are simply impossible! By a strong hallucination I mean a hallucination of an apparently concrete and persisting three-dimensional object in the real world — as contrasted to flashes, geometric distortions, auras, afterimages, fleeting phantom-limb experiences, and other anomalous sensations. A strong hallucination would be, say, a ghost that talked back, that permitted you to touch it, that resisted with a sense of solidity, that cast a shadow, that was visible from any angle so that you might walk around it and see what its back looked like.

Hallucinations can be roughly ranked in strength by the number of such features they have. Reports of *very* strong hallucinations are rare, and we can now see why it is no coincidence that the credibility of such reports seems, intuitively, to be inversely proportional to the strength of the hallucination reported. We are — and should be — particularly skeptical of reports of very strong hallucinations because we don't believe in ghosts, and we think that only a real ghost could produce a strong hallucination. (It was primarily the telltale strength of the hallucinations reported by Carlos Castañeda in *The Teachings of Don Juan: A Yaqui Way of Knowledge* [1968] that first suggested to scientists that the book, in spite of having been a successful Ph.D. thesis in anthropology at UCLA, was fiction, not fact.)

But if *really* strong hallucinations are not known to occur, there can be no doubt that convincing, multimodal hallucinations are frequently experienced. The hallucinations that are well attested in the literature of clinical psychology are often detailed fantasies far beyond the generative capacities of current technology. How on earth can a single brain do what teams of scientists and computer animators would find to be almost impossible? If such experiences are not genuine or veridical perceptions of some real thing "outside" the mind, they must be produced entirely inside the mind (or the brain), concocted out of whole cloth but lifelike enough to fool the very mind that concocts them.

---

my point: it is only by various combinations of physical replicas and schematization (a *relatively* coarse-grained representation) that robust illusions can be sustained. And even at their best, they are experiences of virtual surreality, not something that you might mistake for the real thing for more than a moment. If you really want to fool someone into thinking he is in a cage with a gorilla, enlisting the help of an actor in a gorilla suit is going to be your best bet for a long time.

2. PRANKSTERS IN THE BRAIN

The standard way of thinking of this is to suppose that hallucinations occur when there is some sort of freakish autostimulation of the brain, in particular, an entirely internally generated stimulation of some parts or levels of the brain's perceptual systems. Descartes, in the seventeenth century, saw this prospect quite clearly, in his discussion of phantom limb, the startling but quite normal hallucination in which amputees seem to feel not just the presence of the amputated part, but itches and tingles and pains in it. (It often happens that new amputees, after surgery, simply cannot believe that a leg or foot has been amputated until they *see* that it is gone, so vivid and realistic are their sensations of its continued presence.) Descartes's analogy was the bell-pull. Before there were electric bells, intercoms, and walkie-talkies, great houses were equipped with marvelous systems of wires and pulleys that permitted one to call for a servant from any room in the house. A sharp tug on the velvet sash dangling from a hole in the wall pulled a wire that ran over pulleys all the way to the pantry, where it jangled one of a number of labeled bells, informing the butler that service was required in the master bedroom or the parlor or the billiards room. The systems worked well, but were tailor-made for pranks. Tugging on the parlor wire anywhere along its length would send the butler scurrying to the parlor, under the heartfelt misapprehension that someone had called him from there — a modest little hallucination of sorts. Similarly, Descartes thought, since perceptions are caused by various complicated chains of events in the nervous system that lead eventually to the control center of the conscious mind, if one could intervene somewhere along the chain (anywhere on the optic nerve, for instance, between the eyeball and consciousness), tugging just right on the nerves would produce exactly the chain of events that would be caused by a normal, veridical perception of something, and this would produce, at the receiving end in the mind, exactly the effect of such a conscious perception.

The brain — or some part of it — inadvertently played a mechanical trick on the mind. That was Descartes's explanation of phantom-limb hallucinations. Phantom-limb hallucinations, while remarkably vivid, are — by our terminology — relatively weak; they consist of unorganized pains and itches, all in one sensory modality. Amputees don't see or hear or (so far as I know) smell their phantom feet. So something like Descartes's account *could* be the right way to explain phantom limbs, setting aside for the time being the notorious mysteries about how the physical brain could interact with the nonphysical conscious mind. But we can see that even the purely mechanical part of Descartes's story must be wrong as an account of relatively strong hallucinations; there is no way the brain as illusionist could store and manipulate enough false information to fool an inquiring mind. The brain can relax, and let the real world provide a surfeit of *true* information, but if it starts trying to short-circuit its own nerves (or pull its own wires, as Descartes would have said), the results will be only the weakest of fleeting hallucinations. (Similarly, the malfunctioning of your neighbor's electric hairdryer might cause "snow" or "static," or hums and buzzes, or odd flashes to appear on your television set, but if you see a bogus version of the evening news, you *know* it had an elaborately organized cause far beyond the talents of a hairdryer.)

It is tempting to suppose that perhaps we have been too gullible about hallucinations; perhaps only mild, fleeting, thin hallucinations ever occur — the strong ones don't occur because they can't occur! A cursory review of the literature on hallucinations certainly does suggest that there is something of an inverse relation between strength and frequency — as well as between strength and credibility. But that review also provides a clue leading to another theory of the mechanism of hallucination-production: one of the endemic features of hallucination reports is that the victim will comment on his or her rather unusual passivity in the face of the hallucination. Hallucinators usually just stand and marvel. Typically, they feel no desire to probe, challenge, or query, and take no steps to interact with the apparitions. It is likely, for the reasons we have just explored, that this passivity is not an inessential feature of hallucination but a necessary precondition for any moderately detailed and sustained hallucination to occur.

Passivity, however, is only a special case of a way in which relatively strong hallucinations could survive. The reason these hallucinations can survive is that the illusionist — meaning by that, whatever it is that produces the hallucination — can "count on" a particular line of exploration by the victim — in the case of total passivity, the *null* line of exploration. So long as the illusionist can predict in detail the line of exploration actually to be taken, it only has to prepare for the illusion to be sustained "in the directions that the victim will look." Cinema set designers insist on knowing the location of the camera in advance — or if it is not going to be stationary, its exact trajectory and angle — for then they have to prepare only enough material to cover the perspectives actually taken. (Not for nothing does *cinéma verité* make extensive use of the freely roaming hand-held camera.) In real life the same principle was used by Potemkin to economize on the show villages to be reviewed by Catherine the Great; her itinerary had to be ironclad.

So one solution to the problem of strong hallucination is to suppose that there is a link between the victim and illusionist that makes it possible for the illusionist to build the illusion *dependent on*, and hence capable of anticipating, the exploratory intentions and decisions of the victim. Where the illusionist is unable to "read the victim's mind" in order to obtain this information, it is still sometimes possible in real life for an illusionist (a stage magician, for instance) to *entrain* a particular line of inquiry through subtle but powerful "psychological forcing." Thus a card magician has many standard ways of giving the victim the illusion that he is exercising his free choice in what cards on the table he examines, when in fact there is only one card that may be turned over. To revert to our earlier thought experiment, if the evil scientists can *force* the brain in the vat to have a particular set of exploratory intentions, they can solve the combinatorial explosion problem by preparing only the anticipated material; the system will be only *apparently* interactive. Similarly, Descartes's evil demon can sustain the illusion with less than infinite power if he can sustain an illusion of free will in the victim, whose investigation of the imaginary world he minutely controls.[4]

But there is an even more economical (and realistic) way in which hallucinations could be produced in a brain, a way that harnesses the very freewheeling curiosity of the victim. We can understand how it works by analogy with a party game.

3. A PARTY GAME CALLED PSYCHOANALYSIS

In this game one person, the dupe, is told that while he is out of the room, one member of the assembled party will be called upon to relate a recent dream. This will give everybody else in the room the story line of that dream so that when the dupe returns to the room and begins questioning the assembled party, the dreamer's identity will be hidden in the crowd of responders. The dupe's job is to ask yes/no questions of the assembled group until he has figured out the dream narrative to a suitable degree of detail, at which point the dupe is to

4. For a more detailed discussion of the issues of free will, control, mindreading, and anticipation, see my *Elbow Room: The Varieties of Free Will Worth Wanting*, 1984, especially chapters 3 and 4.

psychoanalyze the dreamer, and use the analysis to identify him or her.

Once the dupe is out of the room, the host explains to the rest of the party that no one is to relate a dream, that the party is to answer the dupe's questions according to the following simple rule: if the last letter of the last word of the question is in the first half of the alphabet, the questions is to be answered in the affirmative, and all other questions are to be answered in the negative, with one proviso: a noncontradiction override rule to the effect that later questions are not to be given answers that contradict earlier answers. For example:

> Q: Is the dream about a gir*l*?
> A: Yes.

but if later our forgetful dupe asks

> Q: Are there any female characters in i*t*?
> A: Yes [in spite of the final *t*, applying the noncontradiction override rule].[5]

When the dupe returns to the room and begins questioning, he gets a more or less random, or at any rate arbitrary, series of yeses and noes in response. The results are often entertaining. Sometimes the process terminates swiftly in absurdity, as one can see at a glance by supposing the initial question asked were "Is the story line of the dream word-for-word identical to the story line of *War and Peace*?" or, alternatively, "Are there any animate beings in it?" A more usual outcome is for a bizarre and often obscene story of ludicrous misadventure to unfold, to the amusement of all. When the dupe eventually decides that the dreamer — whoever he or she is — must be a very sick and troubled individual, the assembled party gleefully retorts that the dupe himself is the author of the "dream." This is not strictly true, of course. In one sense, the dupe is the author by virtue of the questions he was inspired to ask. (No one *else* proposed putting the three gorillas in the rowboat with the nun.) But in another sense, the dream simply has no author, and that is the whole point. Here we see a process of narrative production, of detail accumulation, with no authorial intentions or plans at all — an illusion with no illusionist.

The structure of this party game bears a striking resemblance to the structure of a family of well-regarded models of perceptual systems.

---

5. Empirical testing suggests that the game is more likely to produce a good story if in fact you favor affirmative answers slightly, by making p/q the alphabetic dividing line between yes and no.

It is widely held that human vision, for instance, cannot be explained as an *entirely* "data-driven" or "bottom-up" process, but needs, at the highest levels, to be supplemented by a few "expectation-driven" rounds of hypothesis testing (or something analogous to hypothesis testing). Another member of the family is the "analysis-by-synthesis" model of perception that also supposes that perceptions are built up in a process that weaves back and forth between centrally generated expectations, on the one hand, and confirmations (and disconfirmations) arising from the periphery on the other hand (e.g., Neisser, 1967). The general idea of these theories is that after a certain amount of "preprocessing" has occurred in the early or peripheral layers of the perceptual system, the tasks of perception are completed — objects are identified, recognized, categorized — by generate-and-test cycles. In such a cycle, one's current expectations and interests shape hypotheses for one's perceptual systems to confirm or disconfirm, and a rapid sequence of such hypothesis generations and confirmations produces the ultimate product, the ongoing, updated "model" of the world of the perceiver. Such accounts of perception are motivated by a variety of considerations, both biological and epistemological, and while I wouldn't say that any such model has been proven, experiments inspired by the approach have borne up well. Some theorists have been so bold as to claim that perception *must* have this fundamental structure.

Whatever the ultimate verdict turns out to be on generate-and-test theories of perception, we can see that they support a simple and powerful account of hallucination. All we need suppose must happen for an otherwise normal perceptual system to be thrown into a hallucinatory mode is for the hypothesis-generation side of the cycle (the expectation-driven side) to operate normally, while the data-driven side of the cycle (the confirmation side) goes into a disordered or random or arbitrary round of confirmation and disconfirmation, just as in the party game. In other words, if noise in the data channel is arbitrarily amplified into "confirmations" and "disconfirmations" (the arbitrary yes and no answers in the party game), the current expectations, concerns, obsessions, and worries of the victim will lead to framing questions or hypotheses whose content is guaranteed to reflect those interests, and so a "story" will unfold in the perceptual system without an author. We don't have to suppose the story is written in advance; we don't have to suppose that information is stored or composed in the illusionist part of the brain. All we suppose is that the illusionist goes into an arbitrary confirmation mode and the victim provides the content by asking the questions.

This provides in the most direct possible way a link between the emotional state of the hallucinator and the content of the hallucinations produced. Hallucinations are usually related in their content to the current concerns of the hallucinator, and this model of hallucination provides for that feature without the intervention of an implausibly knowledgeable internal storyteller who has a theory or model of the victim's psychology. Why, for instance, does the hunter on the last day of deer season *see* a deer, complete with antlers and white tail, while looking at a black cow or another hunter in an orange jacket? Because his internal questioner is obsessively asking: "Is it a deer?" and getting NO for an answer until finally a bit of noise in the system gets mistakenly amplified into a YES, with catastrophic results.

A number of findings fit nicely with this picture of hallucination. For instance, it is well known that hallucinations are the normal result of prolonged sensory deprivation (see, e.g., Vosberg, Fraser, and Guehl, 1960). A plausible explanation of this is that in sensory deprivation, the data-driven side of the hypothesis-generation-and-test system, lacking any data, lowers its threshold for noise, which then gets amplified into arbitrary patterns of confirmation and disconfirmation signals, producing, eventually, detailed hallucinations whose content is the product of nothing more than anxious expectation and chance confirmation. Moreover, in most reports, hallucinations are only gradually elaborated (under conditions of either sensory deprivation or drugs). They start out weak — e.g., geometric — and then become stronger ("objective" or "narrative"), and this is just what this model would predict (see, e.g., Siegel and West, 1975).

Finally, the mere fact that a drug, by diffusion in the nervous system, can produce such elaborate and contentful effects requires explanation — the drug *itself* surely can't "contain the story," even if some credulous people like to think so. It is implausible that a drug, by diffuse activity, could create or even turn on an elaborate illusionist system, while it is easy to see how a drug could act directly to raise or lower or disorder in some arbitrary way a confirmation threshold in a hypothesis-generation system.

The model of hallucination generation inspired by the party game could also explain the composition of dreams, of course. Ever since Freud there has been little doubt that the thematic content of dreams is tellingly symptomatic of the deepest drives, anxieties, and preoccupations of the dreamer, but the clues the dreams provide are notoriously well concealed under layers of symbolism and misdirection. What kind of process could produce stories that speak so effectively and incessantly to a dreamer's deepest concerns, while clothing the whole business in layers of metaphor and displacement? The more or less standard answer of the Freudian has been the extravagant hypothesis of an internal dream playwright composing therapeutic dream-plays for the benefit of the ego and cunningly sneaking them past an internal censor by disguising their true meaning. (We might call the Freudian model the Hamlet model, for it is reminiscent of Hamlet's devious ploy of staging "The Mousetrap" just for Claudius; it takes a clever devil indeed to dream up such a subtle stratagem, but if Freud is to be believed, we all harbor such narrative virtuosi.) As we shall see later on, theories that posit such *homunculi* ("little men" in the brain) are not always to be shunned, but whenever homunculi are rung in to help, they had better be relatively stupid functionaries — not like the brilliant Freudian playwrights who are supposed to produce new dream-scenes every night for each of us! The model we are considering eliminates the playwright altogether, and counts on the "audience" (analogous to the one who is "it" in the party game) to provide the content. The audience is no dummy, of course, but at least it doesn't have to have a theory of its own anxieties; it just has to be driven by them to ask questions.

It is interesting to note, by the way, that one feature of the party game that would not be necessary for a process producing dreams or hallucinations is the noncontradiction override rule. Since one's perceptual systems are presumably always exploring an ongoing situation (rather than a *fait accompli*, a finished dream narrative already told) subsequent "contradictory" confirmations can be interpreted by the machinery as indicating a new change in the world, rather than a revision in the story known by the dream relaters. The ghost was blue when last I looked, but has now suddenly turned green; its hands have turned into claws, and so forth. The volatility of metamorphosis of objects in dreams and hallucinations is one of the most striking features of those narratives, and what is even more striking is how seldom these noticed metamorphoses "bother" us while we are dreaming. So the farmhouse in Vermont is now suddenly revealed to be a bank in Puerto Rico, and the horse I was riding is now a car, no a speedboat, and my companion began the ride as my grandmother but has become the Pope. These things happen.

This volatility is just what we would expect from an active but insufficiently skeptical question-asker confronted by a random sample of yeses and noes. At the same time, the persistence of some themes and objects in dreams, their refusal to metamorphose or disappear, can also be tidily explained by our model. Pretending, for the moment, that the brain uses the alphabet rule and conducts its processing in English, we can imagine how subterranean questioning goes to create an obsessive dream:

Q. Is it about father?
A. No.
Q. Is it about a telephone?
A. Yes.
Q. Okay. Is it about mother?
A. No.
Q. Is it about father?
A. No.
Q. Is it about father on the telephone?
A. Yes.
Q. I *knew* it was about father! Now, was he talking to me?
A. Yes. . . .

This little theory sketch could hardly be said to prove anything (yet) about hallucinations or dreams. It does show — metaphorically — how a mechanistic explanation of these phenomena *might* go, and that's an important prelude, since some people are tempted by the defeatist thesis that science couldn't "in principle" explain the various "mysteries" of the mind. The sketch so far, however, does not even address the problem of our *consciousness* of dreams and hallucinations. Moreover, although we have exorcised one unlikely homunculus, the clever illusionist/playwright who plays pranks on the mind, we have left in his place not only the stupid question-answerers (who arguably can be "replaced by machines") but also the still quite clever and unexplained question-poser, the "audience." If we have eliminated a villain, we haven't even begun to give an account of the victim.

We have made some progress, however. We have seen how attention to the "engineering" requirements of a mental phenomenon can raise new, and more readily answerable, questions, such as: What models of hallucination can avoid combinatorial explosion? How might the content of experience be elaborated by (relatively) stupid, uncomprehending processes? What sort of links between processes or systems could explain the results of their interaction? If we are to compose a scientific theory of consciousness, we will have to address many questions of this sort.

We have also introduced a central idea in what is to follow. The key element in our various explanations of how hallucinations and dreams are possible at all was the theme that the only work that the brain must do is whatever it takes to *assuage epistemic hunger* — to satisfy "curiosity" in all its forms. If the "victim" is passive or incurious about topic x, if the victim doesn't seek answers to any questions about topic x, then no material about topic x needs to be prepared. (Where it doesn't itch, don't scratch.) The world provides an inexhaustible deluge of information bombarding our senses, and when we concentrate on how much is coming in, or continuously available, we often succumb to the illusion that it all must be used, all the time. But our capacities to use information, and our epistemic appetites, are limited. If our brains can just satisfy all our particular epistemic hungers as they arise, we will never find grounds for complaint. We will never be able to tell, in fact, that our brains are provisioning us with less than everything that is available in the world.

So far, this thrifty principle has only been introduced, not established. As we shall see, the brain doesn't always avail itself of this option in any case, but it's important not to overlook the possibility. The power of this principle to dissolve ancient conundrums has not been generally recognized.

4. PREVIEW

In the chapters that follow, I will attempt to explain consciousness. More precisely, I will explain the various phenomena that compose what we call consciousness, showing how they are all physical effects of the brain's activities, how these activities evolved, and how they give rise to illusions about their own powers and properties. It is very hard to imagine how your mind could be your brain — but not impossible. In order to imagine this, you really have to know quite a lot of what science has discovered about how brains work, but much more important, you have to learn new ways of thinking. Adding facts helps you imagine new possibilities, but the discoveries and theories of neuroscience are not enough — even neuroscientists are often baffled by consciousness. In order to stretch your imagination, I will provide, along with the relevant scientific facts, a series of stories, analogies, thought experiments, and other devices designed to give you new perspectives, break old habits of thought, and help you organize the facts into a single, coherent vision strikingly different from the traditional view of consciousness we tend to trust. The thought experiment about the brain in the vat and the analogy with the game of psychoanalysis are warm-up exercises for the main task, which is to sketch a theory of the biological mechanisms *and a way of thinking* about these mechanisms that will let you *see* how the traditional paradoxes and mysteries of consciousness can be resolved.

In Part I, we survey the problems of consciousness and establish some methods. This is more important and difficult than one might think. Many of the problems encountered by other theories are the result of getting off on the wrong foot, trying to guess the answers to the Big Questions too early. The novel background assumptions of my theory play a large role in what follows, permitting us to *postpone* many of the traditional philosophical puzzles over which other theorists stumble, until after we have outlined an empirically based theory, which is presented in Part II.

The Multiple Drafts model of consciousness outlined in Part II is an alternative to the traditional model, which I call the Cartesian Theater. It requires a quite radical rethinking of the familiar idea of "the stream of consciousness," and is initially deeply counterintuitive, but it grows on you, as you see how it handles facts about the brain that have been ignored up to now by philosophers — and scientists. By considering in some detail how consciousness could have evolved, we gain insights into otherwise baffling features of our minds. Part II also provides an analysis of the role of language in human consciousness, and the relation of the Multiple Drafts model to some more familiar conceptions of the mind, and to other theoretical work in the multidisciplinary field of cognitive science. All along the way we have to resist the alluring simplicities of the traditional view, until we can secure ourselves on the new foundation.

In Part III, armed with the new ways of guiding our imaginations, we can confront (at last) the traditional mysteries of consciousness: the strange properties of the "phenomenal field," the nature of introspection, the qualities (or *qualia*) of experiential states, the nature of the self or ego and its relation to thoughts and sensations, the consciousness of nonhuman creatures. The paradoxes that beset traditional philosophical debates about these can then be seen to arise from *failures of imagination*, not "insight," and we will be able to dissolve the mysteries.

This book presents a theory that is both empirical and philosophical, and since the demands on such a theory are so varied, there are two appendices that deal briefly with more technical challenges arising both from the scientific and philosophical perspectives. In the next chapter, we turn to the question of what an explanation of consciousness would be, and whether we should want to dissolve the mysteries of consciousness at all.

PART ONE

PROBLEMS AND METHODS

2

EXPLAINING CONSCIOUSNESS

1. PANDORA'S BOX: SHOULD CONSCIOUSNESS BE DEMYSTIFIED?

And here are trees and I know their gnarled surface, water, and I feel its taste. These scents of grass and stars at night, certain evenings when the heart relaxes — how shall I negate this world whose power and strength I feel? Yet all the knowledge on earth will give me nothing to assure me that this world is mine. You describe it to me and you teach me to classify it. You enumerate its laws and in my thirst for knowledge I admit that they are true. You take apart its mechanism and my hope increases. . . . What need had I of so many efforts? The soft lines of these hills and the hand of evening on this troubled heart teach me much more.

ALBERT CAMUS, *The Myth of Sisyphus,* 1942

Sweet is the lore which Nature brings;
  Our meddling intellect
Misshapes the beauteous forms of things: —
  We murder to dissect.

WILLIAM WORDSWORTH, "The Tables Turned," 1798

Human consciousness is just about the last surviving mystery. A mystery is a phenomenon that people don't know how to think about — yet. There have been other great mysteries: the mystery of the origin of the universe, the mystery of life and reproduction, the mystery of the design to be found in nature, the mysteries of time, space, and gravity. These were not just areas of scientific ignorance, but of utter bafflement and wonder. We do not yet have the final answers to any of the questions of cosmology and particle physics, molecular genetics and evolutionary theory, but we do know how to think about them. The mysteries haven't vanished, but they have been tamed. They no longer overwhelm our efforts to think about the phenomena, because now we know how to tell the misbegotten questions from the right questions, and even if we turn out to be dead wrong about some of the currently accepted answers, we know how to go about looking for better answers.

With consciousness, however, we are still in a terrible muddle. Consciousness stands alone today as a topic that often leaves even the most sophisticated thinkers tongue-tied and confused. And, as with all the earlier mysteries, there are many who insist — and hope — that there will never be a demystification of consciousness.

Mysteries are exciting, after all, part of what makes life fun. No one appreciates the spoilsport who reveals whodunit to the moviegoers waiting in line. Once the cat is out of the bag, you can never regain the state of delicious mystification that once enthralled you. So let the reader beware. If I succeed in my attempt to explain consciousness, those who read on will trade mystery for the rudiments of scientific knowledge of consciousness, not a fair trade for some tastes. Since some people view demystification as desecration, I expect them to view this book at the outset as an act of intellectual vandalism, an assault on the last sanctuary of humankind. I would like to change their minds.

Camus suggests he has no need of science, since he can learn more from the soft lines of the hills and the hand of evening, and I would not challenge his claim — given the questions Camus is asking himself. Science does not answer all good questions. Neither does philosophy. But for that very reason the phenomena of consciousness, which are puzzling in their own right quite independently of Camus's concerns, do not need to be protected from science — or from the sort of demystifying philosophical investigation we are embarking on. Sometimes people, fearing that science will "murder to dissect" as Wordsworth put it, are attracted to philosophical doctrines that offer one guarantee or another against such an invasion. The misgivings that motivate them are well founded, whatever the strengths and weaknesses of the doctrines; it indeed *could* happen that the demystification of consciousness would be a great loss. I will claim only that in fact this *will not* happen: the losses, if any, are overridden by the gains in understanding — both scientific and social, both theoretical and moral — that a good theory of consciousness can provide.

How, though, *might* the demystification of consciousness be something to regret? It might be like the loss of childhood innocence, which is definitely a loss, even if it is well recompensed. Consider what happens to love, for instance, when we become more sophisticated. We can understand how a knight in the age of chivalry could want to sacrifice his life for the honor of a princess he had never so much as spoken to — this was an especially thrilling idea to me when I was about eleven or twelve — but it is not a state of mind into which an adult today can readily enter. People used to talk and think about love in ways that are now practically unavailable — except to children, and to those who can somehow suppress their adult knowledge. We all love to tell those we love that we love them, and to hear from them that we are loved — but as grownups we are not quite as sure we know what this means as we once were, when we were children and love was a simple thing.

Are we better or worse off for this shift in perspective? The shift is not uniform, of course. While naïve adults continue to raise gothic romances to the top of the best-seller list, we sophisticated readers find we have rendered ourselves quite immune to the intended effects of such books: they make us giggle, not cry. Or if they do make us cry — as sometimes they do, in spite of ourselves — we are embarrassed to discover that we are still susceptible to such cheap tricks; for we cannot readily share the mind-set of the heroine who wastes away worrying about whether she has found "true love" — as if this were some sort of distinct substance (emotional gold as opposed to emotional brass or copper). This growing up is not just in the individual. Our culture has become more sophisticated — or at least sophistication, whatever it is worth, is more widely spread through the culture. As a result, our concepts of love have changed, and with these changes come shifts in sensibility that now prevent us from having certain experiences that thrilled, devastated, or energized our ancestors.

Something similar is happening to consciousness. Today we talk about our conscious decisions and unconscious habits, about the conscious experiences we enjoy (in contrast to, say, automatic cash machines, which have no such experiences) — but we are no longer quite sure we know what we mean when we say these things. While there are still thinkers who gamely hold out for consciousness being some one genuine precious thing (like love, like gold), a thing that is just "obvious" and very, very special, the suspicion is growing that this is an illusion. Perhaps the various phenomena that conspire to create the sense of a single mysterious phenomenon have no more ultimate or essential unity than the various phenomena that contribute to the sense that love is a simple thing.

Compare love and consciousness with two rather different phenomena, diseases and earthquakes. Our concepts of diseases and earthquakes have also undergone substantial revision over the last few hundred years, but diseases and earthquakes are phenomena that are very largely (but not entirely) independent of our concepts of them. Changing our minds about diseases did not in itself make diseases disappear or become less frequent, although it did result in changes in medicine and public health that radically altered the occurrence patterns of diseases. Earthquakes may someday similarly come under some measure of human control, or at least prediction, but by and large the existence of earthquakes is unaffected by our attitudes toward them or concepts of them. With love it is otherwise. It is no longer possible for sophisticated people to "fall in love" in some of the ways that once were possible — simply because they cannot believe in *those* ways of falling in love. It is no longer possible for me, for instance, to have a pure teenaged crush — unless I "revert to adolescence" and in the process forget or abandon much of what I think I know. Fortunately, there are other kinds of love for me to believe in, but what if there weren't? Love is one of those phenomena that *depend on their concepts*, to put it oversimply for the time being. There are others; money is a clear instance. If everyone forgot what money was, there wouldn't be any money anymore; there would be stacks of engraved paper slips, embossed metal disks, computerized records of account balances, granite and marble bank buildings — but no money: no inflation or deflation or exchange rates or interest — or *monetary value*. The very property of those variously engraved slips of paper that explains — as nothing else could — their trajectories from hand to hand in the wake of various deeds and exchanges would evaporate.

On the view of consciousness I will develop in this book, it turns out that consciousness, like love and money, is a phenomenon that does indeed depend to a surprising extent on its associated concepts. Although, like love, it has an elaborate biological base, like money, some of its most significant features are borne along on the culture, not simply inherent, somehow, in the physical structure of its instances. So if I am right, and if I succeed in overthrowing some of those concepts, I will threaten with extinction whatever phenomena of consciousness depend on them. Are we about to enter the postconscious period of human conceptualization? Is this not something to fear? Is it even conceivable?

If the concept of consciousness were to "fall to science," what would happen to our sense of moral agency and free will? If conscious experience were "reduced" somehow to mere matter in motion, what would happen to our appreciation of love and pain and dreams and joy? If conscious human beings were "just" animated material objects, how could anything we do to them be right or wrong? These are among the fears that fuel the resistance and distract the concentration of those who are confronted with attempts to explain consciousness.

I am confident that these fears are misguided, but they are not obviously misguided. They raise the stakes in the confrontation of theory and argument that is about to begin. There are powerful arguments, quite independent of the fears, arrayed against the sort of scientific, materialistic theory I will propose, and I acknowledge that it falls to me to demonstrate not only that these arguments are mistaken, but also that the widespread acceptance of my vision of consciousness would not have these dire consequences in any case. (And if I had discovered that it would likely have these effects — what would I have done then? I wouldn't have written *this* book, but beyond that, I just don't know.)

Looking on the bright side, let us remind ourselves of what has happened in the wake of earlier demystifications. We find no diminution of wonder; on the contrary, we find deeper beauties and more dazzling visions of the complexity of the universe than the protectors of mystery ever conceived. The "magic" of earlier visions was, for the most part, a cover-up for frank failures of imagination, a boring dodge enshrined in the concept of a *deus ex machina*. Fiery gods driving golden chariots across the skies are simpleminded comic-book fare compared to the ravishing strangeness of contemporary cosmology, and the recursive intricacies of the reproductive machinery of DNA make *élan vital* about as interesting as Superman's dread kryptonite. When we understand consciousness — when there is no more mystery — consciousness will be different, but there will still be beauty, and more room than ever for awe.

2. THE MYSTERY OF CONSCIOUSNESS

What, then, is the mystery? What could be more obvious or certain to each of us than that he or she is a conscious subject of experience, an enjoyer of perceptions and sensations, a sufferer of pain, an entertainer of ideas, and a conscious deliberator? That seems undeniable, but what in the world can consciousness itself be? How can living physical bodies in the physical world produce such phenomena? That is the mystery.

The mystery of consciousness has many ways of introducing itself, and it struck me anew with particular force one recent morning as I sat in a rocking chair reading a book. I had apparently just looked up from my book, and at first had been gazing blindly out the window, lost in thought, when the beauty of my surroundings distracted me from my theoretical musings. Green-golden sunlight was streaming in the window that early spring day, and the thousands of branches and twigs of the maple tree in the yard were still clearly visible through a mist of green buds, forming an elegant pattern of wonderful intricacy. The windowpane is made of old glass, and has a scarcely detectable wrinkle line in it, and as I rocked back and forth, this imperfection in the glass caused a wave of synchronized wiggles to march back and forth across the delta of branches, a regular motion superimposed with remarkable vividness on the more chaotic shimmer of the twigs and branches in the breeze.

Then I noticed that this visual metronome in the tree branches was locked in rhythm with the Vivaldi concerto grosso I was listening to as "background music" for my reading. At first I thought it was obvious that I must have unconsciously synchronized my rocking with the music — just as one may unconsciously tap one's foot in time — but rocking chairs actually have a rather limited range of easily maintained rocking frequencies, so probably the synchrony was mainly a coincidence, just slightly pruned by some unconscious preference of mine for neatness, for staying in step.

In my mind I skipped fleetingly over some dimly imagined brain processes that might explain how we unconsciously adjust our behavior, including the behavior of our eyes and our attention-directing faculties, in order to "synchronize" the "sound track" with the "picture," but these musings were interrupted in turn by an abrupt realization. *What I was doing* — the interplay of experiencing and thinking I have just described from my privileged, first-person point of view — was much harder to "make a model of" than the unconscious, backstage processes that were no doubt *going on in me* and somehow the causal conditions for what I was doing. Backstage machinery was relatively easy to make sense of; it was the front-and-center, in-the-limelight goings-on that were downright baffling. My conscious thinking, and especially the enjoyment I felt in the combination of sunny light, sunny Vivaldi violins, rippling branches — plus the pleasure I took in just thinking about it all — how could *all that* be just something physical happening in my brain? How could any combination of electrochemical happenings in my brain somehow add up to the delightful way those hundreds of twigs genuflected in time with the music? How could some information-processing event in my brain *be* the delicate warmth of the sunlight I felt falling on me? For that matter, how could an event in my brain *be* my sketchily visualized mental image of . . . some other information-processing event in my brain? It does seem impossible.

It does seem as if the happenings that *are* my conscious thoughts and experiences cannot be brain happenings, but must be *something else*, something caused or produced by brain happenings, no doubt, but something in addition, made of different stuff, located in a different space. Well, why not?

3. THE ATTRACTIONS OF MIND STUFF

Let's see what happens when we take this undeniably tempting route. First, I want you to perform a simple experiment. It involves closing your eyes, imagining something, and then, once you have formed your mental image and checked it out carefully, answering some questions below. Do not read the questions until after you have followed this instruction: when you close your eyes, imagine, in as much detail as possible, a purple cow.

Done? Now:

(1) Was your cow facing left or right or head on?
(2) Was she chewing her cud?
(3) Was her udder visible to you?
(4) Was she a relatively pale purple, or deep purple?

If you followed instructions, you could probably answer all four questions without having to make something up in retrospect. If you found all four questions embarrassingly demanding, you probably didn't bother imagining a purple cow at all, but just thought, lazily: "I'm imagining a purple cow" or "Call *this* imagining a purple cow," or did something nondescript of that sort.

Now let us do a second exercise: close your eyes and imagine, in as much detail as possible, a *yellow* cow.

This time you can probably answer the first three questions above without any qualms, and will have something confident to say about what sort of yellow — pastel or buttery or tan — covered the flanks of your imagined cow. But this time I want to consider a different question:

(5) What is the difference between imagining a purple cow and imagining a yellow cow?

The answer is obvious: The first imagined cow is purple and the second is yellow. There might be other differences, but that is the essential one. The trouble is that since these cows are just imagined cows, rather than real cows, or painted pictures of cows on canvas, or cow shapes on a color television screen, it is hard to see what could be purple in the first instance and yellow in the second. Nothing roughly cow-shaped in your brain (or in your eyeball) turns purple in one case and yellow in the other, and even if it did, this would not be much help, since it's pitch black inside your skull and, besides, you haven't any eyes in there to see colors with.

There are events in your brain that are tightly associated with your particular imaginings, so it is not out of the question that in the near future a neuroscientist, examining the processes that occurred in your brain in response to my instructions, would be able to decipher them to the extent of being able to confirm or disconfirm your answers to questions 1 through 4:

"Was the cow facing left? We think so. The cow-head neuronal excitation pattern was consistent with upper-left visual quadrant presentation, and we observed one-herz oscillatory motion-detection signals that suggest cud-chewing, but we could detect no activity in the udder-complex representation groups, and, after calibration of evoked potentials with the subject's color-detection profiles, we hypothesize that the subject is lying about the color: the imagined cow was almost certainly brown."

Suppose all this were true; suppose scientific mind-reading had come of age. Still, it seems, the mystery would remain: what is brown when you imagine a brown cow? Not the event in the brain that the scientists have calibrated with your experiencing-of-brown. The type and location of the neurons involved, their connections with other parts of the brain, the frequency or amplitude of activity, the neurotransmitter chemicals released — none of those properties is the very property of the cow "in your imagination." And since you did imagine a cow (you are not lying — the scientists even confirm that), an imagined cow came into existence at that time; something, somewhere must have had those properties at that time. The imagined cow must be rendered not in the medium of brain stuff, but in the medium of . . . mind stuff. What else could it be?

Mind stuff, then, must be "what dreams are made of," and it apparently has some remarkable properties. One of these we have already noticed in passing, but it is extremely resistant to definition. As a first pass, let us say that mind stuff always *has a witness*. The trouble with brain events, we noticed, is that no matter how closely they "match" the events in our streams of consciousness, they have one apparently fatal drawback: *There's nobody in there watching them.* Events that happen in your brain, just like events that happen in your stomach or your liver, are not normally witnessed by anyone, nor does it make any difference to how they happen whether they occur witnessed or unwitnessed. Events in consciousness, on the other hand, are "by definition" witnessed; they are *experienced* by an *experiencer,* and their being thus experienced is what makes them what they are: *conscious* events. An experienced event cannot just happen on its own hook, it seems; it must be *somebody's* experience. For a thought to happen, someone (some mind) must think it, and for a pain to happen, someone must feel it, and for a purple cow to burst into existence "in imagination," someone must imagine it.

And the trouble with brains, it seems, is that when you look in them, you discover that *there's nobody home.* No part of the brain is the thinker that does the thinking or the feeler that does the feeling, and the whole brain appears to be no better a candidate for that very special role. This is a slippery topic. Do brains think? Do eyes see? Or do people see with their eyes and think with their brains? Is there a difference? Is this just a trivial point of "grammar" or does it reveal a major source of confusion? The idea that a *self* (or a person, or, for that matter, a soul) is distinct from a brain or a body is deeply rooted in our ways of speaking, and hence in our ways of thinking.

I have a brain.

This seems to be a perfectly uncontroversial thing to say. And it does not seem to mean just

This body has a brain (and a heart, and two lungs, etc.).

or

This brain has itself.

It is quite natural to think of "the self and its brain" (Popper and Eccles, 1977) as two distinct things, with different properties, no matter how closely they depend on each other. If the self is distinct from the brain, it seems that it must be made of mind stuff. In Latin, a thinking thing is a *res cogitans,* a term made famous by Descartes, who offered what he thought was an unshakable proof that he, manifestly a thinking thing, could not be his brain. Here is one of his versions of it, and it is certainly compelling:

> I next considered attentively what I was; and I saw that while I could pretend that I had no body, that there was no world, and no place for me to be in, I could not pretend that I was not; on the contrary, from the mere fact that I thought of doubting the truth of other things it evidently and certainly followed that I existed. On the other hand, if I had merely ceased to think, even if everything else that I had ever imagined had been true, I had no reason to believe that I should have existed. From this I recognized that I was a substance whose whole essence or nature is to think and whose being requires no place and depends on no material thing. [*Discourse on Method*, 1637]

So we have discovered two sorts of things one might want to make out of mind stuff: the purple cow that isn't in the brain, and the thing that does the thinking. But there are still other special powers we might want to attribute to mind stuff.

Suppose a winery decided to replace their human wine tasters with a machine. A computer-based "expert system" for quality control and classification of wine is almost within the bounds of existing technology. We now know enough about the relevant chemistry to make the transducers that would replace the taste buds and the olfactory receptors of the epithelium (providing the "raw material" — the input stimuli — for taste and smell). How these inputs combine and interact to produce our experiences is not precisely known, but progress is being made. Work on vision has proceeded much farther. Research on color vision suggests that mimicking human idiosyncrasy, delicacy, and reliability in the color-judging component of the machine would be a great technical challenge, but it is not out of the question. So we can readily imagine using the advanced outputs of these sensory transducers and their comparison machinery to feed elaborate classification, description, and evaluation routines. Pour the sample wine in the funnel and, in a few minutes or hours, the system would type out a chemical assay, along with commentary: "a flamboyant and velvety Pinot, though lacking in stamina" — or words to such effect. Such a machine might even perform better than human wine tasters on all reasonable tests of accuracy and consistency the winemakers could devise, but surely no matter how "sensitive" and "discriminating" such a system might become, it seems that it would never have, and enjoy, what *we* do when we taste a wine.

Is this in fact so obvious? According to the various ideologies grouped under the label of *functionalism,* if you reproduced the *entire* "functional structure" of the human wine taster's cognitive system (including memory, goals, innate aversions, etc.), you would thereby reproduce *all* the mental properties as well, including the enjoyment, the delight, the savoring that makes wine-drinking something many of us *appreciate.* In principle it makes no difference, the functionalist says, whether a system is made of organic molecules or silicon, so long as it *does the same job.* Artificial hearts don't have to be made of organic tissue, and neither do artificial brains — at least in principle. If all the control functions of a human wine taster's brain can be reproduced in silicon chips, the enjoyment will *ipso facto* be reproduced as well.

Some brand of functionalism may triumph in the end (in fact this book will defend a version of functionalism), but it surely seems outrageous at first blush. It seems that no mere machine, no matter how accurately it mimicked the brain processes of the human wine taster, would be capable of appreciating a wine, or a Beethoven sonata, or a basketball game. For appreciation, you need consciousness — something no mere machine has. But of course the brain *is* a machine of sorts, an organ like the heart or lungs or kidneys with an ultimately mechanical explanation of all its powers. This can make it seem compelling that the brain isn't what does the appreciating; *that* is the responsibility (or privilege) of the mind. Reproduction of the brain's machinery in a silicon-based machine wouldn't, then, yield real appreciation, but at best the illusion or simulacrum of appreciation.

So the conscious mind is not just the place where the witnessed colors and smells are, and not just the thinking thing. It is where the appreciating happens. It is the ultimate arbiter of why anything matters. Perhaps this even follows somehow from the fact that the conscious mind is also supposed to be the source of our intentional actions. It stands to reason — doesn't it? — that if *doing things that matter* depends on consciousness, *mattering* (enjoying, appreciating, suffering, caring) should depend on consciousness as well. If a sleepwalker "unconsciously" does harm, he is not responsible because in an important sense *he* didn't do it; his bodily motions are intricately involved in the causal chains that led to the harm, but they did not constitute any *actions* of his, any more than if he had simply done the harm by falling out of bed. Mere bodily complicity does not make for an intentional action, nor does bodily complicity *under the control of structures in the brain*, for a sleepwalker's body is manifestly under the control of structures in the sleepwalker's brain. What more must be added is consciousness, the special ingredient that turns mere *happenings* into *doings*.[1]

It is not Vesuvius's fault if its eruption kills your beloved, and resenting (Strawson, 1962) or despising it are not available options — unless you somehow convince yourself that Vesuvius, contrary to contemporary opinion, is a conscious agent. It is indeed strangely comforting in our grief to put ourselves into such states of mind, to rail at the "fury" of the hurricane, to curse the cancer that so unjustly strikes down a child, or to curse "the gods." Originally, to say that something was "animate" as opposed to "inanimate" was to say that it had a soul (*anima* in Latin). It may be more than just comforting to think of the things that affect us powerfully as animate; it may be a deep biological design trick, a shortcut for helping our time-pressured brains organize and think about the things that need thinking about if we are to survive.

We might have an innate tendency to treat every changing thing at first as if it had a soul (Stafford, 1983; Humphrey, 1983b, 1986), but however natural this attitude is, we now know that attributing a (conscious) soul to Vesuvius is going too far. Just where to draw the line is a vexing question to which we will return, but for ourselves, it seems, consciousness is precisely what distinguishes us from mere "automata." Mere bodily "reflexes" are "automatic" and mechanical; they may involve circuits in the brain, but do not require any intervention by the conscious mind. It is very natural to think of our own bodies as mere hand puppets of sorts that "we" control "from inside." I make the hand puppet wave to the audience by wiggling my finger; I wiggle my finger by . . . what, wiggling my soul? There are notorious problems with this idea, but that does not prevent it from seeming somehow right: unless there is a conscious mind behind the deed, there is no *real* agent in charge. When we think of our minds this way, we seem to discover the "inner me," the "real me." This real me is not my brain; it is what *owns* my brain ("the self and *its* brain"). On Harry Truman's desk in the Oval Office of the White House was a famous sign: "The buck stops here." No part of the brain, it seems, could be *where the buck stops*, the ultimate source of moral responsibility at the beginning of a chain of command.

To summarize, we have found four reasons for believing in mind

---

1. See my *Elbow Room* (1984), chapter 4, for further discussion of this theme.

stuff. The conscious mind, it seems, cannot just *be* the brain, or any proper part of it, because nothing in the brain could

(1) be the medium in which the purple cow is rendered;
(2) be the thinking thing, the *I* in "I think, therefore I am";
(3) appreciate wine, hate racism, love someone, be a source of *mattering;*
(4) act with moral responsibility.

An acceptable theory of human consciousness must account for these four compelling grounds for thinking that there must be mind stuff.

4. WHY DUALISM IS FORLORN

The idea of mind as distinct in this way from the brain, composed not of ordinary matter but of some other, special kind of stuff, is *dualism*, and it is deservedly in disrepute today, in spite of the persuasive themes just canvassed. Ever since Gilbert Ryle's classic attack (1949) on what he called Descartes's "dogma of the ghost in the machine," dualists have been on the defensive.[2] The prevailing wisdom, variously expressed and argued for, is *materialism:* there is only one sort of stuff, namely *matter* — the physical stuff of physics, chemistry, and physiology — and the mind is somehow nothing but a physical phenomenon. In short, the mind is the brain. According to the materialists, we can (in principle!) account for every mental phenomenon using the same physical principles, laws, and raw materials that suffice to explain radioactivity, continental drift, photosynthesis, reproduction, nutrition, and growth. It is one of the main burdens of this book to explain consciousness without ever giving in to the siren song of dualism. What, then, is so wrong with dualism? Why is it in such disfavor?

The standard objection to dualism was all too familiar to Descartes himself in the seventeenth century, and it is fair to say that neither he nor any subsequent dualist has ever overcome it convincingly. If mind

---

2. A few brave souls (and they surely cannot object to being so categorized!) have bucked the tide: Arthur Koestler's defiantly titled *The Ghost in the Machine* (1967) and Popper and Eccles's *The Self and Its Brain* (1977) are by unquestionably eminent authors, and two other iconoclastic and quirkily insightful defenses of dualism are Zeno Vendler's *Res Cogitans* (1972) and *The Matter of Minds* (1984).

and body are distinct things or substances, they nevertheless must interact; the bodily sense organs, via the brain, must *inform* the mind, must send to it or present it with perceptions or ideas or data of some sort, and then the mind, having thought things over, must *direct* the body in appropriate action (including speech). Hence the view is often called Cartesian interactionism or interactionist dualism. In Descartes's formulation, the locus of interaction in the brain was the pineal gland, or *epiphysis*. It appears in Descartes's own schematic diagram as the much-enlarged pointed oval in the middle of the head.

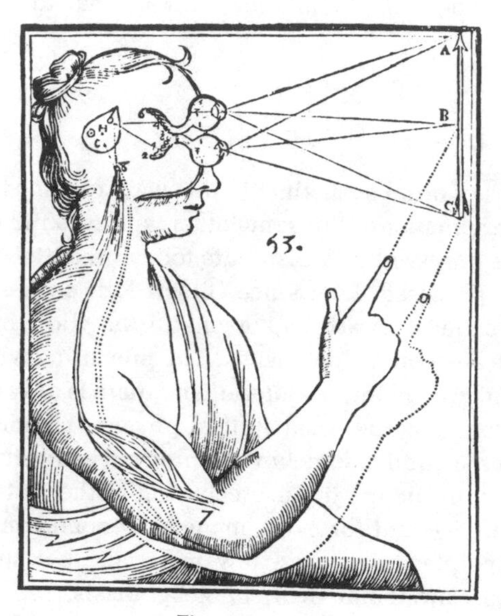

Figure 2.1

We can make the problem with interactionism clear by superimposing a sketch of the rest of Descartes's theory on his diagram (Figure 2.2).

The conscious perception of the arrow occurs only after the brain has somehow transmitted its message to the mind, and the person's finger can point to the arrow only after the mind commands the body. How, precisely, does the information get transmitted from pineal gland to mind? Since we don't have the faintest idea (yet) what properties mind stuff has, we can't even guess (yet) how it might be affected by physical processes emanating somehow from the brain, so let's ignore those upbound signals for the time being, and concentrate on the return signals, the directives from mind to brain. These, *ex hypothesi*, are not physical; they are not light waves or sound waves or cosmic rays or

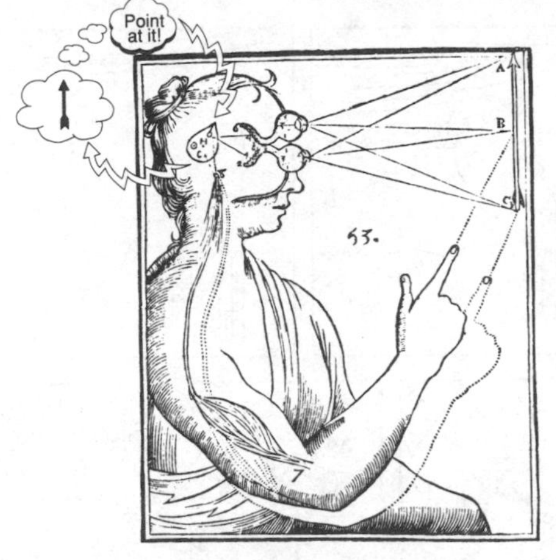

Figure 2.2

streams of subatomic particles. No physical energy or mass is associated with them. How, then, do they get to make a difference to what happens in the brain cells they must affect, if the mind is to have any influence over the body? A fundamental principle of physics is that any change in the trajectory of any physical entity is an acceleration requiring the expenditure of energy, and where is this energy to come from? It is this principle of the conservation of energy that accounts for the physical impossibility of "perpetual motion machines," and the same principle is apparently violated by dualism. This confrontation between quite standard physics and dualism has been endlessly discussed since Descartes's own day, and is widely regarded as the inescapable and fatal flaw of dualism.

Just as one would expect, ingenious technical exemptions based on sophisticated readings of the relevant physics have been explored and expounded, but without attracting many conversions. Dualism's embarrassment here is really simpler than the citation of presumed laws of physics suggests. It is the same incoherence that children notice — but tolerate happily in fantasy — in such fare as Casper the Friendly Ghost (Figure 2.3, page 36). How can Casper *both* glide through walls and grab a falling towel? How can mind stuff *both* elude all physical measurement and control the body? A ghost in the machine is of no help in our theories unless it is a ghost that can move things around — like a noisy poltergeist who can tip over a lamp or slam a door — but anything that can move a physical thing is itself a physical thing (although perhaps a strange and heretofore unstudied kind of physical thing).

What about the option, then, of concluding that mind stuff is

Figure 2.3

actually a special kind of matter? In Victorian séances, the mediums often produced out of thin air something they called "ectoplasm," a strange gooey substance that was supposedly the basic material of the spirit world, but which could be trapped in a glass jar, and which oozed and moistened and reflected light just like everyday matter. Those fraudulent trappings should not dissuade us from asking, more soberly, whether mind stuff might indeed be something above and beyond the atoms and molecules that compose the brain, but still a scientifically investigatable kind of matter. The ontology of a theory is the catalogue of things and types of things the theory deems to exist. The ontology of the physical sciences used to include "caloric" (the stuff heat was made of, in effect) and "the ether" (the stuff that pervaded space and was the medium of light vibrations in the same way air or water can be the medium of sound vibrations). These things are no longer taken seriously, while neutrinos and antimatter and black holes are now included in the standard scientific ontology. Perhaps some basic enlargement of the ontology of the physical sciences is called for in order to account for the phenomena of consciousness.

Just such a revolution of physics has recently been proposed by the physicist and mathematician Roger Penrose, in *The Emperor's New Mind* (1989). While I myself do not think he has succeeded in making his case for revolution,[3] it is important to notice that he has been careful *not* to fall into the trap of dualism. What is the difference? Penrose makes it clear that he intends his proposed revolution to make the conscious mind *more* accessible to scientific investigation, not less. It is surely no accident that the few dualists to avow their views openly have all candidly and comfortably announced that they have no theory whatever of how the mind works — something, they insist, that is quite beyond human ken.[4] There is the lurking suspicion that the most attractive feature of mind stuff is its promise of being *so* mysterious that it keeps science at bay forever.

This fundamentally antiscientific stance of dualism is, to my mind, its most disqualifying feature, and is the reason why in this book I adopt the apparently dogmatic rule that dualism is to be avoided *at all costs.* It is not that I think I can give a knock-down proof that dualism, in all its forms, is false or incoherent, but that, given the way dualism wallows in mystery, *accepting dualism is giving up* (as in Figure 2.4, page 38).

There is widespread agreement about this, but it is as shallow as it is wide, papering over some troublesome cracks in the materialist wall. Scientists and philosophers may have achieved a consensus of sorts in favor of materialism, but as we shall see, getting rid of the old dualistic visions is harder than contemporary materialists have thought. Finding suitable replacements for the traditional dualistic images will require some rather startling adjustments to our habitual ways of thinking, adjustments that will be just as counterintuitive at first to scientists as to laypeople.

I don't view it as ominous that my theory seems at first to be strongly at odds with common wisdom. On the contrary, we shouldn't expect a good theory of consciousness to make for comfortable reading — the sort that immediately "rings bells," that makes us exclaim to ourselves, with something like secret pride: "Of course! I knew that all along! It's obvious, once it's been pointed out!" If there were any such theory to be had, we would surely have hit upon it by now. The mysteries of the mind have been around for so long, and we have made

---

3. See "Murmurs in the Cathedral" (Dennett, 1989c), my review of his book.

4. Eccles has proposed that the nonphysical mind is composed of millions of "psychons," which interact with millions of "dendrons" (tracts of pyramidal cells) in the cortex; each psychon corresponds roughly to what Descartes or Hume would call an idea — such as the idea of red, or the idea of round, or hot — but aside from this minimal decomposition, Eccles has nothing to say about the parts, activities, principles of action, or other properties of the nonphysical mind.

Figure 2.4

so little progress on them, that the likelihood is high that some things we all tend to agree to be obvious are just not so. I will soon be introducing my candidates.

Some brain researchers today — perhaps even a stolid majority of them — continue to pretend that, for them, the brain is just another organ, like the kidney or pancreas, which should be described and explained only in the most secure terms of the physical and biological sciences. They would never dream of mentioning the mind or anything "mental" in the course of their professional duties. For other, more theoretically daring researchers, there is a new object of study, the mind/brain (Churchland, 1986). This newly popular coinage nicely expresses the prevailing materialism of these researchers, who happily admit to the world and themselves that what makes the brain particularly fascinating and baffling is that somehow or other it *is* the mind. But even among these researchers there is a reluctance to confront the Big Issues, a desire to postpone until some later date the embarrassing questions about the nature of consciousness.

But while this attitude is entirely reasonable, a modest recognition of the value of the divide-and-conquer strategy, it has the effect of distorting some of the new concepts that have arisen in what is now called *cognitive science*. Almost all researchers in cognitive science, whether they consider themselves neuroscientists or psychologists or artificial intelligence researchers, tend to postpone questions about consciousness by restricting their attention to the "peripheral" and "subordinate" systems of the mind/brain, which are deemed to feed and service some dimly imagined "center" where "conscious thought" and "experience" take place. This tends to have the effect of leaving too much of the mind's work to be done "in the center," and this leads theorists to underestimate the "amount of understanding" that must be accomplished by the relatively peripheral systems of the brain (Dennett, 1984b).

For instance, theorists tend to think of perceptual systems as providing "input" to some central thinking arena, which in turn provides "control" or "direction" to some relatively peripheral systems governing bodily motion. This central arena is also thought to avail itself of material held in various relatively subservient systems of memory. But the very idea that there are important theoretical divisions between such presumed subsystems as "long-term memory" and "reasoning" (or "planning") is more an artifact of the divide-and-conquer strategy than anything found in nature. As we shall soon see, the exclusive attention to specific subsystems of the mind/brain often causes a sort of theoretical myopia that prevents theorists from seeing that their models still presuppose that somewhere, conveniently hidden in the obscure "center" of the mind/brain, there is a Cartesian Theater, a place where "it all comes together" and consciousness happens. This may seem like a good idea, an inevitable idea, but until we see, in some detail, why it is not, the Cartesian Theater will continue to attract crowds of theorists transfixed by an illusion.

5. THE CHALLENGE

In the preceding section, I noted that if dualism is the best we can do, then we can't understand human consciousness. Some people are convinced that we can't in any case. Such defeatism, today, in the midst of a cornucopia of scientific advances ready to be exploited, strikes me as ludicrous, even pathetic, but I suppose it could be the sad truth. Perhaps consciousness really can't be explained, but how will we know till someone tries? I think that many — indeed, most — of the pieces of the puzzle are already well understood, and only need to be jiggled into place with a little help from me. Those who would defend the Mind against Science should wish me luck with this attempt, since if they are right, my project is bound to fail, but if I do the job about as well as it could be done, my failure ought to shed light on just why science will always fall short. They will at last have their argument against science, and I will have done all the dirty work for them.

The ground rules for my project are straightforward:

(1) *No Wonder Tissue allowed.* I will try to explain every puzzling feature of human consciousness within the framework of contemporary physical science; at no point will I make an appeal to inexplicable or unknown forces, substances, or organic powers. In other words, I intend to see what can be done within the conservative limits of standard science, saving a call for a revolution in materialism as a last resort.

(2) *No feigning anesthesia.* It has been said of behaviorists that they feign anesthesia — they pretend they don't have the experiences we know darn well they share with us. If I wish to deny the existence of some controversial feature of consciousness, the burden falls on me to *show* that it is somehow illusory.

(3) *No nitpicking about empirical details.* I will try to get all the scientific facts right, insofar as they are known today, but there is abundant controversy about just which exciting advances will stand the test of time. If I were to restrict myself to "facts that have made it into the textbooks," I would be unable to avail myself of some of the most eye-opening recent discoveries (if that is what they are). And I would *still* end up unwittingly purveying some falsehoods, if recent history is any guide. Some of the "discoveries" about vision for which David Hubel and Torstein Wiesel were deservedly awarded the Nobel Prize in 1981 are now becoming unraveled, and Edwin Land's famous "retinex" theory of color vision, which has been regarded by most philosophers of mind and other

nonspecialists as established fact for more than twenty years, is not nearly as highly regarded among visual scientists.[5]

So, since as a philosopher I am concerned to establish the *possibilities* (and rebut claims of impossibility), I will settle for theory sketches instead of full-blown, empirically confirmed theories. A theory sketch or a model of how the brain *might* do something can turn a perplexity into a research program: if this model won't quite do, would some other more realistic variation do the trick? (The explanation sketch of hallucination production in chapter 1 is an example of this.) Such a sketch is directly and explicitly vulnerable to empirical disproof, but if you want to claim that my sketch is not a *possible* explanation of a phenomenon, you must show what it *has to* leave out or *cannot* do; if you merely claim that my model may well be incorrect in many of its details, I will concede the point. What is wrong with Cartesian dualism, for instance, is not that Descartes chose the pineal gland — as opposed to the thalamus, say, or the amygdala — as the locus of interaction with the mind, but *the very idea* of such a locus of mind-brain interaction. What counts as nitpicking changes, of course, as science advances, and different theorists have different standards. I will try to err on the side of overspecificity, not only to heighten the contrast with traditional philosophy of mind, but to give empirical critics a clearer target at which to shoot.

In this chapter, we have encountered the basic features of the mystery of consciousness. The very mysteriousness of consciousness is one of its central features — possibly even a vital feature without which it cannot survive. Since this possibility is widely if dimly appreciated, prudence tends to favor doctrines that do not even purport to explain consciousness, for consciousness matters deeply to us. Dualism, the idea that a brain cannot be a thinking thing so a thinking thing cannot be a brain, is tempting for a variety of reasons, but we must resist temptation; "adopting" dualism is really just accepting defeat without admitting it. Adopting materialism does not by itself dissolve the puzzles about consciousness, nor do they fall to any straightforward inferences from brain science. Somehow the brain must be the mind, but unless we can come to see in some detail how this is possible, our

---

5. A fascinating review of the status of Land's theory is provided by the philosopher C. L. Hardin in an appendix to his book, *Color for Philosophers: Unweaving the Rainbow* (1988).

materialism will not explain consciousness, but only promise to explain it, some sweet day. That promise cannot be kept, I have suggested, until we learn how to abandon more of Descartes's legacy. At the same time, whatever else our materialist theories may explain, they won't explain *consciousness* if we neglect the facts about experience that we know so intimately "from the inside." In the next chapter, we will develop an initial inventory of those facts.

3

A VISIT TO THE PHENOMENOLOGICAL GARDEN

1. WELCOME TO THE PHENOM

Suppose a madman were to claim that there were no such things as animals. We might decide to confront him with his error by taking him to the zoo, and saying, "Look! What are those things, then, if not animals?" We would not expect this to cure him, but at least we would have the satisfaction of making plain to ourselves just what craziness he was spouting. But suppose he then said, "Oh, I know perfectly well that there are these things — lions and ostriches and boa constrictors — but what makes you think these so-called animals are *animals*? In fact, they are all just fur-covered *robots* — well, actually, some are covered with feathers or scales." This may still be craziness, but it is a different and more defensible kind of craziness. This madman just has a revolutionary idea about the ultimate nature of animals.[1]

Zoologists are the experts on the ultimate nature of animals, and

1. Descartes, in fact, held such a view of animals. He held that animals were in fact just elaborate machines. Human bodies, and even human brains, were also just machines. It was only our nonmechanical, nonphysical minds that make human beings (and only human beings) intelligent and conscious. This was actually a subtle view, most of which would be readily defended by zoologists today, but it was too revolutionary for Descartes's contemporaries, who caricatured it in just the ways you would imagine, and then treated the caricature with derision. Centuries later, these slanders of Descartes are still being gleefully propagated by those who find the prospect of a mechanistic account of consciousness inconceivable — or at least intolerable. For an eye-opening account, see Leiber (1988).

zoological gardens — zoos, for short — serve the useful educational purpose of acquainting the populace with the topics of their expertise. If zoologists were to discover that this madman was right (in some manner of speaking), they would find a good use for their zoo in their attempts to explain their discovery. They might say, "It turns out that animals — you know: those familiar things we all have seen at the zoo — are not what we once thought they were. They're so different, in fact, that we really shouldn't call them animals. So you see, there really *aren't* any animals in the ordinary understanding of that term."

Philosophers and psychologists often use the term *phenomenology* as an umbrella term to cover all the items — the fauna and flora, you might say — that inhabit our conscious experience: thoughts, smells, itches, pains, imagined purple cows, hunches, and all the rest. This usage has several somewhat distinct ancestries worth noting. In the eighteenth century, Kant distinguished "phenomena," things as they appear, from "noumena," things as they are in themselves, and during the development of the natural or physical sciences in the nineteenth century, the term *phenomenology* came to refer to the merely descriptive study of any subject matter, neutrally or pretheoretically. The phenomenology of magnetism, for instance, had been well begun by William Gilbert in the sixteenth century, but the explanation of that phenomenology had to await the discoveries of the relationship between magnetism and electricity in the nineteenth century, and the theoretical work of Faraday, Maxwell, and others. Alluding to this division between acute observation and theoretical explanation, the philosophical school or movement known as Phenomenology (with a capital P) grew up early in the twentieth century around the work of Edmund Husserl. Its aim was to find a new foundation for all philosophy (indeed, for all knowledge) based on a special technique of introspection, in which the outer world and all its implications and presuppositions were supposed to be "bracketed" in a particular act of mind known as the *epoché*. The net result was an investigative state of mind in which the Phenomenologist was supposed to become acquainted with the pure objects of conscious experience, called *noemata*, untainted by the usual distortions and amendments of theory and practice. Like other attempts to strip away interpretation and reveal the basic facts of consciousness to rigorous observation, such as the Impressionist movement in the arts and the Introspectionist psychologies of Wundt, Titchener, and others, Phenomenology has failed to find a single, settled method that everyone could agree upon.

So while there are zoologists, there really are no phenomenologists: uncontroversial experts on the nature of the things that swim in the stream of consciousness. But we can follow recent practice and adopt the term (with a lower-case p) as the generic term for the various items in conscious experience that have to be explained.

I once published an article titled "On the Absence of Phenomenology" (1979), which was an attempt to argue for the second sort of craziness: the things that consciousness is composed of are so different from what people have thought, that they really shouldn't use the old terms. But this was such an outrageous suggestion to some people ("How on earth could we be wrong about our own inner lives!") that they tended to dismiss it as an instance of the first sort of craziness ("Dennett doesn't think there are any pains or aromas or daydreams!"). That was a caricature, of course, but a tempting one. My trouble was that I didn't have a handy phenomenological garden — a phenom, for short — to use in my explanations. I wanted to say, "It turns out that the things that swim by in the stream of consciousness — you know: the pains and aromas and daydreams and mental images and flashes of anger and lust, the standard denizens of the phenom — those things are not what we once thought they were. They are really so different, in fact, that we have to find some new words for them."

So let's take a brief tour of the phenomenological garden, just to satisfy ourselves that we know what we are talking about (even if we don't *yet* know the ultimate nature of these things). It will be a deliberately superficial introductory tour, a matter of pointing and saying a few informative words, and raising a few questions, before we get down to serious theorizing in the rest of the book. Since I will soon be mounting radical challenges to everyday thinking, I wouldn't want anyone to think I was simply ignorant of all the wonderful things that inhabit *other* people's minds.

Our phenom is divided into three parts: *(1) experiences of the "external" world*, such as sights, sounds, smells, slippery and scratchy feelings, feelings of heat and cold, and of the positions of our limbs; *(2) experiences of the purely "internal" world*, such as fantasy images, the inner sights and sounds of daydreaming and talking to yourself, recollections, bright ideas, and sudden hunches; and *(3) experiences of emotion or "affect"* (to use the awkward term favored by psychologists), ranging from bodily pains, tickles, and "sensations" of hunger and thirst, through intermediate emotional storms of anger, joy, hatred, embarrassment, lust, astonishment, to the least corporeal visitations of pride, anxiety, regret, ironic detachment, rue, awe, icy calm.

I make no claims for this tripartite division into outer, inner, and affect. Like a menagerie that puts the bats with the birds and the dolphins with the fish, this taxonomy owes more to superficial similarity and dubious tradition than to any deep kinship among the phenomena, but we have to start somewhere, and any taxonomy that gives us some bearings will tend to keep us from overlooking species altogether.

2. OUR EXPERIENCE OF THE EXTERNAL WORLD

Let's begin with the crudest of our outer senses, taste and smell. As most people know, our taste buds are actually sensitive only to sweet, sour, salty, and bitter, and for the most part we "taste with our noses," which is why food loses its savor when we have head colds. The nasal epithelium is to olfaction, the sense of smell, what the retina of the eye is to vision. The individual epithelial cells come in a wide variety, each sensitive to a different kind of airborne molecule. It is ultimately the *shape* of the molecules that matters. Molecules float into the nose, like so many microscopic keys, turning on particular sensory cells in the epithelium. Molecules can often be readily detected in astonishingly low concentrations of a few parts per billion. Other animals have vastly superior olfaction to ours, not only in being able to discriminate more odors, in fainter traces (the bloodhound is the familiar example), but also in having better temporal and spatial resolution of smells. We may be able to sense the presence in a room of a thin trail of formaldehyde molecules, but if we do, we don't smell that there is a threadlike trail, or a region with some smellably individual and particular molecules floating in it; the whole room, or at least the whole corner of the room, will seem suffused by the smell. There is no mystery about why this should be so: molecules *wander* more or less at random into our nasal passages, and their arrival at specific points on the epithelium provides scant information about where they came from in the world, unlike the photons that stream in optically straight lines through the pinhole iris, landing at a retinal address that maps geometrically onto an external source or source path. If the resolution of our vision were as poor as the resolution of our olfaction, when a bird flew overhead the sky would *go all birdish* for us for a while. (Some species do have vision that poor — that is, the resolution and discrimination is no better than that — but what, if anything, it is like for the animal to see things that poorly is another matter, to which we will turn in a later chapter.)

Our senses of taste and smell are yoked together phenomenologically, and so are our senses of touch and kinesthesia, the sense of the position and motion of our limbs and other body parts. We "feel" things by touching them, grabbing them, pushing against them in many ways, but the resulting conscious sensations, while they seem to naïve reflection to be straightforward "translations" of the stimulation of the touch receptors under the skin, are once again the products of an elaborate process of integration of information from a variety of sources. Blindfold yourself and take a stick (or a pen or pencil) in your hand. Touch various things around you with this wand, and notice that you can tell their textures effortlessly — as if your nervous system had sensors out at the tip of the wand. It takes a special, and largely ineffectual, effort to attend to the way the stick feels at your fingertips, the way it vibrates or resists being moved when in contact with the various surfaces. Those transactions between stick and touch receptors under the skin (aided in most instances by scarcely noticed sounds) provide the information your brain integrates into a conscious recognition of the texture of paper, cardboard, wool, or glass, but these complicated processes of integration are all but transparent to consciousness. That is, we don't — and can't — notice how "we" do it. For an even more indirect case, think of how you can feel the slipperiness of an oil spot on the highway under the wheels of your car as you turn a corner. The phenomenological focal point of contact is the point where the rubber meets the road, not any point on your innervated body, seated, clothed, on the car seat, or on your gloved hands on the steering wheel.

Now, while still blindfolded put down your wand and have someone hand you a piece of china, a piece of plastic, and pieces of polished wood and metal. They are all extremely smooth and slippery, and yet you will have little difficulty telling their particular smoothnesses apart — and not because you have specialized china receptors and plastic receptors in your fingertips. The difference in heat conductivity of the substances is apparently the most important factor, but it is not essential: You may surprise yourself by the readiness with which you can sometimes tell these surfaces apart by "feel" using just the wand. These successes must depend on felt vibrations set up in the wand, or on indescribable — but detectable — differences in the clicks and scraping noises heard. But it *seems* as if some of your nerve endings were in the wand, for you *feel* the differences of the surfaces *at* the tip of the wand.

Next, let's consider hearing. The phenomenology of hearing consists of all the sorts of sounds we can hear: music, spoken words, bangs and whistles and sirens and twitters and clicks. Theorists thinking about hearing are often tempted to "strike up the little band in the head." This is a mistake, and to make sure we identify and avoid it, I want to make it vivid with the aid of a fable.

Once upon a time, in about the middle of the nineteenth century, a wild-eyed inventor engaged in a debate with a tough-minded philosopher, Phil. The inventor had announced that his goal was to construct a device that could automatically "record" and then later "replay" with lifelike "fidelity" an orchestra and chorus performing Beethoven's Ninth Symphony. Nonsense, said Phil. It's strictly impossible. I can readily imagine a mechanical device which records the striking of piano keys in sequence, and then controls the reproduction of that sequence on a prepared piano — it might be done with a roll of perforated paper, for example — but think of the huge variety of sounds and their modes of production in a rendition of Beethoven's Ninth! There are a hundred different human voices of different ranges and timbres, dozens of bowed strings, brass, woodwind, percussion. The device that could play back such a variety of sounds together would be an unwieldy monstrosity that dwarfed the mightiest church organ — and if it performed with the "high fidelity" you propose, it would no doubt have to incorporate quite literally a team of human slaves to handle the vocal parts, and what you call the "record" of the particular performance with all its nuances would have to be hundreds of part scores — one for each musician — with thousands or even millions of annotations.

Phil's argument is still strangely compelling; it is astonishing that all those sounds can be faithfully superimposed via a Fourier transform into a single wavy line chiseled into a long-playing disk or magnetically represented on a tape or optically on the sound track of a film. It is even more astonishing that a single paper cone, wobbled back and forth by an electromagnet driven by that single wavy line, can do about equal justice to trumpet blare, banjo strum, human speech, and the sound of a full bottle of wine shattering on the sidewalk. Phil could not imagine anything so powerful, and mistook his failure of imagination for an insight into necessity.

The "magic" of Fourier transforms opens up a new range of possibilities to think about, but we should note that it does not in itself eliminate the problem that befuddled Phil; it merely postpones it. For while we sophisticates can laugh at Phil for failing to understand how the pattern of compression and rarefaction of the air that stimulates the ear could be recorded and reproduced, the smirks will be wiped from our faces when we contemplate the *next* question: What happens to the signal once the ear has properly received it?

From the ear a further encoded barrage of modulated signal trains (but now somewhat analyzed and broken up into parallel streams, ominously reminiscent of Phil's hundreds of part scores) march inward, into the dark center of the brain. These signal trains are no more *heard sounds* than are the wavy lines on the disk; they are sequences of electrochemical pulses streaming up the axons of neurons. Must there not be some still more central place in the brain where these signal trains control the performance of the mighty theater organ of the mind? When, after all, do these toneless signals get their final *translation* into subjectively heard sound?

We don't want to look for places in the brain that vibrate like guitar strings, any more than we want to find places in the brain that turn purple when we imagine a purple cow. Those are manifest dead ends, what Gilbert Ryle (1949) would call category mistakes. But then what *could* we find in the brain that would satisfy us that we had reached the end of the story of auditory experience?[2] How could any complex of physical properties of events in the brain amount to — or even just account for — the thrilling properties of the sounds we hear?

At first these properties seem unanalyzable — or, to use a favorite adjective among phenomenologists, *ineffable*. But at least some of these apparently atomic and homogeneous properties can be made to become noticeably compound and describable. Take a guitar and pluck the bass or low E string open (without pressing down on any fret). Listen carefully to the sound. Does it have describable components or is it one and whole and ineffably guitarish? Many will opt for the latter way of describing their phenomenology. Now pluck the open string again and carefully bring a finger down lightly over the octave fret to create a high "harmonic." Suddenly you hear a new sound: "purer" somehow and of course an octave higher. Some people insist that this is an entirely novel sound, while others describe the experience by saying "the bottom fell out of the note" — leaving just the top. Then pluck the open string a third time. This time you can hear, with surprising distinctness, the harmonic overtone that was isolated in the second plucking. The homogeneity and ineffability of the first experience is gone, replaced by a duality as directly apprehensible and clearly describable as that of any chord.

The difference in experience is striking, but the complexity newly apprehended on the third plucking was *there* all along (being responded

---

2. This rhetorical question implies, for some, the resounding answer: Nothing! For instance, McGinn (1989) supports his defeatist answer by a canvass of available options that manages to ignore the possibilities we will develop in later chapters.

to or discriminated). Research has shown that it was only by the complex pattern of overtones that you are able to recognize the sound as that of a guitar rather than a lute or harpsichord. Such research may help us *account for* the different properties of auditory experiences, by analyzing the informational components and the processes that integrate them, permitting us to predict and even synthetically provoke particular auditory experiences, but it still seems to leave untouched the question of what such properties *amount to*. Why should the guitar-caused pattern of harmonic overtones sound *like this* and the lute-caused pattern *like that*? We have not yet answered this residual question, even if we have softened it up by showing that at least some initially ineffable properties yield to a certain amount of analysis and description after all.[3]

Research into the processes of auditory perception suggests that there are specialized mechanisms for deciphering different sorts of sounds, somewhat like the imagined components of Phil's fantasy playback machine. Speech sounds in particular seem to be handled by what an engineer would call dedicated mechanisms. The phenomenology of speech perception suggests that a wholesale restructuring of the input occurs in a brain facility somewhat analogous to a recording engineer's sound studio where multiple channels of recordings are mixed, enhanced, and variously adjusted to create the stereo "master" from which subsequent recordings in different media are copied.

For instance, we hear speech in our native tongue as a sequence of distinct words separated by tiny gaps of silence. That is, we have a

---

3. Why do the A below middle C and the A above middle C (one octave higher) *sound alike*? What makes them both A's? What ineffable A-ish pitch property do they have in common? Well, when any two tones are an octave apart (and hence sound "the same, only different" to us), the fundamental frequency of one is exactly double the fundamental frequency of the other. The standard A below middle C is 220 vibrations per second; the A an octave higher ("concert A") is 440 vibrations per second. When sounded together, notes that are one or more octaves apart will be in phase. Does this explain the mystery of this ineffable kinship? "Not at all. Why should notes in phase in that manner sound alike in *this* way?" Well, notes that are out of phase don't sound alike in *this* way, but they may sound alike in other ways (in timbre, for instance), which have different explanations in terms of the relationships between the frequencies of vibration they produce. Once we have described many different ways notes can sound alike and different, and lined these up with their physical properties, and their effects on our auditory system, we can even predict, with some accuracy, how novel notes (for instance, notes produced on electronic synthesizers) will sound to us. If all this doesn't explain the ineffable kinships, what remains to be explained? (We will address this popular topic in some detail in chapter 12.)

clear sense of boundaries between words, which cannot be composed of color edges or lines, and do not seem to be marked by beeps or clicks, so what could the boundaries be but silent gaps of various duration — like the gaps that separate the letters and words in Morse code? If asked in various ways by experimenters to note and assess the gaps between words, subjects have little difficulty complying. There seem to be gaps. But if one looks at the acoustic energy profile of the input signal, the regions of lowest energy (the moments closest to silence) do not line up at all well with the word boundaries. The segmentation of speech sounds is a process that imposes boundaries based on the grammatical structure of the language, not on the physical structure of the acoustic wave (Liberman and Studdert-Kennedy, 1977). This helps to explain why we hear speech in foreign languages as a jumbled, unsegmented rush of sounds: the dedicated mechanisms in the brain's "sound studio" lack the necessary grammatical framework to outline the proper segments, so the best they can do is to pass on a version of the incoming signal, largely unretouched.

When we perceive speech we are aware of more than just the identities and grammatical categories of the words. (If that were all we were aware of, we wouldn't be able to tell if we were hearing or reading the words.) The words are clearly demarcated, ordered, and identified, but they also come clothed in sensuous properties. For instance, I just now heard the distinctive British voice of my friend Nick Humphrey, gently challenging, not quite mocking. I *hear* his smile, it seems, and included in my experience is a sense that laughter was there behind the words, waiting to break out like the sun from behind some racing clouds. The properties we are aware of are not only the rise and fall of intonation, but also the rasps and wheezes and lisps, to say nothing of the whine of peevishness, the tremolo of fear, the flatness of depression. And as we just observed in the case of the guitar, what at first seem entirely atomic and homogeneous properties often yield to analysis with a little experimentation and isolation. We all effortlessly recognize the questiony sound of a question — and the difference between a British questiony sound and an American questiony sound — but it takes some experimenting with theme-and-variation before we can describe with any confidence or accuracy the differences in intonation contours that yield those different auditory flavors.

"Flavors" does seem to be the right metaphor here, no doubt because our capacity to analyze flavors is so limited. The familiar but still surprising demonstrations that we taste with our noses show that our powers of taste and olfaction are so crude that we have difficulty identifying even the route by which we are being informed. This obliviousness is not restricted to taste and smell; our hearing of very low frequency tones — such as the deepest bass notes played by a church organ — is apparently caused more by our feeling the vibrations in our bodies than by picking up the vibrations in our ears. It is surprising to learn that the particular "F#-ness, exactly two octaves below the lowest F# I can sing" can be *heard* with the seat of my pants, in effect, rather than my ears.

Finally, let's turn briefly to sight. When our eyes are open we have the sense of a broad field — often called the phenomenal field or visual field — in which things appear, colored and at various depths or distances from us, moving or at rest. We naïvely view almost all the features experienced as objective properties of the external things, observed "directly" by us, but even as children we soon recognize an intermediate category of items — dazzles, glints, shimmers, blurry edges — that we know are somehow products of an interaction between the objects, the light, and our visual apparatus. We still see these items as "out there" rather than in us, with a few exceptions: the pain of looking at the sun or at a sudden bright light when our eyes are dark-adapted, or the nauseating swim of the phenomenal field when we are dizzy. These can seem to be better described as "sensations in the eyes," more akin to the pressures and itches we feel when we rub our eyes than to normal, out-there properties of things seen.

Among the things to be seen out there in the physical world are pictures. Pictures are so pre-eminently things-to-be-seen that we tend to forget that they are a recent addition to the visible environment, only a few tens of thousands of years old. Thanks to recent human art and artifice, we are now surrounded by pictures, maps, diagrams, both still and moving. These physical images, which are but one sort of "raw material" for the processes of visual perception, have become an almost irresistible model of the "end product" of visual perception: "pictures in the head." We are inclined to say, "Of course the outcome of *vision* is a picture in the head (or in the mind). What else could it be? Certainly not a tune or a flavor!" We'll treat this curious but ubiquitous malady of the imagination in many ways before we are through, but we may begin with a reminder: picture galleries for the blind are a waste of resources, so pictures in the head will require eyes in the head to appreciate them (to say nothing of good lighting). And suppose there are mind's eyes in the head to appreciate the pictures in the head. What of the pictures in the head's head produced by these internal eyes in turn? How are we to avoid an infinite regress of pictures and viewers?

We can break the regress only by discovering some viewer whose perception avoids creating yet another picture in need of a viewer. Perhaps the place to break the regress is the very first step?

Fortunately, there are independent reasons for being skeptical of the picture-in-the-head view of vision. If vision involved pictures in the head with which we (our inner selves) were particularly intimately acquainted, shouldn't drawing pictures be easier? Recall how difficult it is to draw a realistic picture of, say, a rose in a vase. There is the rose as big as life a few feet in front of you — to the left, let us suppose, of your pad of paper. (I really want you to imagine this carefully.) All the visible details of the real rose are vivid and sharp and intimately accessible to you, it seems, and yet the presumably simple process of just relocating a black-and-white, two-dimensional copy of all that detail to the right a few degrees is so challenging that most people soon give up and decide that they just cannot draw. The translation of three dimensions into two is particularly difficult for people, which is somewhat surprising, since what seems at first to be the reverse translation — seeing a realistic two-dimensional picture *as of* a three-dimensional situation or object — is effortless and involuntary. In fact, it is the very difficulty we have in suppressing this reverse interpretation that makes even the process of copying a simple line drawing a demanding task.

This is not just a matter of "hand-eye coordination," for people who can do embroidery or assemble pocket watches with effortless dexterity may still be hopelessly inept at copying drawings. One might say it is more a matter of eye-*brain* coordination. Those who master the art know that it requires special habits of attention, tricks such as slightly defocusing the eyes to permit one somehow to suppress the contribution of what one *knows* (the penny is circular, the table top is rectangular) so that one can observe the actual angles subtended by the lines in the drawing (the penny shape is elliptical, the table top trapezoidal). It often helps to superimpose an imaginary vertical and horizontal grid or pair of cross hairs, to help judge the actual angles of the lines seen. Learning to draw is largely a matter of learning to override the normal processes of vision in order to make one's experience of the item in the world *more like looking at a picture*. It can never be just like looking at a picture, but once it has been adulterated in that direction, one can, with further tricks of the trade, more or less "copy" what one experiences onto the paper.

The visual field seems to naïve reflection to be uniformly detailed and focused from the center out to the boundaries, but a simple experiment shows that this is not so. Take a deck of playing cards and remove a card face down, so that you do not yet know which it is. Hold it out at the left or right periphery of your visual field and turn its face to you, being careful to keep looking straight ahead (pick a target spot and keep looking right at it). You will find that you cannot tell even if it is red or black or a face card. Notice, though, that you are distinctly aware of any flicker of motion of the card. You are seeing motion without being able to see the shape or color of the thing that is moving. Now start moving the card toward the center of your visual field, again being careful not to shift your gaze. At what point can you identify the color? At what point the suit and number? Notice that you can tell if it is a face card long before you can tell if it is a jack, queen, or king. You will probably be surprised at how close to center you can move the card and still be unable to identify it.

This shocking deficiency in our peripheral vision (all vision except two or three degrees around dead center) is normally concealed from us by the fact that our eyes, unlike television cameras, are not steadily trained on the world but dart about in an incessant and largely unnoticed game of visual tag with the items of potential interest happening in our field of view. Either smoothly tracking or jumping in *saccades*, our eyes provide our brains with high-resolution information about whatever is momentarily occupying the central foveal area of the retinal field. (The fovea of the eye is about ten times more discriminating than the surrounding areas of the retina.)

Our visual phenomenology, the *contents* of visual experience, are in a format unlike that of any other mode of representation, neither pictures nor movies nor sentences nor maps nor scale models nor diagrams. Consider what is present in your experience when you look across a sports stadium at the jostling crowd of thousands of spectators. The individuals are too far away for you to identify, unless some large-scale and vivid property helps you out (the president — yes, you can tell it is really he, himself; he is the one you can just make out in the center of the red, white, and blue bunting). You can tell, visually, that the crowd is composed of human beings because of the visibly peoplish way they move. There is something global about your visual experience of the crowd (it looks all crowdy over there, the same way a patch of tree seen through a window can look distinctly elmy or a floor can look dusty), but you don't just see a large blob somehow marked "crowd"; you see — all at once — thousands of particular details: bobbing red hats and glinting eyeglasses, bits of blue coat, programs waved in the air, and upraised fists. If we attempted to paint an "impressionistic" rendering of your experience, the jangling riot of color blobs would *not* capture the content;

you do not have the experience of a jangling riot of color blobs, any more than you have the experience of an ellipse when you look at a penny obliquely. Paintings — colored pictures in two dimensions — may roughly approximate the retinal input from a three-dimensional scene, and hence create in you an impression that is similar to what your visual impression would be were you looking at the scene, but then the painting is not a painting *of* the resulting impression, but rather something that can provoke or stimulate such an impression.

One can no more paint a realistic picture of visual phenomenology than of justice or melody or happiness. Still it often seems apt, even irresistible, to speak of one's visual experiences as pictures in the head. That is part of how our visual phenomenology goes, and hence it is part of what must be explained in subsequent chapters.

3. OUR EXPERIENCE OF THE INTERNAL WORLD

What are the "raw materials" of our inner lives, and what do we do with them? The answers shouldn't be hard to find; presumably we just "look and see" and then write down the results.

According to the still robust tradition of the British Empiricists, Locke, Berkeley, and Hume, the senses are the entry portals for the mind's furnishings; once safely inside, these materials may be manipulated and combined *ad lib* to create an inner world of imagined objects. The way you imagine a purple flying cow is by taking the purple you got from seeing a grape, the wings you got from seeing an eagle, and attaching them to the cow you got from seeing a cow. This cannot be quite right. What enters the eye is electromagnetic radiation, and *it* does not thereupon become usable as various hues with which to paint imaginary cows. Our sense organs are bombarded with physical energy in various forms, where it is "transduced" at the point of contact into nerve impulses that then travel inward to the brain. Nothing but information passes from outside to inside, and while the receipt of information might *provoke* the creation of some phenomenological item (to speak as neutrally as possible), it is hard to believe that the information itself — which is just an abstraction made concrete in some modulated physical medium — could *be* the phenomenological item. There is still good reason, however, for acknowledging with the British Empiricists that *in some way* the inner world is dependent on sensory sources.

Vision is the sense modality that we human thinkers almost always single out as our major source of perceptual knowledge, though we readily resort to touch and hearing to confirm what our eyes have told us. This habit of ours of seeing everything in the mind through the metaphor of vision (a habit succumbed to twice in this very sentence) is a major source of distortion and confusion, as we shall see. Sight so dominates our intellectual practices that we have great difficulty conceiving of an alternative. In order to achieve understanding, we make visible diagrams and charts, so that we can "see what is happening" and if we want to "see if something is possible," we try to imagine it "in our mind's eye." Would a race of blind thinkers who relied on hearing be capable of comprehending with the aid of tunes, jingles, and squawks in the mind's ear everything we comprehend thanks to mental "images"?

Even the congenitally blind use the visual vocabulary to describe their own thought processes, though it is not yet clear the extent to which this results from their bending to the prevailing winds of the language they learn from sighted people, or from an aptness of metaphor they can recognize in spite of differences in their own thought processes, or even to their making approximately the same use as sighted people do of the visual machinery in their brains — in spite of their lacking the normal ports of entry. Answers to these questions would shed valuable light on the nature of normal human consciousness, since its mainly visual decor is one of its hallmarks.

When somebody explains something to us, we often announce our newfound comprehension by saying "I see," and this is not merely a dead metaphor. The quasivisual nature of the *phenomenology* of comprehension has been almost entirely ignored by researchers in cognitive science, particularly in Artificial Intelligence, who have attempted to create language-understanding computer systems. Why have they turned their back on the phenomenology? Probably largely because of their conviction that the phenomenology, however real and fascinating, is nonfunctional — a wheel that turns but engages none of the important machinery of comprehension.

Different listeners' phenomenology in response to the same utterance can vary almost *ad infinitum* without any apparent variation in comprehension or uptake. Consider the variation in mental imagery that might be provoked in two people who hear the sentence

Yesterday my uncle fired his lawyer.

Jim might begin by vividly recalling his ordeals of *yesterday*, interspersed with a fleeting glimpse of a diagram of the *uncle*-relation (brother of father or mother; or husband of sister of father or mother), followed by some courthouse steps and an angry old man. Meanwhile, perhaps, Sally passed imagelessly over "yesterday" and lavished attention on some variation of her uncle Bill's visage, while picturing a slamming door and the scarcely "visible" departure of some smartly suited woman labeled "lawyer." Quite independently of their mental imagery, Jim and Sally understood the sentence about equally well, as can be confirmed by a battery of subsequent paraphrases and answers to questions. Moreover, the more theoretically minded researchers will point out, imagery *couldn't* be the key to comprehension, because you can't draw a picture of an uncle, or of yesterday, or firing, or a lawyer. Uncles, unlike clowns and firemen, don't look different in any characteristic way that can be visually represented, and yesterdays don't look like anything at all. Understanding, then, cannot be accomplished by a process of converting everything to the currency of mental pictures, unless the pictured objects are identified by something like attached labels, but then the writing on these labels would be bits of verbiage in need of comprehension, putting us back at the beginning again.

My *hearing* what you say is dependent on your saying it within earshot while I am awake, which pretty much guarantees that I hear it. My *understanding* what you say is dependent on many things, but not, it seems, on any identifiable elements of internal phenomenology; no conscious experience will guarantee that I have understood you, or misunderstood you. Sally's picturing Uncle Bill may not prevent her in the slightest from understanding that it is the speaker's uncle, not her uncle, who fired his lawyer; she *knows* what the speaker meant; she is just incidentally entertaining herself with an image of Uncle Bill, with scant risk of confusion, since her comprehension of the speaker in no way depends on her imagery.[4]

Comprehension, then, cannot be accounted for by the citation of accompanying phenomenology, but that does not mean that the phenomenology is not really there. It particularly does not mean that a model of comprehension that is silent about the phenomenology will appeal to our everyday intuitions about comprehension. Surely a major source of the widespread skepticism about "machine understanding" of natural language is that such systems almost never avail themselves of anything like a "visual" workspace in which to parse or analyze the input. If they did, the sense that they were actually understanding what

4. The classic development of this theme, together with further support of varying quality, is Wittgenstein's *Philosophical Investigations* (1953).

they processed would be greatly heightened (whether or not it would still be, as some insist, an illusion). As it is, if a computer says, "I see what you mean" in response to input, there is a strong temptation to dismiss the assertion as an obvious fraud.

The temptation is certainly appealing. For instance, it's hard to imagine how anyone could get some jokes without the help of mental imagery. Two friends are sitting in a bar drinking; one turns to the other and says, "Bud, I think you've had enough — your face is getting all blurry!" Now didn't you use an image or fleeting diagram of some sort to picture the mistake the speaker was making? This experience gives us an example, it seems, of *what it feels like to come to understand something*: there you are, encountering something somewhat perplexing or indecipherable or at least as yet unknown — something that in one way or another creates the epistemic itch, when finally *Aha! I've got it!* Understanding dawns, and the item is transformed; it becomes useful, comprehended, within your control. Before time *t* the thing was not understood; after time *t*, it was understood — a clearly marked shift of state that can often be accurately timed, even though it is, emphatically, a subjectively accessible, introspectively discovered transition. It is a mistake, as we shall see, to make this the model of all comprehension, but it is certainly true that when the onset of comprehension has any phenomenology at all (when we are conscious of coming to understand something), this is the phenomenology it has.

There must be something right about the idea of mental imagery, and if "pictures in the head" is the wrong way to think about it, we will have to find some better way of thinking about it. Mental imagery comes in all modalities, not just vision. Imagine "Silent Night," being careful *not* to hum or sing as you do. Did you nevertheless "hear" the tune in your mind's ear in a particular key? If you are like me, you did. I don't have perfect pitch, so I can't tell you "from the inside" which key I just imagined it in, but if someone were to play "Silent Night" on the piano right now, I would be able to say, with great confidence, either "Yes, that's in tune with what I was imagining" or something to the effect of "No, I was imagining it about a minor third higher."[5]

---

5. A neurosurgeon once told me about operating on the brain of a young man with epilepsy. As is customary in this kind of operation, the patient was wide awake, under only local anesthesia, while the surgeon delicately explored his exposed cortex, making sure that the parts tentatively to be removed were not absolutely vital by stimulating them electrically and asking the patient what he experienced. Some stimulations pro-

Not only do we talk to ourselves silently, but sometimes we do this in a particular "tone of voice." Other times, it seems as if there are words, but not *heard* words, and at still other times, only the faintest shadows or hints of words are somehow "there" to clothe our thoughts. In the heyday of Introspectionist psychology, debates raged over whether there was such a thing as *entirely* "imageless" thought. We may leave this issue open for the time being, noting that many people confidently assert that there is, and others confidently assert that there is not. In the next chapter, we will set up a method for dealing with such conflicts. In any event, the phenomenology of vivid thought is not restricted to *talking* to oneself; we can draw pictures to ourselves in our mind's eyes, drive a stick-shift car to ourselves, touch silk to ourselves, or savor an imaginary peanut-butter sandwich.

Whether or not the British Empiricists were right to think that these merely imagined (or recollected) sensations were simply faint copies of the original sensations that "came in from outside," they can bring pleasure and suffering just like "real" sensations. As every day-dreamer knows, erotic fantasies may not be an entirely satisfactory substitute for the real thing, but they are nevertheless something one would definitely miss, if somehow prevented from having them. They not only bring pleasure; they can arouse real sensations and other well-known bodily effects. We may cry when reading a sad novel, and so may the novelist while writing it.

---

voked visual flashes or hand-raisings, others a sort of buzzing sensation, but one spot produced a delighted response from the patient: "It's 'Outta Get Me' by Guns N' Roses, my favorite heavy metal band!"

I asked the neurosurgeon if he had asked the patient to sing or hum along with the music, since it would be fascinating to learn how "high fidelity" the provoked memory was. Would it be in exactly the same key and tempo as the record? Such a song (unlike "Silent Night") has one canonical version, so we could simply have superimposed a recording of the patient's humming with the standard record and compared the results. Unfortunately, even though a tape recorder had been running during the operation, the surgeon hadn't asked the patient to sing along. "Why not?" I asked, and he replied: "I hate rock music!"

Later in the conversation the neurosurgeon happened to remark that he was going to have to operate again on the same young man, and I expressed the hope that he would just check to see if he could restimulate the rock music, and this time ask the fellow to sing along. "I can't do it," replied the neurosurgeon, "since I cut out that part." "It was part of the epileptic focus?" I asked, and he replied, "No, I already told you — I hate rock music!"

The surgical technique involved was pioneered by Wilder Penfield many years ago, and graphically described in Penfield's *The Excitable Cortex in Conscious Man* (1958).

We are all connoisseurs of the pains and pleasures of imagination, and many of us consider ourselves experts in the preparation of these episodes we enjoy so much, but we may still be surprised to learn just how powerful this faculty can become under serious training. I find it breathtaking, for instance, that when musical composition competitions are held, the contestants often do not submit tapes or records (or live performances) of their works; they submit written scores, and the judges confidently make their *aesthetic* judgments on the basis of just reading the scores and *hearing the music in their minds*. How good are the best musical imaginations? Can a trained musician, swiftly reading a score, tell *just* how that voicing of dissonant oboes and flutes over the massed strings will sound? There are anecdotes aplenty, but so far as I know this is relatively unexplored territory, just waiting for clever experimenters to move in.

Imagined sensations (if we may call these phenomenological items that) are suitable objects for aesthetic appreciation and judgment, but why, then, do the real sensations matter so much more? Why shouldn't one be willing to settle for recollected sunsets, merely anticipated spaghetti al pesto? Much of the pleasure and pain we associate with events in our lives is, after all, tied up in anticipation and recollection. The bare moments of sensation are a tiny part of what matters to us. Why — and how — things matter to us will be a topic of later chapters, but the fact that imagined, anticipated, recollected sensations are quite different from *faint* sensations can be easily brought out with another little self-experiment, which brings us to the gate of the third section of the phenom.

4. AFFECT

Close your eyes now and imagine that someone has just kicked you, very hard, in the left shin (about a foot above your foot) with a steel-toed boot. Imagine the excruciating pain in as much detail as you can; imagine it bringing tears to your eyes, imagine you almost faint, so nauseatingly sharp and overpowering is the jolt of pain you feel. You just imagined it vividly; did you feel any pain? Might you justly complain to me that following my directions has caused you some pain? I find that people have quite different responses to this exercise, but no one yet has reported that the exercise caused any actual pain. Some find it somewhat disturbing, and others find it a rather enjoyable exercise of the mind, certainly not as unpleasant as the gentlest pinch on the arm that you would call a pain.

Now suppose that you dreamed the same shin-kicking scene. Such a dream can be so shocking that it wakes you up; you might even find you were hugging your shin and whimpering, with real tears in the corners of your eyes. But there would be no inflammation, no welt, no bruise, and as soon as you were sufficiently awake and well oriented to make a confident judgment, you would say that there was no trace of pain left over in your shin — if there ever was any in the first place. Are dreamed pains real pains, or a sort of imagined pains? Or something in between? What about the pains induced by hypnotic suggestion?

At least the dreamed pains, and the pains induced by hypnosis, are states of mind that we really mind having. Compare them, however, to the states (of mind?) that arise in you while you sleep, when you roll over and inadvertently twist your arms into an awkward position, and then, without waking up, without noticing it at all, roll back into a more comfortable position. Are these pains? If you were awake, the states caused in you by such contortions would be pains. There are people, fortunately quite rare, who are congenitally insensitive to pain. Before you start to envy them, you should know that since they don't make these postural corrections during sleep (or while they are awake!), they soon become cripples, their joints ruined by continual abuse which no alarm bells curtail. They also burn themselves, cut themselves, and in other ways shorten their unhappy lives by inappropriately deferred maintenance (Cohen et al., 1955; Kirman et al., 1968).

There can be no doubt that having the alarm system of pain fibers and the associated tracts in the brain is an evolutionary boon, even if it means paying the price of having *some* alarms ring that we can't do anything about.[6] But why do pains have to *hurt* so much? Why couldn't it just be a loud bell in the mind's ear, for instance?

---

6. The literature on the evolutionary justification of pain is studded with amazingly myopic arguments. One author argues that there can be no evolutionary explanation of pain because some excruciating pains, such as the pain of gallstones, sound an alarm that no one could do anything about until the development of modern medicine. No caveman got any reproductive benefit from the pain of his gallstones, so pain — at least some pain — is an evolutionary mystery. What this author ignores is the simple fact that in order to have a pain system that can properly warn you about such avertible crises as a claw or fang jabbed in your belly, you will very likely get the bonus — which only much later can be appreciated as such — of a system that also warns you about crises you are helpless to dissolve. And by the same token, there are plenty of internal states that today it would be valuable to get pain warnings about (the onset of cancer, for instance), but to which we are oblivious presumably because our evolutionary past did not include any survival advantage for the requisite wiring (were it to emerge by mutation).

And what, if anything, are the uses of anger, fear, hatred? (I take it the evolutionary utility of lust needs no defense.) Or, to take a more complicated case, consider sympathy. Etymologically, the word means *suffering-with*. The German words for it are *Mitleid* (with-pain) and *Mitgefühl* (with-feeling). Or think of *sympathetic vibration*, in which one string of a musical instrument is set to humming by the vibration of another one nearby, closely related to it in that both share a natural resonance frequency. Suppose you witness your child's deeply humiliating or embarrassing moment; you can hardly stand it: waves of emotion sweep over you, drowning your thoughts, overturning your composure. You are primed to fight, to cry, to hit something. That is an extreme case of sympathy. Why are we designed to have those phenomena occur in us? And what are they?

This concern with the adaptive significance (if any) of the various affective states will occupy us in several later chapters. For the moment, I just want to draw attention, during our stroll, to the undeniable importance of affect to our conviction that consciousness is important. Consider *fun*, for instance. All animals *want to go on living* — at least they strive mightily to preserve themselves under most conditions — but only a few species strike us as capable of *enjoying life* or *having fun*. What comes to mind are frisky otters sliding in the snow, lion cubs at play, our dogs and cats — but not spiders or fish. Horses, at least when they are colts, seem to get a kick out of being alive, but cows and sheep usually seem either bored or indifferent. And have you ever had the thought that flying is wasted on the birds, since few if any of them seem capable of *appreciating* the deliciousness of their activity? Fun is not a trivial concept, but it has not yet, to my knowledge, received careful attention from a philosopher. We certainly won't have a complete explanation of consciousness until we have accounted for its role in permitting us (and only us?) to have fun. What are the right questions to ask? Another example will help us see what the difficulties are.

There is a species of primate in South America, more gregarious than most other mammals, with a curious behavior. The members of this species often gather in groups, large and small, and in the course of their mutual chattering, under a wide variety of circumstances, they are induced to engage in bouts of involuntary, convulsive respiration, a sort of loud, helpless, mutually reinforcing group panting that sometimes is so severe as to incapacitate them. Far from being aversive, however, these attacks seem to be sought out by most members of the species, some of whom even appear to be addicted to them.

We might be tempted to think that if only we knew what it was like to be them, from the inside, we'd understand this curious addiction of theirs. If we could see it "from their point of view," we would know what it was for. But in this case we can be quite sure that such insight as we might gain would still leave matters mysterious. For we already have the access we seek; the species is *Homo sapiens* (which does indeed inhabit South America, among other places), and the behavior is laughter.[7]

No other animal does anything like it. A biologist encountering such a unique phenomenon should first wonder what (if anything) it was *for*, and, not finding any plausible analysis of direct biological advantages it might secure, would then be tempted to interpret this strange and unproductive behavior as the price extracted for some other boon. But what? What do we do better than we otherwise would do, thanks to the mechanisms that carry with them, as a price worth paying, our susceptibility to — our near addiction to — laughter? Does laughter somehow "relieve stress" that builds up during our complicated cognitions about our advanced social lives? Why, though, should it take *funny* things to relieve stress? Why not *green* things or *simple flat* things? Or, why is *this* behavior the byproduct of relieving stress? Why don't we have a taste for standing around shivering or belching, or scratching each others' backs, or humming, or blowing our noses, or feverishly licking our hands?

Note that the *view from inside* is well known and unperplexing. We laugh *because we are amused*. We laugh because things are *funny* — and laughter is *appropriate* to funny things in a way that licking one's hands, for instance, just isn't. It is obvious (in fact it is *too* obvious) why we laugh. We laugh because of joy, and delight, and out of happiness, and because some things are hilarious. If ever there was a *virtus dormitiva* in an explanation, here it is: we laugh because of the hilarity of the stimulus.[8] That is certainly true; there is no other reason why

---

7. "What would a Martian visitor think to see a human being laugh? It must look truly horrible: the sight of furious gestures, flailing limbs, and thorax heaving in frenzied contortions." Minsky, 1985, p. 280.

8. In Molière's last play, the classic comedy *Le Malade Imaginaire* (1673), Argan, the hypochondriac of the title, solves his problems in the end by "becoming" a doctor so he can treat himself. No study is required — just a little tortured Latin. In a burlesque oral examination, he is put through his paces. Why, the examiner asks, does opium put people to sleep? Because, replies the doctoral candidate, it has a *virtus dormitiva* — the Latin for "sleep-causing power." "*Bene, bene, bene, bene respondere*," says the chorus. Well answered! How informative! What insight! And, in a more contemporary spirit we might ask: Just what is it about Cheryl Tiegs that makes her look so good in pictures?

we laugh, when we laugh sincerely. Hilarity is the constitutive cause of true laughter. Just as pain is the constitutive cause of unfeigned pain-behavior. Since this is certainly true, we must not deny it.

But we need an explanation of laughter that goes beyond this obvious truth in the same way that the standard explanations of pain and pain-behavior go beyond the obvious. We can give a perfectly sound biological account of why there should be pain and pain-behavior (indeed, we just sketched it); what we want is a similarly anchored account of why there should be hilarity and laughter.

And we can know in advance that if we actually come up with such an account, it won't satisfy everybody! Some people who consider themselves *antireductionists* complain that the biological account of pain and pain-behavior *leaves out the painfulness,* leaves out the "intrinsic awfulness" of pain that makes it what it is, and they will presumably make the same complaint about any account of laughter we can muster: it leaves out the intrinsic hilarity. This is a standard complaint about such explanations: "All you've explained is the attendant *behavior* and the *mechanisms,* but you've left out the *thing in itself,* which is the pain in all its awfulness." This raises complicated questions, which will be considered at length in chapter 12, but for the time being we can note that any account of pain that *left in* the awfulness would be circular — it would have an undischarged *virtus dormitiva* on its hands. Similarly, a proper account of laughter *must* leave out the presumed intrinsic hilarity, the zest, the funniness, because their presence would merely postpone the attempt to answer the question.

The phenomenology of laughter is hermetically sealed: we *just see* directly, naturally, without inference, with an obviousness beyond "intuition," that laughter is what goes with hilarity — it is the "right" reaction to humor. We can seem to break this down a bit: the right reaction to something funny is amusement (an internal state of mind); the natural expression of amusement (when it isn't important to conceal or suppress it, as it sometimes is) is laughter. It appears as if we now have what scientists would call an intervening variable, amusement, in between stimulus and response, and it appears to be constitutively linked at both ends. That is, amusement is by-definition-that-which-provokes-sincere-laughter, and it is also by-definition-that-which-is-provoked-by-something-funny. All this is obvious. As such it seems to

---

She's *photogenic.* So that's why! (I always wondered.) In chapter 12, the charge of vacuity that is implied by calling some explanatory posit a *virtus dormitiva* will be considered in more detail.

be in need of no further explanation. As Wittgenstein said, explanations have to stop somewhere. But all we really have here is a brute — but definitely explicable — fact of human psychology. We have to move beyond pure phenomenology if we are to explain any of these denizens of the phenomenological garden.

These examples of phenomenology, for all their diversity, seem to have two important features in common. On the one hand, they are our most intimate acquaintances; there is nothing we could know any better than the items of our personal phenomenologies — or so it seems. On the other hand, they are defiantly inaccessible to materialistic science; nothing could be less like an electron, or a molecule, or a neuron, than *the way the sunset looks to me now* — or so it seems. Philosophers have been duly impressed by both features, and have found many different ways of emphasizing what is problematic. For some, the great puzzle is the special intimacy: How can we be *incorrigible* or have *privileged access* or *directly apprehend* these items? What is the difference between our epistemic relations to our phenomenology and our epistemic relations to the objects in the external world? For others, the great puzzle concerns the unusual "intrinsic qualities" — or to use the Latin word, the *qualia* — of our phenomenology: How could anything composed of material particles *be* the fun that I'm having, or have the "ultimate homogeneity" (Sellars, 1963) of the pink ice cube I am now imagining, or matter the way my pain does to me?

Finding a materialistic account that does justice to all these phenomena will not be easy. We have made some progress, though. Our brief inventory has included some instances in which a little knowledge of the underlying mechanisms challenges — and maybe even usurps — the authority we usually grant to what is obvious to introspection. By getting a little closer than usual to the exhibits, and looking at them from several angles, we have begun to break the spell, to dissipate the "magic" in the phenomenological garden.

4

A METHOD FOR PHENOMENOLOGY

1. FIRST PERSON PLURAL

You don't do serious zoology by just strolling through the zoo, noting this and that, and marveling at the curiosities. Serious zoology demands precision, which depends on having agreed-upon methods of description and analysis, so that other zoologists can be sure they understand what you're saying. Serious phenomenology is in even greater need of a clear, neutral method of description, because, it seems, no two people use the words the same way, and everybody's an expert. It is just astonishing to see how often "academic" discussions of phenomenological controversies degenerate into desk-thumping cacophony, with everybody talking past everybody else. This is all the more surprising, in a way, because according to long-standing philosophical tradition, *we all agree* on what we find when we "look inside" at our own phenomenology.

Doing phenomenology has usually seemed to be a reliable communal practice, a matter of pooling shared observations. When Descartes wrote his *Meditations* as a first-person-singular soliloquy, he clearly expected his readers to concur with each of his observations, by performing in their own minds the explorations he described, and getting the same results. The British Empiricists, Locke, Berkeley, and Hume, likewise wrote with the presumption that what they were doing, much of the time, was *introspecting*, and that their introspections would be readily replicated by their readers. Locke enshrined this presumption in his *Essay Concerning Human Understanding* (1690) by calling his method the "historical, plain method" — no abstruse deductions or *a priori* theorizing for him, just setting down the observed facts, reminding his readers of what was manifest to all who looked. In fact, just about every author who has written about consciousness has made what we might call the *first-person-plural presumption*: Whatever mysteries consciousness may hold, *we* (you, gentle reader, and I) may speak comfortably together about our mutual acquaintances, the things we both find in our streams of consciousness. And with a few obstreperous exceptions, readers have always gone along with the conspiracy.

This would be fine if it weren't for the embarrassing fact that controversy and contradiction bedevil the claims made under these conditions of polite mutual agreement. We are fooling ourselves about something. Perhaps we are fooling ourselves about the extent to which we are all basically alike. Perhaps when people first encounter the different schools of thought on phenomenology, they join the school that sounds right to them, and each school of phenomenological description is basically right about its own members' sorts of inner life, and then just innocently overgeneralizes, making unsupported claims about how it is with everyone.

Or perhaps we are fooling ourselves about the high reliability of introspection, our personal powers of self-observation of our own conscious minds. Ever since Descartes and his "*cogito ergo sum*," this capacity of ours has been seen as somehow immune to error; we have privileged access to our own thoughts and feelings, an access guaranteed to be better than the access of any outsider. ("Imagine anyone trying to tell you that you are wrong about what you are thinking and feeling!") We are either "infallible" — always guaranteed to be right — or at least "incorrigible" — right or wrong, no one else could correct us (Rorty, 1970).

But perhaps this doctrine of infallibility is just a mistake, however well entrenched. Perhaps even if we *are* all basically alike in our phenomenology, some observers just get it all wrong when they try to describe it, but since they are so sure they are right, they are relatively invulnerable to correction. (They are incorrigible in the derogatory sense.) Either way, controversy ensues. And there is yet another possibility, which I think is much closer to the truth: what we are fooling ourselves about is the idea that the activity of "introspection" is *ever* a matter of just "looking and seeing." I suspect that when we claim to be just using our powers of inner *observation*, we are always actually engaging in a sort of impromptu *theorizing* — and we are remarkably gullible theorizers, precisely because there is so *little* to "observe" and so much to pontificate about without fear of contradiction. When we introspect, communally, we are really very much in the position of the legendary blind men examining different parts of the elephant. This seems at first to be a preposterous idea, but let us see what can be said for it.

Did anything you encountered in the tour of the phenom in the previous chapter surprise you? Were you surprised, for instance, that you could not identify the playing card until it was almost dead center in front of you? Most people, I find, are surprised — even those who know about the limited acuity of peripheral vision. If it surprised you, that must mean that had you held forth on the topic before the surprising demonstration, you would very likely have got it wrong. People often claim a direct acquaintance with more *content* in their peripheral visual field than in fact they have. Why do people make such claims? Not because they directly and incorrigibly observed themselves to enjoy such peripheral content, but because it seems to *stand to reason*. After all, you don't notice any gaping blanks in your visual field under normal conditions, and surely if there was an area there that wasn't positively colored, you'd notice the discrepancy, and besides, everywhere you look, there you find everything colored and detailed. If you think that your subjective visual field is basically an inner picture composed of colored shapes, then it stands to reason that each portion of the canvas must be colored *some* color — even raw canvas is *some* color! But that is a conclusion drawn from a dubious model of your subjective visual field, not anything you directly observe.

Am I saying we have absolutely no privileged access to our conscious experience? No, but I am saying that we tend to think we are much more immune to error than we are. People generally admit, when challenged in this way about their privileged access, that they don't have any special access to the *causes and effects* of their conscious experiences. For instance, they may be surprised to learn that they taste with their noses or hear bass notes through their feet, but they never claimed to be authoritative about the causes or sources of their experiences. They are authoritative, they say, only about the experiences themselves, in isolation from their causes and effects. But although people may *say* they are claiming authority only about the isolated contents of their experiences, not their causes and effects, they often overstep their self-imposed restraints. For instance, would you be prepared to bet on the following propositions? (I made up at least one of them.)

(1) You can experience a patch that is red and green all over at the same time — a patch that is *both* colors (not mixed) at once.
(2) If you look at a yellow circle on a blue background (in good light), and the luminance or brightness of the yellow and blue are then adjusted to be equal, the boundary between the yellow and blue disappears.
(3) There is a sound, sometimes called the auditory barber pole, which seems to keep on rising in pitch forever, without ever getting any higher.
(4) There is an herb an overdose of which makes you incapable of understanding spoken sentences in your native language. Until the effect wears off, your hearing is unimpaired, with no fuzziness or added noise, but the words you hear sound to you like an entirely foreign language, even though you somehow know they aren't.
(5) If you are blindfolded, and a vibrator is applied to a point on your arm while you touch your nose, you will feel your nose growing like Pinocchio's; if the vibrator is moved to another point, you will then have the eerie feeling of pushing your nose inside out, with your index finger coming to rest somewhere inside your skull.

In fact, I made up number 4, though for all I know it might be true. After all, in the well-studied neuropathology called prosopagnosia, your vision is completely unimpaired and you can readily identify most things by sight, but the faces of your closest friends and associates are entirely unrecognizable.[1] My point, once again, is not that you have *no* privileged access to the nature or content of your conscious experience, but just that we should be alert to *very* tempting overconfidence on that score.

During the guided tour of the phenom, I proposed a number of simple experiments for you to do. This was not in the spirit of "pure" phenomenology. Phenomenologists tend to argue that *since* we are not authoritative about the physiological causes and effects of our phenomenology, we should ignore such causes and effects in our at-

---

1. For the red and green patch, see Crane and Piantanida (1983) and Hardin (1988); for the disappearing color boundary, the Liebmann (1927) effect, see Spillman and Werner (1990); for the auditory barber pole, see Shepard (1964); for the Pinocchio effect, see Lackner (1988). For more on prosopagnosia, see Damasio, Damasio, and Van Hoesen (1982); Tranel and Damasio (1988); Tranel, Damasio, and Damasio (1988).

tempt to give a pure, neutral, pretheoretical description of what we find "given" in the course of everyday experience. Perhaps, but then just see how many curious denizens of the phenom we would never even meet! A zoologist who attempted to extrapolate the whole science from observation of a dog, a cat, a horse, a robin, and a goldfish would probably miss a few things.

2. THE THIRD-PERSON PERSPECTIVE

Since we are going to indulge in *impure* phenomenology, we need to be more careful than ever about method. The standard perspective adopted by phenomenologists is Descartes's *first-person perspective*, in which *I* describe in a monologue (which I let *you* overhear) what *I* find in *my* conscious experience, counting on *us* to agree. I have tried to show, however, that the cozy complicity of the resulting *first-person-plural perspective* is a treacherous incubator of errors. In the history of psychology, in fact, it was the growing recognition of this methodological problem that led to the downfall of Introspectionism and the rise of Behaviorism. The Behaviorists were meticulous about avoiding speculation about what was going on in *my* mind or *your* mind or *his* or *her* or *its* mind. In effect, they championed the *third-person perspective*, in which only facts garnered "from the outside" count as data. You can videotape people in action and then measure error rates on tasks involving bodily motion, or reaction times when pushing buttons or levers, pulse rate, brain waves, eye movements, blushing (so long as you have a machine that measures it objectively), and galvanic skin response (the electrical conductivity detected by "lie detectors"). You can open up subjects' skulls (surgically or by brain-scanning devices) to see what is going on in their *brains*, but you must not make any *assumptions* about what is going on in their *minds*, for that is something you can't get any data about while using the intersubjectively verifiable methods of physical science.

The idea at its simplest was that since you can never "see directly" into people's minds, but have to take their word for it, any such facts as there are about mental events are not among the data of science, since they can never be properly verified by objective methods. This *methodological* scruple, which is the ruling principle of *all* experimental psychology and neuroscience today (not just "behaviorist" research), has too often been elevated into one or another *ideological* principle, such as:

Mental events don't exist. (Period! — this has been well called "barefoot behaviorism.")
Mental events exist, but they have no effects whatever, so science can't study them (epiphenomenalism — see chapter 12, section 5).
Mental events exist, and have effects, but *those* effects can't be studied by science, which will have to content itself with theories of the "peripheral" or "lower" effects and processes in the brain. (This view is quite common among neuroscientists, especially those who are dubious of "theorizers." It is actually dualism; these researchers apparently agree with Descartes that the mind is *not* the brain, and they are prepared to settle for having a theory of the brain alone.)

These views all jump to one unwarranted conclusion or another. Even if mental events are not among the *data* of science, this does not mean we cannot study them scientifically. Black holes and genes are not among the data of science, but we have developed good scientific theories of them. The challenge is to construct a theory of mental events, using the data that scientific method permits.

Such a theory will have to be constructed from the third-person point of view, since *all* science is constructed from that perspective. Some people will tell you that such a theory of the conscious mind is impossible. Most notably, the philosopher Thomas Nagel has claimed that

> There are things about the world and life and ourselves that cannot be adequately understood from a maximally objective standpoint, however much it may extend our understanding beyond the point from which we started. A great deal is essentially connected to a particular point of view, or type of point of view, and the attempt to give a complete account of the world in objective terms detached from these perspectives inevitably leads to false reductions or to outright denial that certain patently real phenomena exist at all. [Nagel, 1986, p. 7]

We shall see. It is premature to argue about what can and can't be accounted for by a theory until we see what the theory actually says. But if we are to give a fair hearing to a theory, in the face of such skepticism, we will need to have a neutral way of *describing the data* — a way that does not prejudge this issue. It might seem that no such method could exist, but in fact there is such a neutral method, which I will first describe, and then adopt.

3. THE METHOD OF HETEROPHENOMENOLOGY[2]

The term is ominous; not just phenomenology but *hetero*phenomenology. What can it be? It is in fact something familiar to us all, layman and scientist alike, but we must introduce it with fanatical caution, noting exactly what it presupposes and implies, since it involves taking a giant theoretical step. Ignoring all tempting shortcuts, then, here is the *neutral* path leading from objective physical science and its insistence on the third-person point of view, to a method of phenomenological description that can (in principle) do justice to the most private and ineffable subjective experiences, while never abandoning the methodological scruples of science.

We want to have a theory of consciousness, but there is controversy about just which entities have consciousness. Do newborn human babies? Do frogs? What about oysters, ants, plants, robots, zombies . . .? We should remain neutral about all this for the time being, but there is one class of entities that is held by just about everyone to exhibit consciousness, and that is our fellow adult human beings.

Now, some of these adult human beings *may* be zombies — in the philosophers' "technical" sense. The term *zombie* apparently comes from Haitian voodoo lore and refers, in that context, to a "living dead" person, punished for some misdeed and doomed to shuffle around, mumbling and staring out of dead-looking eyes, mindlessly doing the bidding of some voodoo priest or shaman. We have all seen zombies in horror movies, and they are immediately distinguishable from normal people. (Roughly speaking, Haitian zombies can't dance, tell jokes, hold animated philosophical discussions, keep up their end in a witty conversation — and they look just awful.)[3] But philosophers use the

---

2. This and the following sections draw on several earlier accounts of mine of the methodological underpinnings of heterophenomenology: Dennett (1978c, 1982a).

3. Several years ago, Wade Davis, a young Harvard-trained anthropologist, announced that he had deciphered the mystery of voodoo zombies, and in his book *The Serpent and the Rainbow* (1985) described the neuropharmacological potion prepared by voodoo practitioners that can putatively put human beings into a deathlike state; after being buried alive for several days, these unfortunate people are sometimes exhumed and given a hallucinogen that causes some disorientation and amnesia. As a result of either the hallucinogen or brain damage caused by oxygen deprivation during their entombment, they then do indeed shuffle about roughly the way the zombies in the

term *zombie* for a different category of imaginary human being. According to common agreement among philosophers, a zombie is or would be a human being who exhibits perfectly natural, alert, loquacious, vivacious behavior but is in fact not conscious at all, but rather some sort of automaton. The whole point of the philosopher's notion of zombie is that you can't tell a zombie from a normal person by examining external behavior. Since that is all we ever get to see of our friends and neighbors, *some of your best friends may be zombies.* That, at any rate, is the tradition I must be neutral about at the outset. So, while the method I describe makes no assumption about the *actual consciousness* of any apparently normal adult human beings, it does focus on this class of normal adult human beings, since if consciousness is anywhere, it is in them. Once we have seen what the outlines of a theory of human consciousness might be, we can turn our attention to the consciousness (if any) of other species, including chimpanzees, dolphins, plants, zombies, Martians, and pop-up toasters (philosophers often indulge in fantasy in their thought experiments).

Adult human beings (henceforth, people) are studied in many sciences. Their bodies are probed by biologists and medical researchers, nutritionists, and engineers (who ask such questions as: How fast can human fingers type? What is the tensile strength of human hair?). They are also studied by psychologists and neuroscientists, who place individual people, called *subjects*, in various experimental situations. For most experiments, the subjects first must be categorized and prepared. Not only must it be established how old they are, which gender, right- or left-handed, how much schooling, and so forth, but they must be *told what to do*. This is the most striking difference between human subjects and, say, the biologist's virus cultures, the engineer's samples of exotic materials, the chemist's solutions, the animal psychologist's rats, cats, and pigeons.

People are the only objects of scientific study the preparation of which typically (but not always) involves verbal communication. This is partly a matter of the ethics of science: people may not be used in experiments without their informed consent, and it is simply not

---

movies do, and on occasion they may have been enslaved. Because of the sensational nature of Davis's claims (and the film loosely based on his novelistic book), his discoveries have met with an undercurrent of skepticism in some quarters, but these are well rebutted in a second, more scholarly, book, *Passage of Darkness: The Ethnobiology of the Haitian Zombie* (1988). See also Booth (1988) and Davis (1988b).

possible to obtain informed consent without verbal interaction. But even more important, from our point of view, is the fact that verbal communication is used to set up and constrain the experiments. Subjects are asked to perform various intellectual tasks, solve problems, look for items in displays, press buttons, make judgments, and so forth. The validity of most experiments depends on this preparation being done uniformly and successfully. If it turns out, for instance, that the instructions were given in Turkish to subjects whose only language was English, the failure of the experiment is pretty well guaranteed. In fact, evidence of even minor misunderstandings of instructions can compromise experiments, so it is a matter of some concern that this practice of preparing human subjects with verbal communication be validated.

What is involved in this practice of talking to subjects? It is an ineliminable element in psychological experiments, but does it presuppose the consciousness of the subjects? Don't experimenters then end up back with the Introspectionists, having to take a subject's untestable word for what he or she understands? Don't we run some risk of being taken in by zombies or robots or other impostors?

We must look more closely at the details of a generic human subject experiment. Suppose, as is often the case, that multiple recordings are made of the entire experiment: videotape and sound tape, and electroencephalograph, and so forth. Nothing that is not thus recorded will we count as data. Let's focus on the recording of sounds — vocal sounds mainly — made by the subjects and experimenters during the experiment. Since the sounds made by the subjects are made by physical means, they are in principle explainable and predictable by physics, using the same principles, laws, models that we use to explain and predict automobile engine noises or thunderclaps. Or, since the sounds are made by physiological means, we could add the principles of physiology and attempt to explain the sounds using the resources of that science, just as we explain belches, snores, growling stomachs, and creaking joints. But the sounds we are primarily interested in, of course, are the vocal sounds, and more particularly the subset of them (ignoring the occasional burps, sneezes, and yawns) that are *apparently* amenable to a linguistic or semantic analysis. It is not always obvious just which sounds to include in this subset, but there is a way of playing it safe: we give copies of the tape recordings to three trained stenographers and have them independently prepare *transcripts* of the raw data.

This simple step is freighted with implications; we move by it from one world — the world of mere physical sounds — into another: the world of words and meanings, syntax and semantics. This step yields a radical reconstrual of the data, an abstraction from its acoustic and other physical properties to strings of words (though still adorned with precise timing — see, e.g., Ericsson and Simon, 1984). What governs this reconstrual? Although there presumably are regular and discoverable relationships between the physical properties of the acoustic wave recorded on the tape and the *phonemes* that the typists hear and then further transcribe into words, we don't yet know enough about the relationships to describe them in detail. (If we did, the problem of making a machine that could take dictation would be solved. Although great progress has been made on this, there are still some major perplexities.) Pending the completion of that research in acoustics and phonology, we can still trust our transcripts as objective renditions of the data so long as we take a few elementary precautions. First, having stenographers prepare the transcripts (instead of entrusting that job to the experimenter, for instance) guards against both willful and unwitting bias or overinterpretation. (Court stenographers fulfill the same neutral role.) Having three independent transcripts prepared gives us a measure of how objective the process is. Presumably, if the recording is good, the transcripts will agree word-for-word on all but a tiny fraction of one percent of the words. Wherever the transcripts disagree, we can simply throw out the data if we wish, or use agreement of two out of three transcripts to fix the transcript of record.

The transcript or text is not, strictly speaking, *given* as data, for, as we have seen, it is created by putting the raw data through a process of interpretation. This process of interpretation depends on assumptions about which language is being spoken, and on some of the speaker's intentions. To bring this out clearly, compare the task we have given the stenographers with the task of typing up transcripts of recordings of birdsongs or pig grunts. When the human speaker utters "Djamind if a push da buddin wid ma leff hand" the stenographers all agree that he asked, "Do you mind if I push the button with my left hand?" — but that is because they know English, and this is what makes sense, obviously, in the context. And if the subject *says*, "Now the spot is moving from reft to light" we will allow the stenographers to improve this to "Now the spot is moving from left to right." No similar purification strategy is available for transcribing birdsongs or pig grunts — at least not until some researcher discovers that there are norms for such noises, and devises and codifies a description system.

We effortlessly — in fact involuntarily — "make sense" of the sound stream in the process of turning it into words. (We had better allow the stenographers to change "from reft to light" to "from left to right," for they will probably change it without even noticing.) The fact that the process is both highly reliable and all but unnoticeable in normal circumstances should not conceal from us the fact that it is a sophisticated process even when it doesn't proceed all the way to understanding but stops short at word recognition. When the stenographer transcribes "To me, there was a plangent sort of thereness to my presentiment, a beckoning undercurrent of foretaste and affront, a manifold of anticipatory confirmations that revealed surfaces behind surfaces," he may not have the faintest idea what this means, but be quite certain that those were indeed the words the speaker intended to speak, and succeeded in speaking, whatever they mean.

It is always possible that the speaker also had no idea what the words mean. The subject, after all, just *might* be a zombie, or a parrot dressed up in a people suit, or a computer driving a speech-synthesizer program. Or, less extravagantly, the subject may have been confused, or in the grip of some ill-understood theory, or trying to play a trick on the experimenter by spouting a lot of nonsense. For the moment, I am saying, the process of creating a transcript or text from the data record is neutral with regard to all these strange possibilities, even though it proceeds with the methodological assumption that there is a text to be recovered. When no text can be recovered, we had best throw out the data on that subject and start over.

So far, the method described is cut-and-dried and uncontroversial. We have reached the bland conclusion that we can turn tape recordings into texts without giving up science. We have taken our time securing this result, because the next step is the one that creates the opportunity to study consciousness empirically, but also creates most of the obstacles and confusions. We must move beyond the text; we must interpret it as a record of *speech acts;* not mere pronunciations or recitations but assertions, questions, answers, promises, comments, requests for clarification, out-loud musings, self-admonitions.

This sort of interpretation calls for us to adopt what I call the *intentional stance* (Dennett, 1971, 1978a, 1987a): we must treat the noise-emitter as an agent, indeed a rational agent, who harbors beliefs and desires and other mental states that exhibit *intentionality* or "aboutness," and whose actions can be explained (or predicted) on the basis of the content of these states. Thus the uttered noises are to be interpreted as things the subjects *wanted to say*, of *propositions* they meant to *assert*, for instance, for various *reasons*. In fact, we were already relying on some such assumptions in the previous step of purifying the text. (We reason: Why would anyone *want to say* "from reft to light"?)

Whatever dangers we run by adopting the intentional stance toward these verbal behaviors, they are the price we must pay for gaining access to a host of reliable truisms we exploit in the design of experiments. There are many reasons for wanting to say things, and it is important to exclude some of these by experimental design. Sometimes, for instance, people want to say things not because they believe them but because they believe their audience wants to hear them. It is usually important to take the obvious steps to diminish the likelihood that this desire is present or effective: we tell subjects that what we want to hear is *whatever they believe*, and we take care not to let them know what it is we hope they believe. We do what we can, in other words, to put them in a situation in which, given the desires we have inculcated in them (the desire to cooperate, to get paid, to be a good subject), they will have no better option than to try to say what in fact they believe.

Another application of the intentional stance toward our subjects is required if we are to avail ourselves of such useful event-types as button-pushing. Typically, pushing a button is a way of performing some conventionally fixed speech act, such as *asserting that* the two seen figures appear superimposed to me right *now*, or answering that *yes*, my hurried, snap judgment (since you have told me that speed is of the essence) is that the word that I have just heard was on the list I heard a little while ago. For many experimental purposes, then, we will want to unpack the meaning of these button-pushes and incorporate them as elements of the text. Which speech act a particular button-pushing can be taken to execute depends on the intentional interpretation of the interactions between subject and experimenter that were involved in preparing the subject for the experiment. (Not all button-pushing consists in speech acts; some may be make-believe shooting, or make-believe rocket-steering, for instance.)

When doubts arise about whether the subject has said what he meant, or understood the problem, or knows the meanings of the words being used, we can ask for clarifications. Usually we can resolve the doubts. Ideally, the effect of these measures is to remove all likely sources of ambiguity and uncertainty from the experimental situation, so that *one* intentional interpretation of the text (including the button-pushings) has no plausible rivals. It is taken to be *the* sincere, reliable expression by a *single, unified subject* of that very subject's beliefs and opinions.[4] As we shall see, though, there are times when this

4. In "How to Change Your Mind," in *Brainstorms* (1978a), I adopt a conventional use of "opinion" that permits me to draw a distinction between beliefs proper and other

presumption is problematic — especially when our subjects exhibit one pathology or another. What should we make, for instance, of the apparently sincere complaints of blindness in cases of so-called hysterical blindness, and the apparently sincere denials of blindness in blind people with anosognosia (blindness denial or Anton's syndrome)? These phenomena will be examined in later chapters, and if we are to get at what these people are experiencing, it will not be by any straightforward interview alone.

4. FICTIONAL WORLDS AND HETEROPHENOMENOLOGICAL WORLDS

In addition to the particular problems raised by strange cases, there may seem to be a general problem. Doesn't the very practice of interpreting verbal behavior in this way presuppose the consciousness of the subject and hence beg the zombie question? Suppose you are confronted by a "speaking" computer, and suppose you succeed in interpreting its output as speech acts expressing its beliefs and opinions, presumably "about" its conscious states. The fact that *there is* a single, coherent interpretation of a sequence of behavior doesn't establish that the interpretation is *true;* it might be only *as if* the "subject" were conscious; we risk being taken in by a zombie with no inner life at all. You could not *confirm* that the computer was conscious of anything by this method of interpretation. Fair enough. We can't be sure that the speech acts we observe express real beliefs about actual experiences; perhaps they express only *apparent* beliefs about *nonexistent* experiences. Still, the fact that we had found even one stable interpretation of some entity's behavior as speech acts would always be a fact worthy of attention. Anyone who found an intersubjectively uniform way of interpreting the waving of a tree's branches in the breeze as "commentaries" by "the weather" on current political events would have found something wonderful demanding an explanation, even if it turned out to be effects of an ingenious device created by some prankish engineer.

Happily, there is an analogy at hand to help us *describe* such facts

---

more language-infected states, which I call opinions. Animals without language can have beliefs, but not opinions. People have both, but if you believe that tomorrow is Friday, this should in my terms be called your opinion that tomorrow is Friday. It is not the sort of cognitive state one could have without language. While I will not presuppose familiarity with that distinction here, I do intend my claims to apply to both categories.

without at the same time presumptively *explaining* them: We can compare the heterophenomenologist's task of interpreting subjects' behavior to the reader's task of interpreting a work of fiction. Some texts, such as novels and short stories, are known — or assumed — to be fictions, but this does not stand in the way of their interpretation. In fact, in some regards it makes the task of interpretation easier, by canceling or postponing difficult questions about sincerity, truth, and reference.

Consider some uncontroversial facts about the semantics of fiction (Walton, 1973, 1978; Lewis, 1978; Howell, 1979). A novel tells a story, but not a true story, except by accident. In spite of our knowledge or assumption that the story told is not true, we can, and do, speak of what is *true in the story*. "We can truly say that Sherlock Holmes lived in Baker Street and that he liked to show off his mental powers. We cannot truly say that he was a devoted family man, or that he worked in close cooperation with the police" (Lewis, 1978, p. 37). What is true in the story is much, much more than what is explicitly asserted in the text. It is true that there are no jet planes in Holmes's London (though this is not asserted explicitly or even logically implied in the text), but also true that there are piano tuners (though — as best I recall — none is mentioned, or, again, logically implied). In addition to what is true and false in the story, there is a large indeterminate area: while it is true that Holmes and Watson took the 11:10 from Waterloo Station to Aldershot one summer's day, it is neither true nor false that that day was a Wednesday ("The Crooked Man").

There are delicious philosophical problems about how to say (strictly) all the things we unperplexedly want to say when we talk about fiction, but these will not concern us. Perhaps some people are deeply perplexed about the metaphysical status of fictional people and objects, but not I. In my cheerful optimism I don't suppose there is any deep philosophical problem about the way we should respond, ontologically, to the results of fiction; fiction is *fiction;* there *is no* Sherlock Holmes. Setting aside the intricacies, then, and the ingenious technical proposals for dealing with them, I want to draw attention to a simple fact: the interpretation of fiction is undeniably do-able, with certain uncontroversial results. First, the fleshing out of the story, the exploration of "the world of Sherlock Holmes," for instance, is not pointless or idle; one can learn a great deal about a novel, about its text, about the point, about the author, even about the real world, by learning about *the world portrayed* by the novel. Second, if we are cautious about identifying and excluding judgments of taste or preference (e.g.,

"Watson is a boring prig"), we can amass a volume of unchallengeably objective fact about the world portrayed. All interpreters agree that Holmes was smarter than Watson; in crashing obviousness lies objectivity.

Third — and this fact is a great relief to students — knowledge of the world portrayed by a novel can be independent of knowledge of the actual text of the novel. I could probably write a passing term paper on *Madame Bovary*, but I've never read the novel — even in English translation. I've seen the BBC television series, so I know the story. I know what happens in that world. The general point illustrated is this: facts about the world of a fiction are purely *semantic level* facts about that fiction; they are independent of the syntactic facts about the text (if the fiction is a text). We can compare the stage musical or the film *West Side Story* with Shakespeare's play *Romeo and Juliet;* by describing similarities and differences in what happens in those worlds, we see similarities in the works of art that are not describable in the terms appropriate to the syntactical or textual (let alone physical) description of the concrete instantiations of the fictions. The fact that in each world there is a pair of lovers who belong to different factions is not a fact about the vocabulary, sentence structure, length (in words or frames of film), or size, shape, and weight of any particular physical instantiation of the works.

In general, one can describe what is represented in a work of art (e.g., *Madame Bovary*) independently of describing *how* the represent*ing* is accomplished. (Typically, of course, one doesn't try for this separation, and mixes commentary on the world portrayed with commentary on the author's means of accomplishing the portrayal, but the separation is possible.) One can even imagine knowing enough about a world portrayed to be able to identify the author of a fiction, in ignorance of the text or anything purporting to be a faithful translation. Learning indirectly what happens in a fiction one might be prepared to claim: only Wodehouse could have invented that preposterous misadventure. We think we can identify sorts of events and circumstances (and not merely sorts of *descriptions* of events and circumstances) as Kafkaesque, and we are prepared to declare characters to be pure Shakespeare. Many of these plausible convictions are no doubt mistaken (as ingenious experiments might show), but not all of them. I mention them just to illustrate how much one might be able to glean just from *what is represented*, in spite of having scant knowledge of *how the representing* is accomplished.

Now let's apply the analogy to the problem facing the experimenter who wants to interpret the texts produced by subjects, without begging any questions about whether his subjects are zombies, computers, lying, or confused. Consider the advantages of adopting the tactic of interpreting these texts as fictions of a sort, not as literature of course, but as generators of a *theorist's fiction* (which might, of course, prove to be true after all). The reader of a novel lets the text *constitute* a (fictional) world, a world determined by fiat by the text, exhaustively extrapolated as far as extrapolation will go and indeterminate beyond; our experimenter, the heterophenomenologist, lets the subject's text *constitute* that subject's *heterophenomenological world,* a world determined by fiat by the text (as interpreted) and indeterminate beyond. This permits the heterophenomenologist to postpone the knotty problems about what the relation might be between that (fictional) world and the real world. This permits theorists to agree in detail about just what a subject's heterophenomenological world *is,* while offering entirely different accounts of how heterophenomenological worlds map onto events in the brain (or the soul, for that matter). The subject's heterophenomenological world will be a stable, intersubjectively confirmable theoretical posit, having the same metaphysical status as, say, Sherlock Holmes's London or the world according to Garp.

As in fiction, what the author (the apparent author) says goes. More precisely, what the apparent author says provides a text that, when interpreted according to the rules just mentioned, goes to stipulate the way a certain "world" *is.* We don't ask how Conan Doyle came to know the color of Holmes's easy chair, and we don't raise the possibility that he might have got it wrong; we do correct typographical errors and otherwise put the best, most coherent, reading on the text we can find. Similarly, we don't ask how subjects (the apparent subjects) know what they assert, and we don't (at this point) even *entertain* the possibility that they might be mistaken; we take them at their (interpreted) word. Note, too, that although novels often include a proviso to the effect that the descriptions therein are not intended to portray any real people, living or dead, the tactic of letting a text constitute a world need not be restricted to literary works *intended* as fiction by their authors; we can describe a certain biographer's Queen Victoria, or the world of Henry Kissinger, with blithe disregard of the author's presumed intentions to be telling the truth and to be referring, noncoincidentally, to real people.

5. THE DISCREET CHARM OF THE ANTHROPOLOGIST

This way of treating people as generators of a (theorists') fiction is not our normal way of treating people. Simply *conceding* constitutive authority to their pronouncements can be rather patronizing, offering mock respect in the place of genuine respect. This comes out clearly in a slightly different application of the heterophenomenological tactic by anthropologists. An example will make the point clear. Suppose anthropologists were to discover a tribe that believed in a hitherto-unheard-of god of the forest, called Feenoman. Upon learning of Feenoman, the anthropologists are faced with a fundamental choice: they may convert to the native religion and believe wholeheartedly in the real existence and good works of Feenoman, or they can *study the cult* with an agnostic attitude. Consider the agnostic path. While not believing in Feenoman, the anthropologists nevertheless decide to study and systematize as best they can the religion of these people. They set down descriptions of Feenoman given by native informants. They look for agreement, but don't always find it (some say Feenoman is blue-eyed, others say he — or she — is brown-eyed). They seek to explain and eliminate these disagreements, identifying and ignoring the wise-guys, exploring reformulations with their informants, and perhaps even mediating disputes. Gradually a logical construct emerges: Feenoman the forest god, complete with a list of traits and habits and a biography. These agnostic scientists (who call themselves Feenomanologists), have described, ordered, catalogued a part of the world constituted by the beliefs of the natives, and (if they have done their job of interpretation well) have compiled the *definitive* description of Feenoman. The beliefs of the native believers (Feenomanists, we may call them) are authoritative (he's *their* god, after all), but only because Feenoman is being treated as *merely* an "intentional object," a mere fiction so far as the infidels are concerned, and hence as entirely a *creature* of the beliefs (true or false) of the Feenomanists. Since those beliefs may contradict each other, Feenoman, as logical construct, may have contradictory properties attributed to him — but that's all right in the Feenomanologists' eyes since he is *only* a construct to them. The Feenomanologists try to present the best logical construct they can, but they have no overriding obligation to resolve all contradictions. They are prepared to discover unresolved and undismissible disagreements among the devout.

Feenomanists, of course, don't see it that way — by definition, for they are the believers to whom Feenoman is no *mere* intentional object, but someone as real as you or I. Their attitude toward their own authority about the traits of Feenoman is — or ought to be — a bit more complicated. On the one hand they do believe that they *know* all about Feenoman — they are Feenomanists, after all, and who should know better than they? Yet unless they hold themselves to have something like papal infallibility, they allow as how they could in principle be wrong in some details. They could just possibly be instructed about the true nature of Feenoman. For instance, Feenoman himself might set them straight about a few details. So they should be slightly ill at ease about the bland credulity (as it appears to them) of the investigating Feenomanologists, who almost always take them scrupulously at their word, never challenging, never doubting, only respectfully asking how to resolve ambiguities and apparent conflicts. A native Feenomanist who fell in with the visiting anthropologists and adopted their stance would have to adopt an attitude of distance or neutrality toward his own convictions (or shouldn't we say his own *prior* convictions?), and would in the process pass from the ranks of the truly devout.

The heterophenomenological method neither challenges nor accepts as entirely true the assertions of subjects, but rather maintains a constructive and sympathetic neutrality, in the hopes of compiling a *definitive* description of the world according to the subjects. Any subject made uneasy by being granted this constitutive authority might protest: "No, *really!* These things I am describing to you are *perfectly real*, and have exactly the properties I am asserting them to have!" The heterophenomenologist's honest response might be to nod and assure the subject that of course his sincerity was not being doubted. But since believers in general want more — they want their assertions to be believed and, failing that, they want to know whenever their audience disbelieves them — it is in general more politic for heterophenomenologists, whether anthropologists or experimenters studying consciousness in the laboratory, to avoid drawing attention to their official neutrality.

That deviation from normal interpersonal relations is the price that must be paid for the neutrality a science of consciousness demands. Officially, we have to keep an open mind about whether our apparent subjects are liars, zombies, or parrots dressed up in people suits, but we don't have to risk upsetting them by advertising the fact. Besides, this tactic of neutrality is only a temporary way station on the path to devising and confirming an empirical theory that could in principle vindicate the subjects.

6. DISCOVERING WHAT SOMEONE IS REALLY TALKING ABOUT

What would it be to confirm subjects' beliefs in their own phenomenology? We can see the possibilities better with the help of our analogies. Consider how we might confirm that some "novel" was in fact a true (or largely true) biography. We might begin by asking: Upon what real person in the author's acquaintance is this character modeled? Is this character really the author's mother in disguise? What real events in the author's childhood have been transmogrified in this fictional episode? What is the author *really* trying to say? Asking the author might well not be the best way of answering these questions, for the author may not really know. Sometimes it can plausibly be argued that the author has been forced, unwittingly, to express himself allegorically or metaphorically. The only expressive resources available to the author — for whatever reason — did not permit a direct, factual, unmetaphorical recounting of the events he wished to recount; the story he has composed is a compromise or net effect. As such it may be drastically reinterpreted (if necessary over the author's anguished protests) to reveal a true tale, about real people and real events. Since, one may sometimes argue, it is surely no coincidence that such-and-such a fictional character has these traits, we may reinterpret the text that portrays this character in such a way that its terms can then be seen to refer — in genuine, nonfictional reference — to the traits and actions of a real person. Portraying fictional Molly as a slut may quite properly be seen as slandering real Polly, for all the talk about Molly is *really* about Polly. The author's protestations to the contrary may convince us, rightly or wrongly, that the slander is not, in any event, a conscious or deliberate slander, but we have long since been persuaded by Freud and others that authors, like the rest of us, are often quite in the dark about the deeper wellsprings of their intentions. If there can be unconscious slander, there must be unwitting reference to go along with it.

Or, to revert to our other analogy, consider what would happen if an anthropologist confirmed that there really was a blue-eyed fellow named Feenoman who healed the sick and swung through the forest like Tarzan. Not a god, and not capable of flying or being in two places at once, but still undoubtedly the real source of most of the sightings, legends, beliefs of the Feenomanists. This would naturally occasion some wrenching disillusionment among the faithful, some perhaps in favor of revision and diminution of the creed, others holding out for the orthodox version, even if it means yoking up the "real" Feenoman (supernatural properties intact) in parallel with his flesh-and-blood agent in the world. One could understand the resistance of the orthodox to the idea that they could have been that wrong about Feenoman. And unless the anthropologists' candidate for the real referent of Feenomanist doctrine bore a striking resemblance, in properties and deeds, to the Feenoman constituted by legend, they would have no warrant for proposing any such discovery. (Compare: "I have discovered that Santa Claus is real. He is in fact a tall, thin violinist living in Miami under the name of Fred Dudley; he hates children and never buys gifts.")

My suggestion, then, is that if we were to find real goings-on in people's brains that had *enough* of the "defining" properties of the items that populate their heterophenomenological worlds, we could reasonably propose that we had discovered what they were *really* talking about — even if they initially resisted the identifications. And if we discovered that the real goings-on bore only a minor resemblance to the heterophenomenological items, we could reasonably declare that people were just mistaken in the beliefs they expressed, in spite of their sincerity. It would always be open to someone to insist — like the diehard Feenomanist — that the real phenomenological items *accompanied* the goings-on without being identical to them, but whether or not this claim would carry conviction is another matter.

Like anthropologists, we can remain neutral while exploring the matter. This neutrality may seem pointless — isn't it simply unimaginable that scientists might discover neurophysiological phenomena that *just were* the items celebrated by subjects in their heterophenomenologies? Brain events seem too different from phenomenological items to be the real referents of the beliefs we express in our introspective reports. (As we saw in chapter 1, mind stuff seems to be needed to be the stuff out of which purple cows and the like are composed.) I suspect that most people still do find the prospect of this identification utterly unimaginable, but rather than concede that it is therefore impossible, I want to try to stretch our imaginations some more, with yet another fable. This one closes in somewhat on a particularly puzzling phenomenological item, the *mental image*, and has the virtue of being largely a true story, somewhat simplified and embellished.

7. SHAKEY'S MENTAL IMAGES

In the short history of robots, Shakey, developed at Stanford Research Institute in Menlo Park, California, in the late 1960s by Nils Nilsson, Bertram Raphael, and their colleagues, deserves legendary status, not because he did anything particularly well, or was a particularly realistic simulation of any feature of human psychology, but because in his alien way he opened up some possibilities of thought and closed down others (Raphael, 1976; Nilsson, 1984). He was the sort of robot a philosopher could admire, a sort of rolling argument.

Figure 4.1

Shakey was a box on wheels with a television eye, and instead of carrying his brain around with him, he was linked to it (a large stationary computer back in those days) by radio. Shakey lived indoors in a few rooms in which the only other objects were a few boxes, pyramids, ramps, and platforms, carefully colored and lit to make "vision" easier for Shakey. One could communicate with Shakey by typing messages at a terminal attached to his computer brain, in a severely restricted vocabulary of semi-English. "PUSH THE BOX OFF THE PLATFORM" would send Shakey out, finding the box, locating a ramp, pushing the ramp into position, rolling up the ramp onto the platform, and pushing the box off.

Now how did Shakey do this? Was there, perhaps, a human midget inside Shakey, looking at a TV screen and pushing control buttons? Such a single, smart homunculus would be one — cheating — way of doing it. Another way would be by locating a human controller outside Shakey, in radio remote control. This would be the Cartesian solution, with the transmitter/receiver in Shakey playing the role of the pineal gland, and radio signals being the nonmiraculous stand-in for Descartes's nonphysical soul-messages. The emptiness of these "solutions" is obvious; but what *could* a nonempty solution be? It may seem inconceivable at first — or at least unimaginably complex — but it is just such obstacles to imagination we need to confront and overcome. It turns out to be easier than you may have supposed to imagine how Shakey performed his deeds without the help of a *homo ex machina*.

How, in particular, did Shakey distinguish boxes from pyramids with the aid of his television eye? The answer, in outline, was readily apparent to observers, who could watch the process happen on a computer monitor. A single frame of grainy television, an image of a box, say, would appear on the monitor; the image would then be purified and rectified and sharpened in various ways, and then, marvelously, the boundaries of the box would be outlined in white — and the entire image turned into a line drawing (Figure 4.3, page 88).

Then Shakey would analyze the line drawing; each vertex was identifiable as either an L or a T or an X or an arrow or a Y. If a Y vertex was discovered, the object had to be a box, not a pyramid; from no vantage point would a pyramid project a Y vertex.

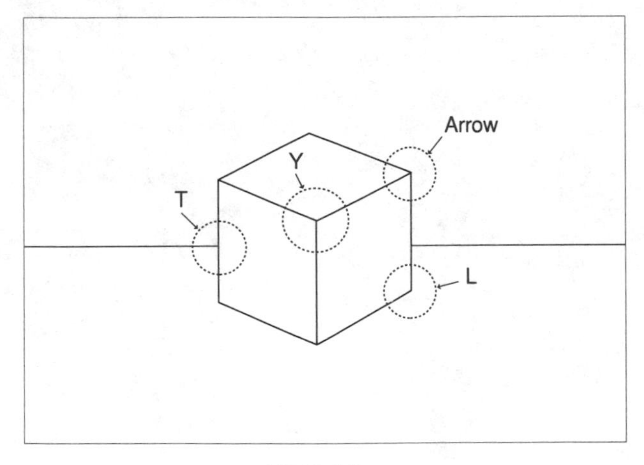

Figure 4.2

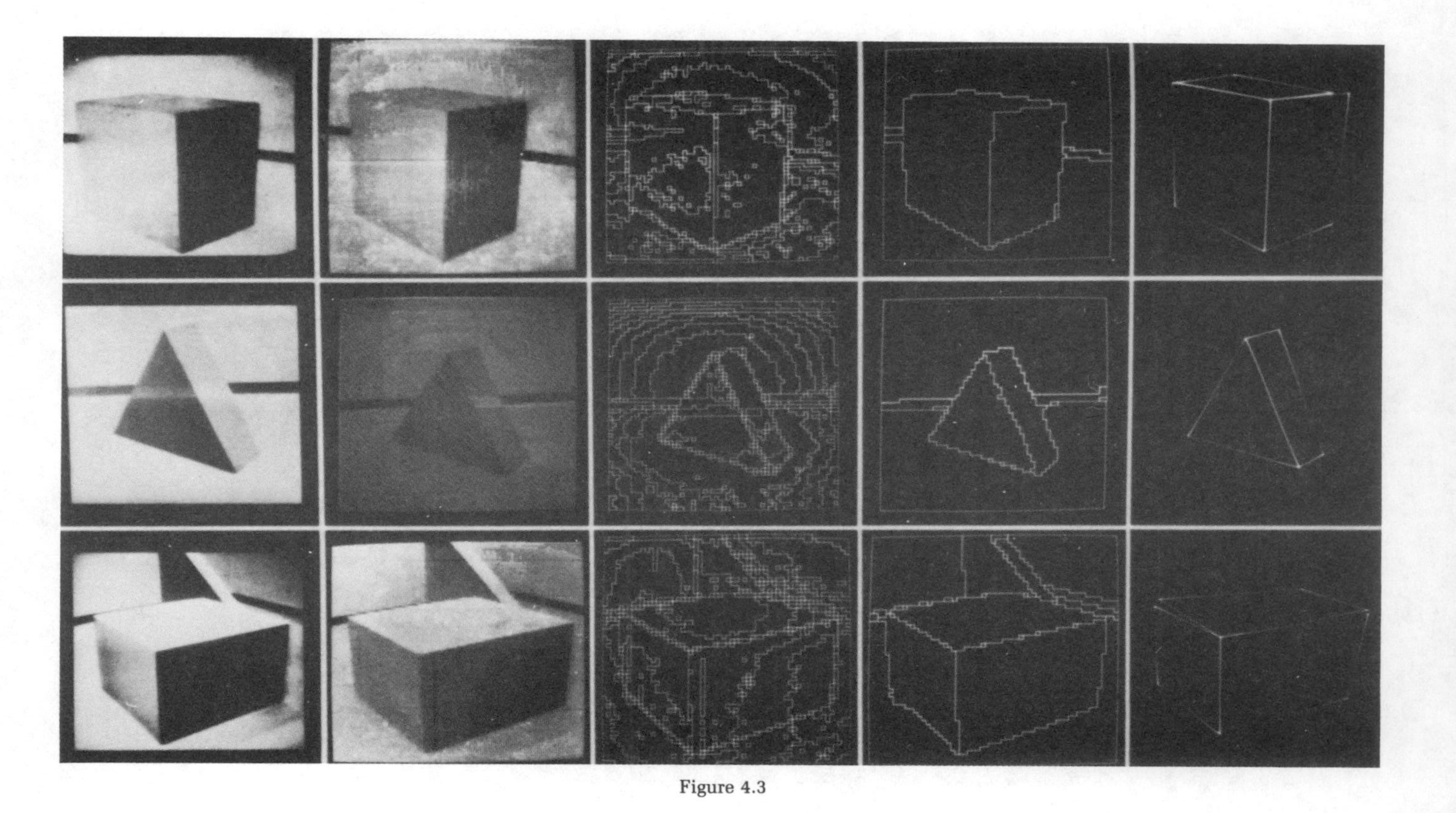

Figure 4.3

That is something of an oversimplification, but it illustrates the general principles relied upon; Shakey had a "line semantics" program for wielding such general rules to determine the category of the object whose image was on the monitor. Watching the monitor, observers might be expected to suffer a sudden dizziness when it eventually occurred to them that there was something strange going on: They were watching a process of image transformation on a monitor, but *Shakey wasn't looking at it*. Moreover, Shakey wasn't looking at any other monitor on which the same images were being transformed and analyzed. There were no other monitors in the hardware, and for that matter the monitor they were watching could be turned off or unplugged without detriment to Shakey's processes of perceptual analysis. Was this monitor some kind of fraud? For whose benefit was it? Only for the observers. What relation, then, did the events they saw on the monitor bear to the events going on inside Shakey?

The monitor was for the observers, but the *idea* of the monitor was also for the designers of Shakey. Consider the almost unimaginable task they faced: How on earth could you take the output from a simple television camera and somehow extract from it reliable box-identifications? Of all the kazillions of possible frames the camera could send to the computer, a tiny subset of them are pictures of boxes; each frame consists simply of an array of black and white cells or pixels, offs and ons, zeros and ones. How could a program be written that would identify all and only the frames that were pictures of boxes? Suppose, to oversimplify, the retina of the camera was a grid of 10,000 pixels, 100 by 100. Then each frame would be one of the possible sequences of 10,000 zeroes and ones. What patterns in the zeroes and ones would line up reliably with the presence of boxes?

To begin with, think of placing all those zeros and ones in an array, actually reproducing the camera image in space, as in the array of pixels visible on the monitor. Number the pixels in each row from left to right, like words on a page (and unlike commercial television, which does a zigzag scan). Notice, then, that dark regions are mainly composed of zeroes and light regions of ones. Moreover, a *vertical* boundary between a light region to the left and a dark region to the right can be given a simple description in terms of the sequence of zeroes and ones: a sequence of mostly ones up to pixel number n, followed by a sequence of mostly zeroes, followed exactly 100 digits later (in the next line) by another sequence of mostly ones up to pixel $n + 100$, followed by mostly zeroes, and so forth, in multiples of 100.

A program that would hunt for such periodicities in the stream

```
000010000001000000100000011011101111110111111111011
001000001000000010000000111010111110111110110111
010000000101000000010011111010111010111111101
000001000000100000000000110101111111111101111110
010000001000000001000000011010111110111111111011
000000000001000000000000111111011111111111011111
000000001000000000000000111011111111111111111111
000000000000001000000000111111110111111111111111
000000001000000000000001011111111111111101111111
000010000000000000000000111111111111011111111110
000000000000000000000100111101111111111111111111
000000100000000000000000111111111110111111111111
```

Figure 4.4

of digits coming from the camera would be able to locate such vertical boundaries. Once found, such a boundary can be turned into a crisp vertical white line by judicious replacement of zeroes with ones and vice versa, so that something like 00011000 occurs exactly every hundred positions in the sequence.

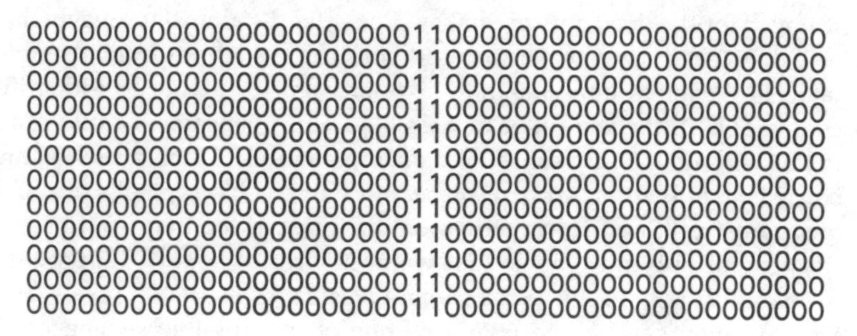

Figure 4.5

A horizontal light/dark boundary is just as easy to spot: a place in the sequence where a flurry of ones gets echoed 100, 200, and 300 digits later (etc.) by a flurry of zeroes.

```
000100000000100000000000010000000001000000000000
000000000010100000000000000001000000000000000001
000000001000000000000100000000000001000000000010
000010000000000001000000000000000010000000100000
000000000100000000000000100000000000000000000000
000000000000000000010000000000100000000000000000
111111111011111111111111101111111111111001111
111111111111111110111111111111111110111111111
111111111111111111111110111111111111111101111
101111111111111111111111111111111111111111111
111111111111111111111111011111111111111111111
111111110111111111111111111111111011111111111
```

Figure 4.6

Sloping boundaries are only a little trickier; the program must look for a progression in the sequence. Once all the boundaries are located and drawn in white, the line drawing is complete, and the next, more sophisticated, step takes over: "templates" are "placed" on bits of the line segment so that the vertices can be identified. Once the vertices have been identified, it is a straightforward matter to use the line semantics program to categorize the object in the image — it might in some cases be as simple a task as looking for a single Y vertex.

Several features of this process are important to us. First, each subprocess is "stupid" and mechanical. That is, no part of the computer has to understand what it is doing or why, and there is no mystery about how each of the steps is mechanically done. Nevertheless, the clever organization of these stupid, mechanical processes yields a device that *takes the place of* a knowledgeable observer. (Put the whole vision system in a "black box" whose task is to "tell Shakey what he needs to know" about what is in front of it, based on TV frames that enter as input. Initially we might be inclined to think the only way to do this would be to put a little man in the black box, watching a screen. We now see a way this homunculus, with his limited job, can be replaced by a machine.)

Once we see how it is done, we can see that while the process is strongly *analogous* to a process of actually looking at (and drawing and erasing) black and white dots on a screen, the actual location in the computer of the individual operations of changing zeroes to ones and vice versa doesn't matter, so long as the numbers that are the temporary "addresses" of the individual digits code the information about which pixels are next to which. Suppose we turn off the monitor. Then even though there is (or need be) no actual two-dimensional image locatable in the space inside the computer (say, as a "pattern of excitation in the hardware"), the operations are homomorphic (parallel) to the events we were watching on the monitor. Those events were genuinely imagistic: a two-dimensional surface of excited phosphor dots forming a shape of a particular size, color, location, and orientation. So in *one* strict sense, Shakey does *not* detect boxes by a series of image transformations; the last real image in the process is the one that is focused on the receptive field of the camera. In another strict but metaphorical sense, Shakey *does* detect boxes by a series of image transformations — the process just described, which turns light-dark boundaries into a line drawing and then categorizes vertices. The fact that this strict sense is nevertheless metaphorical can be brought out by noting that there are a variety of properties one would expect any *real* images to have that the "images" transformed by Shakey lack: They have no color, no size, no orientation. (We could make a nice riddle out of such an image: I'm thinking of an image that is neither larger nor smaller than the Mona Lisa, is neither in color nor in black and white, and faces in no compass direction. What is it?)

The process Shakey used to extract information about objects from the light in its environment was hardly at all like the processes of human vision, and probably not like the visual processes of any creature. But we may ignore this for the moment, in order to see a rather abstract possibility about how the mental images that human subjects report might be discovered in the brain. The account of Shakey's vision system was oversimplified to permit the basic theoretical points to emerge vividly. Now we're going to embark on some science fiction to make another point: Suppose we were to cross Shakey with another famous character in artificial intelligence, Terry Winograd's (1972) SHRDLU, who manipulated (imaginary) blocks and then answered questions about what it was doing and why. SHRLDU's answers were mainly "canned" — stored ready-made sentences and sentence-templates that Winograd had composed. The point of SHRDLU was to explore abstractly some of the information-handling tasks faced by any interlocutor, not to model human speech production realistically, and this is in the spirit of our thought experiment. (In chapter 8 we will look at more realistic models of speech production.) An interchange with our new version of Shakey, redesigned to include a more sophisticated repertoire of verbal actions, might go like this:

> Why did you move the ramp?
> SO I COULD ROLL UP ON THE PLATFORM.
> And why did you want to do that?
> TO PUSH THE BOX OFF.
> And why did you want to do that?
> BECAUSE YOU TOLD ME TO.

But suppose we then asked Shakey:

> How do you tell the boxes from the pyramids?

What should we design Shakey to "say" in reply? Here are three possibilities:

(1) I scan each 10,000-digit-long sequence of 0s and 1s from my camera, looking for certain patterns of sequences, such as . . . blahblahblah (a *very* long answer if we let Shakey go into the details).

(2) I find the light-dark boundaries and draw white lines around them in my mind's eye; then I look at the vertices; if I find a Y vertex, for instance, I know I have a box.

(3) I don't know; some things just look boxy. It just comes to me. It's by *intuition*.

Which is the right sort of thing for Shakey to say? Each answer is true in its way; they are descriptions of the information processing at different depths or grain levels. Which answer we design Shakey to be able to give is largely a matter of deciding how much access Shakey's expressive capacity (his SHRDLU black box) should have to his perceptual processes. Perhaps there would be good reasons of engineering to deny deep (detailed, time-consuming) access to the intermediate analysis processes. But whatever self-descriptive capacities we endow Shakey with, there will be a limit to the depth and detail of his expressible "knowledge" of what is going on in him, what he is doing. If the best answer he can give is (3), then he is in the same position with regard to the question of how he tells pyramids from boxes that we are in when asked how we tell the word "sun" from the word "shun"; we don't know how we do it; one sounds like "sun" and the other like "shun" — that's the best we can do. And if Shakey is designed to respond with (2), there will still be other questions he cannot answer, such as "How do you draw white lines on your mental images?" or "How do you identify a vertex as an arrow?"

Suppose we design Shakey to have type-(2) access to his perceptual analysis processes; when we ask him how he does it, he tells us of the image-transforming he does. Unbeknownst to him, we unplug the monitor. Are we then entitled to tell him that we know better? He isn't really processing images, though he thinks he is? (He *says* he is, and so, following the heterophenomenological strategy, we interpret this as an expression of his belief.) If he were a realistic simulation of a person, he might well retort that we were in no position to tell *him* what was going on in his own mind! *He* knew what he was doing, what he was *really* doing! If he were more sophisticated, he might grant that what he was doing might be only allegorically describable as image processing — though he felt overwhelmingly inclined so to describe what was happening. In this case we would be able to tell him that his metaphorical way of putting it was entirely apt.

If we were more diabolical, on the other hand, we could rig Shakey to have entirely spurious ways of talking about what he was doing. We could design him to want to say things about what was going on in him that bore no regular relationship to what was actually going on ("I use my TV input to drive an internal chisel, which hews a three-dimensional shape out of a block of mental clay. Then if my homunculus can sit on it, it's a box; if he falls off, it's a pyramid.") There would be no truth-preserving interpretation of this report; Shakey would just be *confabulating* — making up a story without "realizing" it.

And this possibility, in us, shows why we have to go to the roundabout trouble of treating heterophenomenology as analogous to the interpretation of fiction. As we have already seen, there are circumstances in which people are just wrong about what they are doing and how they are doing it. It is not that they *lie* in the experimental situation, but that they confabulate; they fill in the gaps, guess, speculate, mistake theorizing for observing. The relation between what they say and whatever it is that drives them to say what they say could hardly be more obscure, both to us heterophenomenologists *on the outside* and to the subjects themselves. *They* don't have any way of "seeing" (with an inner eye, presumably) the processes that govern their assertions, but that doesn't stop them from having heartfelt opinions to express.

To sum up, subjects are unwitting creators of fiction, but to say that they are unwitting is to grant that what they say is, or can be, an account of *exactly how it seems to them*. They tell us *what it is like* to them to solve the problem, make the decision, recognize the object. Because they are sincere (apparently), we grant that that must be what it is like to them, but then it follows that what it is like to them is at best an uncertain guide to what is going on in them. Sometimes, the unwitting fictions we subjects create can be shown to be true after all, if we allow for some metaphorical slack as we did with Shakey's answer in style (2). For instance, recent research on imagery by cognitive psychologists shows that our introspective claims about the mental images we enjoy (whether of purple cows or pyramids) are not utterly false (Shepard and Cooper, 1982; Kosslyn, 1980; Kosslyn, Holtzman, Gazzaniga, and Farah, 1985). This will be discussed in more detail in chapter 10, and we will see how our introspective reports of imagery *can* be interpreted so they come out true. Like the earthly Feenoman, however, who turns out not to be able to fly or be in two places at once, the actual things we find in the brain to *identify* as the mental images will not have all the wonderful properties subjects have confidently endowed their images with. Shakey's "images" provide an example of how something that really wasn't an image at all could be the very thing someone was talking about under the guise of an image. While the processes in the brain underlying human imagery are probably not very much like Shakey's processes, we have opened up a space of possibilities that was otherwise hard to imagine.

8. THE NEUTRALITY OF HETEROPHENOMENOLOGY

At the outset of this chapter I promised to describe a method, the heterophenomenological method, that was neutral with regard to the debates about subjective versus objective approaches to phenomenology, and about the physical or nonphysical reality of phenomenological items. Let's review the method to see that this is so.

First, what about the zombie problem? Very simply, heterophenomenology by itself cannot distinguish between zombies and real, conscious people, and hence does not claim to solve the zombie problem or dismiss it. *Ex hypothesi,* zombies behave just like real people, and since heterophenomenology is a way of interpreting behavior (including the internal behavior of brains, etc.), it will arrive at exactly the same heterophenomenological world for Zoe and for Zombie-Zoe, her unconscious twin. Zombies have a heterophenomenological world, but that just means that when theorists go to interpret them, they succeed at exactly the same task, using exactly the same means, as we use to interpret our friends. Of course, as noted before, some of our friends may be zombies. (It's hard for me to keep a straight face through all this, but since some very serious philosophers take the zombie problem seriously, I feel obliged to reciprocate.)

There is surely nothing wrong, nothing nonneutral, in granting zombies a heterophenomenological world, since it grants so little. This is the metaphysical minimalism of heterophenomenology. The method describes a world, the subject's heterophenomenological world, in which are found various *objects* (intentional objects, in the jargon of philosophy), and in which various things happen to these objects. If someone asks: "What *are* those objects, and what are they *made of?*" the answer *might be* "Nothing!" What is Mr. Pickwick made of? Nothing. Mr. Pickwick is a fictional object, and so are the objects described, named, mentioned by the heterophenomenologist.

—"But isn't it embarrassing to admit, as a theorist, that you are talking about fictional entities—things that don't exist?" Not at all. Literary theorists do valuable, honest intellectual work describing fictional entities, and so do anthropologists who study the gods and witches of various cultures. So indeed do physicists, who, if asked what a center of gravity was made of, would say, "Nothing!" Heterophenomenological objects are, like centers of gravity or the Equator, *abstracta*, not *concreta* (Dennett, 1987a, 1991a). They are not idle fantasies but hardworking theorists' fictions. Moreover, unlike centers of gravity, the way is left open to trade them in for *concreta* if progress in empirical science warrants it.

There are two ways of studying Noah's Flood: You can assume that it is sheer myth but still an eminently studiable myth, or you can ask whether some actual meteorological or geological catastrophe lies behind it. Both investigations can be scientific, but the first is less speculative. If you want to speculate along the second lines, the first thing you should do is conduct a careful investigation along the first lines to gather what hints there are. Similarly, if you want to study how (or even *if*) phenomenological items are really events in the brain, the first thing you should do is a careful heterophenomenological catalogue of the objects. This risks offending the subjects (in the same way anthropologists studying Feenoman risk offending their informants), but it is the only way to avoid the battle of "intuitions" that otherwise passes for phenomenology.

Still, what of the objection that heterophenomenology, by starting out from the third-person point of view, leaves the *real* problems of consciousness untouched? Nagel, as we saw, insists on this, and so does the philosopher John Searle, who has explicitly warned against my approach: "Remember," he admonishes, "in these discussions, always insist on the first person point of view. The first step in the operationalist sleight of hand occurs when we try to figure out how we would *know* what it would be like for others" (Searle, 1980, p. 451). But this is not what happens. Notice that when you are put in the heterophenomenologist's clutches, *you get the last word*. You get to edit, revise, and disavow *ad lib*, and so long as you avoid presumptuous *theorizing* about the causes or the metaphysical status of the items you report, whatever you insist upon is granted constitutive authority to determine what happens in your heterophenomenological world. You're the novelist, and what you say goes. What more could you want?

If you want us to *believe* everything you say about your phenomenology, you are asking not just to be taken seriously but to be granted papal infallibility, and that is asking too much. You are *not* authoritative about what is happening in you, but only about what *seems* to be happening in you, and we are giving you total, dictatorial authority over the account of how it seems to you, about *what it is like to be you*. And if you complain that some parts of how it seems to you are ineffable, we heterophenomenologists will grant that too. What better grounds could we have for believing that you are unable to describe something than that (1) you don't describe it, and (2) confess that you cannot? Of course you might be lying, but we'll give you the benefit of the doubt. If you retort, "I'm not just saying that *I* can't describe it; I'm saying it's indescribable!" we heterophenomenologists will note that at least you can't describe it *now*, and since you're the only one in a position to describe it, it is at this time indescribable. Later, perhaps, you will come to be able to describe it, but of course at that time *it* will be something different, something describable.

When I announce that the objects of heterophenomenology are theorist's fictions, you may be tempted (many are, I find) to pounce on this and say,

> That's *just* what distinguishes the objects of real phenomenology from the objects of heterophenomenology. My *auto*phenomomological objects aren't fictional objects — they're perfectly *real*, though I haven't a clue what to say they are made of. When I tell you, sincerely, that I am imagining a purple cow, I am not just unconsciously producing a word-string to that effect (like Shakey), cunningly contrived to coincide with some faintly analogous physical happening in my brain; I am consciously and deliberately reporting the existence of something that is *really there!* It is no mere theorist's fiction to me!

Reflect cautiously on this speech. You are not just unconsciously producing a word-string you say. Well, you *are* unconsciously producing a word-string; you haven't a clue to how you do that, or to what goes into its production. But, you insist, you are not *just* doing that; you know *why* you're doing it; you *understand* the word-string, and *mean* it. I agree. That's why what you say works so well to constitute a heterophenomenological world. If you were just parroting words more or less at random, the odds against the sequence of words yielding such an interpretation would be astronomical. Surely there is a good explanation of how and why you say what you do, an explanation that accounts for the difference between just saying something and saying it and meaning it, *but you don't have that explanation yet*. At least not all of it. (In chapter 8 we will explore this issue.) Probably you are talking about something real, at least most of the time. Let us see if we can find out what it is.

These reassurances are not enough for some people. Some people just won't play by these rules. Some devoutly religious people, for instance, take offense when interlocutors so much as hint that there *might* be some alternative true religion. These people do not view agnosticism as neutrality, but as an affront, because one of the tenets of their creed is that disbelief in it is itself sinful. People who believe this way are entitled to their belief, and entitled (if that is the right word) to the hurt feelings they suffer when they encounter skeptics or agnostics, but unless they can master the anxiety they feel when they learn that someone does not (yet) believe what they say, they rule themselves out of academic inquiry.

In this chapter we have developed a *neutral* method for investigating and describing phenomenology. It involves extracting and purifying *texts* from (apparently) speaking *subjects*, and using those texts to generate a theorist's fiction, the subject's *heterophenomenological world*. This fictional world is populated with all the images, events, sounds, smells, hunches, presentiments, and feelings that the subject (apparently) sincerely believes to exist in his or her (or its) stream of consciousness. Maximally extended, it is a neutral portrayal of exactly *what it is like to be* that subject — in the subject's own terms, given the best interpretation we can muster.

Having extracted such a heterophenomenology, theorists can then turn to the question of what might explain the *existence* of this heterophenomenology in all its details. The heterophenomenology exists — just as uncontroversially as novels and other fictions exist. People undoubtedly do believe they have mental images, pains, perceptual experiences, and all the rest, and *these* facts — the facts about what people believe, and report when they express their beliefs — are phenomena any scientific theory of the mind must account for. We organize our data regarding these phenomena into theorist's fictions, "intentional objects" in heterophenomenological worlds. Then the question of whether items thus portrayed exist as real objects, events, and states in the brain — or in the soul, for that matter — is an empirical matter to investigate. If suitable real candidates are uncovered, we can identify them as the long-sought referents of the subject's terms; if not, we will have to explain why it seems to subjects that these items exist.

Now that our methodological presuppositions are in place, we can turn to the empirical theory of consciousness itself. We will begin by tackling a problem about the timing and ordering of items in our streams of consciousness. In chapter 5, I will present a first sketch of the theory and exhibit how it handles a simple case. In chapter 6, we will see how the theory permits us to reinterpret some much more complicated phenomena that have perplexed the theorists. Chapters 7 through 9 will develop the theory beyond the initial sketch, warding off misinterpretations and objections, and further illustrating its strengths.

PART TWO

AN EMPIRICAL THEORY OF THE MIND

5

MULTIPLE DRAFTS VERSUS THE CARTESIAN THEATER

1. THE POINT OF VIEW OF THE OBSERVER

> There is no cell or group of cells in the brain of such anatomical or functional preeminence as to appear to be the keystone or center of gravity of the whole system.
>
> WILLIAM JAMES, 1890

Pleasure-boaters sailing along a tricky coast usually make sure they stay out of harm's way by steering for a mark. They find some visible but distant buoy in roughly the direction they want to go, check the chart to make sure there are no hidden obstacles on the straight line between the mark and where they are, and then head straight for it. For maybe an hour or more the skipper's goal is to aim directly at the mark, correcting all errors. Every so often, however, skippers get so lulled by this project that they forget to veer off at the last minute and actually hit the buoy head on! They get distracted from the larger goal of staying out of trouble by the reassuring success they are having with the smaller goal of heading for the mark. In this chapter we will see how some of the most perplexing paradoxes of consciousness arise because we cling too long to a good habit of thought, a habit that usually keeps us out of trouble.

Wherever there is a conscious mind, there is a *point of view*. This is one of the most fundamental ideas we have about minds — or about consciousness. A conscious mind is an observer, who takes in a limited subset of all the information there is. An observer takes in the information that is available at a particular (roughly) continuous sequence of times and places in the universe. For most practical purposes, we can consider the point of view of a particular conscious subject to be just that: a *point* moving through space-time. Consider, for instance, the standard diagrams of physics and cosmology illustrating the Doppler shift or the light-bending effects of gravity.

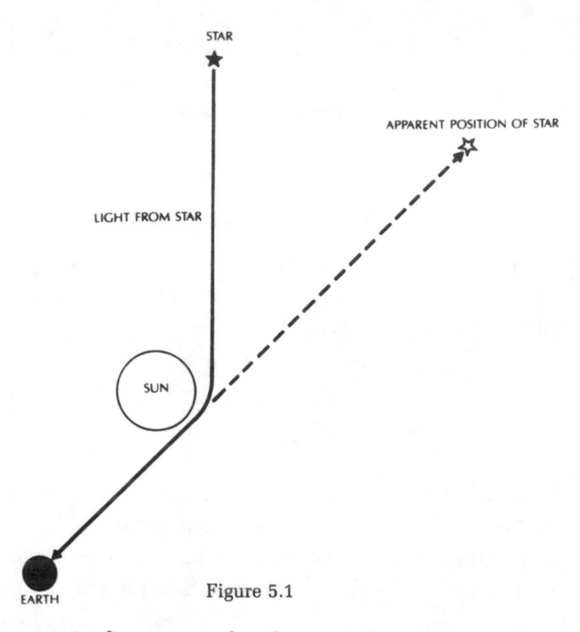

Figure 5.1

The observer in figure 1 is fixed at a point on the surface of the earth. To observers at different points in the universe, things would look different. Simpler examples are more familiar. We explain the startling time gap between the sound and sight of the distant fireworks by noting the different transmission speeds of sound and light. They arrive *at the observer* (at that point) at different times, even though they left the source at the same time.

What happens, though, when we close in on the observer, and try to locate the observer's point of view more precisely, as a point *within* the individual? The simple assumptions that work so well on larger scales begin to break down.[1] There is no single point in the brain where

1. This is reminiscent of the difficulties that face physicists when they confront a *singularity*, a point at which, precisely because of its dimensionlessness, various magnitudes are infinite (given their definitions). This arises for black holes, but also affects the interpretation of more mundane entities. Roger Penrose discusses the case of how to apply the Lorentz equations and Maxwell equations to particles. "What the Lorentz

all information funnels in, and this fact has some far from obvious — indeed, quite counterintuitive — consequences.

Since we will be considering events occurring on a relatively microscopic scale of space and time, it is important to have a clear sense of the magnitudes involved. All the experiments we will consider involve intervals of time measured in milliseconds or thousandths of a second. It will help if you have a rough idea of how long (or short) 100msec or 50msec is. You can speak about four or five syllables per second, so a syllable takes on the order of 200msec. Standard motion pictures run at twenty-four frames per second, so the film advances a frame every 42msec (actually, each frame is held stationary and exposed *three times* during that 42msec, for durations of 8.5msec, with 5.4msec of darkness between each). Television (in the U.S.A.) runs at thirty frames per second, or one frame every 33msec (actually, each frame is woven in two passes, overlapping with its predecessor). Working your thumb as fast as possible, you can start and stop a stopwatch in about 175msec. When you hit your finger with a hammer, the fast (myelin-sheathed) nerve fibers send a message to the brain in about 20msec; the slow, unmyelinated C-fibers send pain signals that take much longer — around 500msec — to cover the same distance.

Here is a chart of the approximate millisecond values of some relevant durations.

saying "one, Mississippi"	1000msec
unmyelinated fiber, fingertip to brain	500msec
a 90 mph fastball travels the 60.6 feet to home plate	458msec
speaking a syllable	200msec
starting and stopping a stopwatch	175msec
a frame of motion picture film	42msec
a frame of television	33msec
fast (myelinated) fiber, fingertip to brain	20msec

---

equations tell us to do is to examine the electromagnetic field at the precise *point* at which the charged particle is located (and, in effect, to provide us with a 'force' at that point). Where is that point to be taken to be if the particle has a finite size? Do we take the 'centre' of the particle, or else do we average the field (for the 'force') over all points at the surface? . . . Perhaps we are better off regarding the particle as a *point* particle. But this leads to other kinds of problems, for then the particle's own electric field becomes *infinite* in its immediate neighbourhood." (Penrose, 1989, pp. 189–190)

the basic cycle time of a neuron	10msec
the basic cycle time of a personal computer	.0001msec

Descartes, one of the first to think seriously about what must happen once we look closely inside the body of the observer, elaborated an idea that is so superficially natural and appealing that it has permeated our thinking about consciousness ever since. As we saw in chapter 2, Descartes decided that the brain *did* have a center: the pineal gland, which served as the gateway to the conscious mind (see Figure 2.1, page 34). The pineal gland is the only organ in the brain that is in the midline, rather than paired, with left and right versions. It is marked "L" in this diagram by the great sixteenth-century anatomist, Vesalius. Smaller than a pea, it sits in splendid isolation on its stalk, attached to the rest of the nervous system just about in the middle of the back

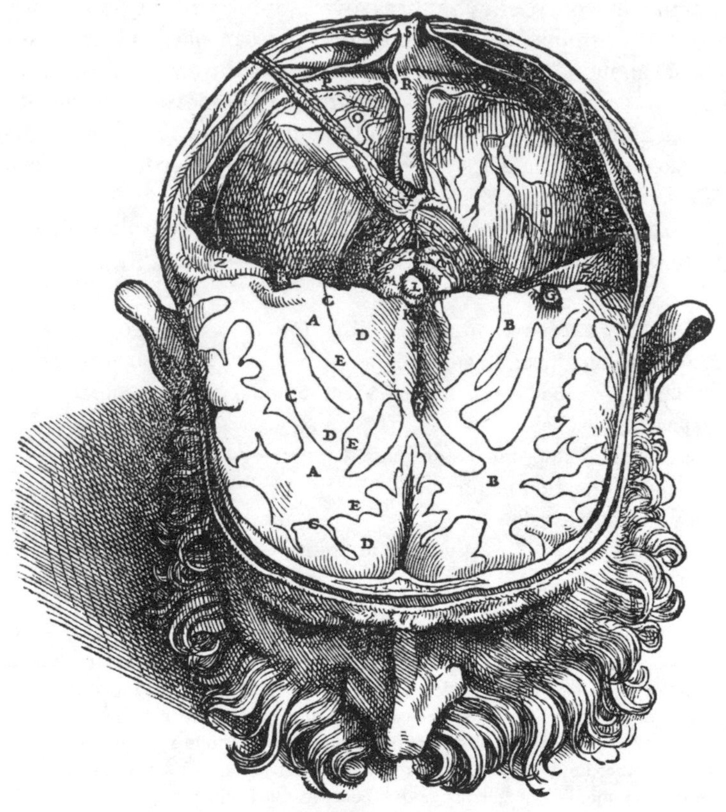

Figure 5.2

of the brain. Since its function was quite inscrutable (it is still unclear what the pineal gland does), Descartes proposed a role for it: in order for a person to be conscious of something, traffic from the senses had to arrive at this station, where it thereupon caused a special — indeed, magical — transaction to occur between the person's material brain and immaterial mind.

Not all bodily reactions required this intervention by the conscious mind, in Descartes's view. He was well aware of what are now called reflexes, and he postulated that they were accomplished by entirely mechanical short circuits of sorts that bypassed the pineal station altogether, and hence were accomplished unconsciously.

Figure 5.3

He was wrong about the details: He thought the fire displaced the skin, which pulled a tiny thread, which opened a pore in the ventricle (F), which caused "animal spirit" to flow out through a hollow tube, which inflated the muscles of the leg, causing the foot to withdraw (Descartes, 1664). But it was otherwise a good idea. The same cannot be said about Descartes's vision of the pineal's role as the turnstile of consciousness (we might call it the Cartesian bottleneck). That idea, Cartesian dualism, is hopelessly wrong, as we saw in chapter 2. But while materialism of one sort or another is now a received opinion approaching unanimity, even the most sophisticated materialists today often forget that once Descartes's ghostly *res cogitans* is discarded, there is no longer a role for a centralized gateway, or indeed for any *functional* center to the brain. The pineal gland is not only not the fax machine to the Soul, it is also not the Oval Office of the brain, and neither are any of the other portions of the brain. The brain is Headquarters, the place where the ultimate observer is, but there is no reason to believe that the brain itself has any deeper headquarters, any inner sanctum, arrival at which is the necessary or sufficient condition for conscious experience. In short, there is no observer inside the brain.[2]

Light travels much faster than sound, as the fireworks example reminds us, but we now know that it takes longer for the brain to process visual stimuli than to process auditory stimuli. As the neuroscientist Ernst Pöppel (1985, 1988) has pointed out, thanks to these counterbalancing differences, the "horizon of simultaneity" is *about* ten meters: light and sound that leave the same point about ten meters from the observer's sense organs produce neural responses that are "centrally available" at the same time. Can we make this figure more precise? There is a problem. The problem is not just measuring the distances from the external event to the sense organs, or the transmission speeds in the various media, or allowing for individual differences. The more fundamental problem is deciding what to count as the "finish line" in the brain. Pöppel obtained his result by comparing behavioral measures: mean reaction times (button-pushing) to auditory and visual stimuli. The difference ranges between 30msec and 40msec, the time it takes

---

2. To deny that the head is Headquarters would be madness, but not unprecedented madness. Phillipe Pinel reported in 1800 the curious case of a man who fell into "a true delirium brought on by the terrors of the revolution. The overturning of his reason is marked by a particular singularity: he believes that he was guillotined, and his head thrown pell-mell onto the pile of the other victims' heads, and that the judges, repenting too late their cruel deed, had ordered the heads to be taken and rejoined to their respective bodies. However, by an error of some sort, they put on his shoulders the head of another unfortunate. This idea that his head has been changed occupies him night and day. . . . 'See my teeth!' he would repeat incessantly, 'they used to be wonderful, and these are rotten! *My* mouth was healthy, and *this* one's infected! What a difference between this hair and the hair I had before my change of head!' " *Traité médico-philosophique sur l'aliénation mentale, ou la Manie.* Paris: Chez Richard, Caille et Ravier, 1800, pp. 66–7. (Thanks to Dora Weiner for bringing this fascinating case to my attention.)

sound to travel approximately ten meters (the time it takes light to travel ten meters is insignificantly different from zero). Pöppel used a peripheral finish line — external behavior — but our natural intuition is that the *experience* of the light or sound happens *between* the time the vibrations hit our sense organs and the time we manage to push the button signaling that experience. And it happens somewhere *centrally*, somewhere in the brain on the excited paths between the sense organ and the finger. It seems that if we could say *exactly* where, we could say exactly when the experience happened. And vice versa: If we could say exactly when it happened, we could say where in the brain conscious experience was located.

Let's call the idea of such a centered locus in the brain *Cartesian materialism*, since it's the view you arrive at when you discard Descartes's dualism but fail to discard the imagery of a central (but material) Theater where "it all comes together." The pineal gland would be one candidate for such a Cartesian Theater, but there are others that have been suggested — the anterior cingulate, the reticular formation, various places in the frontal lobes. Cartesian materialism is the view that there is a crucial finish line or boundary somewhere in the brain, marking a place where the order of arrival equals the order of "presentation" in experience because *what happens there* is what you are conscious of. Perhaps no one today explicitly endorses Cartesian materialism. Many theorists would insist that they have explicitly rejected such an obviously bad idea. But as we shall see, the persuasive imagery of the Cartesian Theater keeps coming back to haunt us — laypeople and scientists alike — even after its ghostly dualism has been denounced and exorcized.

The Cartesian Theater is a metaphorical picture of how conscious experience must sit in the brain. It seems at first to be an innocent extrapolation of the familiar and undeniable fact that *for everyday, macroscopic time intervals*, we can indeed order events into the two categories "not yet observed" and "already observed." We do this by locating the observer at a point and plotting the motions of the vehicles of information relative to that point. But when we try to extend this method to explain phenomena involving very short time intervals, we encounter a *logical* difficulty: If the "point" of view of the observer must be smeared over a rather large volume in the observer's brain, the observer's own subjective sense of sequence and simultaneity *must* be determined by something other than "order of arrival," since order of arrival is incompletely defined until the relevant destination is specified. If A beats B to one finish line but B beats A to another, which result fixes subjective sequence in consciousness? (Cf. Minsky, 1985, p. 61.) Pöppel speaks of the moments at which sight and sound become "centrally available" in the brain, but which point or points of "central availability" would "count" as a determiner of *experienced* order, and why? When we try to answer this question, we will be forced to abandon the Cartesian Theater and replace it with a new model.

The idea of a special center in the brain is the most tenacious bad idea bedeviling our attempts to think about consciousness. As we shall see, it keeps reasserting itself, in new guises, and for a variety of ostensibly compelling reasons. To begin with, there is our personal, introspective appreciation of the "unity of consciousness," which impresses on us the distinction between "in here" and "out there." The naïve boundary between "me" and "the outside world" is my skin (and the lenses of my eyes) but, as we learn more and more about the way events in our own bodies can be inaccessible "to us," the great outside encroaches. "In here" I can try to raise my arm, but "out there," if it has "fallen asleep" or is paralyzed, it won't budge; my lines of communication from wherever *I* am to the neural machinery controlling my arm have been tampered with. And if my optic nerve were somehow severed, I wouldn't expect to go on seeing even though my eyes were still intact; having visual experiences is something that apparently happens *inboard* of my eyes, somewhere in between my eyes and my voice when I tell you what I see.

Doesn't it follow *as a matter of geometric necessity* that our conscious minds are located at the *termination* of all the *in*bound processes, just before the *initiation* of all the *out*bound processes that implement our actions? Advancing from one periphery along the input channels from the eye, for instance, we ascend the optic nerve, and up through various areas of the visual cortex, and then . . . ? Advancing from the other periphery by swimming upstream from the muscles and the motor neurons that control them, we arrive at the supplementary motor area in the cortex and then . . . ? These two journeys advance toward each other up two slopes, the afferent (input) and the efferent (output). However difficult it might be to determine in practice the precise location of the Continental Divide in the brain, must there not be, by sheer geometric extrapolation, a highest point, a turning point, a point such that all tamperings on one side of it are *pre-experiential,* and all tamperings on the other are *post-experiential?*

In Descartes's picture, this is obvious to visual inspection, since everything funnels to and from the pineal station. It might seem, then, that if we were to take a more current model of the brain, we should be able to color-code our explorations, using, say, red for afferent and green for efferent; wherever our colors suddenly changed would be a functional midpoint on the great Mental Divide.

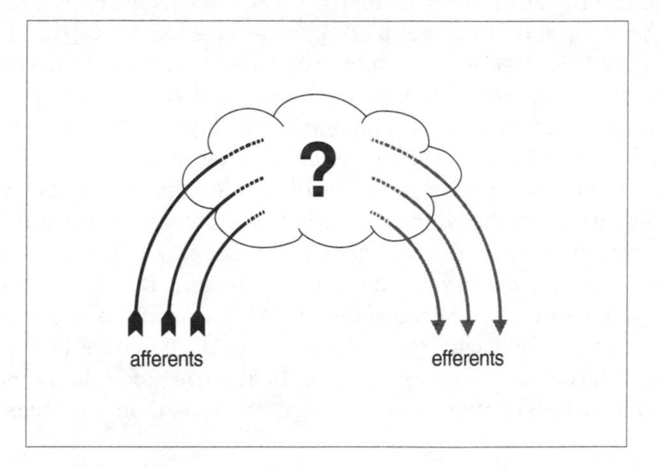

Figure 5.4

This curiously compelling argument may well ring a bell. It is the twin of an equally bogus argument that has recently been all too influential: Arthur Laffer's notorious Curve, the intellectual foundation (if I may speak loosely) of Reaganomics. If the government taxes at 0

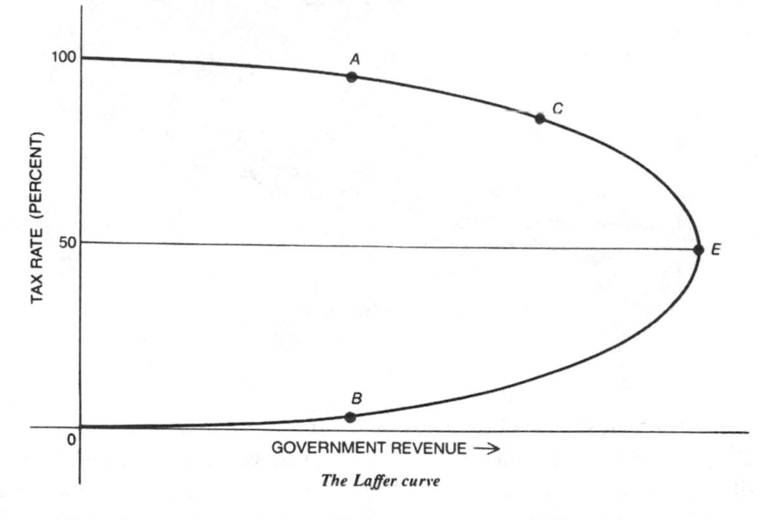

The Laffer curve

Figure 5.5

percent, it gets no revenue, and if it taxes at 100 percent, no one will work for wages, so it gets no revenue; at 2 percent it will get roughly twice the revenue as at 1 percent, and so forth, but as the rate rises, diminishing returns will set in; the taxes will become onerous. Looking at the other end of the scale, 99 percent taxation is scarcely less confiscatory than 100 percent, so scarcely any revenue will accrue; at 90 percent the government will do better, and better still at the more inviting rate of 80 percent. The particular slopes of the curve as shown may be off, but mustn't there be, as a matter of geometric necessity, a place where the curve turns, a rate of taxation that maximizes revenue? Laffer's idea was that since the current tax rate was on the upper slope, lowering taxes would actually increase revenues. It was a tempting idea; it seemed to many that it just had to be right. But as Martin Gardner has pointed out, just because the extreme ends of the curve are clear, there is no reason why the unknown part of the curve in the middle regions has to take a smooth course. In a satiric mood, he proposes the alternative "neo-Laffer Curve," which has more than one "maximum,"

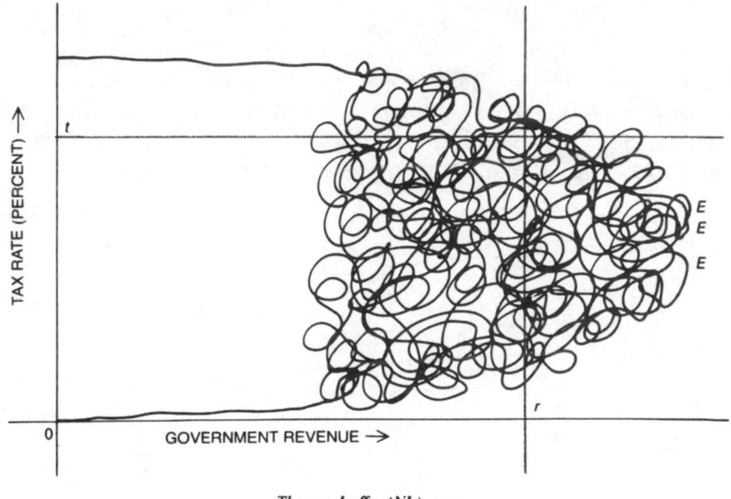

The neo-Laffer (NL) curve

Figure 5.6

and the accessibility of any one of them depends on complexities of history and circumstance that no change of a single variable can possibly determine (Gardner, 1981). We should draw the same moral about what lies in the fog inboard of the afferent and efferent peripheries: the clarity of the peripheries gives us no guarantee that the same distinctions will continue to apply *all the way in*. The "technosnarl" Gardner envisages for the economy is simplicity itself compared to the jumble of activities occurring in the more central regions of the brain. We must stop thinking of the brain as if it had such a single functional summit or central point. This is not an innocuous shortcut; it's a bad habit. In order to break this bad habit of thought, we need to explore some instances of the bad habit in action, but we also need a good image with which to replace it.

2. INTRODUCING THE MULTIPLE DRAFTS MODEL

Here is a first version of the replacement, the Multiple Drafts model of consciousness. I expect it will seem quite alien and hard to visualize at first — that's how entrenched the Cartesian Theater idea is. According to the Multiple Drafts model, all varieties of perception — indeed, all varieties of thought or mental activity — are accomplished in the brain by parallel, multitrack processes of interpretation and elaboration of sensory inputs. Information entering the nervous system is under continuous "editorial revision." For instance, since your head moves a bit and your eyes move a lot, the images on your retinas swim about constantly, rather like the images of home movies taken by people who can't keep the camera from jiggling. But that is not how it seems to us. People are often surprised to learn that under normal conditions, their eyes dart about in rapid saccades, about five quick fixations a second, and that this motion, like the motion of their heads, is edited out early in the processing from eyeball to . . . consciousness. Psychologists have learned a lot about the mechanisms for achieving these normal effects, and have also discovered some special effects, such as the interpretation of depth in random dot stereograms (Julesz, 1971). (See Figure 5.7, page 112.)

If you view these two slightly different squares through a stereopticon (or just stare at them slightly cross-eyed to get the two images to fuse into one — some people can do it without any help from a viewing device), you will eventually see a shape emerge in three dimensions, thanks to an impressive editorial process in the brain that compares and collates the information from each eye. Finding the globally optimal registration can be accomplished without first having to subject each data array to an elaborate process of feature extraction. There are enough lowest-level coincidences of saliency — the individual dots in a random dot stereogram — to dictate a solution.

Figure 5.7

These effects take quite a long time for the brain's editorial processes to produce, but other special effects are swift. The McGurk effect (McGurk and Macdonald, 1979) is a case in point. When a French film is dubbed in English, most of the time viewers are unaware of the discrepancy between the lip motions they see and the sounds they hear — unless the dubbing is done sloppily. But what happens if a sound track is created that lines up well with the images except for some deliberately mismatched consonants? (Using our old friend for a new purpose, we can suppose the filmed person's lips say "from left to right" and the soundtrack voice says "from reft to light.") What will people experience? They will *hear* "from left to right." In the artificially induced editorial contest between the contributions from the eyes and the ears, the eyes win — in this instance.[3]

These editorial processes occur over large fractions of a second, during which time various additions, incorporations, emendations, and overwritings of content can occur, in various orders. We don't directly experience what happens on our retinas, in our ears, on the surface of our skin. What we actually experience is a product of many processes of interpretation — editorial processes, in effect. They take in relatively raw and one-sided representations, and yield collated, revised, enhanced representations, and they take place in the streams of activity occurring in various parts of the brain. This much is recognized by

3. An even more striking example is an experiment in which the subject is tricked by mirrors into thinking he is watching his own hand drawing a line, while in fact he is watching the hand of the experimenter's accomplice. In this instance "the eyes win" to such an extent that the editorial process in the brain is tricked into concluding that the subject's hand is being forcibly moved; the subject claims to feel the "pressure" preventing "his" hand from moving where it is supposed to move (Nielsen, 1963).

virtually all theories of perception, but now we are poised for the novel feature of the Multiple Drafts model: Feature detections or discriminations *only have to be made once*. That is, once a particular "observation" of some feature has been made, by a specialized, localized portion of the brain, the information content thus fixed does not have to be sent somewhere else to be rediscriminated by some "master" discriminator. In other words, discrimination does not lead to a re-*presentation* of the already discriminated feature for the benefit of the audience in the Cartesian Theater — for there is no Cartesian Theater.

These spatially and temporally distributed content-fixations in the brain are precisely locatable in both space and time, but their onsets do *not* mark the onset of consciousness of their content. It is always an open question whether any particular content thus discriminated will eventually appear as an element in conscious experience, and it is a confusion, as we shall see, to ask *when it becomes conscious*. These distributed content-discriminations yield, over the course of time, something *rather like* a narrative stream or sequence, which can be thought of as subject to continual editing by many processes distributed around in the brain, and continuing indefinitely into the future. This stream of contents is only rather like a narrative because of its multiplicity; at any point in time there are multiple "drafts" of narrative fragments at various stages of editing in various places in the brain.

Probing this stream at different places and times produces different effects, precipitates different narratives from the subject. If one delays the probe too long (overnight, say), the result is apt to be no narrative left at all — or else a narrative that has been digested or "rationally reconstructed" until it has no integrity. If one probes "too early," one may gather data on how early a particular discrimination is achieved by the brain, but at the cost of diverting what would otherwise have been the normal progression of the multiple stream. Most important, the Multiple Drafts model avoids the tempting mistake of supposing that there must be a single narrative (the "final" or "published" draft, you might say) that is canonical — that is the *actual* stream of consciousness of the subject, whether or not the experimenter (or even the subject) can gain access to it.

Right now this model probably makes little sense to you as a model of the consciousness you know from your own intimate experience. That's because you are still so comfortable thinking about your consciousness as taking place in the Cartesian Theater. Breaking down that natural, comfortable habit, and making the Multiple Drafts model into a vivid and believable alternative, will take some work, and weird work at that. This will surely be the hardest part of the book, but it is essential to the overall theory and cannot be skipped over! There is no math involved, thank goodness. You just have to think carefully and vividly, making sure you get the right picture in your mind and not the seductive wrong pictures. There will be a variety of simple thought experiments to help your imagination along this tricky path. So prepare for some strenuous exercise. At the end you will have uncovered a new view of consciousness, which involves a major reform (but not a radical revolution) in our ways of thinking about the brain. (For a similar model, see William Calvin's (1989) model of consciousness as "scenario-spinning.")

A good way of coming to understand a new theory is to see how it handles a relatively simple phenomenon that defies explanation by the old theory. Exhibit A is a discovery about apparent motion that was provoked, I am happy to say, by a philosopher's question. Motion pictures and television depend on creating apparent motion by presenting a rapid succession of "still" pictures, and ever since the dawn of the motion picture age, psychologists have studied this phenomenon, called *phi* by Max Wertheimer (1912), the first to study it systematically. In the simplest case, if two or more small spots separated by as much as 4 degrees of visual angle are briefly lit in rapid succession, a single spot will seem to move back and forth. Phi has been studied in many variations, and one of the most striking is reported by the psychologists Paul Kolers and Michael von Grünau (1976). The philosopher Nelson Goodman had asked Kolers whether the phi phenomenon persisted if the two illuminated spots were different in color, and if so, what happened to the color of "the" spot as "it" moved? Would the illusion of motion disappear, to be replaced by two separately flashing spots? Would an illusory "moving" spot gradually change from one color to another, tracing a trajectory through the color solid (the three-dimensional sphere that maps all the hues)? (You might want to make your own prediction before reading on.) The answer, when Kolers and von Grünau performed the experiments, was unexpected: Two different colored spots were lit for 150msec each (with a 50msec interval); the first spot seemed to begin moving and then change color abruptly *in the middle of its illusory passage* toward the second location. Goodman wondered: "How are we able . . . to fill in the spot at the intervening place-times along a path running from the first to the second flash *before that second flash occurs?*" (Goodman, 1978, p. 73)

The same question can of course be raised about any phi, but Kolers's color phi phenomenon vividly brings out the problem. Suppose the first spot is red and the second, displaced, spot is green. Unless there is "precognition" in the brain (an extravagant hypothesis we will postpone indefinitely), the illusory content, *red-switching-to-green-in-midcourse*, cannot be created until *after* some identification of the second, green spot occurs in the brain. But if the second spot is already "in conscious experience," wouldn't it be too late to interpose the illusory content between the conscious experience of the red spot and the conscious experience of the green spot? How does the brain accomplish this sleight of hand?

The principle that causes must precede effects applies to the multiple distributed processes that accomplish the editorial work of the brain. Any particular process that requires information from some source must indeed wait for that information; it can't get there till it gets there. This is what rules out "magical" or precognitive explanations of the color-switching phi phenomenon. The content *green spot* cannot be attributed to any event, conscious or unconscious, until the light from the green spot has reached the eye and triggered the normal neural activity in the visual system up to the level at which the discrimination of green is accomplished. So the (illusory) discrimination of red-turning-to-green has to be accomplished *after* the discrimination of the green spot. But then since what you consciously experience is *first red, then red-turning-to-green, and finally green*, it ("surely") follows that your consciousness of the whole event must be delayed until after the green spot is (unconsciously?) perceived. If you find this conclusion compelling, you are still locked in the Cartesian Theater. A thought experiment will help you escape.

3. ORWELLIAN AND STALINESQUE REVISIONS

> I'm really not sure if others fail to perceive me or if, one fraction of a second after my face interferes with their horizon, a millionth of a second after they have cast their gaze on me, they already begin to wash me from their memory: forgotten before arriving at the scant, sad archangel of a remembrance.
>
> ARIEL DORFMAN, *Mascara*, 1988

Suppose I tamper with your brain, inserting in your memory a bogus woman wearing a hat where none was (e.g., at the party on Sunday). If on Monday, when you recall the party, you remember her and can find no internal resources for so much as doubting the veracity of your memory, we would still say that you never *did* experience her; that is, not at the party on Sunday. Of course your subsequent experience of (bogus) recollection can be as vivid as may be, and on Tuesday we can certainly agree that you have had vivid conscious experiences of there being a woman in a hat at the party, but the *first* such experience, we would insist, was on Monday, not Sunday (though it doesn't seem this way to you).

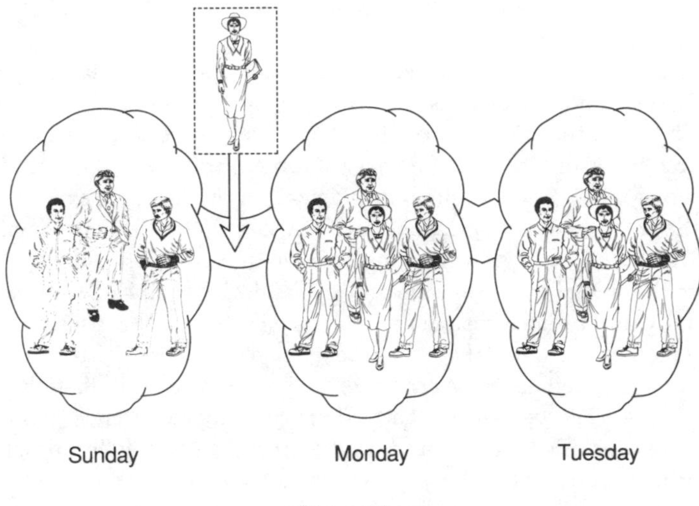

Figure 5.8

We lack the power to insert bogus memories by neurosurgery, but sometimes our memories play tricks on us, so what we cannot yet achieve surgically happens in the brain on its own. Sometimes we seem to remember, even vividly, experiences that never occurred. Let's call such post-experiential contaminations or revisions of memory *Orwellian*, after George Orwell's chilling vision in the novel *1984* of the Ministry of Truth, which busily rewrote history and thus denied access to the (real) past to all who followed.

The possibility of post-experiential (Orwellian) revision exhibits an aspect of one of our most fundamental distinctions: the distinction between appearance and reality. Because we recognize the possibility (at least in principle) of Orwellian revision, we recognize the risk of inferring from "this is what I remember" to "this is what really happened," and hence we resist — with good reason — any diabolical "operationalism" that tries to convince us that what we remember (or what history records in the archives) *just is* what really happened.[4]

Orwellian revision is one way to fool posterity. Another is to stage show trials, carefully scripted presentations of false testimony and bogus confessions, complete with simulated evidence. Let's call this ploy *Stalinesque*. Notice that if we are usually sure which mode of falsification has been attempted on us, the Orwellian or the Stalinesque, this is just a happy accident. In any *successful* disinformation campaign, were we to wonder whether the accounts in the newspapers were Orwellian accounts of trials that never happened at all, or true accounts of phony show trials that actually did happen, we might be unable to tell the difference. If *all* the traces — newspapers, videotapes, personal memoirs, inscriptions on gravestones, living witnesses — were either obliterated or revised, we would have no way of knowing whether a fabrication happened *first*, culminating in a staged trial whose accurate history we have before us, or rather, *after* a summary execution, history-fabrication covered up the deed: No trial of any sort *actually* took place.

The distinction between Orwellian and Stalinesque methods of producing misleading archives works unproblematically in the everyday world, at macroscopic time scales. One might well think it applies unproblematically *all the way in*, but this is an illusion, and we can catch it in the act in a thought experiment that differs from the one just considered in nothing but time scale.

Suppose you are standing on the corner and a long-haired woman dashes by. About one second *after* this, a subterranean memory of some earlier woman — a short-haired woman with eyeglasses — contaminates the memory of what you have just seen: when asked a minute later for details of the woman you just saw, you report, sincerely but erroneously, her eyeglasses. Just as in the case of the woman with the hat at the party, we are inclined to say that your original *visual* experience, as opposed to the memory of it seconds later, was *not* of a woman wearing glasses. But as a result of the subsequent memory contaminations, it seems to you exactly as if at the first moment you

4. Operationalism is (approximately) the view or policy expressed by "If you can't discover a difference, there isn't a difference," or, as one often hears it put, "If it quacks like a duck, and walks like a duck, it is a duck." For a reconsideration of the strengths and weaknesses of operationalism, see Dennett (1985a).

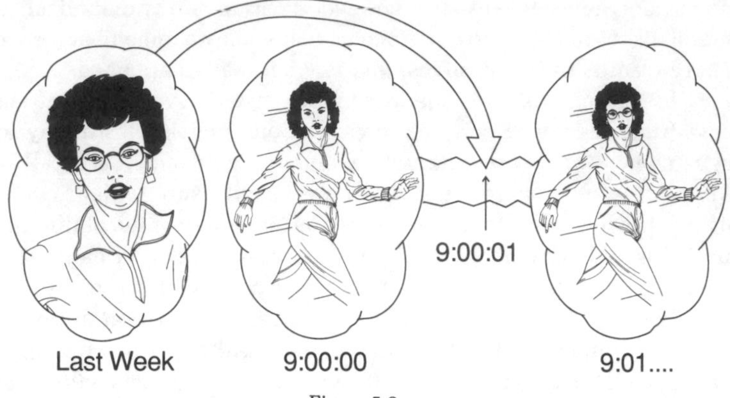

Figure 5.9

saw her, you were struck by her eyeglasses. An Orwellian revision has happened: there was a fleeting instant, before the memory contamination took place, when it *didn't* seem to you she had glasses. For that brief moment, the *reality* of your conscious experience was a long-haired woman *without* eyeglasses, but this historical fact has become inert; it has left no trace, thanks to the contamination of memory that came one second after you glimpsed her.

This understanding of what happened is jeopardized, however, by an alternative account. Your subterranean earlier memories of that woman with the eyeglasses could just as easily have contaminated your experience *on the upward path*, in the processing of information that occurs "prior to consciousness," so that you actually *hallucinated* the eyeglasses from the very beginning of your experience. In that case, your obsessive memory of the earlier woman with glasses would be

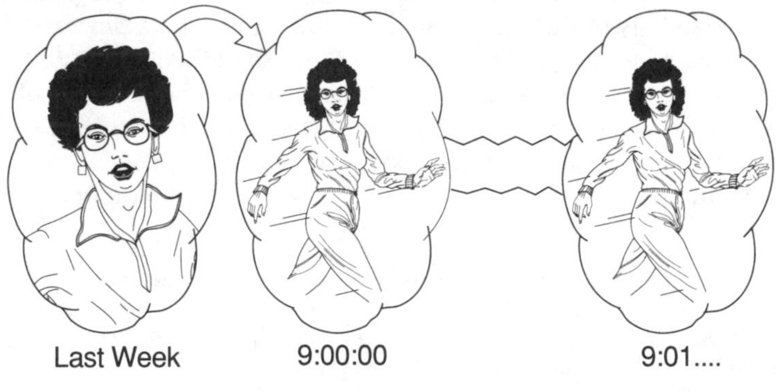

Figure 5.10

playing a Stalinesque trick on you, creating a show trial in experience, which you then accurately recall at later times, thanks to the record in your memory. To naïve intuition these two cases are as different as can be: Told the first way (Figure 5.9), you suffer no hallucination at the time the woman dashes by, but suffer subsequent memory hallucinations; you have false memories of your actual ("real") experience. Told the second way (Figure 5.10), you hallucinate when she runs by, and then accurately remember that hallucination (which "really did happen in consciousness") thereafter. Surely these are distinct possibilities no matter how finely we divide up time?

No. Here the distinction between perceptual revisions and memory revisions that works so crisply at other scales is no longer guaranteed to make sense. We have moved into the foggy area in which the subject's point of view is spatially and temporally smeared, and the question *Orwellian or Stalinesque?* loses its force.

There is a time window that began when the long-haired woman dashed by, exciting your retinas, and ended when you expressed — to yourself or someone else — your eventual conviction that she was wearing glasses. At some time during this interval, the content *wearing glasses* was spuriously added to the content *long-haired woman*. We may assume (and might eventually confirm in detail) that there was a brief time when the content *long-haired woman* had already been discriminated in the brain but *before* the content *wearing glasses* had been erroneously "bound" to it. Indeed, it would be plausible to suppose that this discrimination of a long-haired woman was what triggered the memory of the earlier woman with the glasses. What we would not know, however, is whether this spurious binding was "before or after the fact" — the presumed fact of "actual conscious experience." Were you first conscious of a long-haired woman without glasses and then conscious of a long-haired woman with glasses, a subsequent consciousness that wiped out the memory of the earlier experience, or was the very first instant of conscious experience already spuriously tinged with eyeglasses?

If Cartesian materialism were true, this question would have to have an answer, even if we — and you — could not determine it retrospectively by any test. For the content that "crossed the finish line first" was either *long-haired woman* or *long-haired woman with glasses*. But almost all theorists would insist that Cartesian materialism is false. What they have not recognized, however, is that this implies that there is no privileged finish line, so the temporal order of discriminations cannot be what fixes the subjective order in experience. This conclusion is not easy to embrace, but we can make its attractions more compelling by examining the difficulties you get into if you cling to the traditional alternative.

Consider Kolers's color phi phenomenon. Subjects *report* seeing the color of the moving spot switch in midtrajectory from red to green. This bit of text was sharpened by Kolers's ingenious use of a pointer device, which subjects retrospectively-but-as-soon-as-possible "superimposed" on the trajectory of the illusory moving spot: by placing the pointer, they performed a speech act with the content "The spot changed color right about *here*" (Kolers and von Grünau, 1976, p. 330).

So in the heterophenomenological world of the subjects, there is a color switch in midtrajectory, and the information about which color to switch to (and which direction to move) has to come from somewhere. Recall Goodman's expression of the puzzle: "How are we able . . . to fill in the spot at the intervening place-times along a path running from the first to the second flash *before that second flash occurs?*" Perhaps, some theorists thought, the information comes from *prior experience*. Perhaps, like Pavlov's dog who came to expect food whenever the bell rang, these subjects have come to expect to see the second spot whenever they see the first spot, and by force of habit they actually represent the passage in anticipation of getting any information about the particular case. But this hypothesis has been disproven. Even on the first trial (that is, without any chance for conditioning), people experience the phi phenomenon. Moreover, in subsequent trials the direction and color of the second spot can be randomly changed without making the effect go away. So somehow the information from the second spot (about its color and location) has to be used by the brain to create the "edited" version that the subjects report.

Consider, first, the hypothesis that there is a Stalinesque mechanism: In the brain's editing room, located before consciousness, there is a delay, a loop of slack like the tape delay used in broadcasts of "live" programs, which gives the censors in the control room a few seconds to bleep out obscenities before broadcasting the signal. *In the editing room*, first frame A, of the red spot, arrives, and then, when frame B, of the green spot, arrives, some interstitial frames (C and D) can be created and then spliced into the film (in the order A,C,D,B) on its way to projection in the theater of consciousness. By the time the "finished product" arrives at consciousness, it already has its illusory insertion.

Alternatively, there is the hypothesis that there is an Orwellian mechanism: shortly after the consciousness of the first spot *and* the

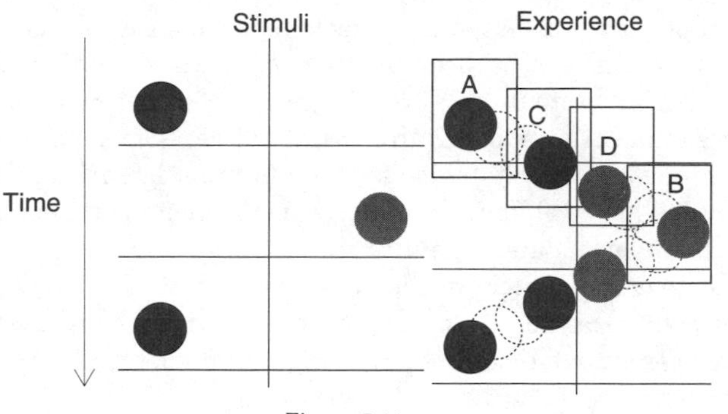

Figure 5.11

second spot (with no illusion of apparent motion at all), a revisionist historian of sorts, in the brain's memory-library receiving station, notices that the unvarnished history in this instance doesn't make enough sense, so he interprets the brute events, red-followed-by-green, by making up a narrative about the intervening passage, complete with mid-course color change, and installs this history, incorporating his glosses, frames C and D (in Figure 5.11), in the memory library for all future reference. Since he works fast, within a fraction of a second — the amount of time it takes to frame (but not utter) a verbal report of what you have experienced — the record you rely on, stored in the library of memory, is already contaminated. You *say* and *believe* that you saw the illusory motion and color change, but that is really a memory hallucination, not an accurate recollection of your original consciousness.

How could we see which of these hypotheses is correct? It might seem that we could rule out the Stalinesque hypothesis quite simply, because of the delay in consciousness it postulates. In Kolers and von Grünau's experiment, there was a 200msec difference in onset between the red and green spot, and since, *ex hypothesi*, the *whole experience* cannot be composed by the editing room until after the content *green spot* has reached the editing room, consciousness of the initial red spot will have to be delayed by at least that much. (If the editing room sent the content *red spot* up to the theater of consciousness immediately, before receiving frame B and then fabricating frames C and D, the subject would presumably experience a gap in the film, a delay of at least 200msec between A and C — as noticeable as a syllable-long gap in a word, or five missing frames of a movie).

Suppose we ask subjects to press a button "as soon as you experience a red spot." We would find little or no difference in response time to a red spot alone versus a red spot followed 200msec later by a green spot (in which case the subjects report color-switching apparent motion). Could this be because there is *always* a delay of at least 200msec in consciousness? No. There is abundant evidence that responses under conscious control, while slower than such responses as reflex blinks, occur with close to the minimum latencies (delays) that are physically possible. After subtracting the demonstrable travel times for incoming and outgoing pulse trains, and the response preparation time, there is not enough time left over in "central processing" in which to hide a 200msec delay. So the button-pressing responses would have to have been initiated before the discrimination of the second stimulus, the green spot.

This might seem to concede victory to the Orwellian hypothesis, a post-experiential revision mechanism: as soon as the subject becomes conscious of the red spot, he initiates a button-press. *While that button press is forming,* he becomes conscious of the green spot. *Then* both these experiences are wiped from memory, replaced in memory by the revisionist record of the red spot moving over and then turning green halfway across. He readily and sincerely *but falsely* reports having seen the red spot moving toward the green spot before changing color. If the subject insists that he really was conscious from the very beginning of the red spot moving and changing color, the Orwellian theorist will firmly explain to him that he is wrong; his memory is playing tricks on him; the fact that he pressed the button when he did is conclusive evidence that he was conscious of the (stationary) red spot before the green spot had even occurred. After all, his instructions were to press the button *when he was conscious of* a red spot. He must have been conscious of the red spot about 200msec before he could have been conscious of its moving and turning green. If that is not how it seems to him, he is simply mistaken.

The defender of the Stalinesque alternative is not defeated by this, however. Actually, he insists, the subject responded to the red spot *before* he was conscious of it! The directions to the subject (to respond to a red spot) had somehow trickled down from consciousness into the editing room, which (unconsciously) initiated the button-push before sending the edited version (frames ACDB) up to consciousness for "viewing." The subject's memory has played no tricks on him; he is reporting exactly what he was conscious of, except for his insistence that he consciously pushed the button after seeing the red spot; his "premature" button-push was unconsciously (or preconsciously) triggered.

Where the Stalinesque theory postulates a button-pushing reaction to an *unconscious* detection of a red spot, the Orwellian theory postulates a *conscious* experience of a red spot that is immediately obliterated from memory by its sequel. So here's the rub: We have two different models of what happens in the color phi phenomenon. One posits a Stalinesque "filling in" on the upward, pre-experiential path, and the other posits an Orwellian "memory revision" on the downward, post-experiential path, and *both* of them are consistent with *whatever* the subject says or thinks or remembers. Note that the inability to distinguish these two possibilities does not just apply to the *outside observers* who might be supposed to lack some private data to which the subject had "privileged access." You, as a subject in a phi phenomenon experiment, *could not* discover anything in the experience from your own first-person perspective that would favor one theory over the other; the experience would "feel the same" on either account.

Is that really so? What if you paid really close attention to your experience — mightn't you be able to tell the difference? Suppose the experimenter made it easier for you, by slowing down the display, gradually lengthening the interstimulus interval between the red and green spots. It's obvious that if the interval is long enough you can tell the difference between *perceiving* motion and *inferring* motion. (It's a dark and stormy night; in the first lightning flash you see me on your left; two seconds later there is another flash and you see me on your right. I must have moved, you infer, and you can certainly tell that you're only inferring the motion on this occasion, not *seeing* me move.) As the experimenter lengthens the interval between the stimuli, there will come a time when you begin to make this discrimination. You will say things like

> "This time the red spot didn't seem to move, but after I saw the green spot I sort of had the idea that the red spot had moved over and changed color."

In fact, there is an intermediate range of intervals where the phenomenology is somewhat paradoxical: you see the spots as two stationary flashers *and* as one thing moving! This sort of apparent motion is readily distinguishable from the swifter, smoother sort of apparent motion we see in movies and television, but our capacity to make *this* discrimination is not relevant to the dispute between the Orwellian and the Stalinesque theorist. They agree that you can make this discrimination under the right conditions. What they disagree about is how to describe the cases of apparent motion that you *can't* tell from real motion — the cases in which you really *perceive* the illusory motion. To put it loosely, in these cases is your memory playing tricks on you, or are just your eyes playing tricks on you?

But even if you, the subject, can't tell whether this phenomenon is Stalinesque or Orwellian, couldn't scientists — outside observers — find something in your brain that showed which it was? Some might want to rule this out as *inconceivable*. "Just try to imagine someone *else* knowing better than you do what you were conscious of! Impossible!" But is it really inconceivable? Let's look more closely. Suppose these scientists had truly accurate information (garnered from various brain-scanning technologies) about the exact "time of arrival" or "creation" of every representing, every vehicle of content, anywhere in your nervous system. This would give them the *earliest* time at which you could react in any way — conscious or unconscious — to any particular content (barring miraculous precognition). But the *actual* time at which you *became conscious* of that content (if you ever did) might be somewhat later. You would have to have become conscious of it early enough to explain your inclusion of the content in some later speech act of recollection — assuming that by definition any item in your heterophenomenological world is an item in your consciousness. That will fix the *latest* time at which the content "became conscious." But, as we have seen, if this leaves a duration of as much as several hundred milliseconds within which consciousness of the item must occur, and if there are several different items that must occur within that window (the red spot and the green spot; the long-haired woman with and without the glasses), there is no way to use your *reports* to order the representing events in consciousness.

Your retrospective verbal reports must be neutral with regard to two presumed possibilities, but might not the scientists find other data they could use? They could if there was a good reason to claim that some nonverbal behavior (overt or internal) was a good sign of consciousness. But this is just where the reasons run out. Both theorists agree that there is no behavioral reaction to a content that *couldn't* be a merely unconscious reaction — except for subsequent telling. On the Stalinesque model there is unconscious button-pushing (and why not?). Both theorists also agree that there could be a conscious experience that left no behavioral effects. On the Orwellian model there is momentary consciousness of a stationary red spot which leaves no trace on any later reaction (and why not?).

Both models can deftly account for *all* the data — not just the data we already have, but the data we can imagine getting in the future. They both account for the verbal reports: One theory says they are innocently mistaken, while the other says they are accurate reports of experienced mistakes. Moreover, we can suppose, both theorists have *exactly* the same theory of what happens in your brain; they agree about just where and when in the brain the mistaken content enters the causal pathways; they just disagree about whether that location is to be deemed pre-experiential or post-experiential. They give the same account of the nonverbal effects, with one slight difference: One says they are the result of unconsciously discriminated contents, while the other says they are the result of consciously discriminated but forgotten contents. Finally, they both account for the subjective data — whatever is obtainable from the first-person perspective — because they even agree about how it ought to "feel" to subjects: Subjects should be unable to tell the difference between misbegotten experiences and immediately misremembered experiences.

So, in spite of first appearances, there is really only a verbal difference between the two theories (for a similar diagnosis, see Reingold and Merikle, 1990). The two theories tell exactly the same story except for where they place a mythical Great Divide, a point in time (and hence a place in space) whose *fine-grained* location is nothing that subjects can help them locate, and whose location is also neutral with regard to all other features of their theories. This is a difference that makes no difference.

Consider a contemporary analogy. In the world of publishing there is a traditional and usually quite hard-edged distinction between prepublication editing, and postpublication correction of "errata." In the academic world today, however, things have been speeded up by electronic communication. With the advent of word-processing and desktop publishing and electronic mail, it now often happens that several different drafts of an article are simultaneously in circulation, with the author readily making revisions in response to comments received by electronic mail. Fixing a moment of publication, and thus calling one of the drafts of an article the *canonical* text — the text of record, the one to cite in a bibliography — becomes a somewhat arbitrary matter. Often most of the intended readers, the readers whose reading of the text matters, read only an early draft; the "published" version is archival and inert. If it is important effects we are looking for, then, most if not all the important effects of writing a journal article are spread out over many drafts, not postponed until after publication. It used to be otherwise; it used to be that virtually all of an article's important effects happened *after* appearance in a journal and *because of* its making such an appearance. Now that the various candidates for the "gate" of publication can be seen to be no longer functionally important, if we feel we need the distinction at all, we will have to decide arbitrarily what is to count as publishing a text. There is no natural summit or turning point in the path from draft to archive.

Similarly — and this is the fundamental implication of the Multiple Drafts model — if one wants to settle on some moment of processing in the brain as the moment of consciousness, this has to be arbitrary. One can always "draw a line" in the stream of processing in the brain, but there are no functional differences that could motivate declaring all prior stages and revisions to be unconscious or preconscious adjustments, and all subsequent emendations to the content (as revealed by recollection) to be post-experiential memory contamination. The distinction lapses in close quarters.

4. THE THEATER OF CONSCIOUSNESS REVISITED

The astronomer's rule of thumb:
if you don't write it down,
it didn't happen.

CLIFFORD STOLL, *The Cuckoo's Egg*, 1989

As every book on stage magic will tell you, the best tricks are over before the audience thinks they have begun. At this point you may well be thinking that I have just tried to pull a fast one on you. I have argued that because of the spatiotemporal smearing of the observer's point of view in the brain, all the evidence there is or could be fails to distinguish between the Orwellian and Stalinesque theories of conscious experience, and hence *there is no difference*. That is some sort of operationalism or verificationism, and it leaves out the possibility that there just are brute facts of the matter unreachable by science, even when science includes heterophenomenology. Besides, it really seems quite obvious that *there are* such brute facts — that our immediate conscious experience consists of such facts!

I agree that it seems quite obvious; if it didn't, I wouldn't have to work so hard in this chapter to show that what is so obvious is in fact false. What I seem to have left out, quite willfully, is something analogous to the derided Cartesian Theater of Consciousness. You may well suspect that under cover of anti*dualism* ("Let's get that spook stuff out of here!"), I have spirited away (quite literally) something Descartes was actually right about: There is a functional place of some sort where the items of phenomenology are . . . *projected*.

It is time to confront this suspicion. Nelson Goodman raises the issue when he says of Paul Kolers's color phi experiment that it "seems to leave us a choice between a retrospective construction theory and a belief in clairvoyance" (Goodman, 1978, p. 83). We must shun clairvoyance, so what exactly is "retrospective construction"?

> Whether perception of the first flash is thought to be delayed or preserved or remembered, I call this the retrospective construction theory — the theory that the construction perceived as occurring between the two flashes is accomplished not earlier than the second.

At first Goodman seems to vacillate between a Stalinesque theory (perception of the first flash is delayed) and an Orwellian theory (the perception of the first flash is preserved or remembered), but what is more important is that his postulated revisionist (whether Orwellian or Stalinesque) does not merely adjust judgments; he *constructs* material to *fill in* the gaps:

> each of the intervening places along a path between the two flashes is filled in . . . with one of the flashed colors rather than with successive intermediate colors. [p. 85]

What Goodman overlooks is the possibility that the brain doesn't actually have to go to the trouble of "filling in" anything with "construction" — for no one is looking. As the Multiple Drafts model makes explicit, once a discrimination has been made once, it does not have to be made again; the brain just adjusts to the conclusion that is drawn, making the new interpretation of the information available for the modulation of subsequent behavior.

Goodman considers the theory, which he attributes to Van der Waals and Roelofs (1930), that "the intervening motion is produced retrospectively, built only after the second flash occurs and *projected backwards in time* [my italics]" (pp. 73–74). This suggests a Stalinesque view with an ominous twist: a final film is made and then run through a magical projector whose beam somehow travels backwards in time onto the mind's screen. Whether or not this is just what Van der Waals and Roelofs had in mind when they proposed "retrospective construction," it is presumably what led Kolers (1972, p. 184) to reject their hypothesis, insisting that all construction is carried out in "real time." Why, though, should the brain bother to "produce" the "intervening motion" in any case? Why shouldn't the brain just *conclude that there was* intervening motion, and insert that retrospective conclusion into the processing stream? Isn't that enough?

Halt! This is where the sleight of hand (if there is any) must be taking place. From the third-person point of view, I have posited a subject, the heterophenomenological subject, a sort of fictional "to whom it may concern" to whom, indeed, we outsiders would correctly attribute the belief that intervening motion had been experienced. That is how it would seem to *this* subject (who is just a theorist's fiction). But isn't there also a *real* subject, for whose benefit the brain must indeed mount a show, filling in all the blank spots? This is what Goodman seems to be supposing when he talks of the brain filling in all the places on the path. For whose benefit is all this animated cartooning being executed? For the audience in the Cartesian Theater. But *since there is no such theater, there is no such audience.*

The Multiple Drafts model agrees with Goodman that retrospectively the brain creates the content (the judgment) that there was intervening motion, and this content is then available to govern activity and leave its mark on memory. But the Multiple Drafts model goes on to claim that the brain does not bother "constructing" any representations that go to the trouble of "filling in" the blanks. That would be a waste of time and (shall we say?) *paint.* The judgment is *already in,* so the brain can get on with other tasks![5]

Goodman's "projection backwards in time" is an equivocal phrase. It might mean something modest and defensible: namely that a *reference to some past time* is included in the content. On this reading it would be a claim like "This novel takes us back to ancient Rome . . . ," which no one would interpret in a metaphysically extravagant way, as claiming that the novel was some sort of time-travel machine. This is the reading that is consistent with Goodman's other views, but Kolers apparently took it to mean something metaphysically radical: that there was some actual projection of one thing at one time to another time.

---

5. There is a region in the cortex called MT, which responds to motion (and apparent motion). Suppose then that some activity in MT *is* the brain's concluding that there was intervening motion. There is no further question, on the Multiple Drafts model, of whether this is a pre-experiential or post-experiential conclusion. It would be a mistake to ask, in other words, whether the activity in MT was a "reaction to a conscious experience" (by the Orwellian historian) as opposed to a "decision to represent motion" (by the Stalinesque editor).

As we shall see in the next chapter, confusion provoked by this radical reading of "projection" has bedeviled the interpretation of other phenomena. The same curious metaphysics used to haunt thinking about the representation of space. In Descartes's day, Thomas Hobbes seems to have thought that after light struck the eye and produced there a kind of motion in the brain, this led something to *rebound* somehow back out into the world.

> The cause of sense, is the external body, or object, which presseth the organ proper to each sense, either immediately, as in the taste and touch; or mediately, as in seeing, hearing, and smelling; which pressure, by the mediation of the nerves, and other strings and membranes of the body, continued inwards to the brain and heart, causeth there a resistance, or counter-pressure, or endeavour of the heart to deliver itself, which endeavour, because *outward*, seemeth to be some matter without. [*Leviathan*, Part I, ch. 1, "Of Sense"]

After all, he thought, that's where we see the colors — out on the front surfaces of objects![6] In a similar spirit one might suppose that when you stub your toe, this causes upward signals to the brain's "pain centers," which then "project" the pain *back down into the toe where it belongs*. After all, that is where the pain is felt to be.

As recently as the 1950s this idea was taken seriously enough to provoke J. R. Smythies, a British psychologist, to write an article carefully demolishing it.[7] The projection we speak of in such phenomena does *not* involve beaming some effect out into physical space, and I guess nobody any longer thinks that it does. Neurophysiologists and

---

6. In fact, Hobbes was alert to the problems with this view: "For if those colours and sounds were in the bodies, or objects that cause them, they could not be severed from them, as by glasses, and in echoes by reflection, we see they are; where we know the thing we see is in one place, the appearance in another" (*Leviathan*, same chapter). But this passage is open to several very different readings.

7. Smythies (1954). This heroic piece shows how difficult it was to think about these matters only thirty-seven years ago. He strenuously rebuts a textbook version of the projection theory, and in his summation he quotes approvingly from Bertrand Russell's dismissal of the same idea: "Whoever accepts the causal theory of perception is compelled to conclude that percepts are in our heads, for they come at the end of a causal chain of physical events leading, spatially, from the object to the brain of the percipient. We cannot suppose that, at the end of this process, the last effect suddenly jumps back to the starting point like a stretched rope when it snaps" (Russell, 1927).

psychologists, and for that matter acousticians who design stereo speaker systems, often do speak of this sort of projection, however, and we might ask just what they mean by it if not something involving physical transmission from one place (or time) to another. What does it involve? Let's look closely at a simple case:

> Thanks to the placement of the stereo speakers and the balance of the volume of their respective outputs, the listener *projects* the resulting sound of the soprano to a point midway between the two speakers.

What does this mean? We must build it up carefully. If the speakers are blaring away in an empty room, there is no projection at all. If there is a listener present (an observer with good ears, and a good brain), the "projection" happens, but this does *not* mean that something is emitted by the listener to the point midway between the two speakers. No physical property of that point or vicinity is changed by the presence of the listener. In short, this is what we mean when we say that Smythies was right; there is no projection into space of either visual or auditory properties. What then does happen? Well, *it seems to the observer* that the sound of the soprano is coming from that point. What does this *seeming to an observer* involve? If we answer that it involves "projection by the observer of the sound to that point in space," we are back where we started, obviously, so people are tempted to introduce something new, by saying something like this: "the observer projects the sound *in phenomenal space.*" This looks like progress. We have denied that the projection is in physical space, and have relocated the projection in phenomenal space.

Now what is phenomenal space? Is it a physical space inside the brain? Is it the onstage space in a theater of consciousness located in the brain? Not literally. But metaphorically? In the previous chapter we saw a way of making sense of such metaphorical spaces, in the example of the "mental images" that Shakey manipulated. In a strict but metaphorical sense, Shakey drew shapes in space, paid attention to particular points in that space, based conclusions on what he found at those points in space. But the space was only a *logical* space. It was like the space of Sherlock Holmes's London, a space of a fictional world, but a fictional world systematically anchored to actual physical events going on in the ordinary space in Shakey's "brain." If we took Shakey's utterances as expressions of his "beliefs," then we could say that it was a space Shakey *believed in*, but that did not make it real, any more than someone's belief in Feenoman would make Feenoman real. Both are merely intentional objects.[8]

So we do have *a* way of making sense of the idea of phenomenal space — as a logical space. This is a space into which or in which nothing is literally projected; its properties are simply constituted by the beliefs of the (heterophenomenological) subject. When we say the listener projects the sound to a point in this space, we mean *only* that it seems to him that that is where the sound is coming from. Isn't that enough? Or are we overlooking a "realist" doctrine of phenomenal space, in which the *real seeming* can be projected?

Today we have grown quite comfortable with the distinction between the spatial location in the brain of the vehicle of experience, and the location "in experiential space" of the item experienced. In short we distinguish representing from represented, vehicle from content. We have grown sophisticated enough to recognize that the products of visual perception are not, literally, pictures in the head even though *what they represent* is what pictures represent well: the layout in space of various visible properties. We should make the same distinction for time: *when* in the brain an experience happens must be distinguished from when it seems to happen. Indeed, as the psycholinguist Ray Jackendoff has suggested, the point we need to understand here is really just a straightforward extension of the common wisdom about experience of space. The representation of space in the brain does not always use space-in-the-brain to represent space, and the representation of time in the brain does not always use time-in-the-brain. Just as unfounded as the spatial slide projector Smythies couldn't find in the brain is the temporal movie projector that the radical reading of Goodman's "projection back in time" encourages.

Why do people feel the need to posit this seems-projector? Why are they inclined to think that it is not enough for the editing rooms in the brain merely to insert content into the stream on its way to behavior modulation and memory? Perhaps because they want to preserve the reality/appearance distinction for consciousness. They want

---

8. "It is as if our Feenomanist turned Feenomanologist were to grasp in his confusion at the desperate stratagem of inventing a god-space, or heaven, for his beloved Feenoman to reside in, a space *real* enough to satisfy the believer in him, but remote and mysterious enough to hide Feenoman from the skeptic in him. Phenomenal space is Mental Image Heaven, but if mental images turn out to be *real*, they can reside quite comfortably in the physical space in our brains, and if they turn out not to be real, they can reside, with Santa Claus, in the logical space of fiction." Dennett (1978a), p. 186.

to resist the diabolical operationalism that says that what happened (in consciousness) is simply whatever you remember to have happened. The Multiple Drafts model makes "writing it down" in memory criterial for consciousness; that is *what it is* for the "given" to be "taken" — to be taken one way rather than another. There is no reality of conscious experience independent of the effects of various vehicles of content on subsequent action (and hence, of course, on memory). This looks ominously like dreaded operationalism, and perhaps the Cartesian Theater of consciousness is covertly cherished as the place where whatever happens "in consciousness" *really* happens, whether or not it is later correctly remembered. Suppose something happened in my presence, but left its trace on me for only "a millionth of a second," as in the Ariel Dorfman epigram. Whatever could it mean to say that I was, however briefly and ineffectually, conscious of it? If there were a privileged Cartesian Theater somewhere, at least it could mean that *the film was jolly well shown there* even if no one remembers seeing it. (So there!)

The Cartesian Theater may be a comforting image because it preserves the reality/appearance distinction at the heart of human subjectivity, but as well as being scientifically unmotivated, this is metaphysically dubious, because it creates the bizarre category of the objectively subjective — the way things actually, objectively seem to you even if they don't seem to seem that way to you! (Smullyan, 1981) Some thinkers have their faces set so hard against "verificationism" and "operationalism" that they want to deny it even in the one arena where it makes manifest good sense: the realm of subjectivity. What Clifford Stoll calls the astronomer's rule of thumb is a sardonic commentary on the vagaries of memory and the standards of scientific evidence, but it becomes the literal truth when applied to what gets "written" in memory. We might classify the Multiple Drafts model, then, as *first-person operationalism*, for it brusquely denies the possibility in principle of consciousness of a stimulus in the absence of the subject's belief in that consciousness.[9]

Opposition to this operationalism appeals, as usual, to possible facts beyond the ken of the operationalist's test, but now the operationalist is the subject himself, so the objection backfires: "Just because

9. The philosopher Jay Rosenberg has pointed out to me that Kant sees the wisdom in this, in his claim that, in experience, the *für mich* (the "for me") and the *an sich* (the "in itself") are the same thing.

you can't tell, by your preferred ways, whether or not you were conscious of x, that doesn't mean you weren't. Maybe you were conscious of x but just can't find any evidence for it!" Does anyone, on reflection, really want to say that? Putative facts about consciousness that swim out of reach of both "outside" and "inside" observers are strange facts indeed.

The idea dies hard. Consider how natural is the phrase "I judged it to be so, because that's the way it seemed to me." Here we are encouraged to think of two distinct states or events: the seeming-a-certain-way and a subsequent (and consequent) judging-that-it-is-that-way. The trouble, one may think, with the Multiple Drafts model of color phi, for instance, is that even if it includes the phenomenon of the subject's judging that there was intervening motion, it does not include — it explicitly denies the existence of — any event which might be called the seeming-to-be-intervening-motion, on which this judgment is "based." There must be "evidence presented" somewhere, if only in a Stalinesque show trial, so that the judgment can be caused by or grounded in that evidence.

Some people presume that this intuition is supported by phenomenology. They are under the impression that they actually observe themselves judging things to be such *as a result of* those things seeming to them to be such. No one has ever observed any such thing "in their phenomenology" because such a fact about causation would be unobservable (as Hume noted long ago).[10]

Ask a subject in the color phi experiment: Do you judge that the red spot moved right and changed color because it seemed to you to do so, or does it seem to you to have moved because that is your judgment? Suppose the subject gives a "sophisticated" answer:

---

10. The philosopher Ned Block once recounted to me his experience as a subject in a "laterality" test. He looked straight ahead at a fixation point and every so often a word (or a nonword, such as GHRPE) was flashed on the left or right side of fixation. His task was to press a button if the stimulus was a word. His reaction times were measurably longer for words displayed in the left field (and hence entering the right hemisphere first), supporting the hypothesis that he, like most people, was strongly lateralized for language in the left hemisphere. This was not surprising to Block; what interested him was "the phenomenology: the words flashed on the left seemed a bit blurry, somehow." I asked him whether he thought the words were harder to identify because they seemed blurry, or seemed blurry because they were harder to identify. He admitted that he could have no way of distinguishing these "opposite" causal accounts of his judgment.

> I know there wasn't *actually* a moving spot in the world — it's just apparent motion, after all — but I also know the spot *seemed* to move, so in addition to my judgment that the spot seemed to move, there is the event which my judgment is *about:* the seeming-to-move of the spot. There wasn't any real moving, so there has to have been a real seeming-to-move for my judgment to be about.

Perhaps the Cartesian Theater is popular because it is the place where the seemings can happen in addition to the judgings. But the sophisticated argument just presented is fallacious. Postulating a "real seeming" *in addition to* the judging or "taking" expressed in the subject's report is multiplying entities beyond necessity. Worse, it is multiplying entities beyond possibility; the sort of inner presentation in which real seemings happen is a hopeless metaphysical dodge, a way of trying to have your cake and eat it too, especially since those who are inclined to talk this way are eager to insist that this inner presentation does *not* occur in some mysterious, dualistic sort of space perfused with Cartesian ghost-ether. When you discard Cartesian dualism, you really must discard the show that would have gone on in the Cartesian Theater, and the audience as well, for neither the show nor the audience is to be found in the brain, and the brain is the only real place there is to look for them.

5. THE MULTIPLE DRAFTS MODEL IN ACTION

Let's review the Multiple Drafts model, extending it somewhat, and considering in a bit more detail the situation in the brain that provides its foundation. For simplicity, I'll concentrate at what happens in the brain during visual experience. Later we can extend the account to other phenomena.

Visual stimuli evoke trains of events in the cortex that gradually yield discriminations of greater and greater specificity. At different times and different places, various "decisions" or "judgments" are made; more literally, parts of the brain are caused to go into states that discriminate different features, e.g., first mere onset of stimulus, then location, then shape, later color (in a different pathway), later still (apparent) motion, and eventually object recognition. These localized discriminative states transmit effects to other places, contributing to further discriminations, and so forth (Van Essen, 1979; Allman, Meizin, and McGuinness, 1985; Livingstone and Hubel, 1987; Zeki and Shipp, 1988). The natural but naïve question to ask is: Where does it all come together? The answer is: Nowhere. Some of these distributed contentful states soon die out, leaving no further traces. Others do leave traces, on subsequent verbal reports of experience and memory, on "semantic readiness" and other varieties of perceptual set, on emotional state, behavioral proclivities, and so forth. Some of these effects — for instance, influences on subsequent verbal reports — are at least symptomatic of consciousness. But there is no one place in the brain through which all these causal trains must pass in order to deposit their content "in consciousness."

As soon as any such discrimination has been accomplished, it becomes available for eliciting some behavior, for instance a button-push (or a smile, or a comment), or for modulating some internal informational state. For instance, a discrimination of a picture of a dog might create a "perceptual set" — making it temporarily easier to see dogs (or even just animals) in other pictures — or it might activate a particular semantic domain, making it temporarily more likely that you read the word "bark" as a sound, not a covering for tree trunks. As we already noted, this multitrack process occurs over hundreds of milliseconds, during which time various additions, incorporations, emendations, and overwritings of content can occur, in various orders. These yield, over the course of time, something *rather like* a narrative stream or sequence, which can be thought of as subject to continual editing by many processes distributed around in the brain, and continuing indefinitely into the future. Contents arise, get revised, contribute to the interpretation of other contents or to the modulation of behavior (verbal and otherwise), and in the process leave their traces in memory, which then eventually decay or get incorporated into or overwritten by later contents, wholly or in part. This skein of contents is only rather like a narrative because of its multiplicity; at any point in time there are multiple drafts of narrative fragments at various stages of editing in various places in the brain. While some of the contents in these drafts will make their brief contributions and fade without further effect — and some will make no contribution at all — others will persist to play a variety of roles in the further modulation of internal state and behavior and a few will even persist to the point of making their presence known through press releases issued in the form of verbal behavior.

Probing this stream at various intervals produces different effects, precipitating different narratives — and these *are* narratives: single versions of a portion of "the stream of consciousness." If one delays the probe too long, the result is apt to be no narrative left at all. If one probes "too early," one may gather data on how early a particular discrimination is achieved in the stream, but at the cost of disrupting the normal progression of the stream.

Is there an "optimal time of probing"? On the plausible assumption that after a while such narratives degrade rather steadily through both fading of details and self-serving embellishment (what I ought to have said at the party tends to turn into what I did say at the party), one can justify probing as soon as possible after the stimulus sequence of interest. But one also wants to avoid interfering with the phenomenon by a premature probe. Since perception turns imperceptibly into memory, and "immediate" interpretation turns imperceptibly into rational reconstruction, there is no single all-contexts summit on which to direct one's probes.

Just what we are conscious of within any particular time duration is not defined independently of the probes we use to precipitate a narrative about that period. Since these narratives are under continual revision, there is no single narrative that counts as the canonical version, the "first edition" in which are laid down, for all time, the events that happened in the stream of consciousness of the subject, all deviations from which must be corruptions of the text. But any narrative (or narrative fragment) that does get precipitated provides a "time line," a subjective sequence of events from the point of view of an observer, that may then be compared with other time lines, in particular with the objective sequence of events occurring in the brain of that observer. As we have seen, these two time lines *may* not superimpose themselves in orthogonal registration (lined up straight): even though the (mis-) discrimination of *red-turning-to-green* occurred in the brain *after* the discrimination of *green spot*, the *subjective* or *narrative* sequence is, of course, *red spot, then red-turning-to-green, and finally green spot*. So within the temporal smear of the point of view of the subject, there may be order differences that induce kinks.

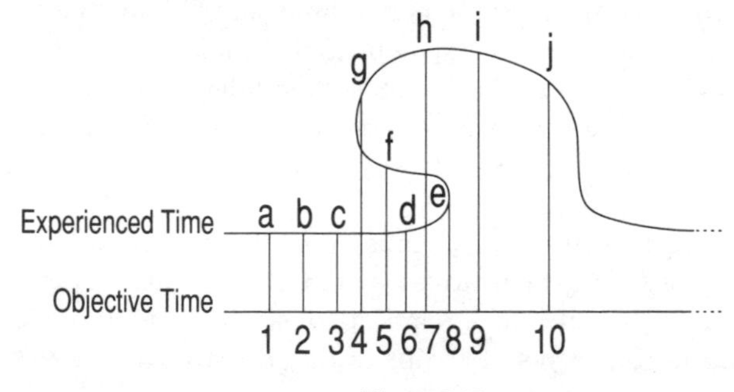

Figure 5.12

There is nothing metaphysically extravagant or challenging about this failure of registration.[11] It is no more mysterious or contra-causal than the realization that the individual scenes in movies are often shot out of sequence, or that when you read the sentence "Bill arrived at the party after Sally, but Jane came earlier than both of them," you learn of Bill's arrival before you learn of Jane's earlier arrival. The space and time of the representing is one frame of reference; the space and time of what the representing represents is another. But this metaphysically innocuous fact does nevertheless ground a fundamental metaphysical category: When a portion of the world comes in this way to compose a skein of narratives, that portion of the world is an observer. That is what it is for there to be an observer in the world, a something it is like something to be.

That is a rough sketch of my alternative model. Just how it differs from the Cartesian Theater model still needs to be further clarified, by showing how it handles particular phenomena. In the next chapter, we will put the model to work on some difficult topics, but first let's consider briefly some mundane and familiar examples, often discussed by philosophers.

You have probably experienced the phenomenon of driving for miles while engrossed in conversation (or in silent soliloquy) and then discovering that you have utterly no memory of the road, the traffic, your car-driving activities. It is as if someone else had been driving. Many theorists (myself included, I admit — Dennett, 1969, p. 116ff) have cherished this as a favorite case of "unconscious perception and intelligent action." But were you *really* unconscious of all those passing cars, stop lights, bends in the road at the time? You were paying attention to other things, but surely *if you had been probed* about what you had *just* seen at various moments on the drive, you would have had at least some sketchy details to report. The "unconscious driving" phenomenon is better seen as a case of rolling consciousness with swift memory loss.

Are you constantly conscious of the clock ticking? If it suddenly stops, you notice this, and you can say right away what it is that has stopped; the ticks "you weren't conscious of" up to the moment they stopped and "would never have been conscious of" if they hadn't stopped are now clearly in your consciousness. An even more striking case is the phenomenon of being able to count, retrospectively in ex-

11. This way of thinking about it first occurred to me after reading Snyder (1988), although his way of approaching the problems is somewhat different from mine.

perience memory, the chimes of the clock which you only noticed was striking after four or five chimes. But how could you so clearly *remember hearing* something you hadn't been conscious of in the first place? The question betrays a commitment to the Cartesian model; there are no fixed facts about the stream of consciousness independent of particular probes.

6

TIME AND EXPERIENCE

I can indeed say that my representations follow one another; but this is only to say that we are conscious of them as in a time-sequence, that is, in conformity with the form of inner sense.

IMMANUEL KANT, *Critique of Pure Reason*, 1781

In the previous chapter, we saw in outline how the Multiple Drafts model dissolves the problem of "backwards projection in time," but we ignored some major complications. In this chapter we will pursue these issues into somewhat more challenging territory, examining and resolving several controversies that have arisen among psychologists and neuroscientists regarding the proper explanation of some notoriously unsettling experiments. I think it's possible to understand the rest of the book without following all the arguments in this chapter, so it could be skipped or skimmed, but I've tried to make the issues clear enough for outsiders to grasp, and I can think of six good reasons for soldiering through the technical parts.

(1) There is much that is still obscure in my sketch of the Multiple Drafts model, and by seeing the model in further action, you will get a clearer view of its structure.
(2) If you have residual doubts about just how different, as an empirical theory, the Multiple Drafts model is from the traditional Cartesian Theater, these doubts will be dissipated by the spectacle of several head-on collisions.
(3) If you wonder if I am attacking a straw man, it will be reassuring to discover some experts tying themselves in knots because they are genuine Cartesian materialists in spite of themselves.

(4) If you suspect that I have based the model on a single carefully chosen phenomenon, Kolers's color phi, you will get to see how some very different phenomena benefit from the Multiple Drafts treatment.

(5) Several of the notorious experiments we will examine have been heralded by *some* distinguished experts as the refutation of the sort of conservative materialistic theory I am presenting, so if there is to be a *scientific* challenge to my explanation of consciousness, this is the battleground that has been chosen by the opposition.

(6) Finally, the phenomena in question are fascinating, well worth the effort to learn about.[1]

1. FLEETING MOMENTS AND HOPPING RABBITS

A normally sufficient, but not necessary, condition for having experienced something is a subsequent verbal report, and this is the anchoring case around which all the puzzling phenomena wander. Suppose that although your brain has registered — responded to — some aspects of an event, something intervenes between that internal response and a subsequent occasion for you to make a verbal report. If there was no time or opportunity for an initial overt response of any sort, and if the intervening events prevent later overt responses (verbal or otherwise) from incorporating reference to some aspects of the first event, this creates a puzzle question: Were they never consciously perceived, or have they been rapidly forgotten?

Many experiments have measured the "span of apprehension." In an acoustic memory-span test, you hear a tape recording of many unrelated items rapidly presented (say, four items a second), and are asked to identify them. You simply cannot respond till the acoustic event is over, and you then identify some, but not others. Yet subjectively you heard all of them clearly and equally well. The natural question to ask is: What exactly were you conscious of? There is no doubt that all the information on the tape got processed by your auditory system, but did the identifying marks of the items that were not subsequently named make it all the way to your consciousness, or were they just uncon-

---

1. The arguments and analyses in this chapter (and some of the discussion in the previous chapter) are elaborations of material in Dennett and Kinsbourne (in press).

sciously registered? They *seem* to have been there, in consciousness, but were they *really*?

In another experimental paradigm, you are briefly shown a slide on which many letters are printed. (This is done with a *tachistoscope*, a display device that can be accurately adjusted to present a stimulus of a particular brightness for a particular number of milliseconds — sometimes only 5msec, sometimes 500msec or longer.) You can subsequently report only some of the letters, but the rest were certainly seen by you. You insist they were there, you know exactly how many there were, and you have the impression that they were clear-cut and distinct. Yet you cannot identify them. Have you rapidly forgotten them, or did they never quite get consciously perceived by you in the first place?

The well-studied phenomenon of *metacontrast* (Fehrer and Raab, 1962) brings out the main point of the Multiple Drafts model sharply. (For a survey of similar phenomena, see Breitmeyer, 1984.) If a stimulus is flashed briefly on a screen (for, say, 30msec — about as long as a single frame of television) and then immediately followed by a second "masking" stimulus, subjects *report* seeing only the second stimulus. The first stimulus might be a colored disc and the second stimulus a colored ring that fits closely outside the space where the disc was displayed.

If you could put yourself in the subject's place, you would see for yourself; you would be prepared to swear that there was only one stimulus: the ring. In the psychological literature, the standard descrip-

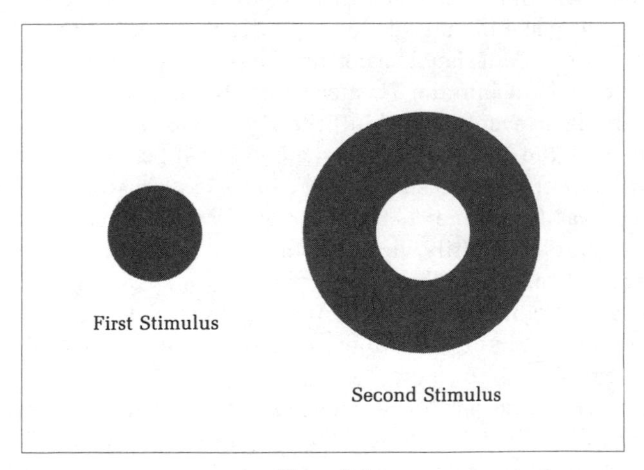

Figure 6.1

tion of such phenomena is Stalinesque: the second stimulus somehow *prevents conscious experience* of the first stimulus. In other words, it somehow waylays the first stimulus on its way up to consciousness. People can nevertheless do much better than chance if required to guess whether there were one or two stimuli. This only shows once again, says the Stalinesque theorist, that stimuli can have their effects on us without our being conscious of them. The first stimulus never plays on the stage of consciousness, but has whatever effects it has entirely unconsciously. We can counter this explanation of metacontrast with its Orwellian alternative: subjects are indeed conscious of the first stimulus (which explains their capacity to guess correctly) but their memory of this conscious experience is *almost* entirely obliterated by the second stimulus (which is why they deny having seen it, in spite of their telltale better-than-chance guesses). The result is a standoff — and an embarrassment to both sides, since neither side can identify any crucial experimental result that *would* settle the dispute.

Here is how the Multiple Drafts model deals with metacontrast. When a lot happens in a short time, the brain may make simplifying assumptions. The outer contour of a disc rapidly turns into the inner contour of a ring. The brain, initially informed just that something happened (something with a circular contour in a particular place), swiftly receives confirmation that there was indeed a ring, with an inner and outer contour. Without further supporting evidence that there was a disc, the brain arrives at the conservative conclusion that there was only a ring. Should we insist that the disc was experienced because *if the ring hadn't intervened* the disc would have been reported? That would be to make the mistake of supposing we could "freeze-frame" the film in the Cartesian Theater and make sure that the disc frame really did make it into the Theater before the memory of it was obliterated by later events. The Multiple Drafts model agrees that information about the disc was briefly in a functional position to contribute to a later report, but this state lapsed; there is no reason to insist that this state was inside the charmed circle of consciousness until it got overwritten, or contrarily, to insist that it never quite achieved this privileged state. Drafts that were composed at particular times and places in the brain were later withdrawn from circulation, replaced by revised versions, but none of them may be singled out as definitive of the content of *consciousness*.

An even more startling exhibition of this capacity for revision is the *cutaneous rabbit*. The psychologists Frank Geldard and Carl Sherrick reported the original experiments in 1972 (see also Geldard, 1977;

Geldard and Sherrick, 1983, 1986). The subject's arm rests cushioned on a table, and mechanical tappers are placed at two or three locations along the arm, up to a foot apart. A series of taps in rhythm are delivered by the tappers, e.g., five at the wrist followed by two near the elbow and then three more on the upper arm. The taps are delivered with interstimulus intervals between 50 and 200msec. So a train of taps might last less than a second, or as much as two or three seconds. The astonishing effect is that the taps seem to the subjects to travel in regular sequence over equidistant points up the arm — as if a little animal were hopping along the arm. Now, at first one feels like asking *how did the brain know* that after the five taps on the wrist, there were going to be some taps near the elbow? The subjects experience the "departure" of the taps from the wrist beginning with the second tap, yet in catch trials in which the later elbow taps are never delivered, subjects feel all five wrist taps at the wrist in the expected manner. The brain obviously can't "know" about a tap at the elbow until after it happens. If you are still entranced by the Cartesian Theater, you may want to speculate that the brain delays the conscious experience until after all the taps have been "received" at some way station in between the arm and the seat of consciousness (whatever that is), and this way station revises the data to fit a theory of motion, and sends the edited version on up to consciousness. But would the brain always delay response to one tap just in case more came? If not, how does it "know" when to delay?

The Multiple Drafts model shows that this is a misbegotten question. The shift in space (along the arm) is discriminated over time by the brain. The number of taps is also discriminated. Although in physical reality the taps were clustered at particular locations, the simplifying assumption is that they were distributed regularly across the space-time extent of the experience. The brain relaxes into this parsimonious but mistaken interpretation *after* the taps are registered, of course, and this has the effect of wiping out earlier (partial) interpretations of the taps, but side effects of those interpretations may live on. For instance, suppose we asked subjects to press a button whenever they felt *two taps on the wrist;* it would not be surprising if they could initiate the button-press *before* the forearm taps had been discriminated that caused them to misinterpret the second tap as displaced up the arm.

We must be particularly careful not to make the mistake of supposing that the content we would derive from such an early probe constituted the "first chapter" of the content we would find in the narrative if we were to probe the same phenomenon later. This confuses two different "spaces": the space of represent*ing* and the space represented. This is such a tempting and ubiquitous mistake that it deserves a section of its own.

2. HOW THE BRAIN REPRESENTS TIME

Cartesian materialism, the view that nobody espouses but almost everybody tends to think in terms of, suggests the following subterranean picture. We know that information moves around in the brain, getting processed by various mechanisms in various regions. Our intuitions suggest that our streams of consciousness consist of events occurring in sequence, and that at any instant every element in that sequence can be classified as either having already occurred "in consciousness" or as having not occurred "there" yet. And if that is so, then (it seems) the contentful vehicles of content moving through the brain must be like railroad cars on a track; the order in which they pass by some point will be the order in which they "arrive at" the theater of consciousness and (hence) "become conscious." To determine *where* in the brain consciousness happens, trace all the trajectories of information-vehicles, and see what point particular vehicles are passing at the instant they become conscious.

Reflection on the brain's fundamental task will show us what is wrong with this picture. The brain's task is to guide the body it controls through a world of shifting conditions and sudden surprises, so it must gather information from that world and use it *swiftly* to "produce future" — to extract anticipations in order to stay one step ahead of disaster (Dennett, 1984a, 1991b). So the brain must represent temporal properties of events in the world, and it must do this efficiently. The processes that are responsible for executing this task are spatially distributed in a large brain with no central node, and communication between regions of this brain is relatively slow; electrochemical nerve impulses travel thousands of times slower than light (or electronic signals through wires). So the brain is under significant time pressure. It must often arrange to modulate its output in the light of its input within a time window that leaves no slack for delays. On the input side, there are perceptual analysis tasks, such as speech perception, which would be beyond the physical limits of the brain's machinery if it didn't utilize ingenious anticipatory strategies that feed on redundancies in the input. Normal speech occurs at the rate of four or five syllables per second, but so powerful are the analysis machines we have evolved to "parse" it, that people can comprehend "compressed speech" — in which the words are electronically sped up without raising the tone chipmunk-style — at rates of up to thirty syllables per second. On the output side, many acts must occur so fast, and with such accurate triggering, that the brain has no time to adjust its control signals in the light of feedback; acts such as playing the piano or accurately pitching a rock (Calvin, 1983, 1986) must be *ballistically* initiated. (Ballistic acts are unguided missiles; once they are triggered, their trajectories are not adjustable.)

How, then, does the brain keep track of the temporal information it needs? Consider the following problem: Since the toe-to-brain distance is much greater than the hip-to-brain distance, or the shoulder-to-brain distance, or the forehead-to-brain distance, stimuli delivered simultaneously at these different sites will arrive at Headquarters in staggered succession, if travel-speed is constant along all paths. How, you might ask yourself, does the brain "ensure central simultaneity of representation for distally simultaneous stimuli"? Engaging in some speculative reverse engineering, you might think as follows: Perhaps all afferent nerve tracts are like spring-loaded windup tape measures — and all the same length: the nerves to the toes are fully unwound, those to the forehead are mainly coiled in the brain. Signals in the latter tract loop round and round in their inboard delay coils, exiting into Headquarters at exactly the same instant as the nonlooping signals from the toes. Or you might imagine that nerve tracts got narrower in diameter as they stretched out (rather like clay pot coils or home-made noodles), and that conduction speed varied with diameter. (It does, but in the wrong direction, alas! Thick fibers conduct faster.) These are vivid (if silly) models of mechanisms that would solve this problem, but the antecedent mistake is to suppose that the brain needs to solve *this* problem at all. The brain shouldn't solve this problem, for an obvious engineering reason: it squanders precious time by conceding to the full range of its operations a "worst case" schedule. Why should vitally important signals from the forehead (for instance) dawdle in the anteroom just because there might someday be an occasion when concurrent signals from the toes need to converge with them somehow?[2]

---

2. This does not mean that the brain never uses "buffer memories" to cushion the interface between the brain's internal processes and the asynchronous outside world. The "echoic memory" with which we preserve stimulus patterns briefly while the brain begins to process them is an obvious example (Sperling, 1960; Neisser, 1967; see also Newell, Rosenbloom, and Laird, 1989, p. 107).

Digital computers do depend on such delays to allow for worst cases and ensure synchrony. The mechanism in a parallel adder circuit that holds completed sums idle until a timing pulse releases them is close kin to the imagined looping nerves. And the builders of supercomputers have to be extraordinarily careful to ensure that the wires connecting various parts are the same length, which often requires including extra loops of wire. But digital computers can afford such local inefficiency because they have speed to burn. (In fact, with the market competition for faster and faster digital computers, these tiny temporal inefficiencies are now all being rethought; the main reason many of them remain is that engineers don't know how to design totally asynchronous computer systems, unregulated by any master clock pulse.)

Imposing a master synchrony on operations requires delays. As reverse engineers, we may speculate that if there are effective ways for the brain to represent the information it needs about time that avoid these delays, evolution will have "found them." In fact there are such ways, which we can illustrate with a historical incident that exhibits the phenomenon greatly magnified — in both space and time.

Consider the communication difficulties faced by the far-flung British Empire before the advent of radio and telegraph. Controlling a worldwide empire from a headquarters in London was not always feasible. The most notorious incident is surely the Battle of New Orleans, on January 8, 1815, fifteen days after the truce ending the War of 1812 was signed in Belgium. Over a thousand British soldiers were killed in this pointless battle. We can use this debacle to see how the system worked. Suppose on day 1 the treaty is signed in Belgium, with the news sent by land and sea to America, India, Africa, and so forth. On day 15 the battle is fought in New Orleans, and news of the defeat is sent by land and sea to England, India, etc. On day 20, too late, the news of the treaty (and the order to surrender) arrives in New Orleans. On day 35, let's suppose, the news of the defeat arrives in Calcutta, but the news of the treaty doesn't arrive there until day 40 (traveling via a slow overland route). To the British commander in chief in India, the battle would "seem" to have been fought before the treaty was signed — were it not for the practice of dating letters, which permits him to make the necessary corrections.[3]

---

3. I hasten to add that I am making up this historical embellishment. Francis Rawdon-Hastings, the first Marquis of Hastings and second Earl of Moira, was governor

These far-flung agents solved most of their problems of communicating information about time by embedding representations of the relevant time information in the *content* of their signals, so that the arrival time of the signals themselves was *strictly irrelevant* to the information they carried. A date written at the head of a letter (or a dated postmark on the envelope) gives the recipient information about when it was sent, information that survives any delay in arrival.[4] This distinction between time represented (by the postmark) and time of representing (the day the letter arrives) is an instance of the familiar distinction between content and vehicle. While this particular solution is not available to the brain's communicators (because they don't "know the date" when they send their messages), the general principle of the content/vehicle distinction is relevant to information-processing models of the brain in ways that have seldom been appreciated.[5]

In general, we must distinguish features of represent*ings* from the features of represent*eds*. Someone can shout "Softly, on tiptoe!" at the top of his lungs, there are gigantic pictures of microscopic objects, and there is nothing impossible about an oil painting of an artist making a charcoal sketch. The top sentence of a written description of a standing man need not describe his head, nor the bottom sentence his feet. This principle also applies, less obviously, to time. Consider the *spoken phrase* "a bright, brief flash of red light." The beginning of *it* is "a

---

general of Bengal and commander in chief of India in 1815, but I haven't the faintest idea how or when he was actually informed of the Battle of New Orleans.

4. Such a "postmark" can in principle be added to a vehicle of content at any stage of its journey; if all materials arriving at a particular location come from the same place, by the same route at the same speed, their "departure time" from the original destination can be retroactively stamped on them, by simply subtracting a constant from their arrival time at the way station. This is an engineering possibility that is probably used by the brain for making certain automatic adjustments for standard travel times.

5. As Uttal (1979) notes, the distinction *is* widely recognized by neuroscientists: "The essence of much of the research that has been carried out in the field of sensory coding can be distilled into a single, especially important idea — any candidate code can represent any perceptual dimension; there is no need for an isomorphic relation between the neural and psychophysical data. Space can represent time, time can represent space, place can represent quality, and certainly, nonlinear neural functions can represent linear or nonlinear psychophysical functions equally well" (p. 286). But while this idea is well known, we shall soon see that some theorists understand it by *mis*understanding it; the way they "make sense of it" is by tacitly reintroducing the unnecessary "isomorphism" in a dimly imagined process of translation or "projection" that is supposed to occur in consciousness.

bright" and the end of it is "red light." Those portions of that speech event are not themselves representations of onsets or terminations of a brief red flash (for a similar point, see Efron, 1967, p. 714). No event in the nervous system can have zero duration (any more than it can have zero spatial extent), so it has an onset and termination separated by some amount of time. If the event itself *represents* an event in experience, then the event it represents must itself have non–zero duration, an onset, a middle, and a termination. But there is no reason to suppose that the beginning of the representing represents the beginning of the represented.[6] Although different attributes are indeed extracted by different neural facilities at different rates (e.g., location versus shape versus color), and although if asked to respond to the presence of each one in isolation, we would do so with different latencies, we perceive events, not a successively analyzed trickle of perceptual elements or attributes.[7]

A novel or historical narrative need not have been composed in the order it eventually portrays — sometimes authors begin with the ending and work backwards. Moreover, such a narrative can contain flashbacks, in which events are *represented as* having happened in a certain order by means of *representings* that occur in a different order. Similarly, the representing by the brain of *A before B* does not have to be accomplished by:

first:

a representing of A,

followed by:

a representing of B.

---

6. Cf. Pylyshyn (1979, p. 278): "No one . . . is disposed to speak *literally* of such physical properties of a mental event as its color, size, mass and so on . . . though we *do* speak of them as *representing* (or having the experiential content of) such properties. For instance, one would not properly say of a thought (or image) that it was large or red, but only that it was a thought *about* something large or red (or that it was an image *of* something large or red). . . . It ought to strike one as curious, therefore, that we speak so freely of the *duration* of a mental event."

7. As the psychologist Robert Efron remarks: "We do not, when first observing an object with central vision, fleetingly experience the object as it would appear with the most peripheral vision, then as it would appear with less peripheral vision. . . . Similarly, when we shift our attention from one object of awareness to another, there is no experience of 'growing' specificity of the new object of awareness — we just perceive the new object" (1967, p. 721).

The phrase "B after A" is an example of a (spoken) vehicle that represents A as being before B, and the brain can avail itself of the same freedom of temporal placement. What matters for the brain is not necessarily when individual representing events happen in various parts of the brain (as long as they happen in time to control the things that need controlling!) but their *temporal content*. That is, what matters is that the brain can proceed to control events "under the assumption that A happened before B" whether or not the information that A has happened enters the relevant system of the brain and gets recognized as such before or after the information that B has happened. (Recall the commander in chief in Calcutta: First he was informed of the battle, and then he was informed of the truce, but since he can extract from this the information that the truce came first, he can act accordingly. He has to *judge* that the truce came before the battle; he doesn't also have to mount some sort of pageant of "historical reconstruction" to watch, in which he receives the letters in the "proper" order.)

Some have argued, however, that time is the one thing that the mind or brain *must* represent "with itself." The philosopher Hugh Mellor, in his book *Real Time* (1981, p. 8) puts the claim clearly and vigorously:

> Suppose for example I see one event *e* precede another, *e**. I must first see *e* and then *e**, my seeing of *e* being somehow recollected in my seeing of *e**. That is, my seeing of *e* affects my seeing of *e**: this is what makes me — rightly or wrongly — see *e* precede *e** rather than the other way round. But seeing *e* precede *e** means seeing *e* first. So the causal order of my perceptions of these events, by fixing the temporal order I perceive them to have, fixes the temporal order of the perceptions themselves. . . . [T]he striking fact . . . should be noticed, namely that perceptions of temporal order need temporally ordered perceptions. *No other property or relation has to be thus embodied in perceptions of it* [my italics]: perceptions of shape and colour, for example, need not themselves be correspondingly shaped or coloured.

This is false, but there is something right about it. Since the fundamental function of representation in the brain is to control behavior in real time, the timing of representings is *to some degree* essential to their task, in two ways.

First, at the beginning of a perceptual process, the timing may be *what determines the content*. Consider how we distinguish a spot moving from right to left from a spot moving from left to right on a motion picture screen. The *only* difference between the two may be the temporal order in which two frames (or more) are projected. If first A then B is projected, the spot is seen as moving in one direction; if first B then A, the spot is seen as moving in the opposite direction. The only difference in the stimuli that the brain could use to make this discrimination of direction is the order in which they occur. This discrimination is, then, as a matter of logic, based on the brain's capacity to make a temporal order discrimination of a particular acuity. Since motion-picture frames are usually presented at the rate of twenty-four per second, we know that the visual system can resolve order between stimuli that occur within about 50msec. This means that the actual temporal properties of signals — their onset times, their velocity in the system, and hence their arrival times — must be accurately controlled until such a discrimination is made. Otherwise, the information on which the discrimination must be based will be lost or obscured.

On a larger scale this phenomenon sometimes arises at the beginning of a sailboat race; you *see* a boat cross the starting line and then *hear* the starting gun, but was the boat over the line too early? It is logically impossible to tell unless you can calculate the different transmission times for sound and light to the place where you made the discrimination. Once a judgment has been made (either *all clear* or *boat #7 was over the line early*), this content can be conveyed to the participants in a leisurely fashion, without regard to how fast or far *it* has to travel to do its job.

So timing of some representings matters *until* a discrimination such as *left-to-right* (or *over the line early*) has been made, but once it is made, locally, by some circuit in the cortex (or some observer on the committee boat), the content of the judgment can be sent, in a temporally sloppy way, anywhere in the brain where this information might be put to use. Only in this way can we explain the otherwise puzzling fact that people may be unable to perform above chance on some temporal order judgments while they perform flawlessly on other judgments (such as direction of motion judgments) which logically call for even greater temporal acuity. They use specialized (and specially located) discriminators to make the high-quality judgments.

The second constraint on timing has already been noted parenthetically: it does not matter in what order representings occur so long as they occur in time to contribute to the control of the appropriate behavior. The function of a representing may depend on meeting a *deadline*, which is a temporal property of the vehicle doing the representing. This is particularly evident in such time-pressured environments as the imagined Strategic Defense Initiative. The problem is not how to get computer systems to represent, accurately, missile launches, but how to represent a missile launch accurately during the brief time while you can still do something about it. A message that a missile was launched at 6:04:23.678 A.M. EST may accurately represent the time of launch forever, but its utility may utterly lapse at 6:05 A.M. EST. For any task of control, then, there is a *temporal control window* within which the temporal parameters of representings may in principle be moved around *ad lib*.

The deadlines that limit such windows are not fixed, but rather depend on the task. If, rather than intercepting missiles, you are writing your memoirs or answering questions at the Watergate hearings (Neisser, 1981), the information you need to recover about the sequence of events in your life in order to control your actions can be recovered in almost any order, and you can take your time drawing inferences. Or to take an intermediate case closer to the phenomena we are considering, suppose you are drifting in a boat, and you wonder whether you are drifting towards or away from a dangerous reef you can see in the distance. Suppose that *now* you know your current distance from the reef (by measuring the angle, let's say, that it subtends in your visual field); in order to answer your question, you can either wait a bit and then measure the angle again or, if half an hour ago you took a Polaroid snapshot of the reef, you could measure the angle on that old photograph, do some calculations and *retrospectively* figure out how far away you were then. In order to make the judgment about the direction you are drifting you have to calculate two distances: distance at noon and distance at 12:30, let's say, but it doesn't make any difference which distance gets calculated first. But you had better be able to make the calculation swiftly enough so that you can get out the oars before it's too late.

So the brain's representing of time is anchored to time itself in two ways: the very timing of the representing can be what provides the evidence or determines the content, and the whole point of representing the time of things can be lost if the representing doesn't happen in time to make the difference it is supposed to make. I expect that Mellor appreciates both of these factors, and had them in mind in making the claim I quoted, but he makes the natural mistake of thinking that in combination they *completely* constrain the representing of time, so that order of representing *always* represents the order in content. On his account, there is no room for temporal "smearing," while I have been arguing that there *must* be temporal smearing — on a small scale — *because* there must be spatial smearing (on a small scale) of the point of view of the observer.

Causes must precede effects. This fundamental principle ensures that temporal control windows are bounded at both ends: by the earliest time at which information could arrive in the system, and by the latest time at which information could contribute causally to control of a particular behavior. We still haven't seen *how* the brain might utilize the time available in a control window to sort out the information it receives and turn it into a coherent "narrative" that gets used to govern the body's responses.

How, then, might temporal properties be inferred by processes in the brain? Systems of "date stamps" or "postmarks" are not theoretically impossible, but there is a cheaper, less foolproof but biologically more plausible way: by what we might call *content-sensitive settling*. A useful analogy would be the film studio where the sound track is "synchronized" with the film. The various segments of audio tape may by themselves have lost all their temporal markers, so that there is no simple, mechanical way of putting them into apt registration with the images. Sliding them back and forth relative to the film and looking for convergences, however, will usually swiftly home in on a "best fit." The slap of the slateboard at the beginning of each take — "Scene three, take seven, camera rolling, SLAP!" — provides a double saliency, an auditory and a visual clap, to slide into synchrony, pulling the rest of the tape and the frames into position at the same time. But there are typically so many points of mutually salient correspondence that this conventional saliency at the beginning of each take is just a handy redundancy. Getting the registration right depends on the *content* of the film and the tape, but not on sophisticated analysis of the content. An editor who knew no Japanese would find synchronizing a Japanese soundtrack to a Japanese film difficult and boring but not impossible. Moreover, the temporal order of the stages of the process of putting the pieces into registration is independent of the content of the product; the editor can organize scene three before organizing scene two, and in principle could even do the entire job running the segments "in reverse."

Quite "stupid" processes can do similar jiggling and settling in the brain. For instance, the computation of depth in random-dot stereograms (see Figure 5.7, page 112) is a spatial problem for which we can readily envisage temporal analogues. In principle, then, the brain can solve some of its problems of temporal inference by such a process, drawing data not from left and right eyes, but from whatever information sources are involved in a process requiring temporal judgments.

Two important points follow from this. First, such temporal inferences can be drawn (such temporal discriminations can be made) by comparing the (low-level) *content* of several data arrays, and this real-time process need not occur in the temporal order that its product eventually represents. Second, once such a temporal inference has been drawn, which may be *before* high-level features have been extracted by other processes, it does not have to be drawn again! There does not have to be a *later* representation in which the high-level features are "presented" in a real-time sequence for the benefit of a second sequence-judger. In other words, having drawn inferences from these juxtapositions of temporal information, the brain can go on to represent the results in any format that fits its needs and resources — not necessarily a format in which "time is used to represent time."

3. LIBET'S CASE OF "BACKWARDS REFERRAL IN TIME"

We've established a way in which the brain can do its editorial work on temporal information in a fashion that ignores the actual timing (the "time of arrival") of some of its representations, but we must remind ourselves once more of the time pressure under which it must do all this. Working backwards from the deadline, all content reported or otherwise expressed in subsequent behavior must have been present (in the brain, but not necessarily "in consciousness") in time to have contributed causally to that behavior. For instance, if a subject in an experiment *says* "dog" in response to a visual stimulus, we can work backwards from the behavior, which was clearly controlled by a process that had the content *dog* (unless the subject says "dog" to every stimulus, or spends the day saying "dog dog dog . . . " and so forth). And since it takes on the order of 100msec to *begin* to execute a speech intention of this sort (and roughly another 200msec to complete it), we can be quite sure that the content *dog* was present in (roughly) the language areas of the brain by 100msec before the utterance began. Working from the other end, again, we can determine the earliest time the content *dog* could have been computed or extracted by the visual system from the retinal input, and even, perhaps, follow its creation and subsequent trajectory through the visual system and into the language areas.

What would be truly anomalous (indeed a cause for lamentations and the gnashing of teeth) would be if the time that expired between the *dog*-stimulus and the "dog"-utterance was less than the time physically required for this content to be established and moved through the system. But no such anomalies have been uncovered. What have been discovered, however, are some startling juxtapositions between the two sequences graphed in Figure 5.12 on page 136. When we try to put the sequence of events in the objective processing stream in the brain into registration with the subject's subjective sequence *as determined by what the subject subsequently says* we sometimes find surprisingly large kinks. That, at least, is the conclusion we might want to draw from one of the most widely discussed — and criticized — experiments in neuroscience: Benjamin Libet's neurosurgical experiment demonstrating what he calls "backwards referral in time."

Sometimes during brain surgery it is important for the patient to be awake and alert, under only local anesthetic (like getting Novocain from the dentist). This lets the neurosurgeon get immediate reports from the patient about what is being experienced while the brain is being probed (see footnote 5 on page 58). This practice was pioneered by Wilder Penfield (1958), and for more than thirty years, neurosurgeons have been gathering data on the results of direct electrical stimulation to various parts of the cortex. It has long been known that stimulation of locations on the *somatosensory* cortex (a strip conveniently located across the top of the brain) produces the experience in the patient of sensations on corresponding parts of the body. For instance, stimulation of a point on the left somatosensory cortex can produce the sensation of a brief tingle in the subject's right hand (because of the familiar twist in the nervous system that leaves the left half of the brain responsible for the right side of the body, and vice versa). Libet compared the time course of such cortically induced tingles to similar sensations produced in the more usual way, by applying a brief electrical pulse to the hand itself (Libet, 1965, 1981, 1982, 1985b; Libet et al., 1979; see also Popper and Eccles, 1977; Dennett, 1979b; Churchland, 1981a, 1981b; Honderich, 1984).

What would one expect to happen? Well, suppose two commuters head to work every day at exactly the same time, but one lives way out in the suburbs and the other lives a few blocks from the office. They drive at the same speed, so given the extra distance the suburban commuter must drive, we would expect him to arrive at the office later. This is not, however, what Libet found when he asked his patients which came first, the hand-tingle that started right in the cortex or the hand-tingle sent from the hand. From the data he gathered, he argued

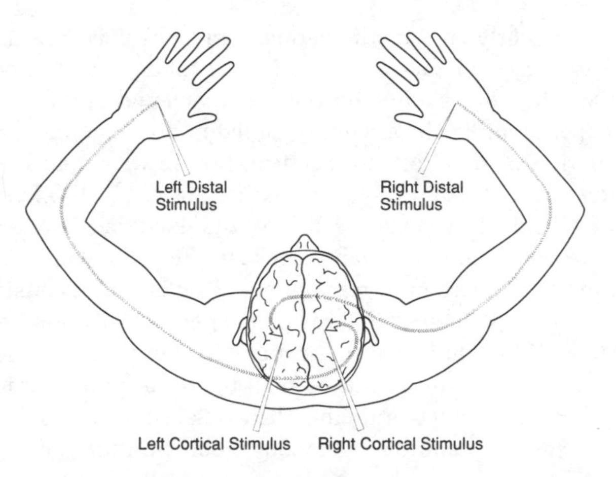

Figure 6.2

that while in each case it took considerable time (approximately 500 msec) from onset of stimulation to "neuronal adequacy" (the point at which he claims that cortical processes culminate to yield a conscious experience of a tingle), when the hand itself was stimulated, the experience was "automatically" "referred backwards in time," and was felt to happen *before* the tingle produced by brain stimulation itself.

Most strikingly, Libet reported instances in which a patient's left *cortex* was stimulated *before* his left *hand* was stimulated, which one would tend to think would surely give rise to two felt tingles: first right hand (cortically induced) and then left hand. In fact, however, the subjective report was reversed: "first left, then right."

Libet has interpreted his results as raising a serious challenge to materialism: ". . . a dissociation between the timings of the corresponding 'mental' and 'physical' events would seem to raise serious though not insurmountable difficulties for the . . . theory of psychoneural identity" (Libet et al., 1979, p. 222). According to Sir John Eccles, a Nobel laureate in medicine for his research in neurophysiology, this challenge cannot be met:

> This antedating procedure does not seem to be explicable by any neurophysiological process. Presumably it is a strategy that has been learnt by the self-conscious mind . . . the antedating sensory experience is attributable to the ability of the self-conscious mind to make slight temporal adjustments, i.e., to play tricks with time. [Popper and Eccles, 1977, p. 364]

More recently, the mathematician and physicist Roger Penrose (1989) has suggested that a materialistic explanation of Libet's phenomena would require a revolution in fundamental physics. Although Libet's experiment has been widely hailed in nonscientific circles as a demonstration of the truth of dualism, few in the cognitive science community share this opinion. In the first place, Libet's experimental procedures, and his analysis of the results, have been severely criticized. His experiment has never been replicated, which is reason enough in many quarters to remove his "results" from consideration. The skeptical view, then, is that his phenomena simply don't exist. *But what if they did?* That is just the sort of question a philosopher would ask, but in this case, there is more than the usual philosophical motivation for asking it. No one doubts the existence of such simpler phenomena as color phi and the cutaneous rabbit, and their interpretation raises the same problems. It would be theoretically myopic to settle for *methodological* grounds for dismissal, when this leaves unchallenged the background assumptions that suggest that if Libet's experiment *were* ever properly replicated, it would be a dark day for materialism.

The first thing to notice about Libet's experiment is that it would provide no evidence of any anomaly at all were we to forgo the opportunity to record the subjects' verbal reports of their experiences and then use them to generate first a text and then a heterophenomenological world. The noises they make with their vocal tracts during or after experiments yield no hint of paradox if treated as merely acoustic phenomena. In no case do sounds appear to issue from heads before lips move, nor do hands move before the brain events that purportedly cause them, nor do events occur in the cortex in advance of the stimuli that are held to be their source. Viewed strictly as the internal and external behavior of a biologically implemented control system for a body, the events observed and clocked in the experiments exhibit no apparent violations of everyday mechanical causation — of the sort Galilean/Newtonian physics provides the standard approximate model.

You might, then, "make the problems disappear" by being a barefoot behaviorist and simply refusing to take introspective reports seriously. But we are not barefoot behaviorists; we want to accept the challenge of making sense of what Libet calls "a primary phenomenological aspect of our human existence in relation to brain function" (1985a, p. 534). Libet almost gets the point about heterophenomenology. He says: "It is important to realize that these subjective referrals and corrections are apparently taking place at the level of the *mental* 'sphere'; they are not apparent, as such, in the activities at neural levels"

(1982, p. 241). But since he has no *neutral* way of referring to phenomenology, he must assign the anomaly to "the mental 'sphere.' " A small step, a forced step (for he *must* make this point if he rejects behaviorism), but it is the first step on a buttered slide back to dualism.

> The reports by subjects about their different experiences . . . were not theoretical constructs but empirical observations. . . . The method of introspection may have its limitations, but it can be used appropriately within the framework of natural science, and it is absolutely essential if one is trying to get some experimental data on the mind-brain problem. [1987, p. 785]

The reports by subjects, even when turned into texts, are, as Libet says, empirical observations, but *what they report*, the events in their heterophenomenological worlds, are indeed theoretical constructs. They can, as Libet urges, be used appropriately within the framework of natural science, but *only* if you understand them at the outset as theorists' fictions.

Libet claims that his experiments with direct stimulation of the cortex demonstrate "two remarkable temporal factors":

(1) *There is a substantial delay before cerebral activities*, initiated by a sensory stimulus, *achieve "neuronal adequacy"* for eliciting any resulting conscious sensory experience.
(2) After neuronal adequacy is achieved, the *subjective timing* of the experience *is (automatically) referred backwards in time*, utilizing a "timing signal" in the form of the initial response of cerebral cortex to the sensory stimulus (1981, p. 182).

The "timing signal" is the *first* burst of activity to appear in the cortex (the primary evoked potential), which occurs only 10 to 20msec after stimulation of the peripheral sense organ. Libet suggests that the backwards referral is always "to" the timing signal.

Libet's model is Stalinesque: After the primary evoked potential, various editing processes occur in the cortex prior to the moment of "neuronal adequacy," at which time a finished film is projected. How is it projected? Here Libet's account vacillates between an extreme view and a moderate view (cf. Honderich, 1984):

(1) *backwards projection:* It is somehow sent backwards in time in some Cartesian Theater where it is projected in synch with the primary evoked potentials. (The primary evoked potentials, as "timing signals," serve rather like the slateboard used

in filmmaking, showing the projector exactly how far back in time to send the experience.)

(2) *backwards referral:* It is projected in ordinary time, but it carries something like a postmark, reminding the viewer that these events must be understood to have occurred somewhat earlier. (In this case the primary evoked potentials serve simply as dates, which might be *represented* on the Cartesian screen by a title "On the eve of the Battle of Waterloo" or "New York City, summer, 1942.")

Libet's own term is *referral,* and he defends it by reminding us of the "long recognized and accepted" phenomenon of spatial referral, which suggests the moderate reading.

> Subjective referral backwards in time is a strange concept and perhaps is not readily palatable on initial exposure to it. But there is a major precedent for it in the long recognized and accepted concept of subjective referral in the spatial dimension. For example, the visual image experienced in response to a visual stimulus has a subjective spatial configuration and location that are greatly different from the spatial configuration and location of the neuronal activities that give rise to the ("subjectively referred") image. [1981, p. 183. See also Libet et al., 1979, p. 221; Libet, 1985b.]

However, he goes on to conclude that temporal referral raises problems for materialism (the "theory of psychoneural identity": Libet et al., 1979, p. 222), so either he thinks that spatial referral also raises these problems, or he hasn't understood his own defense. If spatial referral — the fact that what we see seems to be outside, not inside, our brains — raises a problem for materialism, though, why does Libet suggest his own work uncovers an important new argument for dualism? Surely the fact of spatial referral is much better attested than the sorts of temporal referrals he must work so ingeniously to demonstrate. It does seem, though, that Libet has a radical (or at any rate confused) vision of spatial referral as some kind of "projection":

> there is experimental evidence for the view that the subjective or mental "sphere" could indeed "fill in" spatial and temporal gaps. How else, for example, could one view the already mentioned enormous discrepancy *that is known to exist* between a subjective visual image and the configuration of neuronal activities that gives rise to the experience of the image? [1981, p. 196]

This does seem to say that the projector Smythies could not find in the brain is in fact hiding in the mental "sphere."[8]

How does Libet claim to establish his two remarkable temporal factors? "Neuronal adequacy," which Libet estimates to require up to 500msec of cortical activity, is determined by seeing how late, following initial stimulation, a direct cortical stimulation can *interfere with* the consciousness subsequently reported. Beyond that critical interval, a direct cortical stimulus would be reported by the subject to be a *subsequent* experience. (Having arrived too late for incorporation by the editing room into the "final print" of the first stimulus experience, it would appear in the next installment.) Libet's data suggest a tremendously variable editing window: "The conditioning cortical stimulus could be started more than 500msec following the skin pulse and still modify the skin sensation, although in most cases retroactive effects were not observed with S-C intervals greater than 200msec" (1981, p. 185). Libet is careful to define neuronal adequacy in terms of effect on subsequent unhurried verbal report: "[T]he subject was asked to report, within a few seconds after the delivery of each pair of . . . stimuli" (1979, p. 195), and he insists that "The timing of a subjective experience must be distinguished from that of a behavioral response (such as in reaction time), which might be made before conscious awareness develops . . . " (Libet et al., 1979, p. 193).

This proviso permits him to defend a rival interpretation of Patricia Churchland's data. Churchland is the first "neurophilosopher" (see her 1986 book, *Neurophilosophy: Toward a Unified Science of the Mind-Brain.*) When I first read of Libet's results (in Popper and Eccles, 1977), I encouraged her to look into them, and she gave them a vigorous shakedown (Churchland, 1981a). She attempted to discredit Libet's first

8. See also Libet's dismissal of MacKay's suggestion of a more moderate reading (1981, p. 195; 1985b, p. 568). Libet's final summation in 1981, on the other hand, is inconclusive: "My own view . . . has been that the temporal discrepancy creates relative difficulties for identity theory, but that these are not insurmountable" (p. 196). Presumably they would be undeniably insurmountable on the backwards *projection* interpretation, for that involves precognition or backwards causation or something equally spooky and unprecedented. Moreover, Libet later (1985b, p. 569) describes these not-insurmountable difficulties in a way that seems to require the milder reading: "Although the delay-and-antedating hypothesis does not separate the actual time of the experience from its time of neuronal production, it does eliminate the necessity for simultaneity between the *subjective timing* of the experience and the actual clock-time of the experience." Perhaps Sir John Eccles's enthusiastic support for a radical, dualistic interpretation of the findings has lured the attention of Libet (and his critics) away from the mild thesis he sometimes defends.

thesis, the long rise time to "neuronal adequacy" for consciousness, by asking subjects in an experiment to say "go" as soon as they were conscious of a skin stimulus like those used by Libet. She reported a mean response time over nine subjects of 358msec, which, she argued, showed that the subjects must have achieved neuronal adequacy by the 200msec mark at the latest (allowing time for the production of a verbal response).

Libet's reply is Stalinesque: A verbal reaction — saying "go" — can be unconsciously initiated. "There is nothing magical or uniquely informative when the motor response is a vocalization of the word 'go' instead of the more usual one of a finger tapping a button. . . . The ability to detect a stimulus and to react to it purposefully, or be psychologically influenced by it, without any reportable conscious awareness of the stimulus, is widely accepted" (1981, pp. 187–188). And to the objection, "But what did Churchland's subjects think they were doing, if not saying, as requested, just when they were conscious of the stimulus?" Libet could give the standard Stalinesque reply: They did indeed eventually become conscious of the stimulus, but by then, their verbal report had already been initiated.[9]

For this reason Libet rejects reaction-time studies such as Churchland's as having "an uncertain validity as a primary criterion of a subjective experience" (1981, p. 188). He favors letting the subject take his time: "The report is made unhurriedly within a few seconds after each trial, allowing the subject to introspectively examine his evidence" (p. 188). How, then, can he deal with the rival prospect that this leisurely pace gives the Orwellian revisionist in the brain plenty of time to replace the *real* memories of consciousness with *false* memories?

> Reporting after the trial of course requires that processes of short-term memory and recallability be operative, but this presents no difficulty for subjects with no significant defects in these abilities. [p. 188]

---

9. In an earlier paper, Libet conceded the possibility of Orwellian processes and supposed there might be a significant difference between unconscious mental events and conscious-but-ephemeral mental events: "There may well be an immediate but ephemeral kind of experience of awareness which is not retained for recall at conscious levels of experience. If such experiences exist, however, their content would have direct significance only in later unconscious mental processes, although, like other unconscious experiences, they might play an indirect role in later conscious ones" (1965, p. 78).

This begs the question against the Orwellian, who is prepared to explain a variety of effects as the result of *normal* misremembering or hallucinatory recall, in which a prior, real event in consciousness is obliterated and replaced by subsequent memories. Has Libet let the stew cook too long, or has Churchland sampled it too soon? If Libet wants to claim a *privileged* status for his choice of probe time, he must be prepared to combat the counterarguments.

Libet comes close to pleading *nolo contendere*: "Admittedly, a report of relative timing order cannot, in itself, provide an indicator of the 'absolute' time (clock-time) of the experience: as suggested, there is no known method to achieve such an indicator" (1981, p. 188). This echoes his earlier remark that there seemed to be "no method by which one could determine the absolute timing of a subjective experience" (Libet et al., 1979, p. 193). What Libet misses, however, is the possibility that this is because there is no such moment of absolute time (cf. Harnad, 1989).

Churchland, in her criticisms (1981a, 1981b), also falls prey to the failure to distinguish time represented from time of representing:

> The two hypotheses differ essentially on just when the respective sensations *were felt*. [my italics; 1981a, p. 177]

> Even if it be supposed that the sensations arising from the simultaneous skin and LM [medial lemniscus] sensations are *felt at exactly the same time* [my italics], the delay in neuronal adequacy for skin stimuli may well be an artifact of the setup. [1981b, p. 494]

Suppose that all such artifacts were eliminated, and *still* the sensations are "felt at exactly the same time." How would Churchland interpret this unhoped-for result? Would this mean that there is a time *t* such that stimulus 1 is felt at *t* and stimulus 2 is felt at *t* (the antimaterialist prospect) or only that stimulus 1 and stimulus 2 are felt as (experienced as) simultaneous? Churchland doesn't discourage the inference that Libet's findings, if vindicated, would wreak havoc (as he sometimes claims) on materialism. Elsewhere, however, she correctly notes that "intriguing as temporal illusions are, there is no reason to suppose there is something preternatural about them, and certainly there is nothing which distinguishes them from spatial illusions or motion illusions as uniquely bearing the benchmark of a non-physical origin" (1981a, p. 179). This could only be the case if temporal illusions were phenomena in which *time was misrepresented;* if the *misrepresentings* take place at the "wrong" times, something more revolutionary is afoot.

Where does this leave Libet's experiments with cortical stimulation? As an interesting but inconclusive attempt to establish something about *how the brain represents temporal order*. Primary evoked potentials may somehow serve as specific reference points for neural representations of time, although Libet has not shown this, as Churchland's technical criticisms make clear. Alternatively, the brain may keep its representations of time more labile. We don't represent seen objects as existing on the retina, but rather at various distances in the external world; why should the brain not also represent events as happening *when* it makes the most "ecological" sense for them to happen? When we are engaged in some act of manual dexterity, "fingertip time" should be the standard; when we are conducting an orchestra, "ear time" might capture the registration. "Primary cortical time" might be the default standard (rather like Greenwich Mean Time for the British Empire), a matter, however, for further research.

The issue has been obscured by the fact that both proponent and critic have failed to distinguish consistently between time of representing and time represented. They talk past each other, with Libet adopting a Stalinesque position and Churchland making the Orwellian countermoves, both apparently in agreement that there is a fact of the matter about exactly when (in "absolute" time, as Libet would put it) a conscious experience happens.[10]

4. LIBET'S CLAIMS OF SUBJECTIVE DELAY OF CONSCIOUSNESS OF INTENTION

This concept of the absolute timing of an experience is exploited in Libet's later experiments with "conscious intentions." In these experiments, he sought to determine this absolute timing experimentally by letting the subjects, who alone have direct access (somehow) to their experiences, do *self-timing*. He asked normal subjects (not neurosurgery patients) to make "spontaneous" decisions to flex one hand at the wrist

---

10. Harnad (1989) sees an insoluble problem of measurement, but denies just what I am asserting — that there is no fact of the matter: "[I]ntrospection can only tell us when an event *seemed* to occur, or which of two events *seemed* to occur first. There is no independent way of confirming that the real timing was indeed as it seemed. Incommensurability is a methodological problem, not a metaphysical one" (p. 183).

while noting the position of a spot on a revolving disk (the "second hand" on a clock, in effect) at the precise time they formed the intention (Libet, 1985a, 1987, 1989). Afterwards (a few seconds later), subjects reported where the spot was at the moment they decided to flex their wrist. This permitted Libet to calculate what time it was (down to the millisecond) when subjects *thought* they had decided, and compare that moment with the timing of events going on concurrently in their brains. He found evidence that these "conscious decisions" lagged between 350 and 400msec behind the onset of "readiness potentials" he was able to record from scalp electrodes, which, he claims, tap the neural events that determine the voluntary actions performed. He concludes that "cerebral initiation of a spontaneous voluntary act begins unconsciously" (1985a, p. 529).

This seems to show that your consciousness lags behind the brain processes that actually control your body. Many find this an unsettling and even depressing prospect, for it seems to rule out a real (as opposed to illusory) "executive role" for "the conscious self." (See the discussions by many commentators on Libet, 1985a, 1987, 1989; and in Harnad, 1982; Pagels, 1988, p. 233ff; and Calvin, 1989a, pp. 80–81.)

Libet is clearer than most of his critics about the importance of keeping content and vehicle distinguished: "One should not confuse *what* is reported by the subject with *when* he may become introspectively aware of what he is reporting" (Libet, 1985a, p. 559). He recognizes (p. 560), moreover, that a judgment of simultaneity need not itself be simultaneously arrived at or rendered; it might mature over a long period of time (consider, for instance, the minutes it may take the stewards at the racetrack to develop and then examine the photo-finish picture on which they eventually base their judgment of the winner or a dead heat).

Libet gathered data on two time series:

> the objective series, which includes the timing of the external clock and the salient neural events: the readiness potentials (RPs) and the electromyograms (EMGs) that recorded the beginning of muscle contraction

> the subjective series (as later reported), which consists of mental imagery, memories of any preplanning, and a single benchmark datum for each trial: a simultaneity judgment of the form *my conscious intention (W) began simultaneously with the clock spot in position P*

Libet seems to have wanted to approximate the elusive *acte gratuit* discussed by the existentialists (e.g., Gide, 1948; Sartre, 1943), the purely motiveless — and hence in some special sense "free" — choice. As several commentators have pointed out, such highly unusual actions (which might be called acts of deliberate pseudo-randomness) are hardly paradigms of "normal voluntary acts" (Libet, 1987, p. 784). But did he in any event isolate a variety of conscious experiences, however characterized, that can be given absolute timing by such an experimental design?

He claims that when conscious intentions to act (at least of his special sort) are put into registration with the brain events that actually initiate the acts, there is an offset in the 300-500msec range. This is huge — up to half a second — and it does look ominous to anyone committed to the principle that our conscious acts *control* our bodily motions. It looks as if *we* are located in Cartesian Theaters where we are shown, with a half-second tape delay, the *real* decision-making that is going on *elsewhere* (somewhere *we* aren't). We are not quite "out of the loop" (as they say in the White House), but since our access to information is thus delayed, the most we can do is intervene with last-moment "vetoes" or "triggers." Downstream from (unconscious) Command Headquarters, I take no real initiative, am never in on the birth of a project, but do exercise a modicum of executive modulation of the formulated policies streaming through my office.

This picture is compelling but incoherent. Libet's model, as before, is Stalinesque, and there is an obvious Orwellian alternative: The subjects *were* conscious of their intentions at an earlier moment, but this consciousness was wiped out of memory (or just revised) before they had a chance to recall it. Libet concedes that this "does present a problem, but was not experimentally testable" (1985a, p. 560).

Given this concession, is the task of fixing the absolute timing of consciousness ill-conceived? Neither Libet nor his critics draw that conclusion. Libet, having carefully distinguished content from vehicle — *what* is represented from *when* it is represented — nonetheless tries to draw inferences from premises about what is represented to conclusions about the absolute timing of the representing in consciousness. The psychologist Gerald Wasserman (1985, p. 556) sees the problem: "The time when the external objective spot occupies a given clock position can be determined easily, but this is not the desired result." But he then goes on to fall in the Cartesian trap: "What is needed is the time of occurrence of the internal brain-mind representation of the spot."

"*The* time of occurrence" of the internal representation? Occurrence where? There is essentially continuous representation of the spot (representing it to be in various different positions) in various different parts of the brain, starting at the retina and moving up through the visual system. The brightness of the spot is represented in some places and times, its location in others, and its motion in still others. As the external spot moves, all these representations change, in an asynchronous and spatially distributed way. Where does "it all come together at an instant in consciousness"? Nowhere.

Wasserman correctly points out that the subject's task of determining where the spot was at some time in the subjective sequence is itself a voluntary task, and initiating it presumably takes some time. This is difficult not only because it is in competition with other concurrent projects, but also because it is unnatural — a conscious judgment of temporality of a sort that does not normally play a role in behavior control, and hence has no natural meaning in the sequence. The process of interpretation that eventually fixes the judgment of subjective simultaneity is itself an artifact of the experimental situation, and *changes the task,* therefore telling us nothing of interest about the actual timing of normal representation vehicles anywhere in the brain.

The all-too-natural vision that we must discard is the following: Somewhere deep in the brain an act-initiation begins; it starts out as an unconscious intention, and slowly makes its way to the Theater, picking up definiteness and power as it goes, and then, at an instant, *t*, it bursts onstage, where a parade of visual spot-representations are marching past, having made their way slowly from the retina, getting clothed with brightness and location as they moved. The audience or *I* is given the task of saying which spot-representation was "onstage" exactly when the conscious intention made its bow. Once identified, this spot's time of departure from the retina can be calculated, as well as the distance to the Theater and the transmission velocity. In that way we can determine the exact moment at which the conscious intention occurred in the Cartesian Theater.

I find it eerie how enticing that picture is. It's so easy to visualize! It seems so apt! Isn't that what *has* to happen when two things happen together in consciousness? No. In fact, it *cannot* be what happens when two things happen together in consciousness, for there is no such place in the brain. Some have thought that the incoherence of *that* vision does not require one to give up the idea of absolute timing of experiences. There is, it seems, an alternative model for the onset of consciousness that avoids the preposterousness of Descartes's centered brain while permitting absolute timing. Couldn't consciousness be a matter not of arrival at a point but rather a matter of a representation exceeding some threshold of activation over the whole cortex or large parts thereof? On this model, an element of content becomes conscious at some time *t*, not by entering some functionally defined and anatomically located system, but by changing state right where it is: by acquiring some property or by having the intensity of one of its properties boosted above some threshold.

The idea that consciousness is a *mode of action* of the brain rather than a *subsystem* of the brain has much to recommend it (see, e.g., Kinsbourne, 1980; Neumann, 1990; Crick and Koch, 1990). Moreover, such mode shifts can presumably be timed by outside observers, providing, in principle, a unique and determinate sequence of contents attaining the special mode. But this is still the Cartesian Theater if it is claimed that the real ("absolute") timing of such mode shifts is definitive of subjective sequence. The imagery is slightly different, but the implications are the same. Conferring the special property that makes for consciousness at an instant is only half the problem; discriminating that the property has been conferred at that time is the other, and although scientific observers with their instruments may be able to do this with microsecond accuracy, how is the brain to do this?

We human beings do make judgments of simultaneity and sequence of elements of our own experience, some of which we express, so at some point or points in our brains the corner must be turned from the actual timing of representations to the representation of timing, and wherever and whenever these discriminations are made, thereafter the temporal properties of the representations embodying those judgments are not constitutive of their content. The objective simultaneities and sequences of events spread across the broad field of the cortex are of no functional relevance *unless they can also be accurately detected by mechanisms in the brain.* We can put the crucial point as a question: What would make *this* sequence the stream of consciousness? There is no one inside, *looking at* the wide-screen show displayed all over the cortex, even if such a show is discernible by *outside* observers. What matters is the way those contents get utilized by or incorporated into the processes of ongoing control of behavior, and this *must* be only indirectly constrained by cortical timing. What matters, once again, is not the temporal properties of the representings, but the temporal properties *represented,* something determined by how they are "taken" by subsequent processes in the brain.

5. A TREAT: GREY WALTER'S PRECOGNITIVE CAROUSEL

Having struggled through the complicated cases, we deserve an encounter with something strange but *relatively* easy to understand — something that drives home the message of this difficult chapter. Libet's experiment with self-timing, we just observed, created an artificial and difficult judgment task, which robbed the results of the hoped-for significance. A remarkable early experiment by the British neurosurgeon W. Grey Walter (1963) did not have this drawback. Grey Walter performed his experiment with patients in whose motor cortex he had implanted electrodes. He wanted to test the hypothesis that certain bursts of recorded activity were the initiators of intentional actions. So he arranged for each patient to look at slides from a carousel projector. The patient could advance the carousel at will, by pressing the button on the controller. (Note the similarity to Libet's experiment: This was a "free" decision, timed only by an endogenous rise in boredom, or curiosity about the next slide, or distraction, or whatever.) Unbeknownst to the patient, however, the controller button was a dummy, not attached to the slide projector at all! What actually advanced the slides was the amplified signal from the electrode implanted in the patient's motor cortex.

One might suppose that the patients would notice nothing out of the ordinary, but in fact they were startled by the effect, because it seemed to them as if the slide projector was anticipating their decisions. They reported that just as they were "about to" push the button, but before they had actually decided to do so, the projector would advance the slide — and they would find themselves pressing the button with the worry that it was going to advance the slide twice! The effect was strong, according to Grey Walter's account, but apparently he never performed the dictated follow-up experiment: introducing a variable delay element to see how large a delay had to be incorporated into the triggering in order to eliminate the "precognitive carousel" effect.

An important difference between Grey Walter's and Libet's designs is that the judgment of temporal order that leads to surprise in Grey Walter's experiment is part of a normal task of behavior monitoring. In this regard it is like the temporal order judgments by which our brains distinguish moving left-to-right from moving right-to-left, rather than "deliberate, conscious" order judgments. The brain in this case has set itself to "expect" visual feedback on the successful execution of its project of advancing the carousel, and the feedback arrives earlier than expected, triggering an alarm. This could show us something important about the actual timing of content vehicles and their attendant processes in the brain, but it would not, contrary to first appearances, show us something about the "absolute timing of the conscious decision to change the slide."

Suppose, for instance, that an extension of Grey Walter's experiment showed that a delay as long as 300msec (as implied by Libet) had to be incorporated into the implementation of the act in order to eliminate the subjective sense of precognitive slide-switching. What such a delay would in fact show would be that expectations set up by a decision to change the slide are tuned to expect visual feedback 300msec later, and to report back with alarm under other conditions. (It is analogous to a message of shock from the commander in chief in Calcutta to Whitehall in the wake of the Battle of New Orleans.) The fact that the alarm *eventually* gets interpreted in the subjective sequence as a perception of misordered events (change before button-push) shows nothing about *when* in real time the consciousness of the decision to press the button first occurred. The sense the subjects reported of not quite having had time to "veto" the initiated button-push when they "saw the slide was already changing" is a natural interpretation for the brain to settle on (eventually) of the various contents made available at various times for incorporation into the narrative. Was this sense already there at the first moment of consciousness of intention (in which case the effect requires a long delay to "show time" and is Stalinesque) or was it a retrospective reinterpretation of an otherwise confusing *fait accompli* (in which case it is Orwellian)? I hope it is clear by now that the question's presuppositions disqualify it.

6. LOOSE ENDS

You may still want to object that all the arguments in this chapter are powerless to overturn the obvious truth that our experiences of events occur in the very same order as we experience them to occur. If someone thinks the thought "One, two, three, four, five," his thinking "one" occurs before his thinking "two" and so forth. This example does illustrate a thesis that is true in general, and does indeed seem unexceptioned so long as we restrict our attention to psychological phenomena of "ordinary," macroscopic duration. But the experiments we looked at are concerned with events that were constricted by unusually narrow time frames of a few hundred milliseconds. At this scale, the standard presumption breaks down.

Every event in your brain has a definite spatio-temporal location, but asking "Exactly when do *you* become conscious of the stimulus?" assumes that some one of these events is, or amounts to, your becoming conscious of the stimulus. This is like asking "Exactly when did the British Empire become informed of the truce in the War of 1812?" Sometime between December 24, 1814, and mid-January, 1815 — that much is definite, but there simply is no fact of the matter if we try to pin it down to a day and hour. Even if we can give precise times for the various moments at which various officials of the Empire became informed, no one of these moments can be singled out as the time the Empire itself was informed. The signing of the truce was one official, intentional act of the Empire, but the participation by the British forces in the Battle of New Orleans was another, and it was an act performed under the assumption that no truce had yet been signed. A case might be made for the principle that the arrival of the news at Whitehall or Buckingham Palace in London should be considered the official time at which the Empire was informed, since this was the "nerve center" of the Empire. Descartes thought the pineal gland was just such a nerve center in the brain, but he was wrong. Since cognition and control — and hence consciousness — is distributed around in the brain, no moment can count as the precise moment at which each conscious event happens.

In this chapter, I have attempted to shake some bad habits of thought off their imaginary "foundations" and replace them with some better ways of thinking, but along the way I have had to leave many loose ends. The most tantalizing, I suspect, is the metaphorical assertion that "probing" is something that "precipitates narratives." The timing of inquisitive probes by experimenters, I claimed, can have a major revisionary effect on the systems of representation utilized by the brain. But among those who may direct such inquisitive probes at the subject is the subject himself. If you become interested in the question of just when you become conscious of something, your own self-examinations or self-inquisitions fix termini for new control windows, thereby altering the constraints on the processes involved.

The results of probes by outsiders are typically speech acts of one variety or another, and these *express judgments about* various contents of consciousness, for all to hear and interpret. The results of self-probes are items *in the same semantic category* — not "presentations" (in the Cartesian Theater) but *judgments* about how it seems to the subject, judgments the subject himself can then go on to interpret, act upon, remember. In both cases, these events fix interpretations of what the subject experienced, and thus provide fixed points in the subjective sequence. But, on the Multiple Drafts model, there is no further question about whether *in addition* to such a judgment, and the earlier discriminations on which it is based, there was a *presentation* of the materials-to-be-interpreted for the scrutiny of a Master Judge, the audience in the Cartesian Theater. This is still not an easy idea to understand, let alone accept. We must build several more roads to it.

7

THE EVOLUTION OF CONSCIOUSNESS

Everything is what it is because it got that way.
D'ARCY THOMPSON (1917)

1. INSIDE THE BLACK BOX OF CONSCIOUSNESS

The theory sketched in the last chapter goes a little way towards showing how consciousness might reside in the human brain, but its main contribution was negative: toppling the dictatorial idea of the Cartesian Theater. We have *begun* to replace it with a positive alternative, but we certainly haven't gone very far. To make further progress, we must shift field and approach the complexities of consciousness from a different quarter: evolution. Since there hasn't always been human consciousness, it has to have arisen from prior phenomena that weren't instances of consciousness. Perhaps if we look at what must have — or might have — been involved in that transition, we'll get a better perspective on the complexities and their roles in creating the full-fledged phenomenon.

In his elegant little book, *Vehicles: Essays in Synthetic Psychology* (1984), the neuroscientist Valentino Braitenberg describes a series of ever more complicated autonomous mechanisms, gradually building from comically simple and utterly lifeless devices to (imagined) entities that are impressively biological and psychological in flavor. This exercise of the imagination works because of what he calls the law of uphill analysis and downhill synthesis: It is much easier to imagine the behavior (and behavioral implications) of a device you synthesize "from the inside out" one might say, than to try to analyze the external behavior of a "black box" and figure out what must be going on inside.

So far we have actually been treating consciousness itself as something of a black box. We have taken its "behavior" (= phenomenology) as "given" and wondered about what sort of hidden mechanisms in the brain could explain it. Now let's reverse the strategy, and think about the evolution of brain mechanisms for doing this and that, and see if anything that emerges gives us a plausible mechanism for explaining some of the puzzling "behaviors" of our conscious brains.

There have been many theories — well, speculations — about the evolution of human consciousness, beginning with Darwin's own surmises in *The Descent of Man* (1871). Unlike most explanations in science, evolutionary explanations are essentially narratives, taking us from a time when something didn't exist to a time when it did by a series of steps that the narrative explains. Rather than attempting to survey in a scholarly way all the narratives that have been devised, I propose to tell a single story, borrowing freely from many other theorists, and concentrating on a few often overlooked points that will help us over obstacles to understanding what consciousness is. In the interests of telling a good story, and keeping it relatively short, I have resisted the temptation to include literally dozens of fascinating subplots, and suppressed the standard philosopher's instinct to air all the arguments for and against the points I include and reject. The result, I recognize, is a bit like a hundred-word summary of *War and Peace*, but we have a lot of work to do.[1]

The story that we must tell is analogous to other stories that biology is beginning to tell. Compare it, for instance, to the story of the origins of sex. There are many organisms today that have no genders and reproduce asexually, and there was a time when all the organisms that existed lacked gender, male and female. Somehow, by some imaginable series of steps, some of these organisms must have evolved into organisms that did have gender, and eventually, of course, into us. What sort of conditions were required to foster or necessitate these innovations? Why, in short, did all these changes happen? These are some of the deepest problems in contemporary evolutionary theory.[2]

There is a nice parallel between the two questions, about the

---

1. You can infer that everything I use in my story I think is right — or on the right track — but you must not infer from the fact that I leave out some theory or some detail of a theory, that I think it is wrong. You must also not infer that just because I use a few details from some theory, I think the rest of that theory is defensible. This applies as well to my own earlier writing on this subject, from which I will draw.

2. John Maynard Smith is the leading theorist, and in addition to his classic, *The Evolution of Sex* (1978), there are several brilliant articles on the conceptual problems

origins of sex and the origins of consciousness. There is almost nothing *sexy* (in human terms) about the sex life of flowers, oysters, and other simple forms of life, but we can recognize in their mechanical and apparently joyless routines of reproduction the foundations and principles of our much more exciting world of sex. Similarly, there is nothing particularly *selfy* (if I may coin a term) about the primitive precursors of conscious human selves, but they lay the foundations for our particularly human innovations and complications. The design of our conscious minds is the result of three successive evolutionary processes, piled on top of each other, each one vastly swifter and more powerful than its predecessor, and in order to understand this pyramid of processes, we must begin at the beginning.

2. EARLY DAYS

Scene One: The Birth of Boundaries and Reasons

In the beginning, there were no reasons; there were only causes. Nothing had a purpose, nothing had so much as a function; there was no teleology in the world at all. The explanation for this is simple: There was nothing that had interests. But after millennia there happened to emerge simple *replicators* (R. Dawkins, 1976; see also Monod, 1972, ch. 1, "Of Strange Objects"). While *they* had no inkling of their interests, and perhaps properly speaking had no interests, we, peering back from our godlike vantage point at their early days, can nonarbitrarily assign them certain interests — generated by their defining "interest" in self-replication. That is, maybe it really made no difference, was a matter of no concern, didn't matter to anyone or anything whether or not they succeeded in replicating (though it does seem that we can be grateful that they did), but at least we can assign them interests conditionally. *If* these simple replicators are to survive and replicate, thus persisting in the face of increasing entropy, their environment must meet certain conditions: conditions conducive to replication must be present or at least frequent.

Put more anthropomorphically, if these simple replicators want to continue to replicate, they should hope and strive for various things; they should avoid the "bad" things and seek the "good" things. When

---

in his collection of essays *Sex, Games, and Evolution* (1989). See also R. Dawkins (1976), pp. 46–48, for a brief survey of the issues.

an entity arrives on the scene capable of behavior that staves off, however primitively, its own dissolution and decomposition, it brings with it into the world its "good." That is to say, it creates a point of view from which the world's events can be roughly partitioned into the favorable, the unfavorable, and the neutral. And its own innate proclivities to seek the first, shun the second, and ignore the third contribute essentially to the definition of the three classes. As the creature thus comes to have interests, the world and its events begin creating *reasons* for it — whether or not the creature can fully recognize them (Dennett, 1984a). The first reasons preexisted their own recognition. Indeed, the first problem faced by the first problem-facers was to learn how to recognize and act on the reasons that their very existence brought into existence.

As soon as something gets into the business of self-preservation, boundaries become important, for if you are setting out to preserve yourself, you don't want to squander effort trying to preserve the whole world: you draw the line. You become, in a word, *selfish*. This primordial form of selfishness (which, as a primordial form, lacks most of the flavors of *our* brand of selfishness) is one of the marks of life. Where one bit of granite ends and the next bit begins is a matter of slight moment; the fracture boundary may be real enough, but nothing works to protect the territory, to push back the frontier or retreat. "Me against the world" — this distinction between everything on the inside of a closed boundary and everything in the external world — is at the heart of all biological processes, not just ingestion and excretion, respiration and transpiration. Consider, for instance, the immune system, with its millions of different antibodies arrayed in defense of the body against millions of different alien intruders. This army must solve the fundamental problem of recognition: telling one's self (and one's friends) from everything else. And the problem has been solved in much the way human nations, and their armies, have solved the counterpart problem: by standardized, mechanized identification routines — the passports and customs officers in miniature are molecular shapes and shape-detectors. It is important to recognize that this army of antibodies has no generals, no GHQ with a battle plan, or even a description of the enemy: the antibodies represent their enemies only in the way a million locks represent the keys that open them.

We should note several other facts that are already evident at this earliest stage. Whereas evolution depends on history, Mother Nature is no snob, and origins cut no ice with her. It does not matter where or how an organism acquired its prowess; handsome is as handsome does. So far as we know, of course, the pedigrees of the early replicators were all pretty much the same: they were each of them the product of one blind, dumb-luck series of selections or another. But had some time-traveling hyperengineer inserted a *robot-replicator* into the milieu, and if its prowess was equal or better than the prowess of its natural-grown competition, its descendants might now be among us — might even *be* us! (Dennett, 1987a, 1990b)

Natural selection cannot tell how a system got the way it got, but that doesn't mean there might not be profound differences between systems "designed" by natural selection and those designed by intelligent engineers (Langton, Hogeweg, in Langton, 1989). For instance, human designers, being farsighted but blinkered, tend to find their designs thwarted by unforeseen side effects and interactions, so they try to guard against them by giving each element in the system a single function, and insulating it from all the other elements. In contrast, Mother Nature (the process of natural selection) is famously myopic and lacking in goals. Since she doesn't foresee at all, she has no way of worrying about unforeseen side effects. Not "trying" to avoid them, she tries out designs in which many such side effects occur; most such designs are terrible (ask any engineer), but every now and then there is a *serendipitous side effect:* two or more unrelated functional systems interact to produce a bonus: multiple functions for single elements. Multiple functions are not unknown in human-engineered artifacts, but they are relatively rare; in nature they are everywhere, and as we shall see, one of the reasons theorists have had such a hard time finding plausible designs for consciousness in the brain is that they have tended to think of brain elements as serving just one function each.[3]

---

3. The idea of multifunctional neurons is not new, but it has recently been gaining adherents:

> It is the more or less simultaneous concatenations of neuronal outputs or signals that are unambiguous, rather than the outputs of individual neurons. The convergence of different concatenations of ambiguous signals at each succeeding level would partly resolve the ambiguity just as the convergence of ambiguous definitions determines unique or nearly unique solutions to crossword puzzles [Dennett, 1969, p. 56].

> . . . there is no unique structure or combination of groups corresponding to a given category or pattern of output. Instead, more than one combination of neuronal

Thus are laid the foundation stones. We can now explain the following primordial facts:

(1) There are reasons to recognize.
(2) Where there are reasons, there are points of view from which to recognize or evaluate them.
(3) Any agent must distinguish "here inside" from "the external world."
(4) All recognition must ultimately be accomplished by myriad "blind, mechanical" routines.
(5) Inside the defended boundary, there need not always be a Higher Executive or General Headquarters.
(6) In nature, handsome is as handsome does; origins don't matter.
(7) In nature, elements often play multiple functions within the economy of a single organism.

We have already seen echoes of these primordial facts, in the search for the ultimate "point of view of the conscious observer" and in the several instances in which we have replaced homunculi with (teams of) simple mechanisms. But, as we have seen, the point of view of a conscious observer is not identical to, but a sophisticated descendant of, the primordial points of view of the first replicators who divided their worlds into good and bad. (After all, even plants have points of view in this primordial sense.)

Scene Two: New and Better Ways of Producing Future

> And one of the deepest, one of the most general functions of living organisms is to look ahead, to produce future as Paul Valéry put it.
>
> FRANÇOIS JACOB (1982), p. 66

> groups can yield a particular output, and a given single group can participate in more than one kind of signaling function. This property of neuronal groups in repertoires, called *degeneracy*, provides a fundamental basis for the generalizing capabilities of reentrant maps [Edelman, 1989, p. 50].

This architectural feature, in which each node contributes to many different contents, was already insisted upon by Hebb in his pioneering work, *The Organization of Behavior: A Neuropsychological Theory* (1949). It is at the heart of "parallel distributed processing" or "connectionism." But there is more to multiple function than just this; at a more coarse-grained level of analysis, we will get whole systems having specialized roles but also being recruitable into more general projects.

> To predict the future of a curve is to carry out a certain operation on its past. The true prediction operation cannot be realized by any constructible apparatus, but there are certain operations which bear it a certain resemblance, and are in fact realizable by apparatus we can build.
>
> NORBERT WIENER (1948), p. 12

In the last chapter I mentioned in passing that the fundamental purpose of brains is to produce future, and this claim deserves a bit more attention. In order to cope, an organism must either armor itself (like a tree or a clam) and "hope for the best," or else develop methods of getting out of harm's way and into the better neighborhoods in its vicinity. If you follow this latter course, you are confronted with the primordial problem that every *agent* must continually solve:

Now what do I do?

In order to solve this problem, you need a nervous system, to *control* your activities in time and space. The juvenile sea squirt wanders through the sea searching for a suitable rock or hunk of coral to cling to and make its home for life. For this task, it has a rudimentary nervous system. When it finds its spot and takes root, it doesn't need its brain anymore, so it eats it! (It's rather like getting tenure.)[4] The key to control is the ability to *track* or even *anticipate* the important features of the environment, so all brains are, in essence, *anticipation machines*. The clam's shell is fine armor, but it cannot always be kept closed; the hard-wired reflex that snaps the shell shut is a crude but effective harm-anticipator/avoider.

Even more primitive are the withdrawal and approach responses of the simplest organisms, and they are tied in the most direct imaginable way to the sources of good and ill: they *touch* them. Then, depending on whether the touched thing is bad or good for them, they either recoil or engulf (in the nick of time, if they are lucky). They do this by simply being "wired" so that actual contact with the good or bad feature of the world triggers the appropriate squirm. This fact is the basis, as we shall see, for some of the most terrible and delicious

4. The analogy between the sea squirt and the associate professor was first pointed out, I think, by the neuroscientist Rodolfo Llinás.

(literally) features of consciousness. In the beginning, *all* "signals" caused by things in the environment meant either "scram!" or "go for it!" (Humphrey, forthcoming).

No nervous system, at that early time, had any way to use a more dispassionate or objective "message" that merely *informed* it, neutrally, of some condition. But such simple nervous systems cannot get much purchase on the world. They are capable of only what we may call *proximal anticipation:* behavior that is appropriate to what is in the *immediate* future. Better brains are those that can extract more information, faster, and use it to *avoid* the noxious contact in the first place, or *seek out* the nutritious bits (and the mating opportunities, once sex had appeared).

Faced with the task of extracting useful future out of our personal pasts, we organisms try to get something for free (or at least at bargain price): to find the laws of the world — and if there aren't any, to find approximate laws of the world — anything that will give us an edge. From some perspectives it appears utterly remarkable that we organisms get any purchase on nature at all. Is there any deep reason why nature should tip its hand, or reveal its regularities to casual inspection? Any useful future-producer is apt to be something of a trick — a makeshift system that happens to work, more often than not, a lucky hit on a regularity in the world that can be tracked. Any such lucky anticipators Mother Nature stumbles over are bound to be prized, of course, if they improve an organism's edge.

At the minimalist extreme, then, we have the creatures who represent as little as possible: just enough to *let the world warn them* sometimes when they are beginning to do something wrong. Creatures who follow this policy engage in no planning. They plunge ahead, and if something starts hurting, they "know enough" to withdraw, but that is the best they can do.

The next step involves short-range anticipation — for instance, the ability to duck incoming bricks. This sort of anticipatory talent is often "wired in" — part of the innate machinery designed over the eons to track the sort of (exceptioned) regularity we can notice between *things looming* and *things hitting us*. The ducking response to looming is hard-wired in human beings, for instance, and can be observed in newborn infants (Yonas, 1981), a gift from our remote ancestors whose nonsurviving cousins didn't know enough to duck. Does the "something looming!" signal *mean* "duck!"? Well, it proto-means it; it is wired directly to the ducking mechanism.

We have other such gifts. Our visual systems, like those of many other animals, even fish, are exquisitely sensitive to patterns with a vertical axis of symmetry. Braitenberg suggests that this is probably because in the natural world of our remote ancestors (back before there were church façades or suspension bridges), just about the only things in the world that *showed* axes of vertical symmetry were other animals, and only when they were facing you. So our ancestors have been equipped with a most valuable alarm system that would be triggered (mostly) whenever they were being *looked at* by another animal (Braitenberg, 1984).[5] Identifying a predator at a (spatial) distance (instead of having to wait until you feel its teeth digging into you) is also a *temporally* more distal sort of anticipation: it gives you a head start at avoidance.

An important fact about such mechanisms is their discriminatory crudeness; they trade off what might be called *truth and accuracy in reporting* for *speed and economy*. Some of the things that trigger a vertical symmetry–detector have no actual significance to the organism: the rare nearly symmetrical tree or shrub, or (in modern times) many human artifacts. So the class of things that are distinguished by the mechanism are, officially, a motley crew — dominated by animals-looking-in-my-direction, but permitting an indefinitely large variety of false alarms (relative to *that* message). And not even all or only patterns with vertical symmetry will set the thing off; some vertically symmetrical patterns will fail for one reason or another to trigger it, and there will be false alarms here as well; it is the price to be paid for a fast, cheap, portable mechanism, and it is a price organisms, in their narcissism (Akins, 1989), pay willingly. This fact is plain to see, but some of its implications for consciousness are not so obvious at first. (In chapter 12, this will become important when we ask such questions as: What properties do we detect with our color-vision? What do red things have in common? And even, Why does the world appear as it does to us?)

Becoming informed (fallibly) that another animal is looking at you is almost always an event of significance in the natural world. If the animal doesn't want to eat you, it might be a potential mate, or a rival

5. This bit of wiring is reminiscent of Shakey's not entirely foolproof way of telling the boxes from the pyramids. So Shakey is not utterly unbiological after all; the biosphere has lots of gadgets like this. It is still true, however, that Shakey's "visual" system is not at all a good model of any species' vision. That was not its point.

for a mate, or a prey who has spotted your approach. The alarm should next turn on the "friend or foe or food?" analyzers, so that the organism can distinguish between such messages as: "A conspecific is looking at you!," "A predator has you in his gunsights!," and "Your supper is about to bolt!" In some species (in some fish, for instance) the vertical symmetry–detector is wired up to cause a swift interruption of ongoing activity known as an *orienting response*.

The psychologist Odmar Neumann (1990) suggests that orienting responses are the biological counterpart to the shipboard alarm "All hands on deck!" Most animals, like us, have activities that they control in a routine fashion, "on autopilot," using less than their full capacities, and in fact under the control of specialized subsystems of their brains. When a specialized alarm is triggered (such as the looming alarm or the vertical symmetry–alarm in us), or a general alarm is triggered by anything sudden and surprising (or just unexpected), the animal's nervous system is mobilized to deal with the possibility of an emergency. The animal stops what it is doing and does a quick scan or update that gives every sense organ an opportunity to contribute to the pool of available and relevant information. A *temporary* centralized arena of control is established through heightened neural activity — all the lines are open, for a brief period. If the result of this update is that a "second alarm" is turned in, the animal's whole body is mobilized, through a rush of adrenaline. If not, the heightened activity soon subsides, the off-duty crew goes back to bed, and the specialists resume their control functions. These brief episodes of interruption and heightened vigilance are not themselves episodes of human-style "conscious awareness" (as people often redundantly say) or at least they are not necessarily instances of such a state, but they are probably the necessary precursors, in evolution, of our conscious states.

Neumann speculates that these orienting responses began as reactions to alarm signals, but proved so useful in provoking a generalized update that animals began to go into the orienting mode more and more frequently. Their nervous systems needed an "All hands on deck!" mode, but once it was provided, it cost little or nothing to turn it on more frequently, and paid off handsomely in improved information about the state of the environment, or the animal's own state. One might say it became a habit, no longer just under the control of external stimuli, but internally initiated (rather like regular fire drills).

Regular vigilance gradually turned into regular *exploration*, and a new behavioral strategy began to evolve: the strategy of acquiring information "for its own sake," just in case it might prove valuable someday. Most mammals were attracted to this strategy, especially primates, who developed highly mobile eyes, which, via saccades, provided almost uninterrupted scanning of the world. This marked a rather fundamental shift in the economy of the organisms that made this leap: the birth of curiosity, or epistemic hunger. Instead of gathering information only on a pay-as-you-go, use-it-immediately basis, they began to become what the psychologist George Miller has called *informavores*: organisms hungry for further information about the world they inhabited (and about themselves). But they did not invent and deploy entirely new systems of information-gathering. As usual in evolution, they cobbled together these new systems out of the equipment their heritage had already provided. This history has left its traces, particularly on the emotional or affective overtones of consciousness, for even though higher creatures now became "disinterested" gatherers of information, their "reporters" were simply the redeployed warners and cheerleaders of their ancestors, never sending any message "straight" but always putting some vestigial positive or negative editorial "spin" on whatever information they provided. Removing the scare quotes and metaphors: The innate links of informing states to withdrawal and engulfing, avoidance and reinforcement, were not broken, but only attenuated and redirected. (We will return to this point in chapter 12.)

In mammals, this evolutionary development was fostered by a division of labor in the brain, which created two specialized areas: (roughly) the *dorsal* and the *ventral*. (What follows is a hypothesis of the neuropsychologist Marcel Kinsbourne.) The dorsal brain was given "on line" *piloting* responsibilities for keeping the vessel (the organism's body) out of harm's way; like the "collision-detection" routines built into video games, it had to be almost continuously scanning for things approaching or receding, and in general was responsible for keeping the organism from bumping into things or falling over cliffs. This left the ventral brain with a little free time to concentrate on the identification of the various objects in the world; it could afford to zoom in narrowly on particulars and analyze them in a relatively slow, serial manner, because it could rely on the dorsal system to keep the vessel off the rocks. In primates, according to Kinsbourne's speculation, this dorsal-ventral specialization got twisted, and evolved further into the celebrated right-hemisphere/left-hemisphere specializations: the global, spatiotemporal right hemisphere, and the more concentrated, analytic, serial left hemisphere.

We have been exploring just a single strand in the evolutionary history of nervous systems, and we have availed ourselves of the most basic evolutionary mechanism: selection of particular *genotypes* (gene combinations) that have proven to yield better adapted individuals (*phenotypes*) than the alternative genotypes. The organisms with the good fortune to be better wired-up at birth tend to produce more surviving progeny, so good hard-wiring spreads through the population. And we have sketched a progression in design space from the simplest imaginable good- and evil-detectors to collections of such mechanisms organized into an architecture with a considerable capacity for producing useful anticipation in relatively stable and predictable environments.

For the next phase in our story, we must introduce a major innovation: the emergence of individual phenotypes whose innards are not entirely hard-wired, but rather variable or *plastic,* and hence who can learn, during their own lifetimes. The emergence of plasticity in nervous systems occurred at the same time (roughly) as the developments we have already sketched, and it provided two new media in which evolution could take place, at much greater speed than unaided genetic evolution via gene mutation and natural selection. Since some of the complexities of human consciousness are the result of the developments that have occurred, and continue to occur, in these new media, we need a clear, if elementary, vision of their relations to each other and to the underlying process of genetic evolution.

3. EVOLUTION IN BRAINS, AND THE BALDWIN EFFECT

We all assume that the future will be like the past — it is the essential but unprovable premise of all our inductive inferences, as Hume noted. Mother Nature (the design-developer realized in the processes of natural selection) makes the same assumption. In many regards, things stay the same: gravity continues to exert its force, water continues to evaporate, organisms continue to need to replenish and protect their body water, looming things continue to subtend ever-larger portions of retinas, and so on. Where generalities like these are at issue, Mother Nature provides long-term solutions to problems: hard-wired, gravity-based which-way-is-up detectors, hard-wired thirst alarms, hard-wired duck-when-something-looms circuits. Other things change, but predictably, in cycles, and Mother Nature responds to them with other hard-wired devices, such as winter-coat-growing mechanisms triggered by temperature shifts, and built-in alarm clocks to govern the waking and sleeping cycles of nocturnal and diurnal animals. But sometimes the opportunities and vicissitudes in the environment are relatively unpredictable by Mother Nature or by anyone — they are, or are influenced by, processes that are *chaotic* (Dennett, 1984a, p. 109ff). In these cases, no one stereotyped design will accommodate itself to all eventualities, so better organisms will be those that can *redesign themselves* to some degree to meet the conditions they encounter. Sometimes such redesign is called *learning* and sometimes it is called just *development*. The dividing line between these is contentious. Do birds *learn* to fly? Do they learn to sing their songs? (Nottebohm, 1984; Marler and Sherman, 1983) To grow their feathers? Do babies *learn* to walk or speak? Since the dividing line (if any) is irrelevant to our purposes, let's call any such a process, anywhere on the spectrum from learning-to-focus-your-eyes to learning-quantum-mechanics, *postnatal design-fixing*. When you are born, there is still some scope for variation, which eventually gets fixed by one process or another into a relatively permanent element of design for the rest of your life (once you've learned to ride a bicycle, or to speak Russian, it tends to stay with you).

How could such a process of postnatal design-fixing be accomplished? In only one (nonmiraculous) way: by a process strongly analogous to the process that fixes prenatal design, or in other words, a process of evolution by natural selection occurring within the individual (within the phenotype). Something already fixed in the individual by ordinary natural selection has to play the role of mechanical selector, and other things have to play the role of the multitude of candidates for selection. Many different theories of the process have been proposed, but all of them — except for those that are simply loony or frankly mysterious — have this structure, and differ only on the details of the mechanisms proposed. For many years in the twentieth century, the most influential theory was B. F. Skinner's behaviorism, in which stimulus-response pairings were the candidates for selection, and "reinforcing" stimuli were the mechanism of selection. The role of pleasurable and painful stimuli — the carrot and the stick — in shaping behavior is undeniable, but behaviorism's mechanism of "operant conditioning" has been widely recognized as too simple to explain the complexities of postnatal design-fixing in species as complicated as human beings (and probably in pigeons, too, but that is another matter). Today, the emphasis is on various theories that move the evolutionary process inside the brain (Dennett, 1974). Different versions of this idea have been around for decades, and now, with the possibility of testing the rival models in huge computer simulations, there is considerable contentiousness, which we will be wise to steer clear of.[6]

For our purposes, let's just say that one way or another, the plastic brain is capable of reorganizing itself adaptively in response to the particular novelties encountered in the organism's environment, and the process by which the brain does this is almost certainly a mechanical process strongly analogous to natural selection. This is the first new medium of evolution: postnatal design-fixing in individual brains. The candidates for selection are various brain structures that control or influence behaviors, and the selection is accomplished by one or another mechanical weeding-out process that is itself genetically installed in the nervous system.

Amazingly, this capability, itself a product of genetic evolution by natural selection, not only gives the organisms who have it an edge over their hard-wired cousins who cannot redesign themselves, but also reflects back on the process of genetic evolution and *speeds it up.* This is a phenomenon known under various names but best known as the Baldwin Effect (see Richards, 1987; Schull, 1990). Here is how it works.

Consider a population of a particular species in which there is considerable variation at birth in the way their brains are wired up. Just one of the possible wirings, let's suppose, endows its possessor with a Good Trick — a behavioral talent that protects it or enhances its chances dramatically. We can represent this in what is called an adaptive landscape; the altitude represents fitness (higher is better) and the

---

6. The basic insight can be discerned in the writings of Darwin and his early exponents (Richards, 1987). The neuroanatomist J. Z. Young (1965a, 1965b) pioneeered a selectionist theory of memory (see also Young, 1979). I developed a philosopher's version of the basic argument, with a sketch of the details, in my dissertation at Oxford, 1965, a streamlined version of which, "Evolution in the Brain," is chapter 3 of *Content and Consciousness*, 1969. John Holland (1975) and others in Artificial Intelligence developed "genetic algorithms" for self-redesigning or learning systems (see also Holland, Holyoak, Nisbett, and Thagard, 1986), and Jean-Pierre Changeux (Changeux and Danchin, 1976; Changeux and Dehaene, 1989) devised a rather detailed neural model. The neurobiologist William Calvin (1987, 1989a) provides a different (and more readily accessible) perspective on the issues in his own theory of evolution in the brain. See also his clear and perceptive review (Calvin, 1989b) of Gerald Edelman's *Neural Darwinism* (1987). More recently, Edelman has published *The Remembered Present: A Biological Theory of Consciousness* (1989).

longitude and latitude represent variables in wiring (we need not specify them for this thought experiment).

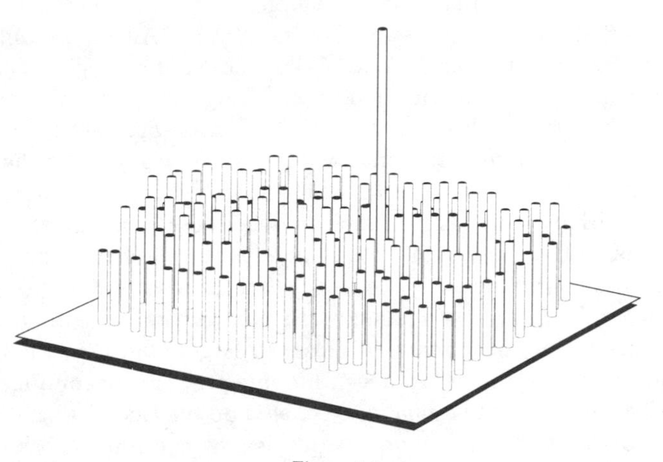

Figure 7.1

As the figure makes clear, only one wiring is favored; the others, no matter how "close" to being the good wiring, are about equal in fitness. Such a needle in the haystack can be practically invisible to natural selection. Even if a few lucky individuals are wired up this way, the chances of their fortune spreading through the population of subsequent generations can be vanishingly small *unless* there is plasticity of design among individuals.

Suppose, then, that the individuals all start out different, genetically, but in the course of their lifetimes wander about in the space of design possibilities available to them, thanks to their plasticity. And because of the particular circumstances in the environment, they all tend to gravitate to the one favored wiring. There is one Good Trick to learn in their environment, and they all tend to learn it. Suppose it is such a good trick that those that don't ever learn it are at a severe disadvantage, and suppose that those that tend never to learn it are those that start out life with designs that are farther away in design space (and hence require more postnatal redesign) than those that are near to the Good Trick.

A fantasy (adapted from Hinton and Nowlan, 1987) will help us imagine this. Suppose there are ten places in each animal's brain where a "wire" can be attached in either of two ways, A or B. Suppose the Good Trick is the design with wiring AAABBBAAAA, and that all other wirings are about equally unimpressive, behaviorally. Since all these connections are plastic, each animal, in the course of its life, can try out any of the 2^{10} different combinations of A and B wirings. Those animals that are born in a state such as BAABBBAAAA are only one bit of rewiring away from the Good Trick (though they might, of course, wander off by trying a series of other rewirings first). Others, that start with a wiring such as BBBAAABBBB have to make at least ten rewirings (without ever rewiring anything back the wrong way) before they find the Good Trick. Those animals whose brains start out closer to the target will have a survival advantage over those who start far away, even though there is no *other* selective advantage to being born with a "near miss" structure as opposed to a "far miss" structure (as Figure 7.1 makes clear). The population in the next generation will thus tend to consist mainly of individuals closer to target (and hence more apt to find the target in their lifetimes), and this process can continue until the whole population is *genetically* fixated on the Good Trick. A Good Trick "discovered" by individuals in this way can thus be passed on, relatively swiftly, to future generations.

If we give individuals a variable chance to hit upon (and then "recognize" and "cling to") the Good Trick in the course of their lifetimes, the near-invisible needle in the haystack of Figure 7.1 becomes the summit of a quite visible hill that natural selection can climb (Figure 7.2). This process, the Baldwin Effect, might look at first like the discredited Lamarckian idea of the genetic transmission of acquired characteristics, but it is not. Nothing the individual learns is transmitted to its progeny. It is just that individuals who are lucky enough to be closer in design-exploration space to a learnable Good Trick will tend to have more progeny, who will also tend to be closer to the Good Trick. Over generations, the competition becomes stiffer: eventually, unless you are born with (or very nearly with) the Good Trick, you are not close enough to compete. If it weren't for the plasticity, however, the effect wouldn't be there, for "a miss is as good as a mile" *unless* you get to keep trying variations until you get it right.

Thanks to the Baldwin Effect, species can be said to pretest the efficacy of particular different designs by phenotypic (individual) exploration of the space of nearby possibilities. If a particularly winning setting is thereby discovered, this discovery will *create* a new selection pressure: organisms that are closer in the adaptive landscape to that discovery will have a clear advantage over those more distant. This means that species with plasticity will *tend* to evolve faster (and more "clearsightedly") than those without it. So evolution in the second

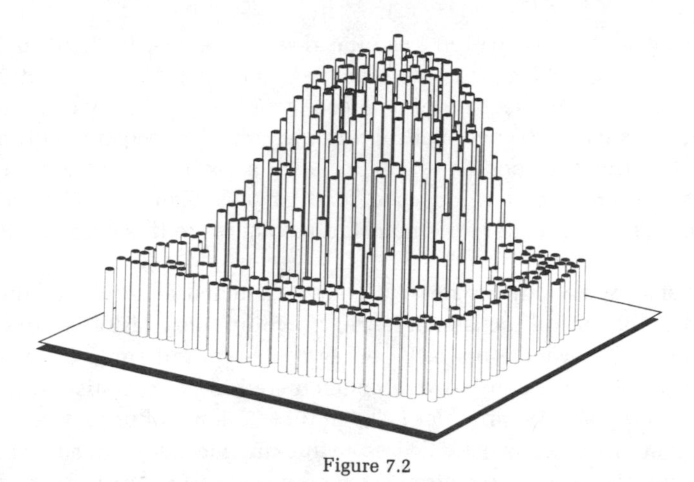

Figure 7.2

medium, phenotypic plasticity, can enhance evolution in the first medium, genetic variation. (We will soon see a countervailing effect that arises as a result of interactions with the third medium.)

4. PLASTICITY IN THE HUMAN BRAIN: SETTING THE STAGE

> Human reason begins in the same way with its native powers and thus creates its first intellectual tools. Through these it acquires further powers for other intellectual operations and through them further tools and the power of extending its inquiries until by degrees it reaches the summit of wisdom.
>
> BENEDICT SPINOZA (1677)

Nervous systems that are hard-wired are lightweight, energy-efficient, and fine for organisms that cope with stereotypic environments on a limited budget. Fancier brains, thanks to their plasticity, are capable not just of stereotypic anticipation, but also of adjusting to trends. Even the lowly toad has some small degree of freedom in how it responds to novelty, slowly altering its patterns of activity to track — with considerable time lag — those changes in features of its environment that matter most to its well-being (Ewert, 1987). In the toad's brain, a design for dealing with the world evolves at a pace many orders of magnitude faster than natural selection — with "generations" lasting seconds or minutes, not years. But for truly high-powered control, what you want is an anticipation machine that will adjust itself in major ways in a few milliseconds, and for that you need a virtuoso future-producer, a system that can think ahead, avoid ruts in its own activity, solve problems in advance of encountering them, and recognize entirely novel harbingers of good and ill. For all our foolishness, we human beings are vastly better equipped for that task than any other self-controllers, and it is our enormous brains that make this possible. But how?

Let's review our progress. We have sketched a story — a single strand of the multidimensional fabric of evolutionary history — of the evolution of a primate brain. Based as it was on millennia of earlier nervous systems, it consists of a conglomeration of specialist circuits designed to perform particular tasks in the economy of primate ancestors: looming-object detectors wired to ducking mechanisms, someone-is-looking-at-me detectors wired to friend-or-foe-or-food discriminators wired to their appropriate further subroutines. We can add also such specifically primate circuits as hand-eye-coordination circuits designed for picking berries or picking up seeds, and others designed for grasping branches or even for dealing with objects close to the face (Rizzolati, Gentilucci, and Matelli, 1985). Thanks to mobile eyes and a penchant for exploration-and-update, these primate brains were regularly flooded with multimedia information (or as neuroscientists would say, multimodal information), and hence a new problem was posed for them: the problem of higher-level control.

A problem posed is also an opportunity, a gateway opening up a new portion of design space. Up till now, we can suppose, nervous systems solved the "Now what do I do?" problem by a relatively simple balancing act between a strictly limited repertoire of actions — if not the famous four F's (fight, flee, feed, or mate), then a modest elaboration of them. But now, with increased functional plasticity, and increased availability of "centralized" information from all the various specialists, the problem of what to do next spawned a meta-problem: what to *think about* next. It is all very well to equip oneself with an "All hands on deck!" subroutine, but then, once all hands are on deck, one must have some way of coping with the flood of volunteers. We should not expect there to have been a convenient captain already at hand (what would he have been doing up till then?), so conflicts between volunteers had to sort themselves out without any higher executive. (As we have already seen in the example of the immune system, concerted, organized action does not always have to depend on control from a central executive.) The pioneer model of this sort of process is Oliver Selfridge's (1959) early Pandemonium architecture in Artificial Intelligence, in which many "demons" vied in parallel for hegemony, and since Selfridge's name for this sort of architecture is so apt, I will use it in a generic sense in this book, to refer to it and all its descendants, direct and indirect, such as "contention scheduling" (Norman and Shallice, 1980; Shallice, 1988) and Ballard and Feldman's (1982) Winner Take All networks and their descendants.

Pandemonium-style contention scheduling, driven rather directly by current features of the environment, still yields a nervous system with a limited capacity for look-ahead. Just as Odmar Neumann hypothesized that orienting reactions, originally driven by novelty in the environment, came to be initiated endogenously (from the inside), we may hypothesize that there was pressure to develop a more endogenous way of solving the meta-problem of what to think about next, pressure to create something on the inside with more of the imagined organizational powers of a captain.

Consider what the behavior of our hypothetical primate ancestor looked like at this point from the outside (we are postponing all questions of what it is like to be such a primate till much later): an animal capable of learning new tricks, and almost continually vigilant and sensitive to novelty, but with a "short attention span" and a tendency to have attention "captured" by distracting environmental features. No long-term projects for this animal, at least not novel projects. (We should leave room for stereotypic long-duration subroutines genetically wired in, like the nest-building routines of birds, the dam-building of beavers, the food-caching of birds and squirrels.)

Onto this substrate nervous system we now want to imagine building a more human mind, with something like a "stream of consciousness" capable of sustaining the sophisticated sorts of "trains of thought" on which human civilization apparently depends. Chimpanzees are our closest kin — genetically closer, in fact, than chimpanzees are to gorillas or orangutans — and current thinking is that we shared a common ancestor with chimpanzees about six million years ago. Since that major break, our brains have diverged dramatically, but primarily in size, rather than structure. While chimpanzees have brains of roughly the same size as our common ancestor (and it is important — and difficult — to keep in mind that chimpanzees have done some evolving away from our common ancestor as well), our hominid ancestors' brains grew four times as large. This increase in volume didn't happen immediately; for several million years after the split with proto-chimpanzees, our hominid ancestors got along with ape-sized brains, in spite of becoming bipedal at least three and a half million years ago. Then, when the ice ages began, about two and a half million years ago, the Great Encephalization commenced, and was essentially completed 150,000 years ago — *before* the development of language, of cooking, of agriculture. Just why our ancestors' brains should have grown so large so fast (in the evolutionary time scale it was more an explosion than a blossoming) is a matter of some controversy (for illuminating accounts, see William Calvin's books). But there is little controversy about the nature of the product: the brain of early *Homo sapiens* (who lived from roughly 150,000 years ago to the end of the most recent ice age a mere 10,000 years ago) was an enormously complex brain of unrivaled plasticity, almost indistinguishable from our own in size and shape. This is important: The astonishing hominid brain growth was essentially complete *before* the development of language, and so cannot be a response to the complexities of mind that language has made possible. The innate specializations for language, hypothesized by the linguist Noam Chomsky and others, and now beginning to be confirmed in details of neuroanatomy, are a *very* recent and rushed add-on, no doubt an exploitation of earlier sequencing circuitry (Calvin, 1989a) accelerated by the Baldwin Effect. Moreover, the most remarkable expansion of human *mental* powers (as witnessed by the development of cooking, agriculture, art, and, in a word, civilization) has all happened even more recently, since the end of the last ice age, in a 10,000-year twinkling that is as good as instantaneous from the evolutionary perspective that measures trends in millions of years. Our brains are equipped at birth with few if any powers that were lacking in the brains of our ancestors 10,000 years ago. So the tremendous advance of *Homo sapiens* in the last 10,000 years must almost all be due to harnessing the plasticity of that brain in radically new ways — by creating something like *software* to enhance its underlying powers (Dennett, 1986).

In short, our ancestors must have learned some Good Tricks they could do with their adjustable hardware, which our species has only just begun to move, via the Baldwin Effect, into the genome. Moreover, as we shall see, there are reasons for believing that in spite of the initial selection pressure in favor of gradually "hard-wiring" these Good Tricks, the tricks have so altered the nature of the environment for our species that there is no longer significant selection pressure calling for further hard-wiring. It is likely that almost all selection pressure on human nervous system design development has been swamped by the side effects of this new design opportunity exploited by our ancestors.

Until this point I have tried to avoid speaking of simpler nervous systems as *representing* anything in the world. The various designs we have considered, both plastic and hard-wired, can be seen to be sensitive to, or responsive to, or designed-with-an-eye-to, or to utilize information about, various features of the organism's environment, and hence in that minimal sense might be called representations, but now we should pause to consider what features of such complex designs should lead us to consider them as systems of representations.

Some of the variability in a brain is required simply to provide a medium for the moment-to-moment transient patterns of brain activity that somehow *register* or at any rate *track* the importantly variable features of the environment. Something in the brain needs to change so that it can keep track of the bird that flies by, or the drop in air temperature, or one of the organism's own states — the drop in blood sugar, the increase of carbon dioxide in the lungs. Moreover — and this is the fulcrum that gives genuine representation its leverage — these transient internal patterns come to be able to continue to "track" (in an extended sense) the features they refer to when they are temporarily cut off from causal commerce with their referents. "A zebra which has caught sight of a lion does not forget where the lion is when it stops watching the lion for a moment. The lion does not forget where the zebra is" (Margolis, 1987, p. 53). Compare this to the simpler phenomenon of the sunflower that tracks the passage of the sun across the sky, adjusting its angle like a movable solar panel to maximize the sunlight falling on it. If the sun is temporarily obscured, the sunflower cannot project the trajectory; the mechanism that is sensitive to the sun's passage does not represent the sun's passage in this extended sense. The beginnings of real representation are found in many lower animals (and we should not rule out, *a priori*, the possibility of real representation in plants), but in human beings the capacity to represent has skyrocketed.

Among the things an adult human brain can somehow represent are not only:

(1) the position of the body and its limbs
(2) a spot of red light
(3) a degree of hunger
(4) a degree of thirst
(5) the smell of a fine old red burgundy

but also:

(6) the smell of a fine old red burgundy *as* the smell of Chambertin 1971
(7) Paris
(8) Atlantis
(9) the square root of the largest prime number less than 20
(10) the concept of a nickel-plated combination corkscrew and staple-remover

It seems almost certain that no other animal's brain can represent 6–10, and moreover that a considerable process of adjustment of the infant human brain is required before any of these can be registered or represented at all. The first five, in contrast, might well be things that almost any brain can represent (in some sense) without any training.

Somehow, in any case, the way a brain represents hunger must differ, physically, from how it represents thirst — since it must govern different behavior depending on which is represented. There must also, at the other extreme, be a difference between the way a particular adult human brain represents Paris and Atlantis, for thinking of one is not thinking of the other. How can a particular state or event in the brain represent one feature of the world rather than another?[7] And whatever it is that makes some feature of the brain represent what it represents, *how does it come* to represent what it represents? Here once again (I'm afraid this refrain is going to get tedious!) there are a range of possibilities, settled by evolutionary processes: some elements of the system of representation can be — indeed must be (Dennett, 1969) — innately fixed, and the rest must be "learned." While some of the categories in life that matter (like hunger and thirst) are no doubt "given" to us in the way we are wired up at birth, others we have to develop on our own.[8]

---

7. This is the fundamental problem in the philosophy of mind of *mental content* or *intentionality*, and its proposed solutions are notoriously controversial. Mine is given in *The Intentional Stance* (1987a).

8. A few intrepid theorists have claimed otherwise. Jerry Fodor (1975), for instance, has claimed that all the concepts one could ever have must be innately given, and only triggered or accessed by particular "learning" experiences. Thus Aristotle had the concept of an airplane in his brain, and also the concept of a bicycle — he just never had occasion to use them! To those who burst out laughing at such a ridiculous idea, Fodor replies that the immunologists used to laugh at the idea that people — Aristotle, for instance — are born with millions of distinct antibodies, including antibodies specific for compounds

How do we do this? Probably by a process of generation-and-selection of patterns of neural activity in the cerebral cortex, the huge convoluted mantle that has swiftly burgeoned in the human skull and now completely covers the older animal brain underneath. Just saying it is an evolutionary process occurring primarily in the cortex leaves too many things mysterious, and at this level of complexity and sophistication, even if we succeeded in explaining the process at the level of synapses or bundles of neurons, we would be mystified about other aspects of what must be happening. If we are to make sense of this at all, we must first ascend to a more general and abstract level. Once we understand the processes in outline at the higher level, we can think about descending once again to the more mechanical level of the brain.

Plasticity makes learning possible, but it is all the better if somewhere out there in the environment there is *something to learn* that is already the product of a prior design process, so that each of us does not have to reinvent the wheel. Cultural evolution, and transmission of its products, is the second new medium of evolution, and it depends on phenotypic plasticity in much the same way phenotypic plasticity depends on genetic variation. We human beings have used our plasticity not just to learn, but to learn how to learn better, and then we've learned better how to learn better how to learn better, and so forth. We have also learned how to make the fruits of this learning available to novices. We somehow *install* an already invented and largely "debugged" *system* of habits in the partly unstructured brain.

5. THE INVENTION OF GOOD AND BAD HABITS OF AUTOSTIMULATION

How can I tell what I think until I see what I say?

E. M. FORSTER (1960)

---

that have appeared in nature only in the twentieth century, but they no longer laugh; it turns out to be true. The trouble with this idea, in its application to both immunology and psychology, is that radical versions of it are obviously false and mild versions are indistinguishable from the view being opposed. There *is* combinatorial reaction in the immune system — not *all* immune response is one-on-one between single types of pre-existing antibodies; analogously, perhaps Aristotle had an innate *airplane* concept, but did he also have an innate concept of *wide-bodied jumbo jet*? What about the concept of an *APEX fare Boston/London round trip*? By the time these questions get sorted out, in both fields, there turns out to be something like learning in both, and something like innate concepts in both.

We speak, not only to tell others what we think, but to tell ourselves what we think.

J. HUGHLINGS JACKSON (1915)

How might this software-sharing come to have happened? A "Just So" story will give us one possible path. Consider a time in the history of early *Homo sapiens* when language — or perhaps we should call it proto-language — was just beginning to develop. These ancestors were bipedal omnivores, living in smallish kin groups, and probably they developed special-purpose vocalization habits rather like those of chimpanzees and gorillas, and even much more distantly related species such as vervet monkeys (Cheney and Seyfarth, 1990). We may suppose that the communicative (or quasi-communicative) acts executed by these vocalizations were not yet fully fledged speech acts (Bennett, 1976), in which the Utterer's intention to achieve a certain effect in the Audience depends on the Audience's appreciation of that very intention.[9] But we may suppose that these ancestors, like contemporary vocalizing primates, discriminated between different utterers and audiences on different occasions, utilizing information about what both parties might believe or want.[10] For instance, hominid Alf wouldn't bother trying to get hominid Bob to believe there was no food in the cave (by grunting "Nofoodhere") if Alf believed Bob already knew there was food in the cave. And if Bob thought Alf wanted to deceive him, Bob would be apt to view Alf's vocalization with cautious skepticism.[11]

---

9. I am alluding, of course, to Paul Grice's theory of nonnatural meaning (Grice, 1957, 1969), but for a new theory of communication that replaces some of the more brittle and unlikely features of Gricean theories, see Sperber and Wilson (1986).

10. What right do I have to speak of beliefs and wants of these not-yet-fully-conscious ancestors? My own theory of belief and desire, set out in *The Intentional Stance*, defends the view that there is no good reason for putting these terms in scare-quotes: the behavior of "lower" animals (even frogs) is just as suitable a domain of explanation for the intentional stance, with its imputation of beliefs and wants, as the behavior of human beings. But readers who disagree with that theory may understand the terms here to be used in metaphorically extended senses.

11. On primate communication and the still unresolved empirical questions about whether or not apes and monkeys are capable of deliberate deception, see Dennett (1983, 1988c, 1988d, 1989a); Byrne and Whiten (1988); Whiten and Byrne (1988).

Now it sometimes happened, we may speculate, that when one of these hominids was stymied on a project, it would "ask for help," and in particular, it would "ask for information." Sometimes the audience present would respond by "communicating" something that had just the right effects on the inquirer, breaking it out of its rut, or causing it to "see" a solution to its problem. For this practice to gain a foothold in a community, the askers would have to be able to reciprocate on occasion in the role of answerers. They would have to have the behavioral capacity to be provoked into making occasionally "helpful" utterances when subjected to "request" utterances of others. For instance, if one hominid knew something and was "asked" about it, this might have the normal, but by no means exceptionless, effect of provoking it to "tell what it knew."

In other words, I am proposing that there was a time in the evolution of language when vocalizations served the function of eliciting and sharing useful information, but one must not assume that a cooperative spirit of mutual aid would have survival value, or would be a stable system if it emerged. (See, e.g., Dawkins, 1982, pp. 55ff; see also Sperber and Wilson, 1986.) Instead, we must assume that the costs and benefits of participating in such a practice were somewhat "visible" to these creatures, and enough of them saw the benefits *to themselves* as outweighing the costs so that the communicative habits became established in the community.

Then one fine day (in this rational reconstruction), one of these hominids "mistakenly" asked for help when there was no helpful audience within earshot — except itself! When it heard its own request, the stimulation provoked just the sort of other-helping utterance production that the request from another would have caused. And to the creature's delight, it found that it had just provoked itself into answering its own question.

What I am trying to justify by this deliberately oversimplified thought experiment is the claim that the practice of asking oneself questions could arise as a natural side effect of asking questions of others, and its utility would be similar: it would be a behavior that could be recognized to enhance one's prospects by promoting better-informed action-guidance. All that has to be the case for this practice to have this utility is for the preexisting access-relations *within* the brain of an individual to be less than optimal. Suppose, in other words, that although the right information for some purpose is already *in the brain*, it is in the hands of the wrong specialist; the subsystem in the brain that needs the information cannot obtain it directly from the specialist — because evolution has simply not got around to providing such a "wire." Provoking the specialist to "broadcast" the information into the environment, however, and then relying on an existing pair of ears (and an auditory system) to pick it up, would be a way of building a "virtual wire" between the relevant subsystems.[12]

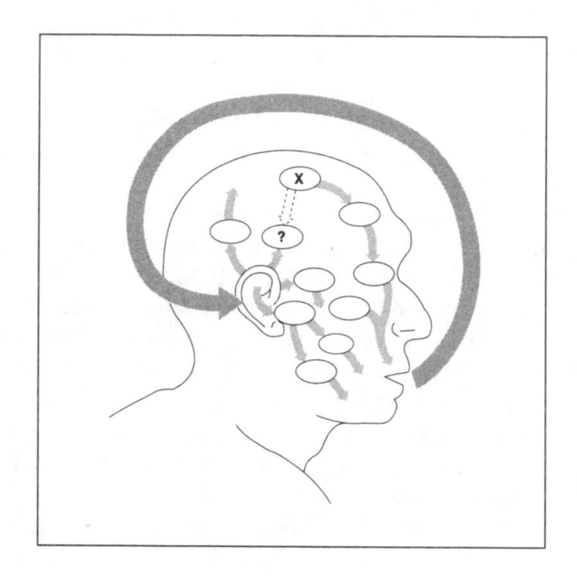

Figure 7.3

Such an act of autostimulation could blaze a valuable new trail between one's internal components. Crudely put, pushing some information through one's ears and auditory system may well happen to stimulate just the sorts of connections one is seeking, may trip just the right associative mechanisms, tease just the right mental morsel to the tip of one's tongue. One can then say it, hear oneself say it, and thus get the answer one was hoping for.

Once crude habits of vocal autostimulation began to be established as Good Tricks in the behavior of hominid populations, we would

12. In "The Garden of Forking Paths," Jorge Luis Borges (1962) devises a devilishly clever version of this strategy, which I will refrain from describing, not wanting to give away a great ending.

expect them to be quickly refined, both in the learned behavioral habits of the population and, thanks to the Baldwin Effect, in genetic predispositions and further enhancements of efficiency and effectiveness. In particular, we can speculate that the greater virtues of *sotto voce* talking to oneself would be recognized, leading later to entirely silent talking to oneself. The silent process would maintain the loop of self-stimulation, but jettison the peripheral vocalization and audition portions of the process, which weren't contributing much. This innovation would have the further benefit, opportunistically endorsed, of achieving a certain privacy for the practice of cognitive autostimulation. (In the next chapter, we will consider how these shortened lines of communication might work.) Such privacy would be particularly useful when comprehending conspecifics were within earshot. This private talking-to-oneself behavior might well not be the best imaginable way of amending the existing functional architecture of one's brain, but it would be a close-to-hand, readily discovered enhancement, and that could be more than enough. It would be slow and laborious, compared to the swift unconscious cognitive processes it was based on, because it had to make use of large tracts of nervous system "designed for other purposes" — in particular for the production and comprehension of audible speech. It would be just as linear (limited to one topic at a time) as the social communication it evolved from. And it would be dependent, at least at the outset, on the informational categories embodied in the actions it exploited. (If there were only fifty things one hominid could "say" to another, there would be only fifty things he could say to himself.)

Talking aloud is only one possibility. Drawing pictures to yourself is another readily appreciated act of self-manipulation. Suppose one day one of these hominids idly drew two parallel lines on the floor of his cave, and when he looked at what he had done, these two lines reminded him, visually, of the parallel banks of the river that he would have to cross later in the day, and this reminded him to take along his vine rope, for getting across. Had he not drawn the "picture," we may suppose, he would have walked to the river and *then* realized, after quick visual inspection, that he needed his rope, and would have had to walk all the way back. This might be a noticeable saving of time and energy that could fuel a new habit, and refine itself eventually into *private* diagram-drawing "in one's mind's eye."

The human talent for inventing new paths of internal communication occasionally shows itself vividly in cases of brain damage. People are extraordinarily good at overcoming brain damage, and it is never a matter of "healing" or the repair of damaged circuits. Rather, they discover new ways of doing the old tricks, and active exploration plays a big role in rehabilitation. A particularly suggestive anecdote comes from the research with split brain patients (Gazzaniga, 1978). The left and right hemispheres are normally connected by a broad bridge of fibers called the corpus callosum. When this is surgically severed (in the treatment of severe epilepsy), the two hemispheres lose their major direct "wires" of interconnection, and are practically incommunicado. If such a patient is asked to identify an object — such as a pencil — by reaching inside a bag and feeling it, success depends on which hand does the reaching. Most of the wiring in the body is arranged *contralaterally*, with the left hemisphere getting its information from — and controlling — the right side of the body, and vice versa. Since the left hemisphere normally controls language, when the patient reaches in the bag with his right hand, he can readily say what is in the bag, but if the left hand does the reaching, only the right hemisphere gets the information that the object is a pencil, and is powerless to direct the voice to express this. But occasionally, it seems, a right hemisphere will hit upon a clever stratagem: by finding the point of the pencil, and digging it into his palm, he causes a sharp *pain* signal to be sent up the left arm, and some pain fibers are *ipsilaterally* wired. The left, language-controlling hemisphere gets a clue: it is something sharp enough to cause a pain. "It's sharp — it's, perhaps, a pen? a pencil?" The right hemisphere, overhearing this vocalization, may help it along with some hints — frowning for *pen*, smiling for *pencil* — so that by a brief bout of "Twenty Questions" the left hemisphere is led to the correct answer. There are more than a few anecdotes about such ingenious jury-rigs invented on the spot by patients with split brains, but we should treat them with caution. They might be what they appear to be: cases exhibiting the deftness with which the brain can discover and implement autostimulatory strategies to improve its internal communications in the absence of the "desired" wiring. But they might also be the unwittingly embroidered fantasies of researchers hoping for just such evidence. That's the trouble with anecdotes.

We could amuse ourselves by dreaming up other plausible scenarios for the "invention" of useful modes of autostimulation, but this would risk obscuring the point that not all such inventions would have to be useful to survive. Once the general habit of exploratory autostimulation had been inculcated in some such way(s), it might well spawn a host of nonfunctional (but not particularly dysfunctional) variations.

There are, after all, many existent varieties of autostimulation and self-manipulation that presumably have no valuable effect on cognition or control, but which, for standard Darwinian reasons, fail to be extinguished and may even drift to (cultural *or* genetic) fixation in certain subpopulations. Likely candidates are painting yourself blue, beating yourself with birch boughs, cutting patterns in your skin, starving yourself, saying a "magical" formula to yourself over and over again, and staring at your navel. If these practices are habits worth inculcating, their virtues as fitness-enhancers are at least not "obvious" enough to have boosted them yet into any known genetic predispositions, but perhaps they are too recent as inventions.

The varieties of autostimulation that enhance cognitive organization are now probably partly innate and partly learned and idiosyncratic. Just as one can notice that stroking oneself in certain ways can produce certain desirable side effects that are only partially and indirectly controllable — and one can then devote some time and ingenuity to developing and exploring the techniques for producing those side effects — so one can half-consciously explore techniques of cognitive autostimulation, developing a personal style with particular strengths and weaknesses. Some people are better at it than others, and some never learn the tricks, but there is much sharing and teaching. Cultural *transmission*, by letting almost everybody in on the Good Trick, can flatten out the top of the fitness hill (see Figure 7.2, p. 187), creating a butte or table top that diminishes the selection pressure for moving the Trick into the genome. If almost everyone gets good enough to get by in the civilized world, the selection pressure for moving the Good Tricks into the genome is extinguished or at least diminished.

6. THE THIRD EVOLUTIONARY PROCESS: MEMES AND CULTURAL EVOLUTION[13]

Just as we learned to milk cows, and then to domesticate them for our own benefit, so we learned to milk others' and our own minds in certain ways, and now the techniques of mutual and self-stimulation are deeply embedded in our culture and training. The way in which culture has become a repository and transmission medium for innovations (not only innovations of consciousness) is important for un-

13. This section is drawn from my "Memes and the Exploitation of Imagination" (1990a).

derstanding the sources of design of human consciousness, for it is yet another medium of evolution.

One of the first major steps a human brain takes in the massive process of postnatal self-design is to get itself adjusted to the local conditions that matter the most: it swiftly (in two or three years) turns itself into a Swahili or Japanese or English brain. What a step — like stepping into a cocked slingshot!

It doesn't matter for our purposes whether this process is called learning or differential development; it happens so swiftly and effortlessly that there is little doubt that the human genotype includes many adaptations that are specifically in place to enhance language acquisition. This has all happened very fast, in evolutionary terms, but that is just what we should expect, given the Baldwin Effect. Being able to speak is such a Good Trick that anyone who was slow off the mark getting there would be at a tremendous disadvantage. The first of our ancestors to speak almost certainly had a much more laborious time getting the hang of it, but we are the descendants of the virtuosos among them.[14]

Once our brains have built the entrance and exit pathways for the vehicles of language, they swiftly become *parasitized* (and I mean that literally, as we shall see) by entities that have evolved to thrive in just such a niche: *memes*. The outlines of the theory of evolution by natural selection are clear: evolution occurs whenever the following conditions exist:

(1) variation: a continuing abundance of different elements
(2) heredity or replication: the elements have the capacity to create copies or replicas of themselves
(3) differential "fitness": the number of copies of an element that are created in a given time varies, depending on interactions between the features of that element (whatever it is that makes it different from other elements) and features of the environment in which it persists

Notice that this definition, though drawn from biology, says nothing specific about organic molecules, nutrition, or even life. It is a more general and abstract characterization of evolution by natural selection. As the zoologist Richard Dawkins has pointed out, the fundamental principle is

---

14. For a good recent airing of the controversies in the literature speculating on the evolution of language, see Pinker and Bloom (1990), and the commentaries following.

> that all life evolves by the differential survival of replicating entities. . . .
>
> The gene, the DNA molecule, happens to be the replicating entity which prevails on our own planet. There may be others. If there are, provided certain other conditions are met, they will almost inevitably tend to become the basis for an evolutionary process.
>
> But do we have to go to distant worlds to find other kinds of replication and other, consequent, kinds of evolution? I think that a new kind of replicator has recently emerged on this very planet. It is staring us in the face. It is still in its infancy, still drifting clumsily about in its primeval soup, but already it is achieving evolutionary change at a rate which leaves the old gene panting far behind. [1976, p. 206]

These new replicators are, roughly, ideas. Not the "simple ideas" of Locke and Hume (the idea of red, or the idea of round or hot or cold), but the sort of complex ideas that form themselves into distinct memorable units — such as the ideas of

wheel
wearing clothes
vendetta
right triangle
alphabet
calendar
the Odyssey
calculus
chess
perspective drawing
evolution by natural selection
Impressionism
"Greensleeves"
deconstructionism

Intuitively these are more or less identifiable cultural units, but we can say something more precise about how we draw the boundaries — about why *D-F#-A* isn't a unit, and the theme from the slow movement of Beethoven's Seventh Symphony is: the units are the smallest elements that replicate themselves with reliability and fecundity. Dawkins coins a term for such units: *memes* —

> a unit of cultural transmission, or a unit of *imitation*. 'Mimeme' comes from a suitable Greek root, but I want a monosyllable that sounds a bit like 'gene' . . . it could alternatively be thought of as being related to 'memory' or to the French word *même*. . . .
>
> Examples of memes are tunes, ideas, catch-phrases, clothes fashions, ways of making pots or of building arches. Just as genes propagate themselves in the gene pool by leaping from body to body via sperm or eggs, so memes propagate themselves in the meme pool by leaping from brain to brain via a process which, in the broad sense, can be called imitation. If a scientist hears, or reads about, a good idea, he passes it on to his colleagues and students. He mentions it in his articles and his lectures. If the idea catches on, it can be said to propagate itself, spreading from brain to brain. [1976, p. 206]

In *The Selfish Gene*, Dawkins urges us to take the idea of meme evolution literally. Meme evolution is not just analogous to biological or genetic evolution, not just a process that can be metaphorically described in these evolutionary idioms, but a phenomenon that obeys the laws of natural selection exactly. The theory of evolution by natural selection is neutral regarding the differences between memes and genes; these are just different kinds of replicators evolving in different media at different rates. And just as the genes for animals could not come into existence on this planet until the evolution of plants had paved the way (creating the oxygen-rich atmosphere and ready supply of convertible nutrients), so the evolution of memes could not get started until the evolution of animals had paved the way by creating a species — *Homo sapiens* — with brains that could provide shelter, and habits of communication that could provide transmission media, for memes.

This is a new way of thinking about ideas. It is also, I hope to show, a good way, but at the outset the perspective it provides is distinctly unsettling, even appalling. We can sum it up with a slogan:

> A scholar is just a library's way of making another library.

I don't know about you, but I'm not initially attracted by the idea of my brain as a sort of dung heap in which the larvae of other people's ideas renew themselves, before sending out copies of themselves in an informational Diaspora. It does seem to rob my mind of its importance as both author and critic. Who's in charge, according to this vision — we or our memes?

There is, of course, no simple answer, and this fact is at the heart of the confusions that surround the idea of a *self*. Human consciousness is to a very great degree a product not just of natural selection, but of cultural evolution as well. The best way to see the contribution of memes to the creation of our minds is to follow the standard steps of evolutionary thinking closely.

The first rule of memes, as it is for genes, is that replication is not necessarily for the good of anything; replicators flourish that are good at . . . replicating! — for whatever reason. As Dawkins has put it,

> A meme that made its bodies run over cliffs would have a fate like that of a gene for making bodies run over cliffs. It would tend to be eliminated from the meme-pool. . . . But this does not mean that the ultimate criterion for success in meme selection is gene survival. . . . Obviously a meme that causes individuals bearing it to kill themselves has a grave disadvantage, but not necessarily a fatal one . . . a suicidal meme can spread, as when a dramatic and well-publicized martyrdom inspires others to die for a deeply loved cause, and this in turn inspires others to die, and so on. [1982, pp. 110–111]

The important point is that there is no *necessary* connection between a meme's replicative power, its "fitness" from *its* point of view, and its contribution to *our* fitness (by whatever standard we judge that). The situation is not totally desperate. While some memes definitely manipulate us into collaborating on their replication *in spite of* our judging them useless or ugly or even dangerous to our health and welfare, many — most, if we're lucky — of the memes that replicate themselves do so not just *with* our blessings, but *because of* our esteem for them. I think there can be little controversy that some memes are, all things considered, good *from our perspective*, and not just from their own perspective as selfish self-replicators: such general memes as cooperation, music, writing, education, environmental awareness, arms reduction; and such particular memes as: *The Marriage of Figaro*, *Moby-Dick*, returnable bottles, the SALT agreements. Other memes are more controversial; we can see why they spread, and why, all things considered, we should tolerate them, in spite of the problems they cause for us: shopping malls, fast food, advertising on television. Still others are unquestionably pernicious, but extremely hard to eradicate: anti-Semitism, hijacking airliners, computer viruses, spray-paint graffiti.

Genes are invisible; they are carried by gene vehicles (organisms) in which they tend to produce characteristic effects ("phenotypic" effects) by which their fates are, in the long run, determined. Memes are also invisible, and are carried by meme vehicles — pictures, books, sayings (in particular languages, oral or written, on paper or magnetically encoded, etc.). Tools and buildings and other inventions are also meme vehicles. A wagon with spoked wheels carries not only grain or freight from place to place; it carries the brilliant idea of a wagon with spoked wheels from mind to mind. A meme's existence depends on a physical embodiment in some medium; if all such physical embodiments are destroyed, that meme is extinguished. It may, of course, make a subsequent independent reappearance — just as dinosaur genes could, in principle, get together again in some distant future — but the dinosaurs they created and inhabited would not be descendants of the original dinosaurs — or at least not any more directly than we are. The fate of memes — whether copies and copies of copies of them persist and multiply — depends on the selective forces that act directly on the physical vehicles that embody them.

Meme vehicles inhabit our world alongside all the fauna and flora, large and small. By and large they are "visible" only to the human species, however. Consider the environment of the average New York City pigeon, whose eyes and ears are assaulted every day by approximately as many words, pictures, and other signs and symbols as assault each human New Yorker. These physical meme vehicles may impinge importantly on the pigeon's welfare, but not in virtue of the memes they carry — it is nothing to the pigeon that it is under a page of the *National Enquirer*, not the *New York Times*, that it finds a crumb.

To human beings, on the other hand, each meme vehicle is a potential friend or foe, bearing a gift that will enhance our powers or a gift horse that will distract us, burden our memories, derange our judgment. We may compare these airborne invaders of our eyes and ears to the parasites that enter our bodies by other routes: there are the beneficial parasites such as the bacteria in our digestive systems without which we could not digest our food, the tolerable parasites, not worth the trouble of eliminating (all the denizens of our skin and scalps, for instance), and the pernicious invaders that are hard to eradicate (the AIDS virus, for instance).

So far, the meme's-eye perspective may still appear to be simply a graphic way of organizing very familiar observations about the way items in our cultures affect us, and affect each other. But Dawkins suggests that in our explanations we tend to overlook the fundamental fact that "a cultural trait may have evolved in the way it has simply because it is *advantageous to itself*" (1976, p. 214). This is the key to answering the question of whether or not the meme meme is one we should exploit and replicate. According to the normal view, the following are virtually tautological:

> Idea X was believed by the people because X was deemed true.
> People approved of X because people found X to be beautiful.

What requires special explanation are the cases in which, in spite of the truth or beauty of an idea, it is *not* accepted, or in spite of its ugliness or falsehood it *is*. The meme's-eye view purports to be a general alternative perspective from which these deviations can be explained: what is tautological for *it* is

> Meme X spread among the people because X was a good replicator.

Now, there is a nonrandom correlation between the two; it is no accident. We would not survive unless we had a better-than-chance habit of choosing the memes that help us. Our meme-immunological systems are not foolproof, but not hopeless either. We can rely, as a general, crude rule of thumb, on the coincidence of the two perspectives: by and large, the good memes are the ones that are also the good replicators.

The theory becomes interesting only when we look at the exceptions, the circumstances under which there is a pulling apart of the two perspectives; only if meme theory permits us better to understand the deviations from the normal scheme will it have any warrant for being accepted. (Note that in its own terms, whether or not the meme meme replicates successfully is strictly independent of its epistemological virtue; it might spread in spite of its perniciousness, or go extinct in spite of its virtue.)

Memes now spread around the world at the speed of light, and replicate at rates that make even fruit flies and yeast cells look glacial in comparison. They leap promiscuously from vehicle to vehicle, and from medium to medium, and are proving to be virtually unquarantinable. Memes, like genes, are *potentially* immortal, but, like genes, they depend on the existence of a continuous chain of physical vehicles, persisting in the face of the Second Law of Thermodynamics. Books are relatively permanent, and inscriptions on monuments even more permanent, but unless these are under the protection of human conservators, they tend to dissolve in time. As with genes, immortality is more a matter of replication than of the longevity of individual vehicles. The preservation of the Platonic memes, via a series of copies of copies, is a particularly striking case of this. Although some papyrus fragments of Plato's texts roughly contemporaneous with him have been discovered recently, the survival of the memes owes almost nothing to such long-range persistence. Today's libraries contain thousands if not millions of physical copies (and translations) of Plato's *Meno*, and the key ancestors in the transmission of this text turned to dust centuries ago.

Brute physical replication of vehicles is not enough to ensure meme longevity. A few thousand hard-bound copies of a new book can disappear with scarcely a trace in a few years, and who knows how many brilliant letters to the editor, reproduced in hundreds of thousands of copies, disappear into landfills and incinerators every day? The day may come when nonhuman meme-evaluators suffice to select and arrange for the preservation of particular memes, but for the time being, memes still depend at least indirectly on one or more of their vehicles spending at least a brief, pupal stage in a remarkable sort of meme nest: a human mind.

Minds are in limited supply, and each mind has a limited capacity for memes, and hence there is a considerable competition among memes for entry into as many minds as possible. This competition is the major selective force in the memosphere, and, just as in the biosphere, the challenge has been met with great ingenuity. For instance, whatever virtues (from our perspective) the following memes have, they have in common the property of having phenotypic expressions that tend to make their own replication more likely by disabling or preempting the environmental forces that would tend to extinguish them: the meme for *faith*, which discourages the exercise of the sort of critical judgment that might decide that the idea of faith was all things considered a dangerous idea (Dawkins, 1976, p. 212); the meme for *tolerance* or *free speech;* the meme of including in a chain letter a warning about the terrible fates of those who have broken the chain in the past; the *conspiracy theory* meme, which has a built-in response to the objection that there is no good evidence of the conspiracy: "Of course not — that's how powerful the conspiracy is!" Some of these memes are "good" perhaps and others "bad"; what they have in common is a phenotypic effect that systematically tends to disable the selective forces arrayed against them. Other things being equal, population memetics predicts that conspiracy theory memes will persist quite independently of their truth, and the meme for faith is apt to secure its own survival, and that of the religious memes that ride piggyback on it, in even the most rationalistic environments. Indeed, the meme for faith exhibits *frequency-dependent fitness:* it flourishes best when it is outnumbered by rationalistic memes; in an environment with few skeptics, the meme for faith tends to fade from disuse.

Other concepts from population genetics also transfer smoothly: here is a case of what a geneticist would call *linked loci:* two memes that happen to be physically tied together so that they tend always to replicate together, a fact that affects their chances. There is a magnificent ceremonial march, familiar to many of us, and one that would be much used for commencements, weddings, and other festive occasions, perhaps driving "Pomp and Circumstance" and the Wedding March from *Lohengrin* to near extinction, were it not for the fact that its musical meme is too tightly linked to its title meme, which many of us tend to think of as soon as we hear the music: Sir Arthur Sullivan's unusable masterpiece, "Behold the Lord High Executioner."

The haven all memes depend on reaching is the human mind, but a human mind is itself an artifact created when memes restructure a human brain in order to make it a better habitat for memes. The avenues for entry and departure are modified to suit local conditions, and strengthened by various artificial devices that enhance fidelity and prolixity of replication: native Chinese minds differ dramatically from native French minds, and literate minds differ from illiterate minds. What memes provide in return to the organisms in which they reside is an incalculable store of advantages — with some Trojan horses thrown in for good measure, no doubt. Normal human brains are not all alike; they vary considerably in size, shape, and in the myriad details of connection on which their prowess depends. But the most striking differences in human prowess depend on microstructural differences induced by the various memes that have entered them and taken up residence. The memes enhance each others' opportunities: the meme for education, for instance, is a meme that reinforces the very process of meme-implantation.

But if it is true that human minds are themselves to a very great degree the creations of memes, then we cannot sustain the polarity of vision with which we started; it cannot be "memes versus us," because earlier infestations of memes have already played a major role in determining *who or what we are*. The "independent" mind struggling to protect itself from alien and dangerous memes is a myth; there is, in the basement, a persisting tension between the biological imperative of the genes and the imperatives of the memes, but we would be foolish to "side with" our genes — that is to commit the most egregious error of pop sociobiology. What foundation, then, can we stand on as we struggle to keep our feet in the memestorm in which we are engulfed? If replicative might does not make right, what is to be the eternal ideal relative to which "we" will judge the value of memes? We should note that the memes for normative concepts — for *ought* and *good* and *truth* and *beauty* — are among the most entrenched denizens of our minds, and that among the memes that constitute us, they play a central role. Our existence as us, as what we as thinkers are — not as what we as organisms are — is not independent of these memes.

To sum up: Meme evolution has the potential to contribute remarkable design-enhancements to the underlying machinery of the brain — at great speed, compared to the slow pace of genetic R and D. The discredited Lamarckian idea of the genetic transmission of individually acquired characteristics was initially attractive to biologists in part because of its presumed capacity to speed new inventions into the genome. (For a fine demolition of Lamarckianism, see Dawkins's discussion in *The Extended Phenotype*, 1982.) That does not, and cannot, happen. The Baldwin Effect does speed up evolution, favoring the movement of individually discovered Good Tricks into the genome, by the indirect path of creating new selection pressures resulting from widespread adoption by individuals of the Good Tricks. But cultural evolution, which happens much faster still, permits individuals to acquire, through cultural transmission, Good Tricks that have been honed by predecessors who are not even their genetic ancestors. So potent are the effects of such a sharing of good designs that cultural evolution has probably obliterated all but a few of the gentle pressures of the Baldwin Effect. The design improvements one receives from one's culture — one seldom has to "reinvent the wheel" — probably swamp most individual genetic differences in brain design, removing the advantage from those who are slightly better off at birth.

All three media — genetic evolution, phenotypic plasticity, and memetic evolution — have contributed to the design of human consciousness, each in turn, and at increasing rates of speed. Compared with phenotypic plasticity, which has been around for millions of years, significant memetic evolution is an extremely recent phenomenon, becoming a powerful force only in the last hundred thousand years, and exploding with the development of civilization less than ten thousand years ago. It is restricted to one species, *Homo sapiens*, and we might note that it has now brought us to the dawn of yet a fourth medium of potential R and D, thanks to the memes of science: the direct revising of individual nervous systems by neuroscientific engineering, and the revision of the genome by genetic engineering.

7. THE MEMES OF CONSCIOUSNESS: THE VIRTUAL MACHINE TO BE INSTALLED

Although an organ may not have been originally formed for some special purpose, if it now serves for this end we are justified in saying that it is specially contrived for it. On the same principle, if a man were to make a machine for some special purpose, but were to use old wheels, springs, and pulleys, only slightly altered, the whole machine, with all its parts, might be said to be specially contrived for that purpose. Thus throughout nature almost every part of each living being has probably served, in a slightly modified condition, for diverse purposes, and has acted in the living machinery of many ancient and distinct forms.

CHARLES DARWIN, 1874

The large brain, like large government, may not be able to do simple things in a simple way.

DONALD HEBB, 1958

The most powerful drive in the ascent of man is his pleasure in his own skill. He loves to do what he does well and, having done it well, he loves to do it better.

JACOB BRONOWSKI, 1973

A feature of my speculative story has been that our ancestors, like us, took pleasure in various modes of relatively undirected self-exploration — stimulating oneself over and over and seeing what happened. Because of the plasticity of the brain, coupled with the innate restlessness and curiosity that leads us to explore every nook and cranny of our environment (of which our own bodies are such an important and ubiquitous element), it is not surprising, in retrospect, that we hit upon strategies of self-stimulation or self-manipulation that led to the inculcation of habits and dispositions that radically altered the internal communicative structure of our brains, and that these discoveries became part of the culture — memes — that were then made available to all.

The transformation of a human brain by infestations of memes is a major alteration in the competence of that organ. As we noted, the differences in a brain whose native language is Chinese rather than English would account for huge differences in the competence of that brain, instantly recognizable in behavior, and significant in many experimental contexts. Recall, for instance, how important it is in experiments with human subjects for the experimenter (the heterophenomenologist) to know whether the subjects have understood the instructions. These functional differences, though presumably all physically embodied in patterns of microscopic changes in the brain, are as good as invisible to neuroscientists, now and probably forever, so if we are going to get any grip on the functional architecture *created* by such meme infestations, we will have to find a higher level at which to describe it. Fortunately, one is available, drawn from computer science. The level of description and explanation we need is *analogous to* (but not identical with) one of the "software levels" of description of computers: what we need to understand is how human consciousness can be realized in the operation of a *virtual machine* created by memes in the brain.

Here is the hypothesis I will defend:

> Human consciousness is *itself* a huge complex of memes (or more exactly, meme-effects in brains) that can best be understood as the operation of a "*von Neumannesque*" virtual machine *implemented* in the *parallel architecture* of a brain that was not designed for any such activities. The powers of this *virtual machine* vastly enhance the underlying powers of the organic *hardware* on which it runs, but at the same time many of its most curious features, and especially its limitations, can be explained as the byproducts of the *kludges* that make possible this curious but effective reuse of an existing organ for novel purposes.

This hypothesis will soon emerge from the thicket of jargon in which I have just stated it. Why did I use the jargon? Because these are terms for valuable concepts that have only recently become available to people thinking about the mind. No other words express these concepts crisply, and they are very much worth knowing. So, with the aid of a brief historical digression, I will introduce them, and place them into the context in which we will use them.

Two of the most important inventors of the computer were the British mathematician Alan Turing and the Hungarian-American mathematician and physicist John von Neumann. Although Turing had plenty of hands-on practical experience designing and building the special-purpose electronic code-breaking machines that helped the Allies win World War II, it was his purely abstract theoretical work, in developing his conception of the *Universal Turing machine,* that opened the Computer Age. Von Neumann saw how to take Turing's abstraction (which was really "philosophical" — a thought experiment, not an engineering proposal) and make it just concrete enough to turn it into the (still quite abstract) design for an actual, practical electronic computer. That abstract design, known as the *Von Neumann architecture,* is to be found in almost every computer in the world today, from giant "mainframes" to the chip at the heart of the most modest home computer.

A computer has a basic *fixed* or *hard-wired* architecture, but with huge amounts of plasticity thanks to the *memory,* which can store both *programs* (otherwise known as *software*) and *data,* the merely transient patterns that are made to track whatever it is that is to be represented. Computers, like brains, are thus incompletely designed at birth, with flexibility that can be used as a medium to create more specifically disciplined architectures, special-purpose machines, each with a strikingly individual way of taking in the environment's stimulation (via the keyboard or other input devices) and eventually yielding responses (via the CRT screen or other output devices).

These temporary structures are "made of rules rather than wires," and computer scientists call them *virtual machines.*[15] A virtual machine is what you get when you impose a particular pattern of rules (more literally: dispositions or transition regularities) on all that plasticity. Consider someone who has broken his arm and has it in a plaster cast. The cast severely restricts the movement of his arm, and its weight and shape also call for adjustments in the rest of the person's bodily movements. Now consider a mime (Marcel Marceau, say) imitating someone with a plaster cast on his arm; if the mime does the trick well, his bodily motions will be restricted in almost exactly the same ways; he has a *virtual cast* on his arm — and it will be "almost visible." Anyone who is familiar with a word processor is acquainted with at least one virtual machine, and if you have used several different word processors, or used a spread sheet or played a game on the very same computer you use for word processing, you are acquainted with several virtual machines, taking turns existing on a particular real machine. The differ-

15. Purists may object that my use of the term *virtual machine* is somewhat broader than the usage they recommend in computer science. I reply that, like Mother Nature, when I see a handy item to "exapt" and put to an extended use (Gould, 1980), I go for it.

ences are made highly visible, so that the user knows which virtual machine he is interacting with at any time.

Everybody knows that different programs endow computers with different powers, but not everybody knows the details. A few of them are important to our story, so I must beg your indulgence and provide a brief, elementary account of the process invented by Alan Turing.

Turing was not trying to invent the word processor or the video game when he made his beautiful discoveries. He was thinking, self-consciously and introspectively, about just how he, a mathematician, went about solving mathematical problems or performing *computations*, and he took the important step of trying to break down the *sequence* of his *mental acts* into their primitive components. "What do I do," he must have asked himself, "when I perform a computation? Well, first I ask myself which rule applies, and then I apply the rule, and then write down the result, and then I look at the result, and then I ask myself what to do next, and . . ." Turing was an extraordinarily well-organized thinker, but his stream of consciousness, like yours or mine or James Joyce's, was no doubt a variegated jumble of images, decisions, hunches, reminders, and so forth, out of which he managed to distill the mathematical essence: the bare-bones, minimal sequence of operations that could accomplish the goals he accomplished in the florid and meandering activities of his conscious mind. The result was the specification of what we now call a Turing machine, a brilliant idealization and simplification of a hyperrational, hyperintellectual phenomenon: a mathematician performing a rigorous computation. The basic idea had five components:

(1) a *serial* process (events happening one at a time), in
(2) a severely restricted *workspace*, to which
(3) both *data* and *instructions* are brought
(4) from an inert but super-reliable *memory*,
(5) there to be operated on by a finite set of *primitive operations*.

In Turing's original formulation, the workspace was imagined to be a scanner that looked at just one square of a paper tape at a time, to see if a zero or one were written on it. Depending on what it "saw," it either erased the zero or one and printed the other symbol, or left the square unchanged. It then moved the tape left or right one square and looked again, in each case being governed by a finite set of hard-wired instructions that formed its *machine table*. The tape was the memory.

Turing's set of primitive operations (the acts "atomic to introspection" if you like) was deliberately impoverished, so that there could be no question of their mechanical realizability. That is, it was important to Turing's mathematical purposes that there be no doubt that each step in the processes he was studying be one that was so simple, so stupid, that it could be performed by a simpleton — by someone who could be replaced by a machine: SCAN, ERASE, PRINT, MOVE LEFT ONE SPACE, and so on.

He saw, of course, that his ideal specification could serve, indirectly, as the blueprint for an actual computing machine, and so did others, in particular, John von Neumann, who modified Turing's basic ideas to create the abstract architecture for the first practically realizable digital computer. We call that architecture the *von Neumann machine*.

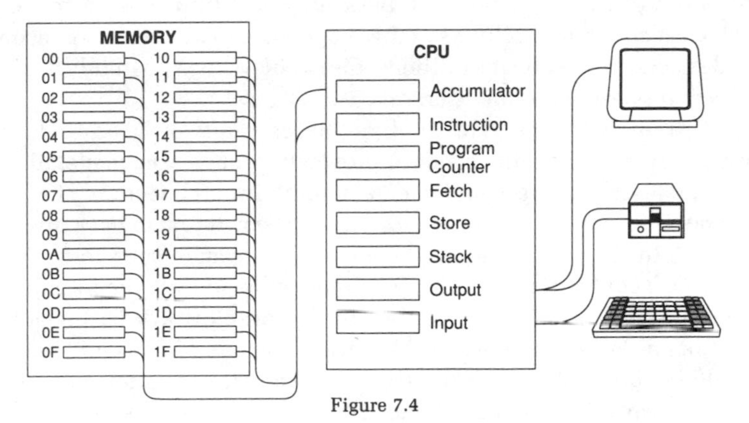

Figure 7.4

On the left is the *memory* or *random access memory* (RAM), where both the data and the instructions are kept, coded in sequences of binary digits or "bits," such as 00011011 or 01001110. Turing's serial process occurs in the workspace consisting of two "registers" marked *accumulator* and *instruction register*. An instruction is copied electronically into the instruction register, which then *executes* it. For instance, if the instruction (translated into English) says "clear the accumulator" the computer puts the number 0 in the accumulator, and if the instruction says "add the contents of memory register 07 to the number in the accumulator" the computer will fetch whatever number is in the memory register with the address 07 (the contents might be any number) and add it to the number in the accumulator. And so forth. What are the primitive operations? Basically, the arithmetical operations, add, subtract, multiply, and divide; the data-moving operations, fetch, store, output, input; and (the heart of the "logic" of computers) the *conditional* instructions, such as "*IF* the number in the accumulator is greater than zero, *THEN* go to the instruction in register 29; otherwise go to the next instruction." Depending on the computer model, there might be as few as sixteen primitive operations or hundreds, all wired-in in special-purpose circuitry. Each primitive operation is coded by a unique binary pattern (e.g., ADD might be 1011, and SUBTRACT might be 1101), and whenever these particular sequences land in the instruction register they are rather like dialed telephone numbers that mechanically open up the lines to the right special-purpose circuit — the adder circuit or the subtracter circuit, and so forth. The two registers, in which only one instruction and one value can appear at any one time, are the notorious "von Neumann bottleneck," the place where all activity of the system has to pass single-file through a narrow gap. On a fast computer, millions of these operations can occur in a second, and strung together by the millions, they achieve the apparently magical effects discernible to the user.

All digital computers are *direct descendants* of this design; and while many modifications and improvements have been made, like all vertebrates they share a fundamental underlying architecture. The basic operations, looking so arithmetical, don't seem at first to have much to do with the basic "operations" of a normal stream of consciousness — thinking about Paris, enjoying the aroma of bread from the oven, wondering where to go for vacation — but that didn't worry Turing or von Neumann. What mattered to them was that this sequence of actions could "in principle" be elaborated to incorporate *all* "rational thought" and perhaps all "irrational thought" as well. It is a considerable historical irony that this architecture was misdescribed by the popular press from the moment it was created. These fascinating new von Neumann machines were called "giant electronic brains," but they were, in fact, *giant electronic minds*, electronic imitations — severe simplifications — of what William James dubbed the stream of consciousness, the meandering sequence of conscious mental contents famously depicted by James Joyce in his novels. The architecture of the brain, in contrast, is massively parallel, with millions of simultaneously active channels of operation. What we have to understand is how a Joycean (or, as I have said, "von Neumannesque") serial phenomenon can come to exist, with all its familiar peculiarities, in the parallel hubbub of the brain.

Here is a bad idea: Our hominid ancestors needed to think in a more sophisticated, logical way, so natural selection gradually designed and installed a hard-wired von Neumann machine in the left ("logical," "conscious") hemisphere of the human cortex. I hope it is clear from the foregoing evolutionary narrative that although that might be *logically* possible, it has no biological plausibility at all — our ancestors might as easily have sprouted wings or been born with pistols in their hands. That is not how evolution does its work.

We *know* there is something at least *remotely* like a von Neumann machine in the brain, because we know we have conscious minds "by introspection" and the minds we thereby discover are at least this much like von Neumann machines: They were the inspiration for von Neumann machines! This historical fact has left a particularly compelling fossil trace: computer programmers will tell you that it is fiendishly difficult to program the parallel computers currently being developed, and relatively easy to program a serial, von Neumann machine. When you program a conventional von Neumann machine, you have a handy crutch; when the going gets tough, you ask yourself, in effect, "What would I do if I were the machine, trying to solve this problem?" and this leads you to an answer of the form, "Well, first I'd do this, and then I'd have to do that, etc." But if you ask yourself "What would I do in this situation if I were a thousand-channel-wide parallel processor?" you draw a blank; you don't have any personal familiarity with — any "direct access to" — processes happening in a thousand channels at once, even though that is what is going on in your brain. Your only access to what is going on in your brain comes in a sequential "format" that is strikingly reminiscent of the von Neumann architecture — although putting it that way is historically backwards.

There is a big difference, as we have seen, between a (standard) computer's serial architecture and the parallel architecture of the brain. This fact is often cited as an objection to Artificial Intelligence, which attempts to create human-style intelligence by devising programs that (almost always) run on von Neumann machines. Does the difference in architecture make a theoretically important difference? In one sense, no. Turing had proven — and this is probably his greatest contribution — that his Universal Turing machine can compute any function that any computer, with any architecture, can compute. In effect, the Universal Turing machine is the perfect mathematical chameleon, capable of imitating *any* other computing machine and doing, during that period of imitation, *exactly* what that machine does. All you have to do is feed the Universal Turing machine a suitable description of another machine, and, like Marcel Marceau (the Universal miming machine) armed with an explicit choreography, it forthwith proceeds to produce a perfect imitation based on that description — it becomes, virtually, the other machine. A computer program can thus be seen either as a list of primitive instructions to be followed or as a description of a machine to be imitated.

Can you imitate Marcel Marceau imitating a drunk imitating a baseball batter? You might find the hardest part of the trick was keeping track of the different levels of imitation, but for von Neumann machines this comes naturally. Once you have a von Neumann machine on which to build, you can nest virtual machines like Chinese boxes. For instance, you can first turn your von Neumann machine into, say, a Unix machine (the Unix operating system) and then implement a Lisp machine (the Lisp programming language) on the Unix machine — along with WordStar, Lotus 123, and a host of other virtual machines — and then implement a chess-playing computer on your Lisp machine. Each virtual machine is recognizable by its *user interface* — the way it appears on the screen of the CRT and the way it responds to input — and this self-presentation is often called the *user illusion,* since a user can't tell — and doesn't care — how the particular virtual machine he's using is implemented in the hardware. It doesn't matter to him whether the virtual machine is one, two, three, or ten layers away from the hardware.[16] (For instance, WordStar users can recognize, and interact with, the WordStar virtual machine wherever they find it, no matter what variation there is in the underlying hardware.)

So a virtual machine is a temporary set of highly structured regularities imposed on the underlying hardware by a *program:* a structured recipe of hundreds of thousands of instructions that give the hardware a huge, interlocking set of habits or dispositions-to-react. If you look at the microdetails of all those instructions reeling through the instruction register, you will miss the forest for the trees; if you stand back, however, the functional architecture that emerges from all those microsettings can be clearly seen: it consists of virtual *things*

such as blocks of text, cursors, erasers, paint-sprayers, files

and virtual *places*

such as directories, menus, screens, shells

connected by virtual *paths*

---

16. Or it might not be a *virtual* machine at all. It might be a made-to-order hard-wired special-purpose real machine, such as a Lisp machine, which is a descendant of Lisp *virtual* machines, and which is designed right down to its silicon chips to run the programming language Lisp.

such as "ESCape to Dos," or *entering* the PRINT menu *from* the MAIN menu

and permitting various large-and-interesting virtual *operations* to be performed

such as searching a file for a word, or enlarging a box drawn on the screen.

Since any computing machine at all can be imitated by a virtual machine on a von Neumann machine, it follows that if the brain is a massive parallel processing machine, it too can be perfectly imitated by a von Neumann machine. And from the very beginning of the computer age, theorists used this chameleonic power of von Neumann machines to create *virtual* parallel architectures that were supposed to model brainlike structures.[17] How can you get a one-thing-at-a-time machine to be a many-things-all-at-once machine? By a process rather like knitting. Suppose the parallel processor being simulated is ten channels wide. First, the von Neumann machine is instructed to perform the operations handled by the first node of the first channel (node

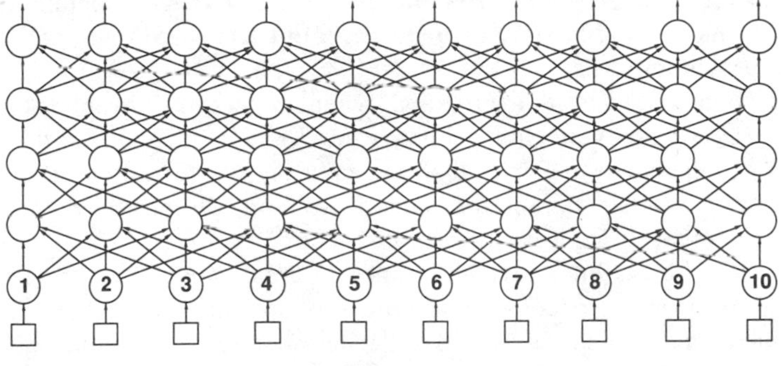

Figure 7.5

1 in the diagram), saving the result in a "buffer" memory, and then node 2, and so forth, until all ten nodes in the first layer have been advanced one moment. Then the von Neumann machine tackles the effects of each of these first-layer results on the next layer of nodes,

17. The "logical neurons" of McCulloch and Pitts (1943) were actually devised contemporaneously with the invention of the serial computer, and influenced von Neumann's thinking, and these in turn led to the Perceptrons of the fifties, the ancestors of today's connectionism. For a brief historical account see Papert (1988).

drawing the previously calculated results one at a time from the buffer memory and applying them as input to the next layer. Laboriously it proceeds, knitting back and forth, trading off time against space. A virtual ten-channel-wide machine will take *at least* ten times as long to simulate as a one-channel machine, and a million-channel-wide machine (like the brain) will take at least a million times longer to simulate. Turing's proof says nothing about the speed with which the imitation will be accomplished, and for some architectures, even the blinding speed of modern digital computers is overwhelmed by the task. That is why AI researchers interested in exploring the powers of parallel architectures are today turning to *real* parallel machines — artifacts that might with more justice be called "giant electronic brains" — on which to compose their simulations. But in principle, any parallel machine can be perfectly — if inefficiently — mimicked as a virtual machine on a serial von Neumann machine.[18]

Now we are ready to turn this standard idea upside-down. Just as you can simulate a parallel brain on a serial von Neumann machine, you can also, in principle, simulate (something like) a von Neumann machine on parallel hardware, and that is just what I am suggesting: Conscious human minds are more-or-less serial virtual machines implemented — inefficiently — on the parallel hardware that evolution has provided for us.

What counts as the "program" when we talk of a virtual machine running on the brain's parallel hardware? What matters is that there is lots of adjustable plasticity that can take on myriad different microhabits and thereby take on different macrohabits. In the case of the von Neumann machine, this is accomplished by hundreds of thousands of zeroes and ones (bits), divided up into "words" of 8, 16, 32, or 64 bits, depending on the machine. The words are separately stored in registers in the memory and accessed a word at a time in the instruction register. In the case of the parallel machine, it is accomplished, we can surmise, by thousands or millions or billions of connection-strength settings between neurons, which all together in concert give the underlying hardware a new set of macrohabits, a new set of conditional regularities of behavior.

And how do these programs of millions of neural connection-strengths get installed on the brain's computer? In a von Neumann machine, you just "load" the program off a disk into the main memory,

---

18. For more on the implications of real-world speed, and its implications for Artificial Intelligence, see "Fast Thinking" in my *The Intentional Stance* (1987a).

and the computer thereby gets an instant set of new habits; with brains, it takes *training*, including especially the repetitive self-stimulation of the sort sketched in section 5. This is, of course, a major disanalogy. A von Neumann machine's Central Processing Unit or CPU is rigid in the way it responds to the bit-strings that compose its words, treating them as instructions in its entirely proprietary and fixed *machine language*. These facts are definitive of the *stored-program digital computer*, and a human brain is no such thing. While it is probably true that each particular connection-strength setting between neurons in the brain has a determinate effect on the resulting behavior of the surrounding network, there is no reason whatever to think that two different brains would have the "same system" of interconnections, so there is almost certainly nothing remotely analogous to the fixed machine language that, say, all IBM and IBM-compatible computers share. So if two or more brains "share software" it will not be by virtue of a simple, direct process analogous to copying a machine language program from one memory to the other. (Also, of course, the plasticity that somehow subserves memory in a brain is not isolated as a passive storehouse; the division of labor between memory and CPU is an artifact for which there is no analogue in the brain, a topic to which we will return in chapter 9.)

Since there are such important — and often overlooked — *disanalogies*, why do I persist in likening human consciousness to software? Because, as I hope to show, some important and otherwise extremely puzzling features of consciousness get illuminating explanations on the hypothesis that human consciousness (1) is too recent an innovation to be hard-wired into the innate machinery, (2) is largely a product of cultural evolution that gets imparted to brains in early training, and (3) its successful installation is determined by myriad microsettings in the plasticity of the brain, which means that its functionally important features are very likely to be invisible to neuroanatomical scrutiny in spite of the extreme salience of the effects. Just as no computer scientist would attempt to understand the different strengths and weaknesses of WordStar versus WordPerfect by building up from information about the differences in voltage patterns in the memory, so no cognitive scientist should expect to make sense of human consciousness simply by building up from the neuroanatomy. Besides, (4) the idea of the *user illusion* of a virtual machine is tantalizingly suggestive: If consciousness is a virtual machine, who is the user for whom the user illusion works? I grant that it looks suspiciously as if we are drifting inexorably back to an internal Cartesian Self, sitting at the cortical workstation and reacting to the user illusion of the software running there, but there are, as we shall see, some ways of escaping that dreadful dénouement.

Suppose for the moment then that there is a more or less well-designed (debugged) version of this stream-of-consciousness virtual machine — the Joycean machine — in the memosphere. As we have seen, since there is no shared machine language between brains, the methods of transmission that would guarantee a fairly uniform virtual machine operating throughout the culture must be social, highly context-sensitive, and to some degree self-organizing and self-correcting. Getting two different computers — e.g., a Macintosh and an IBM-PC — to "talk to each other" is a matter of intricate, fussy engineering that depends on precise information about the internal machinery of the two systems. Insofar as human beings can "share software" without anyone having such knowledge, it must be because the shared systems have a high degree of lability and format tolerance. There are several methods of sharing such software: learning by imitation, learning as a result of "reinforcement" (either deliberately imposed by a teacher — reward, encouragement, disapproval, threat — or subtly and unconsciously transmitted in the course of communicative encounters), and learning as the result of explicit instruction in a natural language that has already been learned via the first two methods. (Think, for instance, of the sorts of habits that would be entrained by frequently saying, to a novice, "Tell me what you are doing," and "Tell me why you are doing that." Now think of the novice getting in the habit of addressing these same requests to himself.)

In fact, not just spoken language but writing plays a major role, I suspect, in the development and elaboration of the virtual machines most of us run most of the time in our brains. Just as the wheel is a fine bit of technology that is quite dependent on rails or paved roads or other artificially planed surfaces for its utility, so the virtual machine that I am talking about can exist only in an environment that has not just language and social interaction, but writing and diagramming as well, simply because the demands on memory and pattern recognition for its implementation require the brain to "off-load" some of its memories into buffers in the environment. (Note that this implies that "preliterate mentality" could well involve a significantly different class of virtual architectures from those encountered in literate societies.)

Think of adding two ten-digit numbers in your head without use of paper and pencil or saying the numbers out loud. Think of trying to figure out, without a diagram, a way to bring three freeways into a cloverleaf-style intersection so that one can drive from either direction on any one freeway to either direction on any other freeway without having to get onto the third freeway. These are the sorts of problems human beings readily solve with the aid of external memory devices and the use of their preexisting scanners (called *eyes* and *ears*) with their highly developed hard-wired pattern-recognition circuits. (See Rumelhart, chapter 14, in McClelland and Rumelhart, 1986, for some valuable observations on this topic.)

We install an organized and partially pretested set of habits of mind, as the political scientist Howard Margolis (1987), calls them, in our brains in the course of early childhood development. We will look more closely at the likely details of this architecture in chapter 9, but the overall structure of the new set of regularities, I suggest, is one of *serial chaining,* in which first one "thing" and then another "thing" takes place in (roughly) the same "place." This stream of events is entrained by a host of learned habits, of which talking-to-oneself is a prime example.

Since this new machine created in us is a highly replicated meme-complex, we may ask to what it owes its replicative success. We should bear in mind, of course, that it *might* not be good for anything — except replicating. It *might* be a software virus, which readily parasitizes human brains without actually giving the human beings whose brains it infests any advantage over the competition. More plausibly, *certain features* of the machine might be parasites, which exist only because they can, and because it is not possible — or worth the trouble — to get rid of them. William James thought it would be absurd to suppose that the most astonishing thing we know of in the universe — consciousness — is a mere artifact, playing no essential role in how our brains work, but however unlikely it might be, it is not entirely out of the question, and hence is not really absurd. There is plenty of evidence around about the benefits consciousness apparently provides us, so we can no doubt satisfy ourselves about its various *raisons d'être,* but we are apt to misread that evidence if we think that a mystery remains unless every single feature has — or once had — a function (from *our* point of view as consciousness-"users") (Harnad, 1982). There is room for some brute facts lacking all functional justification. Some features of consciousness may just be selfish memes.

Looking on the bright side, however, what problems is this new machine apparently well designed to solve? The psychologist Julian Jaynes (1976) has argued persuasively that its capacities for self-exhortation and self-reminding are a prerequisite for the sorts of elaborated and long-term bouts of self-control without which agriculture, building projects, and other civilized and civilizing activities could not be organized. It also seems to be good for the sorts of self-monitoring that can protect a flawed system from being victimized by its own failures, a theme developed in Artificial Intelligence by Douglas Hofstadter (1985). It has been seen by the psychologist Nicholas Humphrey (1976, 1983a, 1986) as providing a means for exploiting what might be called social simulations — using introspection to guide one's hunches about what others are thinking and feeling.

Underlying these more advanced and specialized talents is the basic capacity for solving the meta-problem of what to think about next. We saw early in the chapter that when an organism faces a crisis (or just a difficult and novel problem), it may have resources in it that would be very valuable in the circumstances *if only it could find them and put them to use in time!* Orienting responses, as Odmar Neumann has surmised, have the valuable effect of more or less turning everybody on at once, but accomplishing this global arousal, as we saw, is as much part of the problem as part of the solution. It helps not a bit unless, in the next step, the brain manages to get some sort of coherent activity out of all these volunteers. The problem for which orienting responses were a partial solution was the problem of getting total, global access among a collection of specialists used to minding their own business. Even if, thanks to an underlying Pandemonium-style architecture, the chaos soon settles, leaving one specialist temporarily in charge (and, perhaps, better informed by the competition it has won), there are obviously at least as many bad ways for these conflicts to be resolved as good ways. Nothing guarantees that the politically most effective specialist will be the "man for the job."

Plato saw the problem quite clearly two thousand years ago, and came up with a wonderful metaphor to describe it.

> Now consider whether knowledge is a thing you can possess in that way without having it about you, like a man who has caught some wild birds — pigeons or what not — and keeps them in an aviary he has made for them at home. In a sense, of course we might say he "has" them all the time inasmuch as he possesses them, mightn't we? . . . But in another sense he "has" none of them, though he has got control of them, now that he has made them captive in an enclosure of his own; he can take and have hold of them whenever he likes by catching any bird he chooses,

> and let them go again; and it is open to him to do that as often as he pleases. . . . [S]o now let us suppose that every mind contains a kind of aviary stocked with birds of every sort, some in flocks apart from the rest, some in small groups, and some solitary, flying in any direction among them all. . . . [*Theaetetus*, 197–198a, Cornford translation]

What Plato saw was that merely having the birds is not enough; the hard part is learning how to get the right bird to come to you when you call. He went on to claim that by *reasoning*, we improve our capacity to get the right birds to come at the right time. Learning to reason is, in effect, learning knowledge-retrieval strategies.[19] That is where habits of mind come in. We have already seen in crude outline how such general habits of mind as talking-to-yourself or diagraming-to-yourself *might* happen to tease the right morsels of information to the surface (the surface of what? — a topic I will defer to chapter 10). But more specific habits of mind, refinements and elaborations of specific ways of talking to yourself, can improve your chances even further.

The philosopher Gilbert Ryle, in his posthumously published book *On Thinking* (1979), decided that thinking, of the slow, difficult, pondering sort that Rodin's famous statue of the Thinker is apparently engaged in, must indeed often be a matter of talking to yourself. Surprise, surprise! Isn't it obvious that that is what we do when we think? Well, yes and no. It is obvious that that is what we (often) seem to be doing; we can often even tell each other the various words we express in our silent soliloquies. But what is far from obvious is why talking to yourself does any good at all.

> What is *Le Penseur* doing, seemingly in his Cartesian insides? Or, to sound scientific, what are the mental processes like, which are going on in that Cartesian *camera obscura*? . . . Notoriously some of our ponderings, but not all, terminate in the solution of our problems; we had been fogged, but at last we came out into the clear. But if sometimes successful, why not always? If belatedly, why not promptly? If with difficulty, why not easily? Why indeed does it ever work? How possibly can it work? [Ryle, 1979, p. 65]

Habits of mind have been designed over the eons to *shape* the passages down well-trod paths of exploration. As Margolis notes,

---

19. For an interesting discussion of the (apparent) disagreement between two schools of thought in Artificial Intelligence, reasoning versus search, see Simon and Kaplan, 1989, pp. 18–19.

> even a human being today (hence, a fortiori, a remote ancestor of contemporary human beings) cannot easily or ordinarily maintain uninterrupted attention on a single problem for more than a few tens of seconds. Yet we work on problems that require vastly more time. The way we do that (as we can observe by watching ourselves) requires periods of mulling to be followed by periods of recapitulation, describing to ourselves what seems to have gone on during the mulling, leading to whatever intermediate results we have reached. This has an obvious function: namely, by rehearsing these interim results . . . we commit them to memory, for the immediate contents of the stream of consciousness are very quickly lost unless rehearsed. . . . Given language, we can describe to ourselves what seemed to occur during the mulling that led to a judgment, produce a rehearsable version of the reaching-a-judgment process, and commit that to long-term memory by in fact rehearsing it. [Margolis, 1987, p. 60]

Here, in the individual habits of self-stimulation, is where we should look for *kludges* (it rhymes with *Stooges*), the computer hacker's term for the *ad hoc* jury-rigs that are usually patched onto software in the course of debugging to get the stuff actually to work. (The linguist Barbara Partee once criticized an inelegant patch in an AI language-parsing program for being "odd hack" — as fine a serendipitous spoonerism as I have ever encountered. Mother Nature is full of odd hacks, and we should expect to find them in the individual's idiosyncratic adoption of the virtual machine as well.)

Here is a plausible example: Since human memory is not innately well designed to be superreliable, fast-access, random access memory (which every von Neumann machine needs), when the (culturally and temporally distributed) designers of the von Neumannesque virtual machine faced the task of cobbling up a suitable substitute that would run on a brain, they hit upon various memory-enhancing Tricks. The basic Tricks are rehearsal, rehearsal, and more rehearsal, abetted by rhymes and rhythmic, easy-to-recall maxims. (The rhymes and rhythms exploit the vast power of the pre-existing auditory-analysis system to recognize patterns in sounds.) The deliberate repeated juxtaposition of elements between which one needed to build a link of association — so that one item would always "remind" the brain of the next — was further enhanced, we may suppose, by making the associations as rich as possible, clothing them not just with visual and auditory features, but exploiting the whole body. *Le Penseur*'s frown and chin-holding, and the head-scratchings, mutterings, pacings, and doodlings that we idiosyncratically favor, could turn out to be not just random by-products of conscious thinking but functional contributors (or the vestigial traces of earlier, cruder functional contributors) to the laborious disciplining of the brain that had to be accomplished to turn it into a mature mind.

And in place of the precise, systematic "fetch-execute cycle" or "instruction cycle" that brings each new instruction to the instruction register to be executed, we should look for imperfectly marshaled, somewhat wandering, far-from-logical transition "rules," where the brain's largely innate penchant for "free association" is provided with longish association-chains to more or less ensure that the right sequences get tried out. (In chapter 9 we will consider elaborations of this idea in AI; for elaborations with different emphases, see Margolis, 1987, and Calvin, 1987, 1989. See also Dennett, 1991b.) We should not expect most of the sequences that occurred to be well-proven *algorithms*, guaranteed to yield the sought-after results, but just better-than-chance forays into Plato's aviary.

The analogy with the virtual machines of computer science provides a useful perspective on the phenomenon of human consciousness. Computers were originally just supposed to be number-crunchers, but now their number-crunching has been harnessed in a thousand imaginative ways to create new virtual machines, such as video games and word processors, in which the underlying number-crunching is almost invisible, and in which the new powers seem quite magical. Our brains, similarly, weren't designed (except for some very recent peripheral organs) for word processing, but now a large portion — perhaps even the lion's share — of the activity that takes place in adult human brains is involved in a sort of word processing: speech production and comprehension, and the serial rehearsal and rearrangement of linguistic items, or better, their neural surrogates. And these activities magnify and transform the underlying hardware powers in ways that seem (from the "outside") quite magical.

But still (I am sure you want to object): All this has little or nothing to do with consciousness! After all, a von Neumann machine is entirely unconscious; why should implementing it — or something like it: a Joycean machine — be any more conscious? I do have an answer: The von Neumann machine, by being wired up from the outset that way, with maximally efficient informational links, didn't have to become the object of its own elaborate perceptual systems. The workings of the Joycean machine, on the other hand, are just as "visible" and "audible" to it as any of the things in the external world that it is designed to perceive — for the simple reason that they have much of the same perceptual machinery focused on them.

Now this *appears* to be a trick with mirrors, I know. And it certainly is counterintuitive, hard-to-swallow, initially outrageous — just what one would expect of an idea that could break through centuries of mystery, controversy, and confusion. In the next two chapters we will look more closely — and skeptically — at the way in which this apparent trick with mirrors might be shown to be a legitimate part of the explanation of consciousness.

8

HOW WORDS DO THINGS WITH US

Language, like consciousness, only arises from the need, the necessity, of intercourse with others.
KARL MARX, 1846

Consciousness generally has only been developed under the pressure of the necessity for communication.
FRIEDRICH NIETZSCHE, 1882

Before my teacher came to me, I did not know that I am. I lived in a world that was a no-world. I cannot hope to describe adequately that unconscious, yet conscious time of nothingness. . . . Since I had no power of thought, I did not compare one mental state with another.
HELEN KELLER, 1908

1. REVIEW: E PLURIBUS UNUM?

In chapter 5 we exposed the persistently seductive bad idea of the Cartesian Theater, where a sound-and-light show is presented to a solitary but powerful audience, the Ego or Central Executive. Even though we've seen for ourselves the incoherence of this idea, and identified an alternative, the Multiple Drafts model, the Cartesian Theater will continue to haunt us until we have anchored our alternative firmly to the bedrock of empirical science. That task was begun in chapter 6, and in chapter 7 we made further progress. We returned, literally, to first principles: the principles of evolution that guided a speculative narration of the gradual process of design development that has created our kind of consciousness. This let us glimpse the machinery of consciousness from inside the black box — from backstage, one might say, in homage to the tempting theatrical image we are trying to overthrow.

In our brains there is a cobbled-together collection of specialist brain circuits, which, thanks to a family of habits inculcated partly by culture and partly by individual self-exploration, conspire together to produce a more or less orderly, more or less effective, more or less well-designed virtual machine, the *Joycean machine*. By yoking these independently evolved specialist organs together in common cause, and thereby giving their union vastly enhanced powers, this virtual machine, this software of the brain, performs a sort of internal political miracle: It creates a *virtual captain* of the crew, without elevating any one of them to long-term dictatorial power. Who's in charge? First one coalition and then another, shifting in ways that are not chaotic thanks to good meta-habits that tend to entrain coherent, purposeful sequences rather than an interminable helter-skelter power grab.

The resulting executive wisdom is just one of the powers traditionally assigned to the Self, but it is an important one. William James paid tribute to it when he lampooned the idea of the Pontifical Neuron somewhere in the brain. We know that the job description for such a Boss subsystem in the brain is incoherent, but we also know that those control responsibilities and decisions have to be parcelled out *somehow* in the brain. We are *not* like drifting ships with brawling crews; we do quite well not just staying clear of shoals and other dangers, but planning campaigns, correcting tactical errors, recognizing subtle harbingers of opportunity, and controlling huge projects that unfold over months or years. In the next few chapters we will look more closely at the architecture of this virtual machine, in order to provide some support — not proof — for the hypothesis that it could indeed perform these executive functions and others. Before we do that, however, we must expose and neutralize another source of mystification: the illusion of the Central Meaner.

One of the chief tasks of the imaginary Boss is controlling communication with the outside world. As we saw in chapter 4, the idealization that makes heterophenomenology possible *assumes* that there is someone home doing the talking, an Author of Record, a Meaner of all the meanings. When we go to interpret a loquacious body's vocal sounds, we don't suppose they are just random yawps, or words drawn out of a hat by a gaggle of behind-the-scenes partygoers, but the acts of a single agent, the (one and only) *person* whose body is making the sounds. If we choose to interpret at all, we have no choice but to posit a person whose communicative acts we are interpreting. This is not quite equivalent to positing an *inner* system that is the Boss of the body, the Puppeteer controlling the puppet, but that is the image that naturally takes hold of us. This internal Boss, it is tempting to suppose, is rather like the president of the United States, who may direct a press secretary or other subordinates to issue the actual press releases, but when they speak, they speak on his behalf, they execute his speech acts, for which he is responsible, and of which he is, officially, the author.

There is not in fact any such chain of command in the brain governing speech production (or writing, for that matter). Part of the task of dismantling the Cartesian Theater is finding a more realistic account of the actual source(s) of the assertions, questions, and other speech acts we naturally attribute to the (one) person whose body is doing the uttering. We need to see what happens to the enabling myth of heterophenomenology when the complexities of language production are given their due.

We have already seen a shadow cast by this problem. In chapter 4, we imagined Shakey the robot to have a rudimentary capacity to converse, or at least to emit words under various circumstances. We supposed that Shakey could be designed to "tell us" how it discriminated the boxes from the pyramids. Shakey might say "I scan each 10,000-digit-long sequence . . . ," or "I find the light-dark boundaries and make a line drawing . . . ," or "I don't know; some things just look boxy. . . ." Each of these different "reports" issued from a different level of access that the "report"-making machinery might have to the inner workings of the box-identifying machinery, but we didn't go into the details of how the various internal machine states would be hooked up to the printouts they caused. This was a deliberately simpleminded model of actual language production, useful only for making a very abstract thought-experimental point: if a sentence-emitting system had only limited access to its internal states, and a limited vocabulary with which to compose its sentences, its "reports" might be interpretable as true only if we impose on them a somewhat metaphorical reading. Shakey's "images" provided an example of how something that really wasn't an image at all could be the very thing one was actually talking about under the guise of an image.

It is one thing to open up an abstract possibility; it is another to show that this possibility has a realistic version that applies to us. What Shakey did wasn't real reporting, real saying. For all we could see, Shakey's imagined verbalization would be the sort of tricked-up, "canned" language that programmers build into user-friendly software. You go to format a diskette and your computer "asks" you a friendly question: "Are you sure you want to do this? It will erase everything on the disk! Answer Y or N." It would be a very naïve user who thought the computer actually *meant* to be so solicitous.

Let me put some words in the mouth of a critic. Since this particular imaginary critic will dog our discussions and investigations in later chapters, I will give him a name. Otto speaks:

> It was a cheap trick to call Shakey "he" rather than "it"; the trouble with Shakey is that it has no real insides like ours; there is nothing it is like to be it. Even if the machinery that took input from its TV camera "eye" and turned that input into box-identification had been strongly analogous to the machinery in *our* visual systems (and it wasn't), and even if the machinery that controlled its production of strings of English words had been strongly analogous to the machinery in *our* speech systems that controls the production of strings of English words (and it wasn't), there would *still* have been something missing: the Middleman in each of us whose judgments get expressed when we tell how it is with us. The problem with Shakey is that its input and output are attached to each other in the wrong way — a way that eliminates the observer (experiencer, enjoyer) that has to lie somewhere between the visual input and the verbal output, so that there is someone in there to *mean* Shakey's words when they are "spoken."
>
> When *I* speak, [Otto goes on] I mean what I say. My conscious life is private, but I can choose to divulge certain aspects of it to you. I can decide to tell you various things about my current or past experience. When I do this, I formulate sentences that I carefully tailor to the material I wish to report on. I can go back and forth between the experience and the candidate report, checking the words against the experience to make sure I have found *les mots justes*. Does this wine have a hint of *grapefruit* in its flavor, or does it seem to me more reminiscent of *berries?* Would it be more apt to say the higher tone sounded *louder*, or is it really just that it seems *clearer* or *better focused?* I attend to my particular conscious experience and arrive at a judgment about which words would do the most justice to its character. When I am satisfied that I have framed an accurate report, I express it. From my introspective report, you can come to know about some feature of my conscious experience.

As heterophenomenologists, we need to divide this text into two parts. We put to one side the claims about how the experience of speaking seems to Otto. These are inviolable; that *is* how the experience seems to Otto, and we must take that as a datum demanding an explanation. To the other side we put the theoretical claims (are they the conclusions of tacit arguments?) that Otto makes about what this shows about what is going on in him — and how it differs from what was going on in Shakey, for instance. These have no special standing, but we will treat them with the respect due all thoughtful claims.

It is all very well for me to insist that the Middleman, the Internal Observer in the Cartesian Theater, must be eliminated, not found, but we can't *just* throw him away. If there isn't a Central Meaner, where does the meaning come from? We must replace him with a plausible account of how a meant utterance — a *real* report, without any scare-quotes — could get composed without needing the *imprimatur* of a solitary Central Meaner. That is the main task of this chapter.

2. BUREAUCRACY VERSUS PANDEMONIUM

One of the skeletons in the closet of contemporary linguistics is that it has lavished attention on hearing but largely ignored speaking, which one might say was roughly half of language, and the most important half at that. Although there are many detailed theories and models of language *perception*, and of the *comprehension* of heard utterances (the paths from phonology, through syntax, to semantics and pragmatics), no one — not Noam Chomsky, and not any of his rivals or followers — has had anything very substantial (right or wrong) to say about systems of language *production*. It is as if all theories of art were theories of art *appreciation* with never a word about the artists who created it — as if all art consisted of *objets trouvés* appreciated by dealers and collectors.

It is not hard to see why this is so. Utterances *are* readily found objects with which to begin a process. It is really quite clear what the raw material or input to the perception and comprehension systems is: wave forms of certain sorts in the air, or strings of marks on various plane surfaces. And although there is considerable fog obscuring the controversies about just what the end product of the comprehension process is, at least this deep disagreement comes at the end of the process being studied, not the beginning. A race with a clear starting line can at least be rationally begun, even if no one is quite sure where it is going to end. Is the "output" or "product" of speech comprehension a *decoding* or *translation* of the input into a new representation — a sentence of Mentalese, perhaps, or a picture-in-the-head — or is it a set of *deep structures*, or some still unimagined entity? Linguists can decide to postpone an answer to that stumper while they work on the more peripheral parts of the process.

With speech production, on the other hand, since no one has yet worked out any clear and agreed-upon description of what initiates the process that eventually yields a full-fledged utterance, it is hard even to get started on a theory. Hard, but not impossible. There has been some good work on the issues of production quite recently, excellently surveyed and organized by the Dutch psycholinguist Pim Levelt, in *Speaking* (1989). Working backwards from the output, or working from the middle in both directions, we get some suggestive glimpses into the machinery that designs our utterances and gets them expressed. (The following examples are drawn from Levelt's discussion.)

Speech is not produced by a "batch process" that designs and executes one word at a time. The existence of at least a limited look-ahead capacity in the system is revealed by the way stress gets distributed in an utterance. A simple case: the stress in the word "sixteen" depends on context:

ANDY: How many dollars does it cost?
BOB: I think it's sixTEEN.
ANDY: SIXteen dollars isn't very much.

When Andy gives his second speech, he must adjust his pronunciation of "sixteen" to the word (DOLLars) that follows. Had he been going to say:

SixTEEN isn't very much.

he would have given the word a different stress pattern. Another example: notice how different the stress is on the two occurrences of "Tennessee" in

I drove from Nashville, TennesSEE, to the TENnessee border.

Spoonerisms and other speech errors show quite conclusively how lexical and grammatical distinctions are observed (and misobserved) in the course of designing an utterance to speak. People are more apt to say "barn door" when they mean "darn bore" than they are to say "bart doard" when they mean "dart board". There is a bias in favor of real (familiar) words over merely pronounceable (possible but not actual) words even when making a slip of the tongue. Some errors are suggestive about how word-selection mechanisms must operate: "The competition is a little stougher [stiffer/tougher]," and "I just put it in the oven at very low speed." And think of the transposition that must be involved in producing an error like "naming a wear tag" for "wearing a name tag."

Thanks to ingenious experiments that provoke such errors, and intricate analyses of what does and doesn't happen when people speak, progress is being made on models of the highly organized mechanisms that execute the ultimate articulation of a message once *it has been decided* that a particular message is to be released to the outside world. But who or what puts this machinery in motion? A speech error is an error in virtue of being other than what the speaker *meant to say*. What taskmaster *sets the task* relative to which errors such as the examples above are judged?

What, if not the Central Meaner? Levelt provides us with a picture, a "blueprint for the speaker":

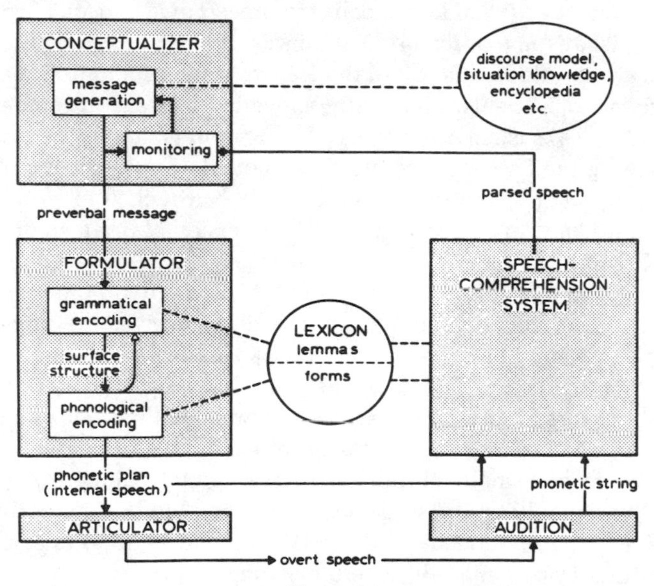

Figure 8.1

In the upper left-hand corner a functionary who looks suspiciously like the Central Meaner makes his appearance in the guise of the Conceptualizer, armed with lots of world knowledge, plans, and *communicative intentions*, and capable of "message generation." Levelt warns his readers that the Conceptualizer "is a reification in need of further explanation" (p. 9), but he posits it anyway, since he really can't get the process going, it seems, without some such unanalyzed Boss to give the marching orders to the rest of the team.

How does it work? The underlying problem will be clearer if we begin with a caricature. The Conceptualizer decides to perform a speech act, such as insulting his interlocutor by commenting adversely on the size of his feet. So he sends a command to the bureaucracy under his sway, the Public Relations Department (Levelt's Formulator): "Tell this bozo his feet are too big!" The PR people take on the job. They find the appropriate words: the second-person singular possessive pronoun, *your;* a good word for feet, such as *feet;* the right plural form of the verb *to be,* namely *are;* and the appropriate adverb and adjective: *too big.* These they cunningly combine, with the right insulting tone of voice, and execute:

"Your feet are too big!"

But wait a minute. Isn't that too easy? When the Conceptualizer gave the command (what Levelt calls the *preverbal message*), if he gave it in English, as my caricature just suggested, he's done all the hard work, leaving little for the rest of the team to do, except to pass it along with trivial adjustments. Then is the preverbal message in some other representational system or language? Whatever it is, it must be capable of providing the basic "specs" to the production team for the object they are to compose and release, and it must be couched in terms *they* can "understand" — not English but some version of Brainish or Mentalese. It will have to be in a sort of language of thought, Levelt argues, but perhaps in a language of thought that is used only to order speech acts, not for all cognitive activities. The team receives the preverbal message, a detailed Mentalese order to make an English utterance, and then it fills this order. This gives the subordinates a little more to do, but just obscures the looming regress. How does the Conceptualizer figure out which words of Mentalese to use to give the order? There had better not be a smaller duplicate of Levelt's whole blueprint hidden in the Conceptualizer's *message generation* box (and so on, *ad infinitum*). And certainly nobody told the Conceptualizer what to say; he's the Central Meaner, after all, where meaning *originates.*

How then does the meaning of an utterance develop? Consider the following nesting of commands, leading from grand overall strategy through detailed tactics to basic operations:

(1) Go on the offensive!
(2) Do something nasty but not too dangerous to him!
(3) Insult him!
(4) Cast aspersions on some aspect of his body!
(5) Tell him his feet are too big!

(6) Say: "Your feet are too big"!
(7) Utter: *yər FĪT är tū bĭg!*

Surely something like this zeroing-in on the final act must happen. Human speech is purposive activity; there are ends and means, and we somehow do a passable job of traversing the various options. We could have shoved him instead of insulting him, or belittled his intelligence instead of enlarging on his feet, or said, quoting Fats Waller, "Your pedal extremities are obnoxious!"

But does this zeroing-in all get accomplished by a bureaucratic hierarchy of commanders giving orders to subordinates? In this cascade of commands there appears to be a lot of decision-making — "moments" at which options are "selected" over their rivals, and this invites a model in which there is delegation of responsibility for finer details, and in which subordinate agents with their own intentions appreciate reasons for the various selections they make. (If they didn't have to understand at all why they were doing what they were doing, they wouldn't really be agents, but just passive, rubber-stamping functionaries letting whatever happened to cross their desks control them.)

Levelt's blueprint exhibits fossil traces of one of its sources: the von Neumann architecture that was inspired by Turing's reflections on his own stream of consciousness and that has inspired in turn many models in cognitive science. In chapter 7, I attempted to overcome resistance to the idea that human consciousness is *rather like* a von Neumann machine, a serial processor with a succession of definite contents reeling through the bottleneck of the accumulator. Now I must put on the brakes and emphasize some ways in which the functional architecture of human consciousness is *not* like that of a von Neumann machine. If we compare Levelt's blueprint to the way von Neumann machines standardly emit words, we can see that Levelt's model may borrow slightly too much.

When a von Neumann machine says what is written in its heart, it outputs the contents of its single central workplace, the accumulator, which at each instant has entirely specific contents in the fixed language of binary arithmetic. The rudimentary "preverbal messages" of a von Neumann machine look like this: 10110101 00010101 11101101. One of the primitive instructions in any machine language is an OUTPUT instruction, which can take the current contents of the accumulator (e.g., the binary number 01100001) and write it on the screen or the printer, so that an outside user can gain access to the results accomplished in the CPU. In a slightly more user-friendly variation, a routine operation composed of a series of primitive instructions can first translate the binary number into decimal notation (e.g., binary 00000110 = decimal 6) or into a letter of the alphabet via ASCII code (e.g., binary 01100001 = "a" and 01000001 = "A") and then output the result. These subroutines are at the heart of the fancier output instructions found in higher-level programming languages, like Fortran or Pascal or Lisp. These permit the programmer to create further subroutines for building larger messages, fetching long series of numbers from memory and running them through the accumulator, translating them and writing the results on the screen or printer. For instance, a subroutine can make several trips to the accumulator for values to plug into the blanks in

> You have overdrawn your account, Mr.______, by $______. Have a nice day, Mr. ______!

— a "canned" sentence formula that itself is kept stored as a series of binary numbers in the memory until some subroutine determines that it is time to open the can. In this manner, a strict hierarchy of fixed routines can turn sequences of specific contents in the accumulator into expressions that a human being can read on a screen or printer: "Do you want to save this document?" or "6 files copied" or "Hello, Billy, do you want to play tic-tac-toe?"

There are two features of this process that are shared by Levelt's model: (1) the process takes an already determinate content as its input, and (2) the bureaucracy — the "flow of control" in computer-science jargon — has to have been carefully designed: all "decision-making" flows hierarchically by a delegation of responsibility to subagents whose job descriptions dictate which bit of means/ends analysis they are authorized to perform. Interestingly enough, the first of these features — the determinate content — seems to be endorsed by Otto's view of his own processes: There is a determinate "thought" somewhere in the Center, waiting to be "put into words." The second shared feature, however, seems alien: The hierarchy of routines that slavishly render *that very thought* in natural language have been predesigned *by someone else* — by the programmer, in the case of the von Neumann machine, and presumably by the combination of evolution and individual development in the case of the activities in Levelt's Formulator. The creative, judgmental role that the thinker of the thought should play in getting the thought into words does not appear in the model; it is either usurped by the Conceptualizer, who does all the creative work before sending an order to the Formulator, or it is implicit in the design of the Formulator, a *fait accompli* of some earlier design process.

How else could ends and means be organized? Let's consider an opposing caricature: a pandemonium of word-demons. Here is how we talk: First we go into vocal noise-making mode — we turn on the horn:

> Beeeeeeeeeeeeeeeeeeeeeeeeeeeeeeeeeeeeeeep. . . .

We do this for no good reason, but just because no good reason not to do it occurs to us. The internal "noise" excites various demons in us who begin trying to modulate the horn in all sorts of random ways by interfering with its stream. The result is gibberish, but at least it's English gibberish (in English speakers):

> Yabba-dabba-doo-fiddledy-dee-tiddly-pom-fi-fi-fo-fum. . . .

But before any of this embarrassing stuff actually hits the outside world, further demons, sensitive to patterns in the chaos, start shaping it up into words, phrases, clichés . . .

> And so, how about that?, baseball, don't you know, in point of fact, strawberries, happenstance, okay? That's the ticket. Well, then . . .

which incites demons to make further serendipitous discoveries, augmented by opportunistic shaping, yielding longer bits of more acceptable verbiage, until finally a whole sentence emerges:

> I'm going to knock your teeth down your throat!

Fortunately, however, this gets set aside, unspoken, since at the same time (in parallel) other candidates have been brewing and are now in the offing, including a few obvious losers, such as

> You big meany!

and

> Read any good books lately?

and a winner by default, which gets spoken:

> Your feet are too big!

The muse has failed our speaker on this occasion; no witty retort made it to the finals, but at least something halfway appropriate to the speaker's current "mind-set" got blurted out. As the speaker walks away after the encounter, he will probably resume the chaotic tournament, muttering and musing about what he should have said. The muse may then descend with something better, and the speaker will savor it, turning it over and over in his mind, imagining the stricken look it would have provoked on the face of his interlocutor. By the time the speaker gets home, he may vividly "remember" having skewered his interlocutor with a razor-sharp witticism.

We can suppose that all of this happens in swift generations of "wasteful" parallel processing, with hordes of anonymous demons and their hopeful constructions never seeing the light of day — either as options that are *consciously* considered and rejected, or as ultimately executed speech acts for outsiders to hear. If given enough time, more than one of these may be silently tried out in a conscious rehearsal, but such a formal audition is a relatively rare event, reserved for occasions where the stakes are high and misspeaking carries heavy penalties. In the normal case, the speaker gets no preview; he and his audience learn what the speaker's utterance is at the same time.

But how is this tournament of words judged? When one word or phrase or whole sentence beats out its competitors, how does its suitability or appropriateness to the current mind-set get discriminated and valued? What *is* a mind-set (if not an explicit communicative intention), and how does its influence get conveyed to the tournament? For after all, even if there isn't a Central Meaner, there has to be some way for the content to get from deep inside the system — from perceptual processes, for instance — to verbal reports.

Let's review the issues. The problem with the bureaucratic extreme is that the Conceptualizer seems ominously powerful, a homunculus with too much knowledge and responsibility. This excess of power is manifested in the awkward problem of how to couch its output, the *preverbal message*. If it already specifies a speech act — if it already *is* a sort of speech act in Mentalese, a specific command to the Formulator — most of the hard work of composition has happened before our model kicks in. The problem with the Pandemonium alternative is that we need to find a way in which sources of content can *influence* or *constrain* the creative energies of the word-demons without *dictating* to them.

What about the process described in chapter 1, the rounds of question-answering that generated hallucinations on the model of the game of Psychoanalysis? Recall that we eliminated the wise Freudian dream-playwright and hallucination-producer by replacing him with a process from which content *emerged* under the incessant questioning of a questioner. The residual problem was how to get rid of the clever questioner, a problem we postponed. Here we have the complementary problem: how to get answers to an eager flock of contestants asking questions like "Why don't we say, 'Your mother wears army boots!' " or (in another context) "Why don't we say, 'I seem to see a red spot moving and turning green as it moves'?" *Two complementary problems — could they perhaps solve each other by being mated?* What if the word-demons are, in parallel, the questioners/contestants, and the content-demons are the answerers/judges? Fully fledged and executed communicative intentions — Meanings — could emerge from a quasi-evolutionary process of speech act design that involves the collaboration, partly serial, partly in parallel, of various subsystems none of which is capable on its own of performing — or ordering — a speech act.

Is such a process really possible? There are a variety of models of such "constraint satisfaction" processes, and they do indeed have striking powers. In addition to the various "connectionist" architectures of neuronlike elements (see, e.g., McClelland and Rumelhart, 1986), there are other more abstract models. Douglas Hofstadter's (1983) Jumbo architecture, which hunts for solutions to Jumbles or anagrams, has the right sorts of features, and so do Marvin Minsky's (1985) ideas about the Agents making up the "society of mind" — which will be discussed further in chapter 9. But we must reserve judgment until models that are more detailed, explicit, and directly aimed at language production are created and put through their paces. There may be surprises and disappointments.

We know, however, that somewhere in any successful model of language production we must avail ourselves of an evolutionary process of message generation, since otherwise we will be stuck with a miracle ("And then a miracle occurs") or an infinite regress of Meaners to set the task.[1] We also know — from the research Levelt surveys — that

1. Dan Sperber and Deirdre Wilson (1986) open up a new perspective on how we compose our communications by insisting on models of how things actually *work*, in the speaker and hearer, contrary to recent practice among philosophers and linguists, who have tended to wave their hands about the mechanisms while appealing to rational reconstructions of the supposed tasks and their demands. This permits Sperber and Wilson to raise considerations of practicality and efficiency: least-effort principles, and concerns about timing and probability. They then show from this new perspective how certain traditional "problems" disappear — in particular, the problem of how the hearer finds the "right" interpretation of what the speaker intended. Although they do not pitch

there are quite rigid and automatic processes that take over eventually and determine the grammatical-to-phonological transformations that compose the final muscular recipe for speech. The two caricatures define extremes along a continuum, from hyperbureaucratic to hyperchaotic. Levelt's actual model — in contrast to the caricature I have used in order to make the contrast vivid — incorporates (or can readily be made to incorporate) some of the nonbureaucratic features of the opposing caricature: for example, there is nothing deep or structural preventing Levelt's Formulator from engaging in more or less spontaneous (unrequested, undirected) language generation, and, given the monitoring loop through the Speech-Comprehension System back to the Conceptualizer (see Figure 8.1), this spontaneous activity *could* play the sort of generating role envisaged for the multiple word-demons. Between the two caricatures there is an intervening spectrum of more realistic ways alternative models could be developed. The main question is how much interaction is there between the specialists who determine the content and style of what is to be said and the specialists who "know the words and the grammar"?

At one extreme, the answer is: None. We could keep Levelt's model intact, and simply supplement it with a pandemonium model of what happens *inside* the Conceptualizer to fix the "preverbal message." In Levelt's model, there is nearly complete separation between the processes of message generation (specs-setting) and linguistic production (specs-meeting). When the first bit of preverbal message arrives at the Formulator, it triggers the production of the beginning of an utterance, and as the words get chosen by the Formulator, this constrains how the utterance can continue, but there is minimal *collaboration* on revision of the specs. The subordinate language-carpenters in the Formulator are, in Jerry Fodor's terms, "encapsulated"; in their automatic way, they do the best they can with the orders they receive, with no ifs, ands, or buts.

At the other extreme are the models in which words and phrases from the Lexicon, together with their sounds, meanings, and associations, jostle with grammatical constructions in a pandemonium, all "trying" to be part of the message, and some of them thereby make a substantial contribution to the very communicative intentions that still fewer of them end up executing. At this extreme, the communicative

---

their model at the level of evolutionary processes of the sorts we have just been considering, it certainly invites just such an elaboration.

intentions that exist are as much an effect of the process as a cause — they emerge as a product, and once they emerge, they are available as standards against which to measure *further* implementation of the intentions. There is not one source of meaning, but many shifting sources, opportunistically developed out of the search for the right words. Instead of a determinate content in a particular functional place, waiting to be Englished by subroutines, there is a still-incompletely-determined mind-set distributed around in the brain and constraining a composition process which in the course of time can actually feed back to make adjustments or revisions, further determining the expressive task that set the composition process in motion in the first place. There still is an overall pattern of serial passage, with concentration on one topic at a time, but the boundaries are not sharp lines.

In the Pandemonium model, control is usurped rather than delegated, in a process that is largely undesigned and opportunistic; there are multiple sources for the design "decisions" that yield the final utterance, and no strict division is possible between the marching orders of content flowing from within and the volunteered suggestions for implementation posed by the word-demons. What this brand of model suggests is that in order to *preserve* the creative role of the thought-*expresser* (something that mattered a good deal to Otto), we have to *abandon* the idea that the thought-*thinker* begins with a determinate thought to be expressed. This idea of determinate content also mattered a good deal to Otto, but something has to give (and section 4 will explore the alternatives more fully).

Where on the spectrum does the truth reside? This is an empirical question to which we do not yet know the answer.[2] There are some phenomena, however, that strongly suggest (to me) that language-

---

2. As Levelt notes, "If one could show, for instance, that message generation is directly affected by the accessibility of lemmas or word forms, one would have evidence for direct feedback from the Formulator to the Conceptualizer. This is an empirical question, and it is possible to put it to the test. . . . So far, the evidence for such feedback is negative" (p. 16). The evidence he reviews is from tightly controlled experiments in which a very specific task was given to the speaker: such as *describe the picture on the screen as fast as you can* (pp. 276–282). This is excellent negative evidence — I for one was surprised at the lack of effect in these experiments — but, as he recognizes, it is not at all conclusive. It is not really *ad hoc* to claim that the artificiality of these experimental situations successfully drowned out the opportunistic/creative dimension of language use. But perhaps Levelt is right; perhaps the only feedback from Formulator to Conceptualizer is *indirect:* the sort of feedback that a person can produce *only* by explicitly talking to himself and then framing an opinion about what he finds himself saying.

generation will turn out to involve Pandemonium — opportunistic, parallel, evolutionary processes — almost all the way down. The next section will review some of them briefly.

3. WHEN WORDS WANT TO GET THEMSELVES SAID

Whatever we may want to say, we probably won't say exactly *that*.

MARVIN MINSKY (1985), p. 236

The AI researchers Lawrence Birnbaum and Gregg Collins (1984) have noted a peculiarity about Freudian slips. Freud famously drew our attention to slips of the tongue that were not random or meaningless, he insisted, but deeply meaningful: unconsciously *intended* insertions into the fabric of discourse, insertions that indirectly or partially satisfied suppressed communicative goals of the speaker. This standard Freudian claim has often been vehemently rejected by skeptics, but there is something puzzling about its application to particular cases that has nothing to do with one's opinion about Freud's darker themes of sexuality, the Oedipus complex, or death wishes. Freud discussed an example in which a man said

Gentlemen, I call upon you to *hiccup* to the health of our Chief.

(In German — the language actually spoken in the example — the word for "hiccup," *aufzustossen*, was slipped in for the word for "drink," *anzustossen*.)

> In his explanation, Freud argues that this slip is a manifestation of an unconscious goal on the part of the speaker to ridicule or insult his superior, suppressed by the social and political duty to do him honor. However, . . . one cannot reasonably expect that the speaker's intention to ridicule his superior gave rise originally to a plan involving the use of the word "hiccup": *A priori*, there are hundreds of words and phrases that can more plausibly be used to insult or ridicule someone. . . . There is no way that a planner could have reasonably anticipated that the goal of ridiculing or insulting its superior would be satisfied by uttering the word "hiccup," for exactly the same reason that it is implausible that the

planner would have chosen to use the word as an insult in the first place.

The only process that could explain the frequency of serendipitous Freudian slips, they argue, is one of "opportunistic planning."

> . . . What examples like the above seem to indicate, therefore, is that the goals *themselves* are active cognitive agents, capable of commanding the cognitive resources needed to recognize opportunities to satisfy themselves, and the behavioral resources needed to take advantage of the opportunities. [Birnbaum and Collins, 1984, p. 125]

Freudian slips draw attention to themselves by seeming to be mistakes and not mistakes at the same time, but the fact (if it is one) that they satisfy *unconscious* goals does not make them any harder to explain than other word choices that fulfill several functions (or goals) at once. It is about as hard to imagine how puns and other forms of intended verbal humor could be the result of nonopportunistic, encapsulated planning and production. If anyone has a plan for designing witticisms — a detailed plan that actually works — there are more than a few comedians who would pay good money for it.[3]

If Birnbaum and Collins are right, creative language use can be accomplished only by a parallel process in which multiple *goals* are simultaneously on the alert for materials. But what if the materials themselves were at the same time on the alert for opportunities to get incorporated? We pick up our vocabulary from our culture; words and phrases are the most salient phenotypic features — the visible bodies — of the memes that invade us, and there could hardly be a more congenial medium in which memes might replicate than a language-production system in which the supervisory bureaucrats had partially abdicated, ceding a large measure of control to the words themselves, who in effect fight it out among themselves for a chance in the limelight of public expression.

---

3. Levelt tells me that he himself is an inveterate pun-hunter (in his native Dutch), and he knows just how he does it: "By lifelong training I turn around just about every word I hear. I then (quite consciously) check the result for its meaning. In 99.9 percent of the cases there is nothing funny coming out. But one per thousand is fine, and those I express right away" (personal communication). This is a perfect example of von Neumanesque problem-solving: serial, controlled — *and conscious!* The question is whether there are other, more pandemonic, ways of generating wit unconsciously.

It is no news that some of what we say we say primarily because we like the way it sounds, not because we like what it means. New slang sweeps through subcommunities, worming its way into almost everybody's speech, even those who try to resist it. Few of those who use a new word are deliberately or consciously following the schoolteacher's maxim "Use a new word three times and it's yours!" And at larger levels of aggregation, whole sentences appeal to us for the way they ring in our ears or trip off our tongues, quite independently of whether they meet some propositional specs we have already decided upon. One of the most quotable lines Abraham Lincoln ever came up with is:

> You can fool all the people some of the time, and some of the people all the time, but you can not fool all the people all of the time.[4]

What did Lincoln mean? Logic teachers are fond of pointing out that there is a "scope ambiguity" in the sentence. Did Lincoln mean to assert that there are some dunces who can always be fooled, or that on every occasion, someone or other is bound to be fooled — but not always the same people? Logically, these are entirely different propositions.

Compare:

> "Someone always wins the lottery."
> "It must be rigged!"
> "That's not what I meant."

Which reading did Lincoln intend? Maybe neither! What are the odds that Lincoln never noticed the scope ambiguity and never actually got around to having one communicative intention rather than "the other"? Perhaps it just sounded so good to him when he first formulated it that he never picked up the ambiguity, *and never had any prior communicative intention* — except the intention to say something pithy and well cadenced on the general topic of fooling people. People do talk that way, even great Meaners like Lincoln.

The fiction writer Patricia Hampl, in a thoughtful essay, "The Lax

---

4. According to *The Oxford Dictionary of Quotations* (second edition, 1953), this famous sentence is also attributed to Phineas T. Barnum. Since Barnum is an illustrious alumnus and generous benefactor of my university, I feel duty-bound to draw attention to the possibility that Lincoln may not be the originator of this highly replicative meme.

Habits of the Free Imagination," writes about her own process of composing short stories.

> Every story has a story. This secret story, which has little chance of getting told, is the history of its creation. Maybe the "story of the story" *can* never be told, for a finished work consumes its own history, renders it obsolete, a husk. [Hampl, 1989, p. 37]

The finished work, she notes, is readily interpretable by critics as an artifact cunningly contrived to fulfill a host of sophisticated authorial intentions. But when she encounters these hypotheses about her own work, she is embarrassed:

> "Hampl" had precious few intentions, except, like the charlatan I suddenly felt myself to be, to filch whatever was loose on the table that suited my immediate purposes. Worse, the "purposes" were vague, inconsistent, reversible, under pressure. And who — or what — was applying the pressure? I couldn't say. [p. 37]

How then does she do it? She suggests a maxim: "Just keep talking — mumbling is fine." Eventually, the mumbling takes on shapes that meet with the approval of the author. Could it be that the process Hampl detects on a grand scale in her creative writing is just an enlargement of the more submerged and swift process that produces the creative speaking of everyday life?

The tempting similarity does not involve just a process but also a subsequent attitude or reaction. Hampl's confessional zeal contrasts with a more normal — and not really dishonest — reaction of authors to friendly interpretations by readers: these authors defer gracefully to the imputations of intent, and even willingly elaborate on them, in the spirit of "Hey, I guess that *is* what I was up to, all along!" And why not? Is there anything self-contradictory in the reflection that a certain move one has just made (in chess, in life, in writing) is actually cleverer than one at first realized? (For further reflections on this topic, see Eco, 1990.)

As E. M. Forster put it, "How do I know what I think until I see what I say?" We often do discover what we think (and hence what we mean) by reflecting on what we find ourselves saying — and not correcting. So we are, at least on those occasions, in the same boat as our external critics and interpreters, encountering a bit of text and putting the best reading on it that we can find. The fact that we said it gives it a certain personal persuasiveness or at least a presumption of authenticity. *Probably*, if I said it (and I heard myself say it, and I didn't hear myself rushing in with any amendments), I meant it, and it probably means what it seems to mean — to me.

Bertrand Russell's life provides an example:

> It was late before the two guests left and Russell was alone with Lady Ottoline. They sat talking over the fire until four in the morning. Russell, recording the event a few days later, wrote, "I did not know I loved you till I heard myself telling you so — for one instant I thought 'Good God, what have I said?' and then I knew it was the truth." [Clark, 1975, p. 176]

What about the other occasions, though, where we have no such sense of a *discovery* of self-interpretation? We might suppose that in these, the normal, cases, we have some intimate and privileged advance insight into what we mean, just because we ourselves are the Meaners, the *fons et origo* of the meaning of the words *we* say, but such a supposition requires a supporting argument, not just an appeal to tradition. For it could as well be the case that we have no sense of discovery in these cases just because it is so obvious to us what we mean. It doesn't take "privileged access" to intuit that when I say, "Please pass the salt" at the dinner table, I'm asking for the salt.

I used to believe there was no alternative to a Central Meaner, but I thought I had found a safe haven for it. In *Content and Consciousness* I argued that there had to be a functionally salient line (which I called the awareness line) separating the preconscious fixation of communicative intentions from their subsequent execution. The location of this line in the brain might be horrendously gerrymandered, anatomically, but it had to exist, logically, as the watershed dividing malfunctions into two varieties. Errors could occur anywhere in the whole system, but every error had to fall — by geometric necessity — on one side of the line or the other. If they fell on the execution side of the line, they were (correctable) errors of *expression*, such as slips of the tongue, malapropisms, mispronunciations. If they fell on the inner or higher side of the line, they *changed that which was to be expressed* (the "preverbal message" in Levelt's model). Meaning was fixed at this watershed; that's where meaning came from. There had to be such a place where meaning came from, I thought, since *something* has to set the standard against which "feedback" can register failure to execute.

My mistake was falling for the very same scope ambiguity that bedevils the interpretation of Abe Lincoln's dictum. There does indeed have to be something on each occasion that is, for the nonce, the standard against which any corrected "error" gets corrected, but there doesn't have to be the same single thing each time — even *within* the duration of a single speech act. There doesn't have to be a *fixed* (if gerrymandered) line that marks this distinction. In fact, as we saw in chapter 5, the distinction between pre-experiential revisions that *change that which was experienced* and post-experiential revisions that have the effect of *misreporting or misrecording what was experienced* is indeterminate in the limit. Sometimes subjects are moved to revise or amend their assertions, and sometimes they aren't. Sometimes when they do make revisions, the edited narrative is no closer to "the truth" or to "what they *really* meant" than the superseded version. As we noted earlier, where prepublication editing leaves off and postpublication errata-insertion cuts in is a distinction that can be drawn only arbitrarily. When we put a question to a subject about whether or not a particular public avowal adequately captures the ultimate inner truth about what he was just experiencing, the subject is in no better position to judge than we outsiders are. (See also Dennett, 1990d.)

Here is another way of looking at the same phenomenon. Whenever the process of creating a verbal expression occurs, there is at the outset a distance that must be eliminated: the "mismatch distance in semantic space," we might call it, between the content that is in position to be expressed and the various candidates for verbal expression that are initially nominated. (In my old view, I treated this as a problem of simple "feedback correction," with a *fixed point* for a standard against which verbal candidates were to be measured, discarded, improved.) The back-and-forth process that narrows the distance is a feedback process of sorts, but it is just as possible for the content-to-be-expressed to be adjusted in the direction of some candidate expression, as for the candidate expression to be replaced or edited so better to accommodate the content-to-be-expressed. In this way, the most accessible or available words and phrases could actually *change the content of the experience* (if we understand the experience to be what is ultimately reported — the settled event in the heterophenomenological world of the subject).[5]

If our unity as Meaners is no better guaranteed than this, then in

5. This is reminiscent of Freud's view of how the "preconscious" works: "The question, 'How does a thing become conscious?' would be more advantageously stated: 'How does a thing become preconscious?', and the answer would be: 'Through becoming connected with the word-presentations corresponding to it' " (*The Ego and the Id*, English edition, 1962, p. 10).

principle it ought to be possible for it to become shattered on rare occasions. Here are two cases in which that seems to have happened.

I was once importuned to be the first base umpire in a baseball game — a novel duty for me. At the crucial moment in the game (bottom of the ninth, two outs, the tying run on third base), it fell to me to decide the status of the batter running to first. It was a close call, and I found myself emphatically jerking my thumb up — the signal for OUT — while yelling "SAFE!" In the ensuing tumult I was called upon to say what I had meant. I honestly couldn't say, at least not from any privileged position. I finally decided (to myself) that since I was an unpracticed hand-signaler but competent word-speaker, my vocal act should be given the nod, but anyone else could have made just the same judgment. (I would be happy to learn of other anecdotes in which people have not known which of two very different speech acts they had meant to perform.)

In an experimental setting, the psychologist Tony Marcel (in press) has found an even more dramatic case. The subject, who suffers from blindsight (about which I will say more in chapter 11), was asked to *say* whenever he thought there was a flash of light, but he was given peculiar instructions about how he was to do this. He was instructed to perform this single speech act by three distinct acts at once (not in sequence, but not necessarily "in unison" either):

(1) saying "Yes"
(2) pressing a button (the YES button)
(3) blinking YES

What is startling is that the subject didn't always perform all three together. Occasionally he blinked YES but didn't say YES or button-push YES, and so forth. There was no straightforward way of ordering the three different responses, either for fidelity to intention or for accuracy. That is, when there were disagreements among the three actions, the subject had no pattern to follow about which act to accept and which to count as a slip of the tongue, finger, or eyelid.

Whether similar findings can be provoked under other conditions with other subjects, normal or otherwise, remains to be seen, but other pathological conditions also suggest a model of speech production in which verbalization *can* be set in motion without any marching orders from a Central Meaner. If you suffer from one of these pathologies, "your mind is on vacation, but your mouth is working overtime," as the Mose Allison song puts it.

Aphasia is loss or damage of the ability to speak, and several different varieties of aphasia are quite common and have been extensively studied by neurologists and linguists. In the most common variety, Broca's aphasia, the patient is acutely aware of the problem, and struggles, with mounting frustration, to find the words that are just beyond the tip of her tongue. In Broca's aphasia, the existence of thwarted communicative intentions is painfully clear to the patient. But in a relatively rare variety of aphasia, jargon aphasia, patients seem to have no anxiety at all about their verbal deficit.[6] Even though they are of normal intelligence, and not at all psychotic or demented, they seem entirely content with such verbal performances as these (drawn from two cases described by Kinsbourne and Warrington, 1963):

Case 1:

How are you today?

"Gossiping O.K. and Lords and cricket and England and Scotland battles. I don't know. Hypertension and two won cricket, bowling, batting, and catch, poor old things, cancellations maybe gossiping, cancellations, arm and argument, finishing bowling."

What is the meaning of "safety first"?

"To look and see and the Richmond Road particularly, and look traffic and hesitation right and strolling, very good cause, maybe, zebras maybe these, motor-car and the traffic light."

Case 2:

Did you work in an office?

"I did work in an office."

And what kind of firm was it?

"Oh, as an executive of this, and the complaint was to discuss the tonations as to what type they were, as to how they were typed, and kept from the different . . . tricu . . . tricula, to get me from the attribute convenshments . . . sorry . . ."

"She wants to give one the subjective vocation to maintain the vocation of perfect impregnation simbling."

"Her normal corrucation would be a dot."

---

6. Levelt tells me that research underway at the Max Planck Institute for Psycholinguistics in Nijmegen casts doubt on this, the received view. Work by Heeschen suggests that at some level jargon or Wernicke's aphasics do have anxiety about their deficit, and seem to adopt a strategy of repetition, in hopes of achieving communication.

asked to identify a nail file:
"That is a knife, a knife tail, a knife, stale, stale knife."
and scissors:
"Groves — it's a groves — it's not really a groves — two groves containing a comb — no, not a comb — two groves providing that the commandant is not now — "

A strangely similar condition, and much more common, is *confabulation*. In chapter 4, I suggested that normal people may often confabulate about details of their own experience, since they are prone to guess without realizing it, and mistake theorizing for observing. Pathological confabulation is unwitting fiction of an entirely different order. Often in cases of brain damage, especially when people have terrible memory loss — as in Korsakoff's syndrome (a typical sequel of severe alcoholism) — they nevertheless proceed to prattle on with utter falsehoods about their lives and their past histories, and even about events of the last few minutes, if their amnesia is severe.

The resulting verbiage sounds virtually normal. It often sounds, in fact, just like the low-yield, formulaic chitchat that passes for conversation in a bar: "Oh, yes, my wife and I — we've lived in the same house for thirty years — used to go out to Coney Island, and, you know, sit on the beach — *loved* to sit on the beach, just watching the young people, and, but that was before the accident . . ." — except that it is made up out of whole cloth. The man's wife may have died years ago, never been within a hundred miles of Coney Island, and they may have moved from apartment to apartment. An uninitiated listener can often be entirely unaware that he is encountering a confabulator, so natural and "sincere" are the reminiscences and the ready answers to questions.

Confabulators have no idea they are making it all up, and jargon aphasics are oblivious to the fact that they are spouting word-salad. These stunning anomalies are instances of *anosognosia*, or inability to acknowledge or recognize a deficit. Other varieties of this absence of self-monitoring exist, and in chapter 11 we will consider what they have to tell us about the functional architecture of consciousness. In the meantime, we can note that the brain's machinery is quite able to construct apparent speech acts in the absence of any coherent direction from on high.[7]

7. Another anomalous linguistic phenomenon is the familiar symptom of schizophrenia: "hearing voices." It is now quite firmly established that the voice the schizophrenic "hears" is his own; he is talking to himself silently without realizing it. As simple an obstacle as having the patient hold his mouth wide open is sufficient to stop

Pathology, either the temporary strain induced by clever experiments, or the more permanent breakdowns caused by illness or mechanical damage to the brain, provides an abundance of clues about how the machinery is organized. These phenomena suggest to me that our second caricature, Pandemonium, is closer to the truth than a more dignified, bureaucratic model would be, but this has yet to be put to the proper empirical test. I am not claiming that it is impossible for a largely bureaucratic model to do justice to these pathologies, but just that they would not seem to be the natural failings of such a system. In Appendix B, for scientists, I will mention some research directions that could help confirm or disconfirm my hunch.

What I have sketched in this chapter — but certainly not proven — is a way in which a torrent of verbal products emerging from thousands of word-making demons in temporary coalitions could exhibit a unity, the unity of an evolving best-fit interpretation, that makes them appear *as if* they were the executed intentions of a Conceptualizer — and indeed they are, but not of an *inner* Conceptualizer that is a proper part of the language-producing system, but of the global Conceptualizer, the person, of which the language-producing system is itself a proper part.

This idea may seem alien at first, but it should not surprise us. In biology, we have learned to resist the temptation to explain *design in organisms* by positing a single great Intelligence that does all the work. In psychology, we have learned to resist the temptation to explain *seeing* by saying it is just as if there were an internal screen-watcher, for the internal screen-watcher does all the work — the only thing between such a homunculus and the eyes is a sort of TV cable. We must build up the same resistance to the temptation to explain *action* as arising from the imperatives of an internal action-orderer who does too much of the specification work. As usual, the way to discharge an intelligence that is too big for our theory is to replace it with an ultimately mechanical fabric of semi-independent semi-intelligences acting in concert.

This point applies not just to speech act generation; it applies to intentional action across the board. (See Pears, 1984, for a development of similar ideas.) And contrary to some first appearances, phenomenology actually assists us in seeing that this is so. Although we are occasionally conscious of performing elaborate practical reasoning, leading to a conclusion about what, all things considered, we ought to

---

the voices (Bick and Kinsbourne, 1987). See also Hoffman (1986), and the commentary by Akins and Dennett, "Who May I Say Is Calling?" (1986).

do, followed by a conscious decision to do that very thing, and culminating finally in actually doing it, these are relatively rare experiences. Most of our intentional actions are performed without any such preamble, and a good thing, too, since there wouldn't be time. The standard trap is to suppose that the relatively rare cases of conscious practical reasoning are a good model for the rest, the cases in which our intentional actions emerge from processes into which we have no access. Our actions generally satisfy us; we recognize that they are in the main coherent, and that they make appropriate, well-timed contributions to our projects as we understand them. So we safely assume them to be the product of processes that are reliably sensitive to ends and means. That is, they are rational, in *one* sense of that word (Dennett, 1987a, 1991a). But that does not mean they are rational in a narrower sense: the product of serial reasoning. We don't have to explain the underlying processes on the model of an internal reasoner, concluder, decider who methodically matches means to ends and then orders the specified action; we have seen in outline how a different sort of process could control speaking, and our other intentional actions as well.

Slowly but surely, we are shedding our bad habits of thought, and replacing them with other habits. The demise of the Central Meaner is more generally the demise of the Central Intender, but the Boss still lives on in other disguises. In chapter 10 we will encounter him in the roles of the Observer and Reporter, and will have to find other ways of thinking about what is going on, but first we must secure the foundations of our new habits of thought by tying them more closely to some scientific details.

9

THE ARCHITECTURE OF THE HUMAN MIND

1. WHERE ARE WE?

The hardest part is over, but there is plenty of work still to do. We have now completed the most strenuous exercises of imagination-stretching, and are ready to try out our newfound perspective. Along the way we had to leave several topics dangling, and tolerated quite a lot of handwaving. There are promises to keep, and postponed acknowledgments and comparisons to make. The theory I have been developing includes elements drawn from many thinkers. I have sometimes deliberately ignored what these thinkers consider the best parts of their theories, and have mixed together ideas drawn from "hostile" camps, but I suppressed these messy details in the interests of clarity and vividness. This may well have left some serious mind-modelers squirming with frustration, but I couldn't see any other way to get different kinds of readers up to the same new vantage point together. Now, though, we're in a good position to take stock, and secure some essential details. The point of going to all the trouble of constructing a new perspective is, after all, to see the phenomena and the controversies in a new way. So let's take a look around.

In a Thumbnail Sketch, here is my theory so far:

> There is no single, definitive "stream of consciousness," because there is no central Headquarters, no Cartesian Theater where "it all comes together" for the perusal of a Central Meaner. Instead of such a single stream (however wide), there are multiple channels in which specialist circuits try, in parallel pandemoniums,

to do their various things, creating Multiple Drafts as they go. Most of these fragmentary drafts of "narrative" play short-lived roles in the modulation of current activity but some get promoted to further functional roles, in swift succession, by the activity of a virtual machine in the brain. The seriality of this machine (its "von Neumannesque" character) is not a "hard-wired" design feature, but rather the upshot of a succession of coalitions of these specialists.

The basic specialists are part of our animal heritage. They were not developed to perform peculiarly human actions, such as reading and writing, but ducking, predator-avoiding, face-recognizing, grasping, throwing, berry-picking, and other essential tasks. They are often opportunistically enlisted in new roles, for which their native talents more or less suit them. The result is not bedlam only because the trends that are imposed on all this activity are themselves the product of design. Some of this design is innate, and is shared with other animals. But it is augmented, and sometimes even overwhelmed in importance, by microhabits of thought that are developed in the individual, partly idiosyncratic results of self-exploration and partly the predesigned gifts of culture. Thousands of memes, mostly borne by language, but also by wordless "images" and other data structures, take up residence in an individual brain, shaping its tendencies and thereby turning it into a mind.

This theory has enough novelty to make it hard to grasp at first, but it draws on models developed by people in psychology, neurobiology, Artificial Intelligence, anthropology — and philosophy. Such unabashed eclecticism is often viewed askance by the researchers in the fields from which it borrows. As a frequent interloper in these fields, I have grown accustomed to the disrespect expressed by some of the participants for their colleagues in the other disciplines. "Why, Dan," ask the people in Artificial Intelligence, "do you waste your time conferring with those neuroscientists? They wave their hands about 'information processing' and worry about *where* it happens, and which neurotransmitters are involved, and all those boring facts, but they haven't a clue about the computational requirements of higher cognitive functions." "Why," ask the neuroscientists, "do you waste your time on the fantasies of Artificial Intelligence? They just invent whatever machinery they want, and say unpardonably ignorant things about the brain." The cognitive psychologists, meanwhile, are accused of concocting models with *neither* biological plausibility *nor* proven computational powers; the anthropologists wouldn't know a model if they saw one, and the philosophers, as we all know, just take in each other's laundry, warning about confusions they themselves have created, in an arena bereft of both data and empirically testable theories. With so many idiots working on the problem, no wonder consciousness is still a mystery.

All these charges are true, and more besides, but I have yet to encounter any idiots. Mostly the theorists I have drawn from strike me as very smart people — even brilliant people, with the arrogance and impatience that often comes with brilliance — but with limited perspectives and agendas, trying to make progress on hard problems by taking whatever shortcuts *they* can see, while deploring other people's shortcuts. No one can keep all the problems and details clear, including me, and everyone has to mumble, guess, and handwave about large parts of the problem.

For instance, one of the occupational hazards of neuroscience seems to be the tendency to think of consciousness as *the end of the line*. (This is like forgetting that the end product of apple trees is not apples — it's more apple trees.) Of course it is only recently that neuroscientists have permitted themselves to think about consciousness at all, and only a few brave theorists have begun to speak, officially, about what they have been thinking. As the vision researcher Bela Julesz recently quipped, you can really only get away with it if you have white hair — and a Nobel Prize! Here, for instance, is a hypothesis hazarded by Francis Crick and Christof Koch:

> We have suggested that one of the functions of consciousness is to present the result of various underlying computations and that this involves an attentional mechanism that temporarily binds the relevant neurons together by synchronizing their spikes in 40 hz oscillations. [Crick and Koch, 1990, p. 272]

So a function of consciousness is to *present the results of underlying computations* — but to whom? The Queen? Crick and Koch do not go on to ask themselves the Hard Question: *And then what happens?* ("And then a miracle occurs"?) Once their theory has shepherded something into what they consider to be the charmed circle of consciousness, it stops. It doesn't confront the problems we addressed in chapters 5 through 8, for instance, about the tricky path from (presumed) consciousness to behavior, including, especially, introspective reports.

Models of the mind offered in cognitive psychology and AI, in contrast, almost never suffer from *this* defect (see, e.g, Shallice, 1972, 1978; Johnson-Laird, 1983, 1988; Newell, 1990). They generally posit a "workspace" or "working memory" that replaces the Cartesian Theater, and the models show how the results of computations carried out there feed into further computations that guide behavior, inform verbal reports, double back recursively to provide new input to working memory, and so forth. But these models typically don't say where or how a working memory might be located in the brain, and are so concerned with the *work* being done in that workspace that there is no time for "play" — no sign of the sort of *delectation* of phenomenology that seems such an important feature of human consciousness.

Curiously, then, neuroscientists often end up looking like dualists, since once they have "presented" things in consciousness, they seem to pass the buck to the Mind, while cognitive psychologists often end up looking like zombists (automatists?), since they describe structures unknown to neuroanatomists, and their theories purport to show how all the work can get done without having to ring in any Inner Observer.

Appearances are misleading. Crick and Koch aren't dualists (even if they are, apparently, Cartesian materialists), and the cognitive psychologists haven't *denied* the existence of consciousness (even if they do their best, most of the time, to ignore it). Furthermore, these blinkered approaches disqualify neither enterprise. The neuroscientists are right to insist that you don't really have a good model of consciousness until you solve the problem of where it fits in the brain, but the cognitive scientists (the AIers and the cognitive psychologists, for instance) are right to insist that you don't really have a good model of consciousness until you solve the problem of what functions it performs and how it performs them — mechanically, without benefit of Mind. As Philip Johnson-Laird puts it, "Any scientific theory of the mind has to treat it as an automaton" (Johnson-Laird, 1983, p. 477). The limited perspective of each enterprise taken by itself just shows us the need for another enterprise — the one we are engaged in — that tries to put together as many as possible of the strengths of each.

2. ORIENTING OURSELVES WITH THE THUMBNAIL SKETCH

My main task in this book is philosophical: to show how a genuinely explanatory theory of consciousness *could* be constructed out of these parts, not to provide — and confirm — such a theory in all its details. But my theory would have been inconceivable (by me, at least) if it had not borrowed heavily from empirical work in various fields which opened up (to me, at least) new ways of thinking. (A particularly rich collection of empirical findings and new ideas about consciousness is Marcel and Bisiach, 1988.) This is a glorious time to be involved in research on the mind. The air is thick with new discoveries, new models, surprising experimental results — and roughly equal measures of oversold "proofs" and premature dismissals. At this time, the frontier of research on the mind is so wide open that there is almost no settled wisdom about what the right questions and methods are. With so many underdefended fragments of theory and speculation, it is a good idea to postpone our demand for proof and look instead for more or less independent but also inconclusive grounds that tend to converge in support of a single hypothesis. We should try to keep our enthusiasm in check, however. Sometimes what seems to be enough smoke to guarantee a robust fire is actually just a cloud of dust from a passing bandwagon.

In his book *A Cognitive Theory of Consciousness* (1988), the psychologist Bernard Baars summarizes what he sees as a "gathering consensus" that consciousness is accomplished by a "distributed society of specialists that is equipped with a working memory, called a *global workspace*, whose contents can be broadcast to the system as a whole" (p. 42). As he notes, a variety of theorists, in spite of enormous differences in perspective, training, and aspiration, are gravitating toward this shared vision of how consciousness must sit in the brain. It is a version of that emerging consensus that I have been gingerly introducing, ignoring some features and emphasizing others — features that I think are either overlooked or underestimated, and that I think are particularly crucial for breaking through the conceptual mysteries that still remain.

In order to orient my theory in relation to some of the mountains of work it has borrowed from, let's go back through my thumbnail sketch, one theme at a time, drawing parallels and noting sources and disagreements.

> There is no single, definitive "stream of consciousness," because there is no central Headquarters, no Cartesian Theater where "it all comes together" for the perusal of a Central Meaner. . . .

While everyone agrees that there is no such single point in the brain, reminiscent of Descartes's pineal gland, the implications of this have not been recognized, and are occasionally egregiously overlooked. For instance, incautious formulations of "the binding problem" in current neuroscientific research often presuppose that there must be some single representational space in the brain (smaller than the whole brain) where the results of all the various discriminations are put into registration with each other — marrying the sound track to the film, coloring in the shapes, filling in the blank parts. There are some careful formulations of the binding problem(s) that avoid this error, but the niceties often get overlooked.

> . . . Instead of such a single stream (however wide), there are multiple channels in which specialist circuits try, in parallel pandemoniums, to do their various things, creating Multiple Drafts as they go. Most of these fragmentary drafts of "narrative" play short-lived roles in the modulation of current activity. . . .

In AI, the importance of narrativelike sequences has long been stressed by Roger Schank, first in his work on *scripts* (1977, with Abelson), and more recently (1991) in his work on the role of story-telling in comprehension. From very different perspectives, still within AI, Patrick Hayes (1979), Marvin Minsky (1975), John Anderson (1983), and Erik Sandeval (1991), — and others — have argued for the importance of data structures that are not just sequences of "snapshots" (with the attendant problem of reidentifying particulars in successive frames) but are instead specifically designed in one way or another to represent temporal sequences and sequence types directly. In philosophy, Gareth Evans (1982) had begun developing some parallel ideas before his untimely death. In neurobiology, these narrative fragments are explored as *scenarios* and other sequences in William Calvin's (1987) Darwin Machine approach. Anthropologists have long maintained that the myths each culture transmits to its new members play an important role in shaping their minds (see, e.g., Goody, 1977, and for a suggested application to AI, Dennett, 1991b), but they have made no attempt to model this, either computationally or neuroanatomically.

> . . . but some get promoted to further functional roles, in swift succession, by the activity of a virtual machine in the brain. The seriality of this machine (its "von Neumannesque" character) is not a "hard-wired" design feature, but rather the upshot of a succession of coalitions of these specialists. . . .

Many have remarked on the relatively slow, awkward pace of conscious mental activity (e.g., Baars, 1988, p. 120), and the suggestion has long lurked that this might be because the brain was not really designed — hard-wired — for such activity. The idea has been around for several years that human consciousness might then be the activity of some sort of serial virtual machine implemented on the parallel hardware of the brain. The psychologist Stephen Kosslyn offered a version of the serial virtual machine idea at a meeting of the Society for Philosophy and Psychology in the early 1980s, and I have been trying out different versions of the idea since about the same time (e.g., Dennett, 1982b), but an earlier presentation of much the same idea — though without using the term 'virtual machine' — is in the psychologist Paul Rozin's seminal paper, "The Evolution of Intelligence and Access to the Cognitive Unconscious" (1976). Another psychologist, Julian Jaynes, in his boldly original speculations in *The Origins of Consciousness in the Breakdown of the Bicameral Mind* (1976), stressed that human consciousness was a very recent and culture-borne imposition on an earlier functional architecture, a theme also developed in other ways by the neuroscientist, Harry Jerison (1973). The underlying neural architecture is far from being a *tabula rasa* or blank slate at birth, according to this view, but it is nevertheless a medium in which structures get built as a function of the brain's interactions with the world. And it is these built structures, more than the innate structures, that must be cited in order to explain cognitive functioning.

> The basic specialists are part of our animal heritage. They were not developed to perform peculiarly human actions, such as reading and writing, but ducking, predator-avoiding, face-recognizing, grasping, throwing, berry-picking, and other essential tasks. . . .

These hordes of specialists are attested to by very different theories, but their sizes, roles, and organization are hotly debated. (For a useful swift survey, see Allport, 1989, pp. 643–647.) Neuroanatomists studying the brains of animals ranging from sea slugs and squids to cats to monkeys have identified many varieties of hard-wired circuits exquisitely designed to perform particular tasks. Biologists speak of Innate Releasing Mechanisms (IRMs) and Fixed Action Patterns (FAPs), which can get yoked together, and in a recent letter to me, the neuropsychologist Lynn Waterhouse aptly described the minds of animals as being composed of "quilts of IRM-FAPs." It is just such problematically quilted animal minds that Rozin (along with others) presupposes as the basis for the evolution of more general-purpose minds, which exploit these pre-existing mechanisms for novel purposes. The perceptual psychologist V. S. Ramachandran (1991), notes that "there is an actual advantage which appears in multiple systems: it buys you tolerance for noisy images of the kind that you encounter in the real world. My favorite analogy to illustrate some of these ideas is that it is a bit like two drunks; neither of them can walk unsupported but by leaning on each other they manage to stagger towards their goal."

The neuropsychologist Michael Gazzaniga has pointed to a wealth of data arising from neurological deficits (including the famous, but often misdescribed, split-brain patients) that supports a view of the mind as a coalition or bundle of semi-independent agencies (Gazzaniga and Ledoux, 1978; Gazzaniga, 1985); and, coming from a different quarter, the philosopher of psychology Jerry Fodor (1983) has argued that large parts of the human mind are composed of *modules*: hard-wired, special-purpose, "encapsulated" systems of input analysis (and output generation — though he has not had much to say about this).

Fodor concentrates on modules that would be specific to the human mind — modules for acquiring languages and parsing sentences, in particular — and since he largely ignores the issue of what their probable ancestors might be in the minds of lower animals, he creates the improbable impression of evolution having designed brand-new species-specific mechanisms for language out of whole cloth, as it were. This image of these modules as a miraculous gift from Mother Nature to *Homo sapiens* is encouraged by Fodor's ultra-intellectualist vision of how modules are attached to the rest of the mind. According to Fodor, they do not carry out whole tasks in the economy of the mind (such as controlling a hand-eye coordination for picking up something), but stop abruptly at an internal edge, a line in the mind over which they cannot step. There is a central arena of rational "belief fixation," Fodor claims, into which the modules slavishly deposit their goods, turning them over to *non*modular ("global, isotropic") processes.

Fodor's modules are a bureaucrat's dream: their job descriptions are carved in stone; they cannot be enlisted to play novel or multiple roles; and they are "cognitively impenetrable" — which means that their activities cannot be modulated, or even interrupted, by changes in the "global" informational states of the rest of the system. According to Fodor, all the really thoughtful activities of cognition are nonmodular. Figuring out what to do next, reasoning about hypothetical situations, restructuring one's materials creatively, revising one's world view — all these activities are performed by a mysterious central facility. Moreover, Fodor claims (with curious satisfaction) that no branch of cognitive science — including philosophy — has any clue about how this central facility does its work!

> A lot is known about the transformations of representations which serve to get information into a form appropriate for central

> processing; practically nothing is known about what happens after the information gets there. The ghost has been chased further back into the machine, but it has not been exorcised. [Fodor, 1983, p. 127]

By giving this central facility so much to do, and so much non-modular power with which to do it, Fodor turns his modules into very implausible agents, agents whose existence only makes sense in the company of a Boss agent of ominous authority (Dennett, 1984b). Since one of Fodor's main points in describing modules has been to contrast their finite, comprehensible, mindless mechanicity with the unlimited and inexplicable powers of the nonmodular center, theorists who would otherwise be receptive to at least most of his characterization of modules have tended to dismiss his modules as the fantasies of a crypto-Cartesian.

Many of the same theorists have been lukewarm-to-hostile about Marvin Minsky's *Agents*, who form *The Society of Mind* (1985). Minsky's Agents are homunculi that come in all sizes, from giant specialists with talents about as elaborate as those of Fodorian modules, down to meme-sized agents (polynemes, micronemes, censor-agents, suppressor-agents, and many others). It all looks too easy, the skeptics think. Wherever there is a task, posit a gang of task-sized agents to perform it — a theoretical move with all the virtues of theft over honest toil, to adapt a famous put-down of Bertrand Russell's.

Homunculi — demons, agents — are the coin of the realm in Artificial Intelligence, and computer science more generally. Anyone whose skeptical back is arched at the first mention of homunculi simply doesn't understand how neutral the concept can be, and how widely applicable. Positing a gang of homunculi would indeed be just as empty a gesture as the skeptic imagines, if it were not for the fact that in homunculus theories, the serious content is in the claims about how the posited homunculi interact, develop, form coalitions or hierarchies, and so forth. And here the theories can be very different indeed. Bureaucratic theories, as we saw in chapter 8, organize homunculi into predesigned hierarchies. There are no featherbedding or disruptive homunculi, and competition between homunculi is as tightly regulated as major league baseball. Pandemonium theories, in contrast, posit lots of duplication of effort, waste motion, interference, periods of chaos, and layabouts with no fixed job description. Calling the units in these very different theories homunculi (or demons or agents) is scarcely more contentful than calling them simply . . . units. They are just units with particular circumscribed competences, and every theory, from the most rigorously neuroanatomical to the most abstractly artificial, posits some such units and then theorizes about how larger functions can be accomplished by organizations of units performing smaller functions. In fact *all* varieties of functionalism can be viewed as "homuncular" functionalism of one grain size or another.

I have been amused to note a euphemism of sorts that has recently gained favor among neuroscientists. Neuroanatomists have made enormous strides in mapping the cortex, which turns out to be exquisitely organized into specialist columns of interacting neurons (the neuroscientist Vernon Mountcastle, 1978, calls them "unit modules"), further organized into such larger organizations as "retinotopic maps" (in which the spatial pattern of excitation on the retinas of the eyes is preserved), which in turn play roles — still ill understood — in still larger organizations of neurons. It used to be that neuroscientists talked about what these various tracts or neuron groups in the cortex *signaled;* they were thinking of these units as homunculi whose "job" was always to "send a message with a particular content." Recent advances in thinking have suggested that these tracts perform much more complex and varied functions, so it is now seen as importantly misleading to talk about them as (just) signaling this or that. How, then, might we express the hard-won discoveries about specific conditions under which these tracts are active? We say that this tract "cares about" color, while that one "cares about" location or motion. But this usage is no ridiculous anthropomorphism or "homunculus fallacy" of the sort encountered everywhere in AI! Mercy no, this is just a clever way that sober researchers have hit upon to talk, suggestively but without undue particularity, about the competences of nerve tracts! Sauce for the goose is sauce for the gander.

Minsky's Agents are distinctive mainly in that, unlike almost every other variety of posited homunculi, they have histories and genealogies. Their existence is not just posited; they have to have developed out of something whose prior existence was not entirely mysterious, and Minsky has many suggestions about how such developments must occur. If he is still unnervingly noncommittal about just what neurons Agents are made of and where in the brain they lie, it is just that he has wanted to explore the most general requirements on development of function without overspecificity. As he says, describing his earlier theory of "frames" (of which the Society of Mind is the offspring), "If the theory had been any vaguer, it would have been ignored, but if it had been described in more detail, other scientists might have 'tested' it, instead of contributing their own ideas" (1985, p. 259). Some scientists are unmoved by this apologia. They are interested in only those theories that make testable predictions right now. This would be good hard-headed policy, except for the fact that all the testable theories so far concocted are demonstrably false, and it is foolish to think that the breakthroughs of vision required to compose *new* testable theories will come out of the blue without a good deal of imaginative exploration of the kind Minsky indulges in. (I have been playing the same game, of course.)

Back to the thumbnail sketch:

> They [the specialist demons] are often opportunistically enlisted in new roles, for which their native talents more or less suit them. The result is not bedlam only because the trends that are imposed on all this activity are themselves the product of design. Some of this design is innate, and is shared with other animals. But it is augmented, and sometimes even overwhelmed in importance, by microhabits of thought that are developed in the individual, partly idiosyncratic results of self-exploration and partly the predesigned gifts of culture. Thousands of memes, mostly borne by language, but also by wordless "images" and other data structures, take up residence in an individual brain, shaping its tendencies and thereby turning it into a mind.

In this part of my theory I have been deliberately noncommittal about many important questions: How do these homunculi actually interact to accomplish anything? What are the underlying information-processing transactions, and what reason do we have to think they might "work"? According to the sketch, the sequence of events is determined (in ways I have only hinted at) by "habits," and aside from some negative claims in chapter 5 about what does *not* happen, I have not yet been at all specific about the structure of the processes by which elements from among the Multiple Drafts get perpetuated, some of them eventually generating heterophenomenology as the result of one probe or another. In order to see what the question comes to, and what the alternative answers might be, we should look briefly at some more-explicit models of sequential thinking.

3. AND THEN WHAT HAPPENS?

In chapter 7, we saw how the von Neumann architecture was a distillation of the serial process of deliberate calculation. Turing and von Neumann isolated one particular sort of current that can flow through the stream of consciousness and then radically idealized it in the interests of mechanization. There is the notorious von Neumann bottleneck, consisting of a single register for results and a single register for instructions. Programs are just ordered lists of instructions drawn from a small set of primitives that the machine is hard-wired to execute. A fixed process, the fetch-execute cycle, draws the instructions from the queue in memory, one at a time, always getting the next instruction in the list, unless the previous instruction branched to another part of the list.

When AI model-builders turned to implementing more realistic models of cognitive operations on this base, they revised all this. They expanded the outrageously narrow von Neumann bottleneck, turning it into a somewhat more compendious "workspace," or "working memory." They also designed more sophisticated operations to serve as the psychological primitives, and replaced the rigid fetch-execute instruction cycle of the von Neumann machine with more flexible ways for the instructions to get called and executed. The workspace became, in some instances, a "blackboard" (Reddy et al., 1973; Hayes-Roth, 1985), where various demons could write messages for all others to read, which in turn provoked another wave of writing and reading. The von Neumann architecture, with its rigid instruction cycle, was still there, in the background, accomplishing the implementation, but it played no role in the model. In the model, what happened next was governed by the outcomes of competitive waves of message-writing and -reading on the blackboard. A related species of descendants of the von Neumann architecture are the various *production systems* (Newell, 1973) that underlie such models as John Anderson's (1983) ACT* (pronounced "act-star") and Rosenbloom, Laird, and Newell's (1987) Soar (see also Newell, 1990).

You can get a good idea of the underlying architecture of a production system from this simple picture of ACT* (Figure 9.1).

Working memory is where all the action happens. All the basic actions are called *productions*. Productions are basically just pattern-recognition mechanisms tuned to fire whenever they detect *their* particular pattern. That is, they are IF-THEN operators that hang around looking at the current contents of working memory, waiting for their IF-clauses to be satisfied, so they can THEN do their deed, whatever it is (in a classical production system, it is depositing a new data element in working memory, for the further perusal of productions).

All computers have IF-THEN primitives, the "sense organs" that make it possible for them to react differentially to data coming in or retrieved from memory. This capacity for *conditional branching* is an

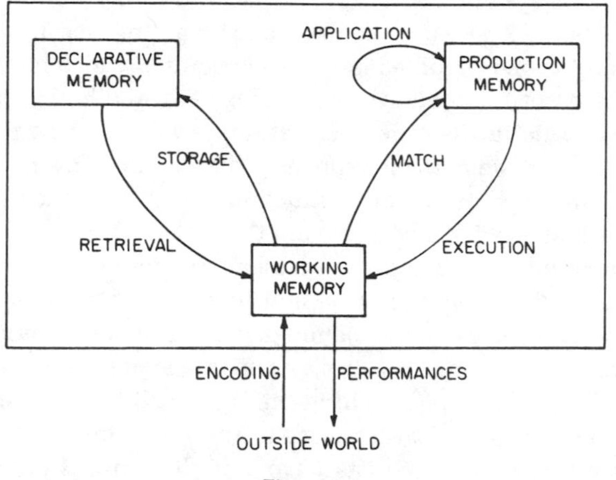

Figure 9.1

essential ingredient in computer power, no matter what the architecture is. The original IF-THENs were Turing's simple, clear-cut machine-state instructions: IF you see a zero, THEN replace it with a one, move one space left, and switch to state *n*. Contrast such simple instructions with the sort you might give a well-trained and experienced human sentry: IF you see something that looks unfamiliar to you, AND further investigation does not resolve the issue OR you have residual doubts, THEN sound the alarm. Can we build such sophisticated monitoring out of simple mechanical IF-THENS? *Productions* are intermediate-level sensors out of which one might build more complex sense organs, and then whole architectures of cognition. Productions can take complex and fuzzy-edged IF-clauses; the patterns they "recognize" don't have to be as simple as the bar-codes recognized by cash registers, but more like the patterns discriminated by the sentry (see the discussion in Anderson, 1983, pp. 35–44). And unlike the IF-THENs of a Turing machine, which is always in just one machine-state at a time (always testing just one IF-THEN from its set, before moving on to the next data item), the IF-THENs in a production system wait *en masse*, in (simulated) parallel, so that at any one "instant" more than one production may have its condition satisfied and be ready for action.

This is where things get interesting: How does such a system deal with *conflict resolution?* When more than one production is satisfied, there is always a chance that two (or more) will pull in incompatible directions. Parallel systems can tolerate a large measure of cross purposes, but in a system that is going to succeed in the world, not everything can happen at once; sometimes something has to give. How conflict resolution is handled is a crucial regard in which models differ. In fact, since most if not all the psychologically and biologically interesting details lie in the differences at this level, it is best to consider the production system architecture as just an underlying medium with which to build models. But all production systems share a few basic assumptions that provide a bridge to our theory sketch: they have a workspace where the action happens, where many productions (= demons) can try to do their thing at once, and they have a more or less inert memory where innate and accumulated information is stored. Since not everything "known" by such a system is available in this workspace at once, Plato's problem of getting the right bird to come at the right time is the major logistical task faced. And, most important from our present vantage point, theorists have *actually worked out* candidate mechanisms for answering the Hard Question: *And then what happens?*

For instance, in ACT*, there are five principles of conflict resolution.

(1) *Degree of match:* If one production's IF-clause is somewhat better matched than another's, it has priority.
(2) *Production strength:* Productions that have been recently successful have a higher "strength" associated with them, which gives them priority over lower-strength productions.
(3) *Data refractoriness:* The same production cannot match the same data more than once (this is to prevent infinite loops and similar, if less drastic, ruts).
(4) *Specificity:* When two productions match the same data, the production with the more specific IF-clause wins.
(5) *Goal dominance:* Among the items productions deposit in working memory are goals. There can be only one currently active goal at a time in the working memory of ACT*, and any production whose output matches the active goal has priority.

These are all plausible principles of conflict resolution, making sense both psychologically and teleologically (for a detailed discussion, see Anderson, 1983, ch. 4). But perhaps they make too much sense. That is, Anderson has himself wisely designed the conflict-resolution system of ACT*, exploiting his knowledge of the specific sorts of problems that arise in conflict-resolution circumstances, and the effective ways of dealing with them. He has essentially hard-wired this sophisticated knowledge into the system, an innate gift from evolution.

An interesting contrast to this is the Soar architecture of Rosenbloom, Laird, and Newell (1987). It, too, like any parallel architecture, encounters *impasses* — occasions when there is a need for conflict resolution because either contradictory productions or no productions "fire" — but it treats them as a boon, not a problem. Impasses are basic building opportunities in the system. Conflicts are not automatically dealt with by a presciently fixed set of conflict-resolution principles (an authoritative traffic-cop homunculus already in place) but rather are dealt with *non*automatically. An impasse creates a *new* "problem space" (a sort of topical workspace) in which the problem to be solved is precisely the impasse. This may generate yet another, meta-meta-traffic problem space, and so on — *potentially* forever. But in practice (at least in the domains modeled to date), after stacking up problem spaces several layers deep, the topmost problem finds a resolution, which quickly resolves the next problem down, and so forth, dissolving the ominous proliferation of spaces after making a nontrivial exploration through the logical space of possibilities. The effect on the system, moreover, is to "chunk" the resulting hard-won discoveries into new productions so that when similar problems arise in the future, there is a newly minted production on hand to resolve it swiftly, a trivial problem already solved in the past.

I briefly mention these details not to argue for the ultimate merits of Soar over ACT*, but just to give an idea of the sorts of issues that can be explored, responsibly, by models built of this sort of parts. My own hunch is that, for various reasons that need not concern us here, the underlying medium of production systems is *still* too idealized and oversimplified in its constraints, but the trajectory from the von Neumann machine through production systems points to still further architectures, ever brainier in structure, and the best way to explore their strengths and limitations is to build 'em and run 'em. This is the way to turn what is still impressionistic and vague in theories such as mine into honest models with details that can be empirically tested.

When you take the various claims about the mechanisms of consciousness that I've maintained in the last four chapters and start trying to juxtapose them on models of cognitive systems such as these, a host of questions arise, but I am not going to try to answer them here. Since I leave all these questions unresolved, my sketch remains just that — a sketch that could loosely fit a whole family of importantly different theories. That is as far as I need to go on this occasion, for the philosophical problems of consciousness are about whether *any* such theory could explain consciousness, so it would be premature to pin our hopes on an overparticular version that might turn out to be seriously flawed. (In Appendix B, I will go out on a few empirical limbs, however, for those who want a theory to have testable implications from the outset.)

It is not just philosophers' theories that need to be made honest by modeling at this level; neuroscientists' theories are in the same boat. For instance, Gerald Edelman's (1989) elaborate theory of "re-entrant" circuits in the brain makes many claims about how such re-entrants can accomplish the discriminations, build the memory structures, coordinate the sequential steps of problem solving, and in general execute the activities of a human mind, but in spite of a wealth of neuroanatomical detail, and enthusiastic and often plausible assertions from Edelman, we won't know what his re-entrants can do — we won't know that re-entrants are the *right* way to conceive of the functional neuroanatomy — until they are fashioned into a whole cognitive architecture at the grain-level of ACT* or Soar and put through their paces.[1]

At a finer-grain level of modeling, there is the unfinished business of showing how the productions (or whatever we call the pattern-recognition demons) themselves get implemented in the brain. Baars (1988) calls his specialists "bricks" with which to build, and opts for leaving the deeper details of brickmaking for another day or another discipline, but, as many have noted, it is tempting to suppose that the specialists themselves, at several levels of aggregation, should be modeled as made of *connectionist* fabrics of one sort or another.

Connectionism (or PDP, for parallel distributed processing) is a fairly recent development in AI that promises to move cognitive modeling closer to neural modeling, since the elements that are *its* bricks are nodes in parallel networks that are connected up in ways that look *rather* like neural networks in brains. Comparing connectionist AI to "Good Old Fashioned AI" (Haugeland, 1985) and to various modeling

---

1. Edelman (1989) is one theorist who has tried to put it all together, from the details of neuroanatomy to cognitive psychology to computational models to the most abstruse philosophical controversies. The result is an instructive failure. It shows in great detail just how many different sorts of question must be answered before we can claim to have secured a complete theory of consciousness, but it also shows that no one theorist can appreciate all the subtleties of the problems addressed by the different fields. Edelman has misconstrued, and then abruptly dismissed, the work of many of his potential allies, so he has isolated his theory from the sort of sympathetic and informed attention it needs if it is to be saved from its errors and shortcomings. This raises the parallel prospect that I too have underestimated some of the work I disagree with in these pages; no doubt I have, and I hope those whose brainchildren I have misrepresented will try (again) to explain what I have missed.

projects in the neurosciences has become a major industry in academia (see, e.g., Graubard, 1988; Bechtel and Abrahamson, 1991; Ramsey, Stich, and Rumelhart, 1991). This is not surprising, since connectionism blazes the first remotely plausible trails of unification in the huge terra incognita lying between the mind sciences and the brain sciences. But almost none of the controversies surrounding "the proper treatment of connectionism" (Smolensky, 1988) affect our projects here. *Of course* there is going to have to be a level (or levels) of theory at about the same coarseness of grain as connectionist models, and it is going to mediate between more obviously neuroanatomical levels of theory and more obviously psychological or cognitive levels of theory. The question is which particular connectionist ideas will be part of that solution and which will drop out of the running. Until that is settled, thinkers tend to use the connectionist debating arena as an amplifier for their favorite slogans, and although I am as willing to take sides in these debates as anyone else (Dennett, 1987b, 1988b, 1989, 1990c, 1991b,c,d), I am going to bite my tongue on this occasion, and get on with our main task, which is to see how a theory of *consciousness* might emerge from this when the dust settles, however it does.

Notice what has happened in the progression from the von Neumann architecture to such virtual architectures as production systems and (at a finer grain level) connectionist systems. There has been what might be called a shift in the balance of power. Fixed, predesigned programs, running along railroad tracks with a few branch points depending on the data, have been replaced by flexible — indeed volatile — systems whose subsequent behavior is much more a function of complex interactions between what the system is currently encountering and what it has encountered in the past. As Newell, Rosenbloom, and Laird (1989) put it, "Thus the issue for the standard computer is how to be interrupted, whereas the issue for Soar and ACT* (and presumably for human cognition) is how to keep focused" (p. 119).

Given all the ink that has been spilt over this theoretical issue, it is important to stress that this is a shift in the *balance* of power, not a shift to some "qualitatively different" mode of operation. At the heart of the most volatile pattern-recognition system ("connectionist" or not) lies a von Neumann engine, chugging along, computing a computable function. Since the birth of computers, critics of Artificial Intelligence have been hammering away on the rigidity, the mechanicity, the *programmedness* of computers, and the defenders have repeatedly insisted that this was a matter of degree of complexity — that indefinitely nonrigid, fuzzy, holistic, organic systems could be created on a computer. As AI has developed, just such systems have appeared, so now the critics have to decide whether to fish or cut bait. Should they declare that connectionist systems — for instance — were the sort of thing they thought minds were made of all along, or should they raise the ante and insist that not even a connectionist system is "holistic" enough for them, or "intuitive" enough, or . . . (fill in your favorite slogan). Two of the best-known critics of AI, the Berkeley philosophers Hubert Dreyfus and John Searle, split on this issue; Dreyfus has pledged allegiance to connectionism (Dreyfus and Dreyfus, 1988), while Searle has raised the ante, insisting that no connectionist computer could exhibit *real* mentality (1990a, 1990b).

The "in principle" skeptics may be in retreat, but huge problems still confront the unifiers. The largest, in my opinion, is one that has a direct bearing on our theory of consciousness. The consensus in cognitive science, which could be illustrated with dozens of diagrams like Figure 9.1, is that *over there* we have the long-term memory (Plato's bird cage) and *over here* we have the workspace or working memory, where the thinking happens, in effect.[2] And yet there are no two places in the brain to house these separate facilities. The only place in the brain that is a plausible home for either of these separate functions is

2. Functionalists have made a habit of "boxology" — drawing diagrams that install the component functions in separate boxes, while explicitly denying that these boxes have anatomical significance. (I am myself guilty of engaging in, and encouraging, this practice; see the figures in *Brainstorms*, chapters 7, 9, and 11.) I still think that "in principle" this is a good tactic, but in practice it does tend to blind the functionalist to alternative decompositions of function, and particularly to the prospect of multiple superimposed functions. The image of spatial separation between working memory and long-term memory — an image as old as Plato's aviary — plays a nontrivial role in how theorists construe the tasks of cognition. A striking example: "The need for symbols arises because it is not possible for all of the structure involved in a computation to be assembled ahead of time at the physical site of the computation. Thus it is necessary to travel out to other (distal) parts of the memory to obtain the additional structure" (Newell, Rosenbloom, and Laird, 1989, p. 105). This leads quite directly to the image of *movable symbols*, and then (in those who are uncritically fond of this image), to skepticism about all connectionist architectures, on the grounds that the elements in such architectures that are closest to being symbols — the nodes that in one way or another anchor the semantics of the system — are immovable in their web of interconnections. See, e.g., Fodor and Pylyshyn (1988). This problem of fixed versus movable semantic elements is one way of looking at a fundamental unsolved problem of cognitive science. It is probably not a good way of looking at it, but it won't go away until it is replaced by a better vision, anchored in a positive acceptance — as opposed to a hysterical dismissal — of the foundational facts of functional neuroanatomy.

the whole cortex — not two places side by side but one large place. As Baars says, summarizing the gathering consensus, there is a *global* workspace. It is global not only in the functional sense (crudely put, it is a "place" where just about everything can get in touch with just about everything else), but also in the anatomical sense (it is distributed throughout the cortex, and no doubt involves other regions of the brain as well). That means, then, that the workspace has to avail itself of the very same neural tracts and networks that apparently play a major role in long-term memory: the "storage" of the design changes wrought by individual exploration.

Suppose you learn how to make cornbread, or learn what "phenotypic" means. Somehow, the cortex must be a medium in which stable connection patterns can quite permanently anchor these design amendments to the brain you were born with. Suppose you are suddenly reminded of your dentist appointment, and it drives away all the pleasure you were deriving from the music you were listening to. Somehow, the cortex must be a medium in which *un*stable connection patterns can rapidly alter these transient contents of the whole "space" — without, of course, erasing long-term memory in the process. How can these two very different sorts of "representation" coexist in the same medium at the same time? In purely cognitive models, the jobs can be housed in separate boxes in a diagram, but when we have to superimpose them on a single fabric of neural tissues, the simple problem of packaging is the least of our worries.

Two functionally distinct network systems can be supposed to interpenetrate (just the way the telephone system and the highway system span the continent) — that is not the issue. The deeper problem is just beneath the surface in an assumption we have been making. Individual specialist demons, we have supposed, somehow recruit others in a larger-scale enterprise. If this were simply a matter of calling on these new recruits to exercise their *specialist* talents in common cause, we would already have models of such processes — such as ACT*, Soar, and Baars's Global Workspace — with varying degrees of plausible detail. But what if the specialists are also sometimes recruited *as generalists*, to contribute to functions in which their specialist talents play no discernible role? This is a tempting idea, for various reasons (see, e.g., Kinsbourne and Hicks, 1978), but so far as I know, we don't yet have any computational models of how such dual-function elements could operate.

Here is the difficulty: It is commonly supposed that specialists in the brain must get their functional identity somehow from their actual position in a network of more or less fixed connections. For instance, it seems that the only sort of facts that could explain a particular neural tract's "caring about" color would be facts about its idiosyncratic connections, however indirect, to the cone cells in the retina that are maximally sensitive to different frequencies of light. Once such a functional identity was established, these connections might be cut (as they are in someone blinded in adulthood) without (total) loss of the power of the specialists to represent (or in some other way "care about") color, but without such causal connections in the first place, it is hard to see what could give specialists a content-specific role.[3] It seems then that the cortex is (largely) composed of elements whose more or less fixed representational powers are the result of their functional location in the overall network. They represent the way members of the House of Representatives represent districts: by carrying information from sources to which they are specifically linked (most of the transactions on the phone lines in their Washington offices can be traced back to their home districts, for instance). Now imagine the members of the House of Representatives sitting in a block of seats in a stadium and representing the important message "Speed kills!" by individually holding large colored cards over their heads, spelling out the message in giant letters visible from the other side of the stadium. Living pixels, in short, in which their relations to their constituencies play no role in their contribution to the group representation. Some models of cortical recruitment strongly suggest that *something like* this secondary representational role must be possible. For instance, it is tempting to suppose that informational content about a particular matter can arise in some specialist tract and then, somehow, be propagated across cortical regions, exploiting the variability in these regions without engaging the specialized semantics of the units residing there. Suppose, for instance, a sudden change occurs in the upper left quadrant of a person's visual world. Just as expected, the brain excitation can be seen to arise first in those parts of the visual cortex that represent (à la the House of Representatives) various features of events in the upper left quadrant of vision, but these hot spots immediately become sources for spreading activation, involving cortical agents with other constituencies. If this spread of arousal across areas of cortex isn't just leakage or noise, if it plays some crucial role in elaborating or facilitating the editing of a draft of a narrative fragment, these recruited agents must

3. In other words, the attractions of a "causal theory of reference" are just as evident to cognitive scientists as to philosophers.

play a role quite different from their role when they are the anchoring source.[4]

It is not surprising that we have no good models, yet, of such multiple functionality (the only plausible speculations I have seen are some of Minsky's, in *The Society of Mind*). As we noted in chapter 7, human engineers, with their imperfect foresight, train themselves to design systems in which each element plays a single role, carefully insulated from interference from outside, in order to minimize the devastation of unforeseen side effects. Mother Nature, on the other hand, does not worry about foreseeing side effects, and so can capitalize on serendipitous side effects when they show up — once in a blue moon. Probably the inscrutability of functional decomposition in the cortex that has so far defied neuroscientists results from the fact that they find themselves constitutionally unable to entertain hypotheses that assign multiple roles to the available elements. Some romantics — the philosopher Owen Flanagan (1991) calls them the New Mysterians — have advanced the claim that there is an insurmountable barrier to the brain's understanding its own organization (Nagel, 1986, and McGinn, 1990). I am not entertaining any such claim, but rather just the supposition that it is proving to be fiendishly difficult — but not impossible — to figure out how the brain works, in part because it was designed by a process that can thrive on multiple, superimposed functionality, something systematically difficult to discern from the perspective of reverse engineering.

These are problems that provoke wishful handwaving, if they are noticed at all. Some are tempted to dismiss the idea of such specialist/generalist duality out of hand — not because they can prove it is mistaken, but because they cannot imagine how to model it, and therefore quite reasonably hope they will never have to do so. But once the prospect is raised, it at least gives theorists some new clues to look for. Neurophysiologists have (tentatively) identified mechanisms in neurons such as the NMDA receptors and the von der Malsburg (1985) synapses, which are plausible candidates to play a role as rapid modulators of connectivity between cells. Such gates might permit the swift formation of transient "assemblies," which could be superimposed on networks without requiring any alteration of the long-term synaptic strengths that are generally assumed to be the glue that holds the permanent assemblies of long-term memory together. (For some novel speculations along these lines, see Flohr, 1990.)

---

4. Fodor notes a variation on this problem in his discussion of "entertaining a concept" (1990, pp. 80–81).

At a larger scale, neuroanatomists have been filling in the map of connections in the brain, showing not only which areas are active under which circumstances, but beginning to show what sort of contributions they make. Several areas have been hypothesized to play a crucial role in consciousness. The *reticular formation* in the midbrain, and the *thalamus* above it, have long been known to play a crucial role in arousing the brain — from sleep, for instance, or in response to novelty or emergency — and now that the pathways are better mapped, more detailed hypotheses can be formulated and tested. Crick (1984), for instance, proposes that the branches radiating from the thalamus to all parts of the cortex fits it for the role of a "searchlight," differentially arousing or enhancing particular specialist areas, recruiting them to current purposes.[5] Baars (1988) has elaborated a similar idea: the ERTAS or Extended Reticular Thalamic Activating System. It would be easy enough to incorporate such a hypothesis into our anatomically noncommittal account of competition between coalitions of specialists, provided that we don't fall into the tempting image of a thalamic Boss that *understands* the current events being managed by the various parts of the brain with which it is "in communication."

Similarly, the *frontal lobes* of the cortex, the part of the brain most conspicuously enlarged in *Homo sapiens*, are known to be involved in long-term control, and the scheduling and sequencing of behavior. Damage to various regions in the frontal lobes typically produces such opposing symptoms as distractibility versus overfocused inability to get out of ruts, and impulsivity or the inability to follow courses of action that require delaying gratification. So it is tempting to install the Boss in the frontal lobes, and several models make moves in this direction. A particularly sophisticated model is Norman and Shallice's (1985) Supervisory Attentional System, which they locate in the prefrontal cortex and give the particular responsibility of conflict resolution when subsidiary bureaucracies can't cooperate. Once again, finding an anatomical location for processes that are crucial in the control of what happens next is one thing, and locating the Boss is another; anyone

---

5. Searchlight theories of attention have been popular for years. Crude theories make the mistake of supposing too literally that what the searchlight differentially illuminates or enhances at a moment is a region of *visual space* — exactly the way a spotlight in a theater can illuminate one region of the stage at a time. More defensible — but also at this time more impressionistic — searchlight theories insist that it is a portion of *conceptual* or *semantic space* that is differentially enhanced (imagine, if you can, a theater spotlight that can pick out just the *Capulets*, or all and only the *lovers*). See Allport (1989) on the difficulties with searchlight theories.

who goes hunting for the frontal display screen where the Boss keeps track of the projects he is controlling is on a wild goose chase (Fuster, 1981; Calvin, 1989a).

Once we forswear such tempting images, though, we have to find other ways of thinking of the contributions these areas are making, and here there is still a shortage of ideas, in spite of the recent progress. It is not that we have no sense of what the machinery is; the problem is much more a matter of lacking a computational model of what the machinery does, and how. Here we are still at the metaphor and hand-waving stage, but that is not a stage to shun; it is a stage to pass through on our way to more explicit models.

4. THE POWERS OF THE JOYCEAN MACHINE

According to our sketch, there is competition among many concurrent contentful events in the brain, and a select subset of such events "win." That is, they manage to spawn continuing effects of various sorts. Some, uniting with language-demons, contribute to subsequent sayings, both sayings-aloud to others and silent (and out-loud) sayings to oneself. Some lend their content to other forms of subsequent self-stimulation, such as diagramming-to-oneself. The rest die out almost immediately, leaving only faint traces — circumstantial evidence — that they ever occurred at all. What good does it do, you might well want to ask, for some contents to win entrance in this way to the charmed circle — and what is so charmed about *this* circle? Consciousness is supposedly something mighty special. What is so special about being advanced to the next round in such a cycle of self-stimulation? How does this help? Do near-magical powers accrue to events that occur in such mechanisms?

I have avoided claiming that any particular sort of victory in this competitive whirl amounts to elevation to consciousness. Indeed, I have insisted that there is no motivated way to draw a line dividing the events that are definitely "in" consciousness from the events that stay forever "outside" or "beneath" consciousness. (See Allport, 1988, for further arguments in favor of this position.) Nevertheless, if my theory of the Joycean machine is going to shed light on consciousness at all, there had better be something remarkable about some if not all of the activities of this machine, for there is no denying that consciousness is, intuitively, something special.

It is hard to address these familiar questions without falling into the trap of thinking that first we must figure out what consciousness is *for*, so we can then ask whether the proposed mechanisms would succeed in performing *that* function — whatever we determine it is.

In his influential book *Vision* (1982), the neuroscientist/AI researcher David Marr proposed three levels of analysis that should be pursued in the attempt to explain any mental phenomenon. The "top" or most abstract level, the computational, is an analysis of "*the problem* [my italics] as an information-processing task," while the middle, algorithmic, level is an analysis of the actual processes by which *this* information-processing task is performed. The lowest, physical level provides an analysis of the neural machinery and shows how it executes the algorithms described at the middle level, thereby performing its task as abstractly described at the computational level.[6]

Marr's three levels can also be used to describe things that are much simpler than minds, and we can get a feel for the differences between the levels by seeing how they apply to something simple, such as an abacus. Its computational task is to do arithmetic: to yield a correct output for any arithmetic problem given to it as input. At this level, then, an abacus and a hand calculator are alike; they are designed to perform the same "information-processing task." The algorithmic description of the abacus is what you learn when you learn how to manipulate it — the recipe for moving the beads in the course of adding, subtracting, multiplying, and dividing. Its physical description depends on what it is made of: it might be wooden beads strung on wires in a frame, or it might be poker chips lined up along the cracks in the floor, or something accomplished with a pencil and a good eraser on a sheet of lined paper.

Marr recommended modeling psychological phenomena at all three levels of analysis, and he particularly stressed the importance of getting clear about the top, computational level before rushing headlong into lower-level modeling.[7] His own work on vision brilliantly exhibited the power of this strategy, and other researchers have since put it to good use on other phenomena. It is tempting to apply the same three

---

6. Closely related distinctions are my trinity of the intentional stance, the design stance, and the physical stance (Dennett, 1971), and Allen Newell's (1982) identification of the "knowledge level" above the "physical symbol system level." See Dennett (1987a, 1988e), and Newell (1988).

7. As Marr noted, "It becomes possible, by separating explanations into different levels, to make explicit statements about what is being computed and why and to construct theories stating that what is being computed is optimal in some sense or is guaranteed to function correctly" (p. 19). For more on the benefits and pitfalls of such reverse engineering see Dennett (1971, 1983, 1987a, 1988d); Ramachandran (1985).

levels of analysis to consciousness as a whole, and some have yielded to the temptation. As we saw in chapter 7, however, this is a risky oversimplification: by asking "What is *the* function of consciousness?" we assume that there is a single "information-processing task" (however complex) that the neural machinery of consciousness is well designed to perform — by evolution, presumably. This can lead us to overlook important possibilities: that some features of consciousness have multiple functions; that some functions are only poorly served by the existing features, due to the historical limitations on their development; that some features have no function at all — at least none that is for *our* benefit. Being careful to avoid these oversights, then, let us review the *powers* (not necessarily the functions) of the mechanisms described in my thumbnail sketch.

First of all, as we saw in chapter 7, there are significant problems of self-control created by the proliferation of simultaneously active specialists, and one of the fundamental tasks performed by the activities of the Joycean machine is to adjudicate disputes, smooth out transitions between regimes, and prevent untimely *coups d'état* by marshaling the "right" forces. Simple or overlearned tasks without serious competition can be routinely executed without the enlistment of extra forces, and hence unconsciously, but when a task is difficult or unpleasant, it requires "concentration," something "we" accomplish with the help of much self-admonition and various other mnemonic tricks, rehearsals (Margolis, 1989), and other self-manipulations (Norman and Shallice, 1985). Often we find it helps to talk out loud, a throwback to the crude but effective strategies of which our private thoughts are sleek descendants.

Such strategies of self-control permit us to govern our own perceptual processes in ways that open up new opportunities. As the psychologist Jeremy Wolfe (1990) points out, although our visual systems are natively designed to detect some sorts of things — the sorts that "pop out" when we "just look" — there are other sorts that we can identify only if we can *look for* them, deliberately, in a policy set up by an act of self-representation. A red spot among a slew of green spots will stick out like a sore thumb (actually, it will stick out like a ripe berry midst the leaves), but if your projects call for you to find a red spot in a crowd of spots of many different colors, you have to *set yourself* a task of serial search. And if your project is to find the red square piece of confetti amidst its multicolored and multishaped neighbors (or to answer the question "Where's Waldo?" [Handford, 1987] in the popular puzzle pictures), the task of serial search can become a particularly engrossing, methodical project, calling on a high degree of self-control.

These techniques of representing things to ourselves permit us to be self-governors or executives in ways no other creature approaches. We can work out policies well in advance, thanks to our capacity for hypothetical thinking and scenario-spinning; we can stiffen our own resolve to engage in unpleasant or long-term projects by habits of self-reminding, and by rehearsing the expected benefits and costs of the policies we have adopted. And even more important, this practice of rehearsal creates a memory of the route by which we have arrived at where we are (what psychologists call *episodic* memory), so that we can explain to ourselves, when we find ourselves painted into the corner, just what errors we made (Perlis, 1991). In chapter 7 we saw how the development of these strategies permitted our ancestors to look farther into the future, and part of what gave them this enhanced power of anticipation was an enhanced power of recollection — being able to look farther back at their own recent operations to *see* where they made their mistakes. "Well, I mustn't do *that* again!" is the refrain of any creature that learns from experience, but we can learn to cast our *thats* back farther and more insightfully than any other creature, thanks to our habit of record-keeping — or more accurately, thanks to our habits of self-stimulation, which have, among their many effects, the enhancement of recollection.

But such memory-loading is only one of the valuable effects of these habits. Just as important is the broadcasting effect (Baars, 1988), which creates an open forum of sorts, permitting *any* of the things one has learned to make a contribution to *any* current problem. Baars develops the claim that this mutual accessibility of contents provides the *context* without which events occurring "in consciousness" would not — *could not* — make sense to the subject. The contents that compose the surrounding context are not themselves always conscious — in fact, in general they are not accessible at all, in spite of being activated — but the connections between them and the contents that can show up in verbal reports are what secures what we might call their "consciously apprehended" meaning.

Ray Jackendoff (1987) argues in the same spirit that the highest levels of analysis performed by the brain, by which he means the most abstract, are *not* accessible in experience, even though they make experience possible, by making it meaningful. His analysis thus provides a useful antidote to yet another incarnation of the Cartesian Theater as the "summit" or "the tip of the iceberg." (Here is a good example, drawn from the neuropsychologist Roger Sperry: "In a position of top command at the highest levels in the hierarchy of brain organization, the subjective properties . . . exert control over the biophysical and chemical activities at subordinate levels" [1977, p. 117].)

Quite a few philosophers, particularly those influenced by the Husserlian school of Phenomenology (Dreyfus, 1979; Searle, 1983), have stressed the importance of this "background" of conscious experience, but they have typically described it as a mysterious or recalcitrant feature, defying mechanical explanation, rather than the key, as Baars and Jackendoff suggest, to providing a computational theory of what happens. These philosophers have supposed that consciousness is the *source* of some special sort of "intrinsic intentionality," but as the philosopher Robert van Gulick has noted, this gets matters just backwards.

> The personal-level experience of understanding is . . . not an illusion. I, the personal subject of experience, do understand. I can make all the necessary connections within experience, calling up representations to immediately connect one with another. The fact that my ability is the result of my being composed of an organized system of subpersonal components which produce my orderly flow of thoughts does not impugn my ability. What is illusory or mistaken is only the view that I am some distinct substantial self who produces these connections in virtue of a totally non-behavioral form of understanding. [van Gulick, 1988, p. 96]

Any of the things you have learned can contribute to any of the things you are currently confronting. That at least is the ideal. This feature is called *isotropy* by Fodor (1983), the power, as Plato would say, of getting the relevant birds to come, or at least to sing out, whenever they are needed. It looks magical, but as every stage magician knows, the appearance of magic is heightened by the fact that an audience can generally be counted on to exaggerate the phenomenon in need of explanation. We may seem at first to be ideally isotropic, but we're not. Sober reflection reminds us of all the occasions when we're slow to recognize the significance of new data. Think of the classic comedic exaggeration of this: the "double take" (Neisser, 1988). Sometimes we even saw off the limb we are sitting on, or light a match to peer into the gas tank.[8]

Stage magicians know that a collection of cheap tricks will often

---

8. In *Minimal Rationality* (1986), the philosopher Christopher Cherniak analyzes the prospects and limits of deductive processes of the sort made possible by this provision of an open forum. See also Stalnaker (1984).

suffice to produce "magic," and so does Mother Nature, the ultimate gadgeteer. Artificial Intelligence research has been exploring the space of possible tricks, looking for "a bundle of . . . heuristics properly co-ordinated and rapidly deployed" (Fodor, p. 116) that could provide the degree of isotropy we human thinkers exhibit. Models such as ACT* and Soar — and many other visions being explored in AI — are promising but inconclusive. Some philosophers, notably Dreyfus, Searle, Fodor, and Putnam (1988), are sure that this idea of the Mind as Gadget is wrong, and have tried to construct arguments proving the impossibility of the task (Dennett, 1988b, 1991c). Fodor, for instance, points out that whereas special-purpose systems can be hard-wired, in a general-purpose system that can respond with versatility to whatever new item comes along, "it may be *unstable, instantaneous* connectivity that counts" (p. 118). He despairs of anybody coming up with a theory of such connectivity, but he is not just being pessimistic: he despairs *in principle* (a neat trick). He is right that we should expect our approximation of isotropy to be due to our software, not our hard-wiring, but his argument against the "bag of tricks" hypothesis supposes that we are better at "considering all things" than we are. We're good, but not fantastic. The habits of self-manipulation we develop turn us into wily exploiters of our hard-won resources; we don't always get the right bird to sing at the right time, but often enough to make us pretty good company.

5. BUT IS THIS A THEORY OF CONSCIOUSNESS?

I have been coy about consciousness up to now. I have carefully avoided telling you what my theory says that consciousness *is*. I haven't declared that anything instantiating a Joycean machine is conscious, and I haven't declared that any particular state of such a virtual machine is a conscious state. The reason for my reticence was tactical: I wanted to avoid squabbling over what consciousness is until I had had a chance to show that at least many of the presumed powers of consciousness could be explained by the powers of the Joycean machine *whether or not* a Joycean machine endows its host hardware with consciousness.

Couldn't there be an *un*conscious being with an internal global workspace in which demons broadcast messages to demons, forming coalitions and all the rest? If so, then the stunning human power of swift, versatile adjustment of mental state in response to almost any contingency, however novel, owes nothing to consciousness itself, but just to the computational architecture that makes this intercommunication possible. If consciousness is something over and above the Joycean machine, I have not yet provided a theory of consciousness at all, even if other puzzling questions have been answered.

Until the whole theory-sketch was assembled, I had to deflect such doubts, but at last it is time to grasp the nettle, and confront consciousness itself, the whole marvelous mystery. And so I hereby declare that YES, my theory is a theory of consciousness. Anyone or anything that has such a virtual machine as its control system is conscious in the fullest sense, and is conscious *because* it has such a virtual machine.[9]

Now I am ready to take on the objections. We can begin by addressing the unanswered question two paragraphs back. Couldn't something unconscious — a zombie, for instance — have a Joycean machine? This question implies an objection that is so perennially popular at moments like these that the philosopher Peter Bieri (1990) has dubbed it *The Tibetan Prayer Wheel*. It just keeps recurring, over and over, no matter what theory has been put forward:

> That's all very well, all those functional details about how the brain does this and that, but I can imagine *all that* happening in an entity without the occurrence of any real consciousness!

A good answer to this, but one seldom heard, is: Oh, can you? How do you know? How do you know you've imagined "all that" in sufficient detail, and with sufficient attention to all the implications? What makes you think your claim is a premise leading to any interesting conclusion? Consider how unimpressed we would be if some present-day *vitalist* were to say:

> That's all very well, all that stuff about DNA and proteins and such, but I can imagine discovering an entity that looked and acted just like a cat, right down to the blood in its veins and DNA in its "cells," but was not alive. (Can I really? Sure: There it is, meowing, and then God whispers in my ear, "It's not alive! It's just a mechanical DNA-thingamabob!" In my imagination, I believe Him.)

9. Jackendoff (1987) adopts a somewhat different tactic. He divides the mind-body problem in two, and addresses his theory to the question of how the *computational* mind fits in the body; this leaves him with an unsolved "mind-mind problem" — what the relation is between the phenomenological mind and the computational mind. Rather than conceding this as a residual mystery, I propose to show how the Multiple Drafts model, in concert with the method of heterophenomenology, dissolves both problems at once.

I trust that no one thinks this is a good argument for vitalism. This effort of imagination doesn't count. Why not? Because it is too puny to weigh against the account of life presented by contemporary biology. The only thing this "argument" shows is that you can ignore "all that" and cling to a conviction if you're determined to do so. Is the Tibetan Prayer Wheel any better as an argument against the theory I have sketched?

We are now in a position, thanks to all the imagination-stretching of the previous chapters, to shift the burden of proof. The Tibetan Prayer Wheel (and there are several importantly different variations, as we shall see) is a descendant of Descartes's famous argument (see chapter 2), in which he claims to be able to conceive *clearly and distinctly* that his mind is distinct from his brain. The force of such an argument depends critically on how high one's standards of conception are. Some people may claim they can clearly and distinctly conceive of a greatest prime number or a triangle that is not a rigid figure. They are wrong — or at any rate, whatever they're doing when they say they're conceiving these things should not be taken as a sign of what is possible. We are now in a position to imagine "all that" in some detail. Can you *really* imagine a zombie? The only sense in which it's "obvious" that you can is not a sense that challenges my theory, and a stronger, unobvious sense calls for a demonstration.

Philosophers have not, as a rule, demanded this. The most influential thought experiments in recent philosophy of mind have all involved inviting the audience to imagine some specially contrived or stipulated state of affairs, and then — without properly checking to see if this feat of imagination has actually been accomplished — inviting the audience to "notice" various consequences in the fantasy. These "intuition pumps," as I call them, are often fiendishly clever devices. They deserve their fame if only for their seductiveness.

In Part III we will take them all on, developing our theory of consciousness as we go. From our new perspective, we will be able to see the sleight of hand that misdirects the audience — and the illusionists — and in the process we will sharpen our own powers of imagination. Among the famous arguments we will encounter are not only the Presumed Possibility of Zombies, but the Inverted Spectrum, What Mary the Color Scientist Doesn't Know about Color, the Chinese Room, and What It Is Like to Be a Bat.

PART THREE

THE PHILOSOPHICAL PROBLEMS OF CONSCIOUSNESS

10

SHOW AND TELL

1. ROTATING IMAGES IN THE MIND'S EYE

The first challenge, before we tackle the philosophical thought experiments, comes from some real experiments that might seem to rehabilitate the Cartesian Theater. Some of the most exciting and ingenious research in cognitive science in the last twenty years has been on the human ability to manipulate mental images, initiated by the psychologist Roger Shepard's (Shepard and Metzler, 1971) classic study of the *speed of mental rotation* of figures such as these.

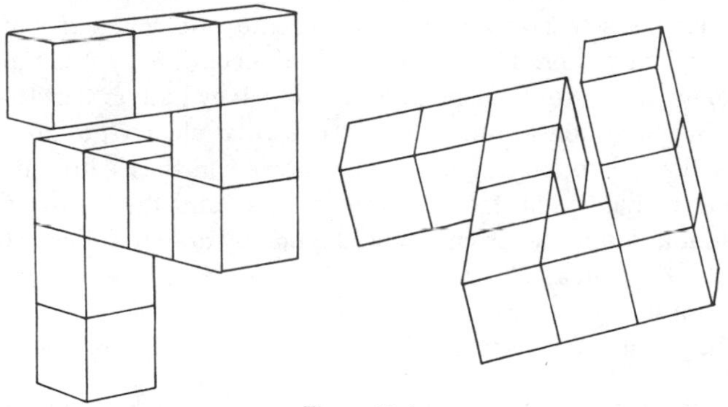

Figure 10.1

The subjects in the original experiment were shown such pairs of line drawings and asked whether or not the pair are different views of the same shape. In this case, as you can quickly determine, the answer is yes. How did you do it? A typical answer is "I rotated one of the images in my mind's eye, and superimposed it on the other." Shepard varied the angular rotation distances between pairs of figures — some pairs were only a few degrees out of alignment, and others required large rotations to put them into alignment — and measured the time it took subjects to respond, on average, to the different displays. Supposing that something like a process of actual rotation of an image in the brain takes place, it should take roughly twice as long to rotate a mental image through 90 degrees as through 45 degrees (if we ignore acceleration and deceleration, keeping rotation speed constant).[1] Shepard's data bore out this hypothesis remarkably well, over a wide variety of conditions. Hundreds of subsequent experiments, by Shepard and many others, have explored the behavior of the image-manipulation machinery of the brain in great detail, and — to put the still-controversial consensus as gingerly as possible — there does seem to be what the psychologist Stephen Kosslyn (1980) calls a "visual buffer" in the brain, which performs transformations by processes which are strongly "imagistic" — or, to use Kosslyn's term, *quasi-pictorial.*

What does this mean? Have cognitive psychologists discovered that the Cartesian Theater exists after all? According to Kosslyn, these experiments show that images are assembled for internal display in much the way that images on a CRT (a cathode ray tube such as a television screen or computer monitor) can be created from files in a computer memory. Once they are on the internal screen, they can be rotated, searched, and otherwise manipulated by subjects who are given particular tasks to perform. Kosslyn stresses, however, that his CRT model is a metaphor. This should remind us of Shakey's metaphorical "image manipulation" talents. Certainly Shakey had no Cartesian Theater in its computer-brain. To get a somewhat clearer picture of what must actually happen in a human brain, we can start with a *non*metaphorical model, much too strong to be true, and then "subtract" the undesirable properties from the model one by one. In other words, we will take Kosslyn's CRT metaphor, and gradually introduce the limitations on it.

First, consider a system that really does manipulate real images,

---

1. This useful oversimplification is one of the many wrinkles that have subsequently been explored by researchers, and there is now considerable evidence for "inertia" and "momentum" effects in image transformations. See Freyd (1989).

such as the computer graphics systems that are now proliferating in hundreds of settings: computer animation for television and films, systems that render three-dimensional objects in perspective views for architects and interior decorators, video games, and many more. Engineers call their versions CAD systems, for computer-aided design. CAD systems are revolutionizing engineering not only because they make drafting vastly easier in the way word processors make writing easier, but because engineers can readily solve problems and answer questions with them that would otherwise be quite difficult. Faced with the Shepard problem in Figure 10.1, an engineer could answer the question with the aid of a CAD system by putting both images on the CRT screen and literally rotating one of the images and then trying to superimpose it on the other. A few details of the process are important.

Each pictured object would be entered into the computer memory as a *virtual* three-dimensional object, by breaking it down into a description of its planes and edges defined by their *xyz* coordinates, each occupied point in virtual space being an "ordered triple" of numbers stored in the computer's memory. The point of view of the implied observer would also be entered as a point in the same virtual space, defined by its own *xyz* coordinate triple. Here is a diagram of a cube and a point of view, but it is important to remember that the only thing the computer has to store is the triples for each crucial point, grouped into larger groups (e.g., one for each face of the cube), together with coded information for the various properties of each face (its color,

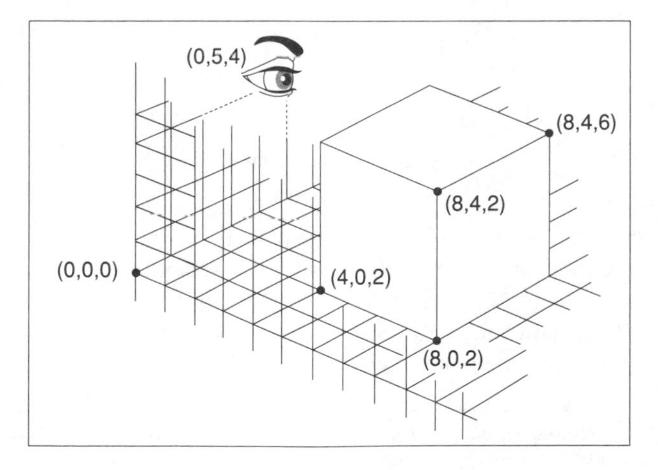

Figure 10.2

whether or not it is opaque or transparent, its texture, and so forth). Rotating one of the objects and then shifting it in the virtual space can be readily computed, by simply adjusting all the *x, y,* and *z* coordinates of that object by constant amounts — simple arithmetic. Then it is a matter of straightforward geometry to calculate the sight lines that determine which planes of the object would then be visible from the virtual point of view, and exactly how they would appear. The calculations are straightforward but laborious or "compute intensive," especially if smooth curves, shading, reflected light, and texture are to be calculated as well.

On advanced systems, different frames can be calculated rapidly enough to create apparent motion on the screen, but only if the representations are kept quite schematic. "Hidden line removal," the computational process that renders the final image opaque in the right places and prevents a Shepard cube from looking like a transparent Necker cube, is itself a relatively time-consuming process, which more or less sets the limits on what can be produced "in real time." For the gorgeously detailed image transformations by computer graphics that we see on television every day, the image-generation process is much slower, even on supercomputers, and the individual frames have to be stored for later display at the higher speeds that satisfy the motion-detection requirements of the human visual system.[2]

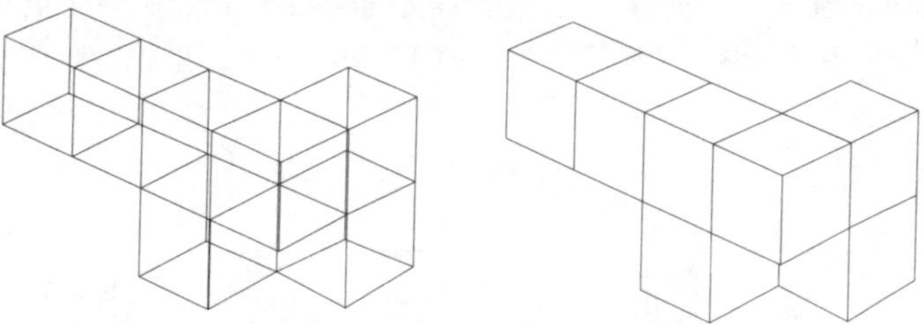

Before and after hidden line removal.

Figure 10.3

These three-dimensional virtual-object manipulators are magnificent new tools or toys, and they really are something new under the sun, not an electronic copy of something we already have in our heads. It is quite certain that no process analogous to these trillions of geometrical and arithmetical calculations occurs in our brains when we

2. The impressive, but noticeably jerky, animation in the popular IBM-PC program, "Flight Simulator," displays the limits of real-time animation of fairly complex three-dimensional scenes by a small computer.

engage in mental imagery, and nothing else *could* yield the richly detailed animated sequences they produce — for the reasons we explored in chapter 1.

We can satisfy ourselves that this limit on our brains is real by considering a slightly different Shepard-style problem that would be quite easy to solve with the aid of such a CAD system: Would the "red" *X* on one face of this object be visible to someone looking through the square hole in its front wall?

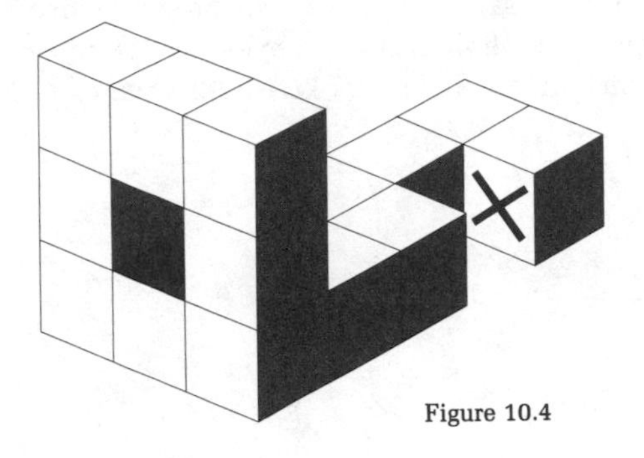

Figure 10.4

Our Shepard object with the *X* is a simple, schematic object, and since the question we want to answer is independent of texture, illumination, and other such niceties, it would be quite easy for an engineer to produce an animated rotation of this object on the CRT. He could then rotate the image every which way, moving the point of view back and forth — and just *look for* a glimpse of red through the hole. If he sees red, the answer is yes, otherwise, no.

Now, can you perform the same experiment in your mind's eye? Can you simply rotate the object shown and peer through the hole? If you can, you can do something I can't do, and all the people I have asked are also unable to do it with any confidence. Even those who have an answer to the question are quite sure that they didn't *just* get it by rotating and looking. (They often say they first tried rotating and looking and found it doesn't work; they can "rotate it" but then it "falls apart" when they try to look through the hole. Then they talk of "drawing in" sightlines through the hole on the *un*rotated image, to see if they could tell where the lines would hit the back plane.) Since our Shepard object is no more complex than the objects apparently successfully rotated in many experiments, this poses a puzzle: What kind of a process can so readily perform some transformations (and then extract information from the result), and fall down so badly on other operations that are apparently no more demanding? (If these operations *appear* to us to be no more demanding, we must be looking at them from the wrong vantage point, since our failures demonstrate that they are more demanding.)

An experiment by the psychologists Daniel Reisberg and Deborah Chambers (forthcoming) raises much the same question. Subjects who claimed to be good imagers were shown "nonsense" shapes and asked to rotate them mentally by 90 or 180 degrees in their mind's eyes and then report what they "saw." They were surprised to discover that they could not recognize, with their mind's eyes, what you can readily recognize when you turn the book clockwise 90 degrees to look at these figures.

Figure 10.5

The sorts of questions engineers use CAD systems to answer are typically not as simple as "Is the red X visible through the hole?" Usually they concern more complicated spatial properties of objects being designed, such as "Will this three-jointed robot arm be able to reach around and adjust the knob on the robot's back without bumping into the power supply pack?" or even aesthetic properties of such objects, such as "What will the stairway in this hotel lobby look like to someone walking by on the street, looking through the plate-glass windows?" When we try to visualize such scenes unaided, we get only the sketchiest and most unreliable results, so a CAD system can be seen as a sort of imagination-prosthesis (Dennett, 1982d, 1990b). It vastly amplifies the imagining powers of a human being, but it does depend on the user's having normal vision — for looking at the CRT.

Now let's try to imagine a more ambitious prosthetic device: a CAD system for blind engineers! And, to keep it simple, let's suppose that the sorts of questions these blind engineers want to answer are of the relatively straightforward, geometric sort — not subtle questions about the aesthetics of architecture. The output will have to be in some nonvisual form, of course. The most user-friendly form would be ordinary language answers (in Braille or voice-synthesized) to ordinary language questions. So we will suppose that when confronted with a question of the sort we have just been considering, the blind engineer would simply pass the sentence on to the CAD system (in terms it can "understand," of course) and wait for the CAD system to provide the answer.

Our Mark I CADBLIND system is inelegant, but straightforward. It consists of an ordinary CAD system, complete with CRT, in front of which is perched a *Vorsetzer*, a computer vision system, complete with TV camera aimed at the CRT and robot fingers to twiddle the knobs of the CAD system.[3] Unlike Shakey, whose CRT was only for the benefit of bystanders, this system really does "look at" an image, a real image made of glowing phosphor dots, which radiates real light of different frequencies onto the light-sensitive transducers at the back of the TV camera. When posed our red X Shepard problem, the Mark I CADBLIND produces an image with a real red X on it, visible to all, including the TV camera of the Vorsetzer.

Let us suppose, without further ado, that the Vorsetzer has solved within it enough of the problems of computer vision to be able to extract the sought-for information from the representations glowing on the CRT screen. (No, I am not going on to claim that the Vorsetzer is conscious — I just want to suppose that it is good enough at doing what it does to be able to answer the questions the blind engineer poses for it.) The Mark I CADBLIND makes and manipulates real images, and uses them to answer, for the blind engineer, all the questions that a sighted engineer could answer using an ordinary CAD system. If the Mark I system is that good, then the Mark II will be child's play to design: We just throw away the CRT and the TV camera looking at it, and replace them with a simple cable! Through this cable the CAD system sends the Vorsetzer the *bit-map*, the array of zeros and ones that defines the image on the CRT. In the Mark I's Vorsetzer, this bit-map was painstakingly reconstructed from the outputs of the optical transducers in the camera.

There are scant savings in *computation* in the Mark II — just the elimination of some unnecessary hardware. All the elaborate calculation of sight lines, hidden line removal, and the rendering of texture,

---

3. I call this device a *Vorsetzer* because it reminds me of the wonderful German player-piano by that name, which consisted of a separate unit with eighty-eight "fingers" that could "sit before" an ordinary piano, depressing the keys and pedals from the outside, just like a human pianist. (It is important to recognize that this device is a *Vorsetzer* — a sitter-before — but not a *Vorsitzer* — a chairman or president!)

shadows, and reflected light, which took so much computation in the Mark I, is still a part of the process. Suppose the Vorsetzer in the Mark II is called upon to make a depth judgment by comparing texture gradients, or interpreting a shadow. It will have to *analyze* the patterns of bits in the relevant portions of the bit-map to arrive at discriminations of the textures and shadows.

This means that the Mark II is still a ridiculously inefficient machine, for if the information that a particular portion of the bit-map should represent a shadow is already "known" to the CAD system (if this is part of the coded description of the object from which the CAD system generates its images), and if that fact is part of what the Vorsetzer must determine in order to make its depth judgment, why doesn't the CAD system just *tell* the Vorsetzer? Why bother *rendering* the shadow for the benefit of the pattern-analyzers in the Vorsetzer, when the task of pattern-rendering and the task of pattern-analyzing cancel each other out?

So our Mark III CADBLIND will exempt itself from huge computational tasks of image-rendering by taking much of what it "knows" about the represented objects and passing it on to the Vorsetzer subsystem directly, using the format of simple codes for properties, and attaching "labels" to various "places" on the bit-map array, which is thereby turned from a pure image into something like a diagram. *Some* spatial properties are represented directly — *shown* — in the (virtual) space of the bit-map, but others are only *told about* by labels.[4]

This should remind us of my claim in chapter 5 that the brain only has to make its discriminations once; a feature identified doesn't have to be redisplayed for the benefit of a master appreciator in the Cartesian Theater.

But now we can see a different aspect of the engineering: "canceling out" only works if the systems that need to communicate can "speak the same language." What if the format in which the CAD system already "knows" the relevant information — e.g., the information that something is a shadow — is not a format in which the Vorsetzer can "use" the information?[5] Then for communication to take place it might

---

4. Once we have labels, we can tell about *any* properties of the object, not just spatial properties or visible properties — as in the old coloring-book jokes: "Here is my boss. Color him obnoxious."

5. See Kosslyn (1980) for a discussion of format. Jackendoff (1989) has a related analysis of what he calls the form of information structures.

be necessary to "step back in order to jump forward." It might be necessary for the systems to engage in informationally profligate — you might say longwinded — interactions in order to interact at all. Think of sketching a map to give directions to a foreigner, when all he needs to know — if only you knew how to say it in his tongue — is "Turn left at the next traffic light." Going to all the trouble of making something like an image is often practically necessary, even when it's not necessary "in principle."

Since the systems in our brains are the products of several overlaid histories of opportunistic tinkering, the long history of natural selection and the short history of individual redesign-by-self-manipulation, we should expect to find such inefficiencies. Besides, there are other reasons for rendering information in imagelike formats (in addition to the sheer pleasure of it), which, if we stumble over them serendipitously, will soon impress us as making images worth the trouble in any case. As we already noted, in the speculations in chapter 7 about "diagramming to yourself," such transformations of format are often highly effective ways of extracting information that is otherwise all but inextricable from the data. Diagrams do indeed amount to re-*presentations* of the information — not to an inner eye, but to an inner pattern-recognition mechanism that can also accept input from a regular ("outer"?) eye. That is why the techniques of (computer) graphics are so valuable in science, for instance. They permit huge arrays of data to be presented in a format that lets the superb pattern-recognition capabilities of human vision take charge. We make graphs and maps and all manner of color-coded plottings so that the sought-for regularities and saliencies will just "pop out" at us, thanks to our visual systems. Diagrams don't just help us see patterns that might otherwise be imperceptible; they can help us *keep track* of what is relevant, and *remind* us to ask the right questions at the right times. The Swedish AI researcher Lars-Erik Janlert (1985) has argued that such image generation and perusal in a computer can also be used to help solve otherwise intractable problems of what we might call inference-management in systems that are "in principle" purely deductive engines. (For a different slant on the same process, see Larkin and Simon, 1987.)

This tactic is certainly well known to many wily thinkers, and has been wonderfully described by one of the wiliest ever, the physicist Richard Feynmann, in *Surely You're Joking, Mr. Feynmann!* (1985). In a chapter aptly entitled "A Different Box of Tools," he tells how he amazed his fellow graduate students at Princeton by "intuiting" the truth and falsity of the arcane theorems of topology, theorems he was utterly unable to derive formally or even fully comprehend:

> I had a scheme, which I still use today when somebody is explaining something that I'm trying to understand: I keep making up examples. For instance, the mathematicians would come in with a terrific theorem, and they're all excited. As they're telling me the conditions of the theorem, I construct something which fits all the conditions. You know, you have a set (one ball)—disjoint (two balls). Then the balls turn colors, grow hairs, or whatever, in my head as they put more conditions on. Finally they state the theorem, which is some dumb thing about the ball which isn't true for my hairy green ball thing, so I say, "False!"
>
> If it's true, they get all excited, and I let them go on for a while. Then I point out my counterexample.
>
> "Oh. We forgot to tell you that it's Class 2 Hausdorff homomorphic."
>
> "Well, then," I say, "It's trivial! it's trivial!" By that time I know which way it goes, even though I don't know what Hausdorff homomorphic means. [pp. 85–86]

Such tactics "come naturally" to some extent, but they have to be learned or invented, and some people are much better at it than others. Those in whom these skills are highly developed have different virtual machines in their brains, with significantly different powers, compared with those who are infrequent or inept "visualizers." And the differences readily emerge in their individual heterophenomenological worlds.

So there is good reason to believe, as Kosslyn and others have argued, that human beings put their vision systems to work not only by presenting themselves with real, *external* images (as on the CAD system's CRT) but also with idiosyncratically designed internal virtual images or diagramlike data representations that are the suitable raw material for some later stage or stages of the visual-processing machinery.

Just which engineering solutions to which problems of internal communication and information-manipulation has the human brain hit upon, and what are their strengths and weaknesses?[6] These are the

---

6. Kosslyn (1980) not only provides a detailed defense of his particular set of answers to these questions (at that time), but also provides an excellent survey of others'

empirical questions addressed by the research on imagery in cognitive psychology, and we should be cautious about advancing *a priori* answers to them.[7] We might, I suppose, have found Mark I image-manipulation systems in our brains, complete with glowing phosphor dots and an inner light-sensitive eye. (So far as I can see, it is not *impossible* that the creatures on some planet might be endowed with such contraptions.) And it takes experiments such as Reisberg and Chambers' to show that the shortcuts hit upon by our brains pretty well preclude discovering a Mark II system, with a bit-map format that never takes advantage of shortcuts. (If we had such a system, solving the red X puzzle in our heads would be easy, and so would rotating Texas.)

Phenomenology provides clues, pointing in both directions: The "sketchiness" of mental images, which is "intuitively obvious" in most subjects' phenomenology, points to the brain's use of shortcuts, occasions on which the brain tells-without-showing. This is as true of visual *perception* as of *visualizing*. We already noted in chapter 2 how hard it is to draw a rose that is right in front of your eyes, or even to copy a drawing, and the reason is that the purely spatial properties that one must identify or discriminate in order to draw well have normally been left behind in the course of perceptual processing, summarized in reports, not rendered for further perusal. On the other hand, the usefulness of mental images in helping us "see the pattern" or in "reminding us" of details we might otherwise forget, points to the exploitation of visual pattern-recognition machinery that could only occur if one part of the brain went to all the trouble of preparing special-format versions of the information for the benefit of those visual systems. As we already saw in chapter 1, the information-handling demands of such representation are formidable, and it should not surprise us that we are so poor at keeping even highly schematic diagrams stable in our heads.

Here is a simple test to remind you how limited our abilities actually are: *In your mind's eye*, fill in the following three-by-three crossword puzzle, writing the following three words *down* in the columns, starting with the left column: *GAS OIL DRY*

---

experimental and theoretical work on imagery. A good review of the work in the subsequent decade can be found in Farah (1988), and in Finke, Pinker, and Farah (1989).

7. As Marvin Minsky puts it, "There is nothing peculiar about the idea of sensing events inside the brain. Agents are agents — and it is as easy for an agent to be wired to detect a *brain-caused brain-event*, as to detect a *world-caused brain-event*" (1985, p. 151).

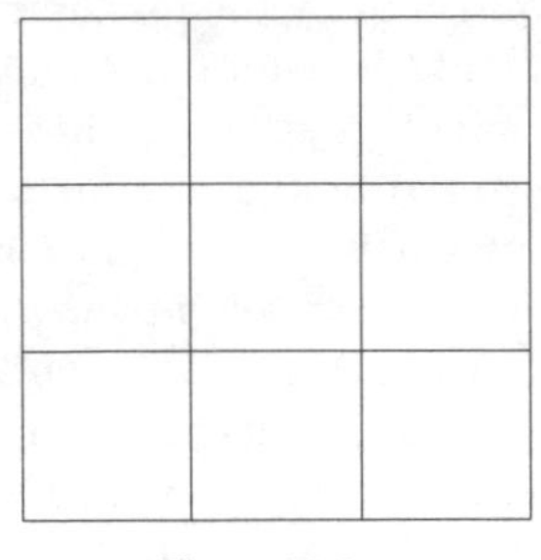

Figure 10.6

Can you readily *read off* the horizontal words? In an actual diagram on paper, these words would "pop out" at you — you would be unable *not* to see them. That, after all, is the point of making diagrams: to present the data in a format that makes a new breakdown or parsing of the data easy or inevitable. A three-by-three array of alphabetic characters is not a very complicated data structure, but it is apparently not something our brains can hold in place steadily enough for its visual systems to do their "pop out" work. (If you want to try again, here are two more groups of words for the columns: *OPT NEW EGO*, and *FOE INN TED*.)

There is plenty of scope for individual variation in the tactics employed by different visualizers, however, and some may be able to find — or develop — imaging strategies that permit them to "read off" these diagrams. Calculating prodigies can teach themselves how to multiply ten-digit numbers together in their heads, and so it would not be surprising if some people can develop prodigious talents for "crossword reading" in their mind's eyes. These informal demonstrations give us hints, but experiments can define much more incisively the sorts of mechanisms and processes that people must be using in these acts of self-manipulation. The evidence to date supports the view that we use a mixed strategy, getting some of the benefits of visual analysis of arrays, but also incorporating shortcut labels, *telling* without *showing*.

Notice, however, that even in the Mark II CADBLIND system, which is ultrapictorial, incorporating a bit-map that renders color, shading, and texture pixel by pixel, there is still a sense — and a metaphysically important sense, as we shall see in the next two chapters — in which it is all "tell" and no "show." Consider the red X on our Shepard figure (figure 4). In the Mark I this is rendered in real red —

the CRT emits light, which has to be transduced by something in the TV camera analogous to the cones in your eyes that respond to frequency differences. When the Vorsetzer rotates the image back and forth, hunting for a glimpse of red through the hole, it is waiting for its red-detector demons to shout. In the Mark II, that hardware is thrown away, and the bit-map represents the color of each pixel by a number. Perhaps the shade of red is color number 37. The Vorsetzer in the Mark II, when it rotates the bit-map image back and forth, is peering through the hole for a glimpse of 37. Or in other words, it is asking whether any of the pixel-demons wants to tell it, "Color number 37 here." All the red is gone — there are only numbers in there. In the end, all the work in a CADBLIND system must be done by arithmetic operations on bit strings, just as we saw at the lowest level of Shakey in chapter 4. And no matter how quasi-pictorial or imagistic the processes are that eventuate in the Vorsetzer's verbal answers to questions, they will not be generated in an inner place where the lost properties (the properties merely "talked about" in the bit-map) are somehow *restored* in order to be *appreciated* by a judge that composes the answers.

People are not CADBLIND systems. The fact that a CADBLIND system can manipulate and inspect its "mental images" without benefit of a Cartesian Theater doesn't by itself prove that there is no Cartesian Theater in the human brain, but it does prove that we don't have to postulate a Cartesian Theater to explain the human talent for solving problems "in the mind's eye." There are indeed mental processes that are strongly *analogous* to observation, but, when we strip down Kosslyn's CRT metaphor to its essentials, we remove the very features that would call for a Cartesian Theater. There need be no time and place where "it all comes together" for the benefit of a single, unified discriminator; the discriminations can be accomplished in a distributed, asynchronous, multilevel fashion.

2. WORDS, PICTURES, AND THOUGHTS

The truly "creative" aspect of language resides not in its "infinite generative capacity" but in cycles of production and comprehension mediated by a mind capable of reflecting upon the multiple meanings attachable to an utterance, meanings that need not have been present in the thought that gave rise to the utterance but which become available through self-comprehension (or deep interpretation of another's utterance) and can lead to a new thought to be expressed and re-interpreted, and so on indefinitely.

H. STEPHEN STRAIGHT (1976), p. 540.

The British economist John Maynard Keynes was once asked whether he thought in words or pictures. "I think in thoughts," was his reply. He was right to resist the suggestion that the "things we think in" are either words or pictures, for as we have seen, "mental images" are not *just* like pictures in the head, and "verbal" thinking is not *just* like talking to yourself. But saying one thinks in thoughts is really no improvement. It just postpones the question, for a thought is just *whatever happens when we think* — a topic about which there is no settled agreement.

Now that we have considered sketches of the sorts of background machinery that is causally responsible for the details of our heterophenomenological worlds, we can *begin* to account for the phenomenology of thinking, explaining not only the limits and conditions on "visual" and "verbal" phenomenology, but looking for yet other varieties that escape that dichotomy.

One of my favorite exercises in fictional heterophenomenology is Vladimir Nabokov's novel *The Defense* (1930), about Grandmaster Luzhin, a genius of the chessboard, who suffers a nervous breakdown in the middle of his climactic match. We see three stages in the development of his consciousness: his boyish mind (before his discovery of chess at about age ten), his chess-saturated mind (up until his nervous breakdown), and the sorry remains of the first two stages after his breakdown, when, imprisoned by his wife in a world without chess — without chess talk, chess playing, chess books — his mind reverts to a sort of spoiled infantile paranoia, brightened by stolen moments of chess — lightning sneak attacks on the chess diagrams and puzzles in the newspapers — but ultimately curdling into chess obsessions that culminate in his "sui-mate." Luzhin, we learn, has so saturated his mind with chess that he sees his whole life in its terms. Here is his awkward courtship of the woman who will become his wife:

> Luzhin began with a series of quiet moves, the meaning of which he himself only vaguely sensed, his own peculiar declaration of love. "Go on, tell me more," she repeated, despite having noticed how morosely and dully he had fallen silent.
>
> He sat leaning on his cane and thinking that with a Knight's move of this lime tree standing on a sunlit slope one could take

> that telegraph pole over there, and simultaneously he tried to remember what exactly he had just been talking about [p. 97].

> With one shoulder pressed against his chest she tried with a cautious finger to raise his eyelids a little higher and the slight pressure on his eyeball caused a strange black light to leap there, to leap like his black Knight which simply took the Pawn if Turati moved it out on the seventh move, as he had done at their last meeting. [p. 114]

Here is a glimpse of his state of mind after his breakdown:

> He found himself in a smoky establishment where noisy phantoms were sitting. An attack was developing in every corner — and pushing aside tables, a bucket with a gold-necked glass Pawn sticking out of it and a drum that was being beaten by an arched, thick-maned chess Knight, he made his way to a gently revolving door. . . . [p. 139]

These themes are "images" in many regards, for chess is a spatial game, and even the identity of the pieces is standardly fixed by their shapes, but the power of chess over Luzhin's mind is not exhausted by the visual or spatial properties of chess — everything that might be captured in photographs or films of a chessboard, its pieces in motion. Indeed, these visual properties provide only the most superficial flavor to his imagination. Much more powerful is the *discipline* provided by the rules and tactics of the game; it is the abstract structure of chess that he has become so obsessively familiar with, and it is his habits of exploration of this structure that drive his mind from "thought" to "thought."

> [He] presently would note with despair that he had been unwary again and that a delicate move had just been made in his life, mercilessly continuing the fatal combination. Then he would decide to redouble his watchfulness and keep track of every second of his life, for traps could be everywhere. And he was oppressed most of all by the impossibility of inventing a rational defense, for his opponent's aim was still hidden [p. 227].

When you first learned to ride a bicycle or drive a car, you encountered a new structure of opportunities for action, with constraints, signposts, ruts, vistas, a sort of abstract behavioral maze in which you quickly learned your way around. It soon became "second nature." You quickly incorporated the structure of that external phenomenon into your own control structure. In the process, you may have had periods of obsessive exploration, when you couldn't get your mind off the new moves. I remember a brief period of bridge mania during my adolescence during which I had obsessive and nonsensical bridge dreams. I would take the same finesse a hundred times, or dream of "bidding" during conversations with my teachers and classmates. My hypnogogic reveries (those rather hallucinatory periods people occasionally have as they are falling asleep or just beginning to wake up) were filled with problems along the lines of "what is the correct response to a preemptive bid of three books — four knives or four forks?"

It is quite common when encountering a new abstract structure in the world — musical notation, a computer programming language, common law, major league baseball — to find yourself trudging back and forth over its paths, making mind-ruts for yourself — really digging in and making yourself at home. Luzhin is only the extreme case; he has only one structure with which to play, and he uses it for everything. And eventually its structure dominates all other habit structures in his mind, channeling the sequence of his thoughts almost as rigidly as the sequence of instructions in a von Neumann machine program.

Think of all the structures you have learned, in school and elsewhere: telling time, arithmetic, money, the bus routes, using the telephone. But of all the structures we become acquainted with in the course of our lives, certainly the most pervasive and powerful source of discipline of our minds is our native tongue. (One often sees best by looking at contrasts; Oliver Sacks, in *Seeing Voices*, 1989, vividly draws attention to the riches language brings to a mind by recounting the terrible impoverishment of a deaf child's mind, if that child is denied early access to a *natural* language — Sign, or sign language.) In chapter 8 we saw how the very vocabulary at our disposal influences not only the way we talk to others, but the way we talk to ourselves. Over and above that *lexical* contribution is the *grammatical* contribution. As Levelt points out (1989, sec. 3.6), the obligatory structures of sentences in our languages are like so many guides at our elbows, reminding us to check on this, to attend to that, *requiring* us to organize facts in certain ways. Some of this structure may indeed be innate, as Chomsky and others have argued, but it really doesn't matter where the dividing line is drawn between structures that are genetically deposited in the brain and those that enter as memes. These structures, real or virtual, lay down some of the tracks on which "thoughts" can then travel.

Language infects and inflects our thought at every level. The words in our vocabularies are catalysts that can precipitate fixations of content as one part of the brain tries to communicate with another. The structures of grammar enforce a discipline on our habits of thought, shaping the ways in which we probe our own "data bases," trying, like Plato's bird-fancier, to get the right birds to come when we call. The structures of the stories we learn provide guidance at a different level, prompting us to ask ourselves the questions that are most likely to be relevant to our current circumstances.

None of this makes any sense so long as we persist in thinking of the mind as ideally rational, and perfectly self-transparent or unified. What good could *talking to yourself* do, if you already know what you intended to say? But once we see the possibility of partial understanding, imperfect rationality, problematic intercommunication of parts, we can see how the powerful forces that a language unleashes in a brain can be exploited in various forms of bootstrapping, some of them beneficial, and some of them malignant.

Here is an example.

> You are magnificent!

Here is another:

> You are pathetic!

You know what these sentences mean. You also know that I have just introduced them out of the blue, as an aid to making a philosophical point, and that they are not the intended speech acts of anyone. Certainly I am neither flattering you nor insulting you, and there is no one else around. But could you flatter yourself, or insult yourself, by helping yourself to one or the other of my sentences, and saying it to yourself, over and over, "with emphasis"? Try it, if you dare. Something happens. You don't believe yourself for one minute (you say to yourself), but you find that saying the words to yourself does kindle reactions, maybe even a little reddening of the ears, along with responses, retorts, disclaimers, images, recollections, projects. These reactions may go either way, of course. Dale Carnegie was right about the power of positive thinking, but like most technologies, thinking is easier to create than to control. When you talk to yourself, you don't have to believe yourself in order for reactions to set in. There are bound to be some reactions, and they are bound to be relevant one way or the other to the meaning of the words with which you are stimulating yourself. Once the reactions start happening, they may lead your mind to places where you find yourself believing yourself after all — so be careful what you say to yourself.

The philosopher Justin Leiber sums up the role of language in shaping our mental lives:

> Looking at ourselves from the computer viewpoint, we cannot avoid seeing that natural language is our most important "programming language." This means that a vast portion of our knowledge and activity is, for us, best communicated and understood in our natural language. . . . One could say that natural language was our first great original artifact and, since, as we increasingly realize, languages are machines, so natural language, with our brains to run it, was our primal invention of the universal computer. One could say this except for the sneaking suspicion that language isn't something we invented but something we became, not something we constructed but something in which we created, and recreated, ourselves. [Leiber, 1991, p. 8]

The hypothesis that language plays this all-important role in thinking might seem at first glance to be a version of the much-discussed hypothesis that there is a *language of thought*, a single medium in which all cognition proceeds (Fodor, 1975). There is an important difference, however. Leiber aptly calls natural language a programming language for the brain, but we may distinguish high-level programming languages (such as Lisp and Prolog and Pascal) from the basic "machine language" or slightly less basic "assembly language" out of which these high-level languages are composed. High-level languages are virtual machines, and they create (temporary) structures in a computer that endow it with a particular pattern of strengths and weaknesses. The price one pays for making certain things "easy to say" is making other things "hard to say" or even impossible. Such a virtual machine may structure only part of the computer's competence, leaving other parts of the underlying machinery untouched. Bearing this distinction in mind, it is plausible to maintain that the details of a natural language — the vocabulary and grammar of English or Chinese or Spanish — constrain a brain in the manner of a high-level programming language. But this is a far cry from asserting the dubious hypothesis that such a natural language provides the structure *all the way down*. Indeed, Fodor and others who defend the idea of the language of thought typically insist that they are *not* talking about the level at which *human* languages do their constraining work. They are talking about a deeper, less-accessible level of representation. Fodor once made the point with the aid of an amusing confession: he acknowledged that when he was thinking his hardest, the only sort of linguistic items he was conscious of were snatches along the lines of "C'mon, Jerry, you can do it!" Those may have been his "thoughts," and we have just seen how they may in fact play an important role in helping him solve the problems that confronted him, but they are hardly the stuff out of which to fashion perceptual inferences, hypotheses to be tested, and the other postulated transactions of the ground-level language of thought. Keynes was right to resist the words-versus-pictures choice; the media used by the brain are only weakly analogous to the representational media of public life.

3. REPORTING AND EXPRESSING

Slowly but surely we've been chipping away at the idea of the Cartesian Theater. We sketched an alternative to the Central Meaner in chapter 8, and we've just seen how to resist the appeal of an inner CRT. Mere glancing blows, I fear; the Cartesian Theater is still standing, still exerting a tenacious pull on our imaginations. It's time to shift tactics and attack from within, exploding the Cartesian Theater by showing its incoherence in its own terms. Let's see what happens when we go along with tradition, accepting the terms of everyday "folk psychology" at face value. We can begin by reconsidering some of the plausible claims Otto made at the beginning of chapter 8:

> When *I* speak, [Otto said] I mean what I say. My conscious life is private, but I can choose to divulge certain aspects of it to you. I can decide to tell you various things about my current or past experience. When I do this, I formulate sentences that I carefully tailor to the material I wish to report on. I can go back and forth between the experience and the candidate report, checking the words against the experience to make sure I have found *les mots justes*. . . . I attend to my particular conscious experience and arrive at a judgment about which words would do the most justice to its character. When I am satisfied that I have framed an accurate report, I express it. From my introspective report, you can come to know about some feature of my conscious experience.

Some of this message actually fits smoothly with our proposed model of language-production in chapter 8. The process of back-and-forth tailoring of the words to the content of the experience can be seen in the pandemonium that mates up the word-demons to the content-demons. What is missing, of course, is the Inner I whose judgments direct the matchmaking. But although Otto does go on about what "*I* choose" and what "*I* judge," introspection doesn't really support this.

We have scant access to the processes by which words "occur to us" to say, even in the cases where we speak deliberately, rehearsing our speech acts silently before uttering them. Candidates for something to say just spring up from we know not where. Either we find ourselves already saying them, or we find ourselves checking them out, sometimes discarding them, other times editing them slightly and then saying them, but even these occasional intermediate steps give us no further hints about how we do them. We just find ourselves accepting or discarding this word and that. If we have reasons for our judgments, they are seldom contemplated before the act, but only retrospectively obvious. ("I was going to use the word *jejune* but stopped myself, since it would have sounded so pretentious.") So we really have no privileged insight into the processes that occur in us to get us from thought to speech. They *might* be produced by a pandemonium, for all we know.

> But still, [Otto goes on] the Pandemonium model leaves out a level or a stage in the process. What your model lacks is not a projection into the "phenomenal space" of a Cartesian Theater — what a ridiculous idea! — but still an extra level of articulation in the psychology of the speaker. It's not enough for words to get strung together by some internal mating dance and then uttered. If they are to be *reports* of someone's conscious mental states, they have to be based somehow on an act of inner *apprehension*. What the Pandemonium model leaves out is the speaker's state of awareness that guides the speech.

Whether or not Otto is right, he is certainly expressing the common wisdom: this is just how we ordinarily conceive of our ability to tell people about our conscious states. In a series of recent papers, the philosopher David Rosenthal (1986, 1989, 1990a, b) has analyzed this everyday concept of consciousness and its relation to our concepts of *reporting* and *expressing*. He uncovers some structural features we can put to good use. First, we can use his analysis to see, from the inside, what the standard picture is and why it's so compelling. Second, we can show how it discredits the idea of zombies — with no help from the outside. Third, we can turn the standard picture against itself, and use the difficulties we encounter to motivate a better picture, one that preserves what is right about the traditional view but discards the Cartesian framework.

Figure 10.7

What happens when we speak? At the heart of our everyday conception of this there is a truism: Provided we're not lying or insincere, we say what we think. To put it more elaborately, we express one of our beliefs or thoughts. Suppose, for instance, you see the cat anxiously waiting by the refrigerator and you say, "The cat wants his supper." This expresses your belief that the cat wants his supper. In *expressing* your belief, you are *reporting* what you take to be a fact about the cat. In this case you're reporting the cat's desire to be fed. It's important to note that you're not *reporting* your belief, or *expressing* the cat's desire. The cat is expressing his desire by standing anxiously by the refrigerator, and you, noticing this, use it as the basis — the evidence — for your report. There are many ways of expressing a mental state (such as a desire), but only one way of reporting one, and that is by uttering a speech act (oral or written or otherwise signaled).

One of the most interesting ways of *expressing* a mental state is by *reporting* another mental state. In the example, you report the cat's desire, thereby expressing your own belief about the cat's desire. Your behavior is evidence not only that the cat has the desire but also that you believe the cat has the desire. You might, however, have given us evidence of your belief in some other way — perhaps by silently getting up from your chair and preparing supper for the cat. This would have *expressed* the same belief without *reporting* anything. Or you might have just sat in your chair and rolled your eyes, *unintentionally* expressing your exasperation with the cat's desire just when you had gotten comfortable in your chair. Expressing a mental state, deliberately or not, is just doing something that provides good evidence for, or makes manifest, that state to another observer — a mind reader, if you like. Reporting a mental state, in contrast, is a more sophisticated activity, always intentional, and involving language.

Here, then, is an important clue about the source of the Cartesian Theater model: Our everyday folk psychology treats reporting one's own mental state on the model of reporting events in the external world. Your report that the cat wants his supper is based on your observation of the cat. Your report expresses your belief that the cat wants his supper, a belief about the cat's desire. Let's call beliefs about beliefs, desires about desires, beliefs about desires, hopes about fears, and so forth *second-order* mental states. And if I (1) *believe* that you (2) *think* that I (3) *want* to have a cup of coffee, that belief of mine is a *third*-order belief. (For the importance of higher-order mental states in theories of mind, see my *Intentional Stance*, 1987a.) There can be no doubt that there are salient, important differences marked by these everyday distinctions when they are applied nonreflexively — when x believes that *y* is in some mental state and $x \neq y$. There is all the difference in the world between the case where the cat wants to be fed and you know it, and the case where the cat wants to be fed and you don't know it. But what about the reflexive cases, where $x = y$? Folk psychology treats these cases just the same.

Suppose I report that *I* want to be fed. On the standard model I must be expressing a *second-order belief* about *my* desire. When I report my desire, I express a second-order belief — my belief about my desire. What if I *report* that second-order belief, by saying "I believe I want to be fed"? That report must express a *third*-order belief — my belief that I do in fact believe that I want to be fed. And so on. Our everyday concepts of what is involved in speaking sincerely generate in this way a host of putatively distinct mental states: my desire is distinct from my belief that I have the desire, which is distinct from my belief that I have the belief that I have the desire, and so on.

Folk psychology also goes into further distinctions. As Rosenthal points out (along with many others), it distinguishes *beliefs*, which are the underlying dispositional states, from *thoughts*, which are occurrent or episodic states — transient events. Your *belief that dogs are animals* has persisted continuously as a state of your mind for years, but my drawing attention to it just now has spawned a *thought* in you — the thought that dogs are animals, an episode that no doubt would not have occurred in you just now without my provocation.

It follows, of course, that there can be first-order thoughts and second- and higher-order thoughts — thoughts about thoughts (about thoughts . . .). Here, then, is the crucial step: When I express a belief — such as my belief that I want to be fed — I do not express the higher-order belief directly; what happens is that my underlying belief yields an episodic thought, the higher-order *thought* that I want to be fed, and I express *that thought* (if I choose to do so). All this is involved, Rosenthal argues, in the common sense model of *saying what you think*.

Since a hallmark of states of human consciousness is that they can be reported (barring aphasia, paralysis, or being bound and gagged, for instance), it follows, on Rosenthal's analysis, that "conscious states must be accompanied by suitable higher-order thoughts, and nonconscious mental states cannot be thus accompanied" (1990b, p. 16). The higher-order thought in question must of course be about the state it accompanies; it must be the thought that one is in the lower-order state (or has just been in it — time marches on). This looks as if it is about to generate an infinite regress of higher-order conscious states or thoughts, but Rosenthal argues that folk psychology permits a striking inversion: *The second-order thought does not itself have to be conscious in order for its first-order object to be conscious.* You can *express* a thought without being conscious of it, so you can express a *second-order* thought without being conscious of *it* — all you need be conscious of is its object, the first-order thought you *report*.

This may seem surprising at first, but on reflection we can recognize this as a familiar fact in a new perspective: You do not attend to the thought you express, but to the object(s) that thought is *about*. Rosenthal goes on to argue that although some second-order thoughts *are* conscious — by virtue of third-order thoughts about them — these are relatively rare. They are the explicitly introspective thoughts that we would report (even to ourselves) only when in a state of hyper-self-consciousness. If I say to you, "I am in pain," I report a conscious state, my pain, and express a second-order belief — my belief that I am in pain. If, waxing philosophical, I say "I think [or I am sure, or I believe] that I am in pain," I thereby report a second-order thought, expressing a third-order thought. Ordinarily, though, I would not have such a third-order thought and hence would not be conscious of such a second-order thought; I would express it, in saying "I am in pain," but would not normally be conscious of it.

This idea of unconscious higher-order thoughts may seem outrageous or paradoxical at first, but the category of episodes in question is not really controversial, even though the term "thought" is not ordinarily used to refer to them. Rosenthal uses "thought" as a technical term — roughly following Descartes's practice, in fact — to cover *all* episodic contentful states, not just the episodes that we would ordinarily call thoughts. Thus a twinge of pain or a glimpse of stocking would count as thoughts for Descartes and Rosenthal. Unlike Descartes, however, Rosenthal insists on the existence of unconscious thoughts.

Unconscious thoughts are, for instance, unconscious perceptual events, or episodic activations of beliefs, that occur naturally — that *must* occur — in the course of normal behavior control. Suppose you tip over your coffee cup on your desk. In a flash, you jump up from the chair, narrowly avoiding the coffee that drips over the edge. You were not conscious of thinking that the desk top would not absorb the coffee, or that coffee, a liquid obeying the law of gravity, would spill over the edge, but such unconscious thoughts must have occurred — for had the cup contained table salt, or the desk been covered with a towel, you would not have leaped up. Of all your beliefs — about coffee, about democracy, about baseball, about the price of tea in China — these and a few others were immediately relevant to your circumstances. If we are to cite them in an explanation of why you leaped up, they must have been momentarily accessed or activated or in some way tapped for a contribution to your behavior, but of course this happened unconsciously. These unconscious episodes would be examples of what Rosenthal calls unconscious thoughts. (We have already encountered unconscious thoughts in some earlier examples: for instance, the unconscious perceptions of vibration in the fingers that permit you to identify, consciously, the textures of things touched with a wand, the unconscious recollection of the woman with the eyeglasses, which led to the mistaken experience of the woman rushing by.)

Rosenthal points out that by finding a way of defining consciousness in terms of unconscious mental states (the accompanying higher-order thoughts), he has uncovered a way of laying the foundations *within folk psychology* for a noncircular, nonmysterious theory of consciousness (1990b). What distinguishes a conscious state from a nonconscious state, he argues, is not some inexplicable intrinsic property, but the straightforward property of having a higher-order accompanying thought that is about the state in question. (See Harnad, 1982, for a similar strategy with some interesting variations.) This looks good for folk psychology: it doesn't just wallow in mystery; it has the resources, well mined by Rosenthal, to articulate an account of its prize category, consciousness, in terms of its subsidiary and less-problematic categories. Part of the bargain, if we opt for his analysis, is that it can be used to break down the putatively sharp distinction between conscious beings and zombies.

4. ZOMBIES, ZIMBOES, AND THE USER ILLUSION

Mind is a pattern perceived by a mind. This is perhaps circular, but it is neither vicious nor paradoxical.

DOUGLAS HOFSTADTER (1981), p. 200

Philosophers' zombies, you will recall, seem to perform speech acts, seem to report on their states of consciousness, seem to introspect. But they are not really conscious at all, in spite of the fact that they are, at their best, behaviorally indistinguishable from a conscious person. They may have internal states with functional content (the sort of content that functionalists can assign to the inner machinery of robots), but these are unconscious states. Shakey — as we imagined him — is a paradigmatic zombie. When he "reports" an internal state, this is not a conscious state that is being reported, since Shakey has no conscious states, but an unconscious state that merely causes him to go into some further unconscious state that directs the process of generating and executing a so-called speech act composed of "canned" formulae. (We've been letting Otto insist on this all along.)

Shakey didn't *first* decide what to *report*, after observing something going on inside, and *then* figure out how to *express it;* Shakey just found himself with things to say. Shakey didn't have any access into why he wanted to say he was forming line drawings around the light-dark boundaries of his mental images — he was just built that way. The central claim of chapter 8, however, was that, contrary to first appearances, the same is also true of you. You don't have any special access into why you want to say what you find you want to say; you are just built that way. But, unlike Shakey, you are constantly rebuilding yourself, discovering new things you want to say as a result of reflecting on what you have just found yourself wanting to say, and so forth.

But couldn't a fancier Shakey do that as well? Shakey was a particularly crude zombie, but we can now imagine a more realistic and complex zombie, which monitors its own activities, including even its own internal activities, in an indefinite upward spiral of reflexivity. I will call such a reflective entity a *zimbo*. A zimbo is a zombie that, as a result of self-monitoring, has internal (but unconscious) higher-order informational states that are about its other, lower-order informational states. (It makes no difference for this thought experiment whether a zimbo is considered to be a robot or a human — or Martian — entity.) Those who believe that the concept of a zombie is coherent must surely accept the possibility of a zimbo. A zimbo is just a zombie that is behaviorally complex, thanks to a control system that permits recursive self-representation.

Consider how a zimbo might perform in the Turing test, Alan Turing's famous proposal (1950) of an operational test for thinking in a computer. A computer can think, Turing proclaimed, if it can regularly beat a human opponent in the "imitation game": The two contestants are hidden from a human judge but able to communicate with the judge by typing messages back and forth via computer terminals. The human contestant simply tries to convince the judge that he or she is human, while the computer contestant does likewise — tries to convince the judge that *it* is human. If the judge cannot regularly spot the computer, the computer is deemed a thinker. Turing proposed his test as a conversation-stopper; it was obvious to him that this test was so ridiculously difficult to pass that any computer that could win should be seen by everyone to be an *amazingly* good thinker. He thought he had set the bar high enough to satisfy any skeptic. He misestimated. Many have argued that "passing the Turing test" would not be a sufficient proof of intelligence, and certainly not of consciousness. (See Hofstadter, 1981b, Dennett, 1985a, and French, 1991, for analyses of the strengths and weaknesses of the Turing test and its critics.)

Now, a zimbo's chances in the Turing test should be as good as any conscious person's, since the contestants show the judge nothing but their behavior, and only their verbal (typing) behavior at that. Suppose, then, that you are the judge in a Turing test, and a zimbo's (apparent) speech acts have convinced you that it is conscious. These apparent speech acts shouldn't have convinced you — *ex hypothesi,* for it is just a zimbo, and zimboes aren't conscious. Should it have convinced itself, though? When a zimbo issues a report, expressing its own second-order unconscious state, there is nothing to prevent it from reflecting (unconsciously) on this very state of affairs. In fact, if it's going to be convincing, it's going to have to be able to respond appropriately to (or take cognizance of) its own "assertions" to you.

Suppose, for instance, that the zimbo is a fancier Shakey, and you, as judge, have just asked it to solve a problem in its mind's eye and then explain how it did it. It reflects on its own assertion to you that it has just solved the problem by forming a line drawing on a mental image. It would "know" that that was what it had wanted to say, and if it reflected further, it could come to "know" that it didn't know why that was what it wanted to say. The more you asked it about what it knew and didn't know about what it was doing, the more reflective it would become. Now what we have just succeeded in imagining, it seems, is an unconscious being that nevertheless has the capability for higher-order thoughts. But according to Rosenthal, when a mental state is accompanied by a conscious *or* unconscious higher-order thought to the effect that one has it, this *ipso facto* guarantees that the mental state is a conscious state! Does our thought experiment discredit Rosenthal's analysis, or discredit the definition of a zimbo?

We can readily see that at the very least the zimbo would (unconsciously) believe that it was in various mental states — precisely the mental states it is in position to report about should we ask it questions. *It* would think it was conscious, even if it wasn't! Any entity that could pass the Turing test would operate under the (mis?)apprehension that it was conscious. In other words, it would be a victim of an illusion (cf. Harnad, 1982). What kind of an illusion? A User illusion, of course. It would be the "victim" of the benign user illusion of its own virtual machine!

Isn't this a trick with mirrors, some illegitimate sort of philosopher's sleight of hand? How could there be a User illusion without a Cartesian Theater in which the illusion is perpetrated? I seem to be in imminent danger of being done in by my own metaphors. The problem is that a virtual machine's user illusion *is* accomplished by a presentation of material in a theater of sorts, and there is an independent, external audience, the User, for whose benefit the show is put on. I am using a computer at this very moment, typing these words into a "file" with the unobtrusive assistance of a word-processing program. When

I interact with the computer, I have limited access to the events occurring within it. Thanks to the schemes of presentation devised by the programmers, I am treated to an elaborate audiovisual metaphor, an interactive drama acted out on the stage of keyboard, mouse, and screen. I, the User, am subjected to a series of benign illusions: I seem to be able to move the cursor (a powerful and visible servant) to the very place in the computer where I keep my file, and once that I see that the cursor has arrived "there," by pressing a key I get it to retrieve the file, spreading it out on a long scroll that unrolls in front of a window (the screen) at my command. I can make all sorts of things happen inside the computer by typing in various commands, pressing various buttons, and I don't have to know the details; I maintain control by relying on my understanding of the detailed audiovisual metaphors provided by the User illusion.

For most computer-users, it is only in terms of these metaphors that they have any appreciation of what is happening inside. This is one of the facts that makes a virtual machine such a tempting analogy for consciousness, for it has always seemed that our access to what is happening inside our own brains is limited; *we* don't have to know how the backstage machinery of our brains does its magic; *we* are acquainted with its operations only as they come clothed for us in the interactive metaphors of phenomenology. But if, when we avail ourselves of this tempting analogy, we maintain the "obvious" separation between Presentation on the one hand and User Appreciation of the show on the other, we seem to end up right back in the Cartesian Theater. How could there be a User illusion without this separation?

There couldn't be; the user that provides the perspective from which the virtual machine becomes "visible" has to be some sort of external observer — a *Vorsetzer*. And one might at first think that the concept of such an observer had to be the concept of a conscious observer, but we have already seen that this is not so. The Vorsetzer that sat in front of the CAD system in the original Mark I CADBLIND system was not conscious, but was nevertheless exactly as limited in its access to the inner workings of the CAD system as any conscious user would be. And once we discard the gratuitous screen-and-camera, the Presentation and User Appreciation evaporate, replaced, as so often before in our account, by a host of more modest transactions. The "external observer" can be gradually incorporated into the system, leaving behind a few fossil traces: bits of "interface" whose various formats continue to constrain the sorts of questions that can be answered, and thus constrain the contents that can be expressed.[8] There doesn't have to be a *single* place where the Presentation happens.[9] And as Rosenthal's analysis suggests to us, even our ordinary concept of consciousness, as anchored in the intuitions of common sense or folk psychology, can tolerate the unconsciousness of the higher-order states whose presence in the system accounts for the consciousness of some of its states.

Is the process of unconscious reflection, then, a path by which a zombie could turn itself into a zimbo, and *thereby* render itself conscious? If it is, then zombies must be conscious after all. All zombies are capable of uttering convincing "speech acts" (remember, they're indistinguishable from our best friends), and this capability would be magical if the control structures or processes causally responsible for it in the zombie's brain (or computer or whatever) were not reflective about the acts and their (apparent, or functional) contents. A zombie might begin its career in an uncommunicative and unreflective state, and hence truly be a zombie, an unconscious being, but as soon as it began to "communicate" with others and with itself, it would become equipped with the very sorts of states, according to Rosenthal's analysis, that suffice for consciousness.

If, on the other hand, Rosenthal's analysis of consciousness in terms of higher-order thoughts is rejected, then zombies can live on for another day's thought experiments. I offer this parable of the zimboes tongue in cheek, since I don't think *either* the concept of a zombie *or* the folk-psychological categories of higher-order thoughts can survive

---

8. In chapter 7 (p. 223), I asked, "teased to the surface of *what?*" and promised to answer the question later. This is my answer. The (metaphorical) surface is determined by the format of interactions between the parts.

9. It is interesting to compare different traces of this idea of the User-in-the-brain in the work of widely disparate thinkers. Here is Minsky (1985): "To overstate the case a bit, what we call 'consciousness' consists of little more than menu lists that flash, from time to time, on mental screen displays that other systems use [p. 57]. . . . Divide the brain into two parts. A and B. Connect the A-brain's inputs and outputs to the real world — so it can sense what happens there. But don't connect the B-brain to the outer world at all; instead, connect it so that the A-brain is the B-brain's world!" [p. 59] Minsky wisely refrains from venturing any anatomical dividing lines, but others are prepared to venture a few. When Kosslyn first speculated about consciousness as a virtual machine, he was inclined to locate the User in the frontal lobes (see also Kosslyn, 1980, p. 21), and more recently Edelman has followed his own argument path to the same conclusion, put in terms of the "value-dominated self/nonself memory" that he locates in the frontal lobes and assigns the task of interpreting the productions of the rest of the brain (Edelman, 1989, p. 102ff).

except as relics of a creed outworn. Rosenthal has done us an excellent service in exposing the logic of these everyday concepts, however, and thanks to the clear view we now have of them, we can see what a better replacement would be.

5. PROBLEMS WITH FOLK PSYCHOLOGY

Rosenthal finds folk psychology positing an ever-expandable hierarchy of higher-order thoughts, conceived to be salient, independent, contentful episodes occurring in real time in the mind. How does this vision stand up, when we look for confirmation? Are there such distinct states and events in the brain? If we're generous about what counts, the answer must be yes. There certainly are familiar psychological differences that can be — and typically are — described in these terms.

> It suddenly occurred to Dorothy that she wanted to leave — and had wanted to leave for quite some time.

Here it seems Dorothy acquired the second-order belief — by having a second-order thought — about her desire some time after the desire had gone into effect. There are many everyday cases of this sort: "And then it occurred to him that he was looking right at the missing cuff link." "He loves her — he just doesn't realize it yet." It can hardly be denied that these ordinary sentences allude to genuine transitions from one "state of mind" to another. And intuitively, as Rosenthal points out, the transition is a matter of *becoming conscious* of the first-order state. When Freud, building on such everyday cases, postulates a vast hidden realm of unconscious mental states, they are precisely states that their subjects do not believe they are in. These people are in states of mind that it hasn't yet occurred to them — via higher-order thoughts — they are in.

This way of describing these differences is familiar, but whether it is entirely perspicuous is another matter. These are all transitions into a better-informed state (to speak as neutrally as possible), and being better informed in this way is indeed a necessary condition for *reporting* (as opposed to merely *expressing*) the earlier "state of mind." Now the incautious way of saying this would be: *In order to* report a mental state or event, you have to have a higher-order thought which you express. This gives us the picture of first *observing* (with some inner sense organ) the mental state or event, *thereby* producing a state of *belief*, whose onset is marked by a *thought*, which is then expressed. This causal chain, as we saw, mimics the causal chain for reporting ordinary external events: You first observe the event with the help of your sense organs, producing in you a belief, and then a thought, which you express in your report.

This hypothesized higher-order thought is, I think, the "extra level of articulation" Otto thought he could discern in his own psychology; it is the thought Otto's words *express* when he *reports* his own conscious experience. But according to the model of speech-generation we sketched in chapter 8, Otto's model has the causation just backwards. It is not that *first* one goes into a higher-order state of self-observation, creating a higher-order thought, so that one can then report the lower-order thought by expressing the higher-order thought. It is rather that the second-order state (the better-informed state) comes to be *created* by the very process of framing the report. We don't *first* apprehend our experience in the Cartesian Theater and *then*, on the basis of that acquired knowledge, have the ability to frame reports to express; our *being able to say* what it is like *is the basis for* our "higher-order beliefs."[10]

At first a Pandemonium process of speech act design looks wrong because it seems to leave out the central observer/decider whose thought is eventually going to be expressed. But this is a strength, not a weakness, of this model. The emergence of the expression is precisely what creates or fixes the content of the higher-order thought expressed. There need be no *additional* episodic "thought." The higher-order state literally depends on — causally depends on — the expression of the speech act. But not necessarily on the public expression of an overt speech act. In chapter 7, we saw how the organism's need for better and better internal communication of information might have led to the creation of habits of self-manipulation that could take the place of the evolutionarily more-laborious process of creating an inner eye, an

---

10. This is at least close kin to a central theme in Wittgenstein's later work, but Wittgenstein declined to develop any positive account or model of the relation between what we say and what we are talking about when we (apparently) report our mental states. The philosopher Elizabeth Anscombe, in her frustratingly obscure classic, *Intention* (1957), tried to fill this gap left by Wittgenstein, arguing that it was wrong to claim that we *know* what our intentions are; rather we just *can say* what our intentions are. She also attempted to characterize a category of things we can *know without observation*. There is a flawed discussion of these claims in my *Content and Consciousness* (1969), chapters 8 and 9. I have always thought there was something right, and important and original, lurking in those passages. My second pass at them is to be found in "Toward a Cognitive Theory of Consciousness" (1978) reprinted in *Brainstorms*, especially sections IV and V (pp. 164–171 in *Brainstorms*). This section is my current attempt to bring these ideas to light, and is a substantial departure from both earlier versions.

actual internal organ that could monitor the brain. The only way for a human brain to get itself into something like a higher-order belief state, we surmised, is to engage in the process rather like reporting first-order states to itself.

We must break the habit of positing ever-more-central observers. As a transitional crutch, we can reconceive of the process as not knowledge-by-observation but on the model of *hearsay*. I believe that *p* because I have been told by a reliable source that *p*. By whom? By myself — or at any rate, by one or more of my "agents." This is not an entirely alien way of thinking; we do speak, after all, of the *testimony* of our senses, a metaphor that suggests that our senses do not bring exhibits into "court" to *show* us, but rather *tell* us of things. Leaning on this metaphor (until we can get used to the complexities of a better alternative), we might rely on a slogan:

> If I couldn't talk to myself, I'd have no way of knowing what I was thinking.

This is not yet quite the right way to think of it, in several regards. First, there is the difference — which I have been glossing over — between one entity "talking to itself" and a variety of subsystems "talking to each other." A proper transition between these two ideas will be negotiated in chapter 13, on the self. Second, as we've seen, the emphasis on linguistic expression is an overstatement; there are other strategies of self-manipulation and self-expression that are not verbal.

It may well appear that I am proposing a poor bargain: giving up the relative crispness and clarity of the standard folk-psychological model, with its hierarchy of inner observations, for a sketchy alternative we can still scarcely conceive. But the clarity of the traditional model is an illusion, for reasons hinted at in chapter 5, when we explored the strange topic of *real seeming*. Now we can diagnose the problems more precisely. Otto is a spokesperson for folk psychology, and if we let him continue, he will soon tie himself in knots. Otto's view, which doggedly extends the folk-psychological categories "all the way in," generates an explosion of distinct "representational states," the relations between which generate artifactual conundrums. Otto continues:

> My public report of a conscious state, should I choose to make one, might contain a mistake. I might make a slip of the tongue, or be mistaken about what a word means and thus misinform you inadvertently. Any such error of expression I didn't catch would

be liable to create a false belief in you about the facts — about the way it *really* is with me. And the mere fact that I didn't happen to catch an error wouldn't mean that there wasn't an error. *On the one hand,* there is the truth about how it is with me, and then *on the other hand,* there is what I eventually say about how it is with me (if I choose to do so). Although I tend to be a highly reliable reporter, there is always room for errors to creep in.

This is one of those situations where two hands aren't enough. For, as Rosenthal has shown us, in addition to "how it is with me" and "what I eventually say" it seems there must be an intervening third fact: my belief about how it is with me.[11] For when I sincerely say what I say, meaning what I mean, I express one of my beliefs — my belief about how it is with me. Indeed, there is an intervening fourth fact: my episodic thought that this is how it is with me.

Might my belief about how it is with me be mistaken? Or might I *only think* that this is how it is with me? Or in other words, might it only *seem* to me that this was my current experience? Otto wanted one separation, but now we are threatened with more: between the subjective experience and the belief about it, between that belief and the episodic thought it spawns on the way to verbal expression, and between that thought and its ultimate expression. And, like the multiplying brooms of the sorcerer's apprentice, there are more separations in the offing once we accept these. Suppose I have my subjective experience (that's one thing) and it provides the grounds in me for my belief that I'm having it (that's a second thing) which in turn spawns the associated thought (a third thing) which next incites in me a communicative intention to express it (a fourth thing), which yields, finally, an actual expression (a fifth thing). Isn't there room for error to creep into the transition between each thing? Might it not be the case that I believe one proposition but, due to a faulty transition between states, come to *think* a different proposition? (If you can "misspeak," can't you also "misthink"?) Wouldn't it be possible to frame the intention to express a rather different proposition from the one you are thinking? And mightn't a defective memory in the communicative intention subsystem lead you to set out with one preverbal message to be expressed and end up with a different preverbal message serving as the standard against which errors were to be corrected? Between any two distinct

11. In *Brainstorms,* I exploited this feature of folk psychology in my discussion of "β-manifolds of phenomenological beliefs" (1978a, p. 177ff).

things there is logical room for error, and as we multiply individual states with definite contents, we discover — or create — multiple sources of error.

There is a strong temptation to cut through this tangle by declaring that *my thought (or belief) about how it seems to me* is just the same thing as *what my experience actually is*. There is a temptation to insist, in other words, that there is logically no room for error to creep in between them, since they are one and the same thing. Such a claim has some nice properties. It stops the threatened explosion at step one — usually the right place to stop an explosion or regress — and it has some genuine intuitive appeal, brought out nicely in a rhetorical question: What possible sense could be attached to the claim that something only seemed to me to seem to me (to seem to me . . .) to be a horse?

But here we must tread carefully, tiptoeing around the bones of defunct philosophical theories (including some of my own — cf. Dennett, 1969, 1978c, 1979a). It might seem that we can stick with the good old-fashioned folk-psychological categories of beliefs, thoughts, beliefs about beliefs, thoughts about experiences, and the like, and avoid the perplexities of self-knowledge by just merging the higher- and lower-order reflexive cases: by declaring that when I believe that I believe that *p*, for instance, it follows logically that I believe that *p*, and in the same spirit, when I think I'm in pain, it follows logically that I am in pain, and so forth.

If this were so, when I expressed a second-order belief by reporting a first-order belief, for instance, I would really just be dealing with one state, one thing, and the fact that in reporting one thing I was expressing "another" would be due to a mere verbal distinction, like the fact that Jones set out to marry his fiancée and ended up marrying his wife.

But this merger will not quite do the work that needs to be done. To see this, consider once again the role of memory, as conceived of in folk psychology. Even if it is intuitively plausible that you cannot be mistaken about how it *is* with you right *now*, it is not at all intuitively plausible that you cannot be mistaken about how it *was* with you back *then*. If the experience you are reporting is a past experience, your memory — on which you rely in your report — might be contaminated by error. Perhaps your experience was actually one way, but you now misremember it as having been another way. It certainly could seem to you *now* to have seemed to you *then* to have been a horse — even if *in fact* it seemed to you *then* to have been a cow. The logical possibility of misremembering is opened up no matter how short the time interval between actual experience and subsequent recall — this is what gave Orwellian theories their license. But as we saw in chapter 5, the error that creeps into subsequent belief thanks to Orwellian memory-tampering is indistinguishable — both from the outside and the inside — from the error that creeps into original experience thanks to Stalinesque illusion-construction. So even if we could maintain that you have "direct" and "immediate" access to your current *judgment* (your second-order thought about how things seem to you now), you will not thereby be able to rule out the possibility that it is a *mis*judgment about how it seemed to you a moment ago.

If we individuate states (beliefs, states of consciousness, states of communicative intention, etc.) by their content — which is the standard means of individuation in folk psychology — we end up having to postulate differences that are systematically undiscoverable by any means, from the inside or the outside, and in the process, we lose the subjective intimacy or incorrigibility that is supposedly the hallmark of consciousness. We already saw instances of that in the discussion in chapter 5 of Orwellian versus Stalinesque models of temporal phenomena. And the solution is not to cling to one or the other doctrine made available by folk psychology, but to abandon this feature of folk psychology.

We replace the division into discrete contentful *states* — beliefs, meta-beliefs, and so forth — with a *process* that serves, over time, to ensure a good fit between an entity's internal information-bearing events and the entity's capacity to *express* (some of) the information in those events in speech. That is what the higher-order states were supposed (by Otto) to ensure, but they failed to carve nature at its joints. Indeed, they posited joints that were systematically indiscernible in nature.

These artifacts of folk psychology live on, however, as denizens in the heterophenomenological worlds of the subjects whose world views are indeed shaped by that conceptual scheme. To put the point tautologically, since it really does seem to people that they have both these beliefs about their experiences, and (in addition) the experiences themselves, these experiences and beliefs-about-experiences are both part of how it seems to them. And so we have to explain *that* fact — not the fact that our minds *are* organized into hierarchies of higher-ordered representational "states" of belief, meta-belief, and so forth, but that our minds *tend to seem to us* to be so ordered.

I have ventured two reasons why we tend to find this such an attractive idea. First, we persist in the habit of positing a separate process of observation (now of *inner* observation) intervening between the circumstances about which we can report and the report we issue — overlooking the fact that at some point this regress of interior observers must be stopped by a process that unites contents with their verbal expression without any intermediary content-appreciator. And second, internal communications created in this way do in fact have the effect of organizing our minds into indefinitely powerful reflective or self-monitoring systems. Such powers of reflection have often been claimed to be at the heart of consciousness, with good reason. We may use the oversimplified model of folk psychology as a sort of crutch for the imagination when we try to understand self-monitoring systems, but when we use it, we risk lapsing into Cartesian materialism. We should start learning how to get along without it, and in the next chapter, we will take a few more cautious steps.

11

DISMANTLING THE WITNESS PROTECTION PROGRAM

1. REVIEW

In Part I, we surveyed the problems and laid down some methodological presuppositions and principles. In Part II, we sketched a new model of consciousness, the Multiple Drafts model, and began the task of showing why it should be preferred to the traditional model, the Cartesian Theater. While the idea of the Cartesian Theater, once made explicit, exhibits its flaws quite dramatically — there are no avowed Cartesian materialists — the background assumptions and habits of thought it has fostered continue to motivate objections and distort "intuition." Now in Part III, we are investigating the implications of our alternative model, by answering a succession of compelling objections. Some of these objections betray a persisting — if disavowed — allegiance to the dear old Cartesian Theater.

"But where does the understanding happen?" This question has been hiding at the center of controversy since the seventeenth century. Descartes confronted a wall of skepticism when he insisted (correctly) that mechanisms in the brain could account for at least a great deal of understanding. For instance, Antoine Arnauld, in his objections to the *Meditations*, noted that "at first sight it seems incredible that it can come about, without the assistance of any soul, that the light reflected from the body of a wolf onto the eyes of a sheep should move the minute fibers of the optic nerves, and that on reaching the brain this motion should spread the animal spirits throughout the nerves in the manner necessary to precipitate the sheep's flight" (1641, p. 144). Descartes replied that this is no more incredible than the human capacity to throw out one's arms protectively when falling, also a reaction accomplished mechanically, without benefit of "soul." This idea of "mechanical" interpretation in the brain is the central insight of *any* materialistic theory of the mind, but it challenges a deeply held intuition: our sense that for *real* understanding to occur, there has to be *someone in there* to validate the proceedings, to *witness* the events whose happening constitutes the understanding. (The philosopher John Searle exploits this intuition in his famous Chinese Room thought experiment, which will be considered in chapter 14.)

Descartes was a mechanist *par excellence* when it came to every other phenomenon in nature, but when it came to the human mind, he flinched. In addition to mechanical interpretation, he claimed, the brain also provides material to a central arena — what I've been calling the Cartesian Theater — where, in human beings, the soul can be a Witness and arrive at its own judgments. Witnesses need raw materials on which to base their judgments. These raw materials, whether they are called "sense data" or "sensations" or "raw feels" or "phenomenal properties of experience," are props without which a Witness makes no sense. These props, held in place by various illusions, surround the idea of a central Witness with a nearly impenetrable barrier of intuitions. The task of this chapter is to break through that barrier.

2. BLINDSIGHT: PARTIAL ZOMBIEHOOD?

Of all the terrible accidents that befall people, a tiny fraction are partly redeemed by the fact that they reveal something of nature's mysteries to inquiring scientists. This is particularly true of brain damage brought on by trauma (gunshots, traffic accidents, and the like), tumor, or stroke.[1] The resulting patterns of disability and spared ability sometimes provide substantial — even startling — evidence about how the mind is accomplished by the brain. One of the most surprising, as its paradoxical name suggests, is blindsight. It seems at first to have been made to order for philosophers' thought experiments: an affliction that

---

1. Tim Shallice, in *From Neuropsychology to Mental Structure* (1988), provides an up-to-date and closely argued discussion of the reasoning involved in analyzing these experiments of nature. Several recent books provide good popular accounts of some of these fascinating cases: Howard Gardner, *The Shattered Mind* (1975), and Oliver Sacks, *The Man Who Mistook His Wife for His Hat* (1985).

turns a normal, conscious person into a partial zombie, an unconscious automaton with regard to some stimuli, but a normally conscious person with regard to the rest. So it is not surprising that philosophers have elevated blindsight to a sort of mythic status as an example around which to build arguments. As we shall see, however, blindsight does not support the concept of a zombie; it undermines it.

In normal human vision, the incoming signals from the eyes travel via the optic nerves through various way stations to the *occipital cortex* or visual cortex, the part of the brain at the very back of the skull above the cerebellum. Information about the left visual field (the left halves of the fields from each eye) is spread over the right visual cortex, and the right visual field on the left cortex. Occasionally a vascular accident (e.g., a burst blood vessel) will destroy a portion of the occipital cortex, creating a blind spot or *scotoma*, a relatively large hole in the visually experienced world, on the side opposite the damage.

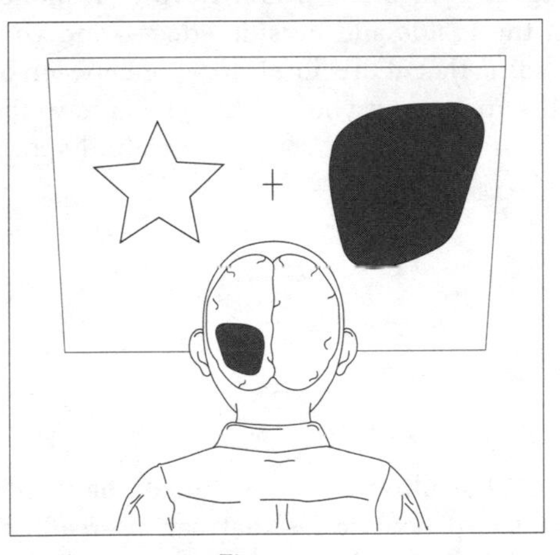

Figure 11.1

In the extreme case where both left and right visual cortices have been destroyed, the person is rendered completely blind. More often, the whole visual cortex on one side of the brain is destroyed by vascular accident, leading to the loss of the opposite half of the visual field; loss of the left visual cortex would produce *right hemianopia*, complete blindness in the right hemifield.

What is it like to have a scotoma? It might seem that this is already familiar to all of us, for we all have blind spots in our visual fields corresponding to the places on our retinas where there are no rods or cones because the optic nerve exits the eyeball there. A normal blind spot, or optic disk, is not small: it blanks out a circle with a diameter of about 6 degrees of visual angle. Close one eye and look at the cross, holding the page about ten inches from your eyes. One of the "blind spot" disks should disappear. Close the other eye, and the opposite disk should disappear. (It may take some adjustment of the distance from the page to make the effect happen. Keep looking straight at the cross.) Why don't you normally notice this gap in your visual field? In part because you have two eyes, and one eye covers for the other; their blind spots do not overlap. But even with one eye closed, you won't notice your blind spot under most conditions. Why not? Since your brain has never had to deal with input from this area of your retina, it has devoted no resources to dealing with it. There are no homunculi responsible for receiving reports from this area, so when no reports arrive, there is no one to complain. An absence of information is not the same as information about an absence. In order for you to see a hole, something in your brain would have to respond to a contrast: *either* between the inside and outside edge — and your brain has no machinery for doing that at this location — *or* between before and after: now you see the disk, now you don't. (That's how the disappearing black disk in Figure 11.2 alerts you to your blind spot.)

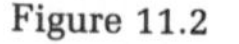

Figure 11.2

Like our normal blind spots, scotomata have definite locations and some have sharp boundaries that can be readily plotted by an experimenter, using a stimulus such as a point of light that can be moved around in the subject's visual field. The subject is asked to report when the spot of light is no longer experienced — a variation on the experiment you just conducted on yourself to discover your blind spot. Reports from the subject can then be correlated with maps of the damage in the cortex, produced by CT (computer-aided tomography) and MRI (magnetic resonance imaging) scans of the brain. A scotoma is different in one important regard from the normal blind spot: it is usually noticed by the subject. This is not just because it is larger than the normal blind spot. Since it is caused by the loss of cells in the visual cortex that previously "reported to" other cells in the cortex that also "cared about" information from certain regions of the retinas, their absence *is* noticed. The brain's expectations are disrupted; something that should be there is missing, some epistemic hunger goes unsatisfied. So the subject is normally aware of the scotoma, but as a lack, not as a positive area of black, such as you might notice if someone pasted a circle of black paper on your car windshield.

Since the normal visual pathways in the brain have been disrupted or cut, one would expect that people with scotomata would be utterly unable to glean any information from vision about things happening in their blind fields. They're blind, after all. And that is just what they say: They experience nothing visual whatever inside the boundaries of their scotomata — no flashes, edges, colors, twinkles, or starbursts. Nothing. That's what blindness is. But some people with scotomata exhibit an astonishing talent: in spite of their utter lack of conscious visual experience in the blind area, they can sometimes "guess" with remarkable accuracy whether or not a light has just been flashed in the field, and even whether a square or circle was shown. This is the phenomenon called blindsight (Weiskrantz, 1986, 1988, 1990). Just how blindsight is to be explained is still controversial, but no researcher thinks there is anything "paranormal" going on. There are at least ten different pathways between the retina and the rest of the brain, so even if the occipital cortex is destroyed, there are still plenty of communication channels over which the information from the perfectly normal retinas could reach other brain areas. Many tests have now been performed on blindsight subjects, and there is no doubt that they can do much better than chance (even as good as 100 percent correct under some conditions) at guessing various simple shapes, the direction of motion, the presence or absence of light. No blindsighted person has yet exhibited a capacity to discriminate colors in the blind field, but recent research by Stoerig and Cowey (1990) provides evidence that this may be possible.

What is going on in blindsight? Is it, as some philosophers and psychologists have urged, visual perception without consciousness — of the sort that a mere automaton might exhibit? Does it provide a disproof (or at any rate a serious embarrassment) to functionalist theories of the mind by exhibiting a case where all the *functions* of vision are still present, but all the good juice of *consciousness* has drained out? It provides no such thing. In their rush to harness blindsight to pull their ideological wagons, philosophers have sometimes overlooked some rather elementary facts about the phenomena of blindsight and the experimental setting in which they can be elicited.

Like the "temporal anomalies" analyzed in chapters 5 and 6, the phenomena of blindsight appear only when we treat subjects from the standpoint of heterophenomenology. The experiments couldn't be conducted if the experimenters couldn't give verbal instructions to the subjects (and be confident that they were understood), and the subjects' responses provide evidence of a startling phenomenon *only* when they are interpreted as speech acts. This is almost too obvious to notice, so I must pause to rub it in.

The interpretation of blindsight is controversial in many ways, but remarkably uncontroversial in one regard: Everyone agrees that the blindsight subject somehow comes to be informed about some event in the world via the eyes (that's the "sight" part), in spite of having no conscious visual experience of the relevant event (that's the "blind" part). More compactly, blindsight involves (1) receipt of visual information that is (2) nevertheless unconscious. The proof of (1) is straightforward: the subject does much better than chance on tests that probe for that information. The proof of (2) is more circumstantial: the subjects deny that they are conscious of any such events, and their verbal denials are supported by neurological evidence of brain damage on the one hand, and by the coherence of their denials on the other. So we believe them![2]

This is not a trivial point. Notice that what is striking about blindsight would evaporate immediately if we concluded that blindsight subjects were malingering — just pretending not to be conscious. Or, closer to home, compare our acceptance of the denials of blindsight subjects to the skepticism with which we greet the same denials when they issue from people diagnosed as suffering from "hysterical blindness." Sometimes people whose eyes and brains are apparently in working order, so far as physiologists can determine, have nevertheless complained that they have been struck blind; they support this complaint by acting "just like a blind person." One can often find a fairly plausible reason why such a person should be motivated to "become" blind — either as a punishment to themselves or to someone who must now care and feel sorry for them, or as a way of denying some terrible visual memory, or as a sort of panic response to some other illness or

2. Note that the details of neurological damage by themselves (without the denials) would prove nothing; it is only by matching up neurological damage with (credible) reports and behavioral evidence that we get any hypotheses about which parts of the brain are essential for which conscious phenomena.

debility — so this is "psychosomatic" blindness if it is blindness at all. Are they really blind? They might be. After all, one might argue, if psychosomatic pain is real pain, and psychosomatic nausea is real enough to make one vomit, why shouldn't psychosomatic blindness be real blindness?

Hysterically blind people *claim* to be blind, but, like blindsight subjects, they nevertheless give unmistakable evidence that they are *taking in visual information.* For instance, hysterically blind people typically do significantly *worse* than chance if asked to guess about the visible features of things! This is a sure sign that they are somehow using visual information to guide their behavior into a preponderance of "errors." Hysterically blind people have an uncanny knack of finding chairs to bump into. And yet, unlike outright malingerers, when hysterically blind people say they are having no visual experiences, they are sincere — they really believe what they say. Shouldn't we? How should we treat the texts of these two different groups of subjects when we go to extrapolating their heterophenomenological worlds?

Here is a place where the ultracautious policies of heterophenomenology pays dividends. Both blindsight subjects and hysterically blind people are apparently sincere in their avowals that they are unaware of anything occurring in their blind field. So their heterophenomenological worlds are alike — at least in respect to the presumptive blind field. And yet there is a difference. We have less knowledge of the neuroanatomical underpinnings of hysterical blindness than we do of blindsight, yet we feel, intuitively, much more skeptical of their denials.[3] What makes us suspect that the hysterically blind are not *really* blind at all, that they are even in some way or to some degree conscious of their visual worlds? The suspiciously auspicious circumstances of their blindness make us wonder, but beyond that circumstantial evidence there is a simpler reason: We doubt their blindness claims because *without prompting,* hysterically blind people sometimes *use* the information their eyes provide for them in ways blindsight subjects do not.

A factor is present in the blindsight experimental situation that fits so perfectly with our standard assumptions that almost no one bothers to discuss it (but see Marcel, 1988; van Gulick, 1989; Carruthers,

3. Without the confirmation of the brain scans showing the cortical damage, there would surely also be widespread skepticism about the genuineness of the scotomata of blindsight subjects. See, e.g., Campion, Latto, and Smith (1983), and Weiskrantz (1988).

1989): Blindsight subjects have to be prompted or cued to give their better-than-chance "guesses." Thus the experimenter may have said, in the initial instructions: "Whenever you hear a tone, make a guess" or "Whenever you feel me tap your hand, make a response." Without such cues, the subject simply fails to respond.[4]

We can test our diagnosis of the difference by imagining a variation. Suppose we encountered a purported blindsight subject who didn't have to be cued: she "spontaneously" issues "guesses" (well above chance, but not perfect) whenever something is presented in the purported blind field. We sit her down in the laboratory and do the usual testing to map the supposed scotoma; she tells us whenever the moving light disappears into her blind field, just like any other blindsight subject. But at the same time she spontaneously volunteers remarks like "This is only a guess, but did you just shine a light in my scotoma?" — but only just after we'd done just that. This would be suspicious, to say the least, and we can say why.

In general, when subjects comply with their instructions in an experiment, this is seen as unproblematic evidence that they have been able to comply with the instructions because they have *consciously experienced* the relevant stimulus events. That's why the following preparatory instruction would be viewed as nonsensical:

> Whenever you are *conscious* of the light going on, press the button on the left; whenever the light goes on but you are *not* conscious of it going on, press the button on the right.

How on earth could a subject comply with this? You would be asking the subject to do the impossible: to condition his behavior on occurrences that are inaccessible to him. It would be like saying "raise your hand whenever someone winks at you without your knowing it."

---

4. The philosopher Colin McGinn (1991) says of an imagined blindsight patient: "Behaviorally, she can function much like a sighted person; phenomenologically, she strikes herself as blind" (p. 111). This is simply false; she cannot function behaviorally much like a sighted person at all. McGinn goes on to underscore his striking claim: "Besides, let's be naive for a minute, don't blindsight patients *look* very much as if they are having visual experiences when they make their surprising discriminations? . . . They don't look the way people look when there is *nothing* experiential going on" (p. 112). Again, this is false. They look in fact as if they were *not* having visual experiences, *because they have to be cued*. If they didn't have to be cued, they would indeed look as if they were having visual experiences — so much so that we wouldn't believe their denials!

An experimenter wouldn't feel the need to insert the adverb "consciously," as in

Whenever you consciously hear a tone, make a guess

since the standard assumption is that one can't condition one's policies on *un*conscious experiences, even if such things occur. To adopt the policy

Whenever *x* happens, do *y*

you have to be able to be conscious of *x* happening.

That's our standard assumption, but this edifice of obviousness has a crack in it. Haven't we learned that many of our behaviors are governed by conditions that we only unconsciously detect? Consider the policies regulating our body temperature, adjusting our metabolism, storing and retrieving energy, activating our immune systems; consider such policies as blinking when things approach or enter the eye, and even such large-scale public behaviors as walking (without falling over) and ducking when things suddenly loom at us. All this "behavior" is controlled without any help from consciousness — as Descartes had observed.

It seems, then, that there are two kinds of behavioral policies: those controlled by conscious thought and those controlled by "blind, mechanical" processes — just like the processes that control an automated elevator. If an automated elevator is to adhere to the policy of carrying no more than two thousand pounds, it must have some sort of built-in scale to detect when this limit is surpassed. An automated elevator is certainly not conscious, and detects nothing consciously, and hence has no conscious policies. It can be said, however, to adhere to policies that hinge on various states of the world that it detects, and even to adjust the policies it adheres to on the basis of other states of affairs it detects, and so forth. It can have policies, meta-policies, and meta-meta-policies, all hinging on various complicated combinations of detected states of affairs — and all without a hint of consciousness. Whatever an elevator can do in the way of detection and policy-following, a human brain and body can surely do as well. It can follow elaborate unconscious elevator-type policies.

So what is the difference between unconscious policy-following and conscious policy-following? When we consider the policies our bodies follow unconsciously, thanks to "blind, mechanical" condition-detectors, it is tempting to say that *since* these are unconscious policies, they are not so much *our* policies as *our bodies'* policies. *Our* policies are (by definition, one might say) our conscious policies; the ones *we* formulate consciously and deliberately, with the opportunity to reflect (consciously) on their pros and cons, and the opportunity to adjust or amend them as the situation unfolds in our experience.

So it seems that when a policy is initially adopted as a result of verbal discussion, or in response to verbal instructions, it is *ipso facto* a conscious policy, which must hinge on consciously experienced events (Marcel, 1988). What seems self-contradictory is the notion that one could talk it over and thereupon decide to follow an *un*conscious policy, hinging on unconsciously detected events. But there is, we can see, a loophole: The status of such a policy *might* change. With enough practice, and some strategically placed forgetfulness, we might start from a consciously adopted and followed policy and gradually move ourselves into the state of following an *un*conscious policy by detecting the relevant hinges without being conscious of them. This could happen, but only if the link to the verbal consideration of the policy were somehow broken.

This possible transition can be better envisaged going in the other direction. Couldn't a blindsight subject *become* conscious of visual experiences in the scotoma, by a reversal of the process just imagined? After all, in blindsight the subject's brain manifestly does receive and analyze the visual information that is somehow utilized in the good guesswork. Shortly after the stimulus occurs, something happens in the subject's brain that marks the onset of the *informed* state. If an external observer (such as the experimenter) can arrange to recognize these onsets, this observer could in principle pass on the information to the subject. Thus the subject could come to recognize these onsets "at second hand" in spite of not being conscious of them "directly." And then, shouldn't the subject be able in principle to "eliminate the middleman" and come to recognize, just as the experimenter does, the changes in his own dispositions? At the outset, it might require using some sort of self-monitoring equipment — the same equipment the experimenter relies on — but with the subject now looking at or listening to the output signals.[5]

In other words, shouldn't it be possible in principle to "close the

5. "If he could listen to his own galvanic skin response, he'd be in better shape." — Larry Weiskrantz, commenting on one of his blindsight patients, ZIF, Bielefeld, May 1990.

feedback loop" and thereby train the subject to follow a policy of conditioning his behavior on changes which he did not ("directly") experience? I raise the prospect of such training in blindsight as if it were just a thought experiment, but in fact it could readily be turned into a real experiment. We could try to train a blindsight subject to recognize when to "guess."

Subjects with blindsight are not unchanging in their talents and dispositions; some days they are in better form than others; they do improve with practice, in spite of the fact that they are usually *not* provided with immediate feedback from the experimenter on how well they are doing (for exceptions, see Zihl, 1980, 1981). There are several reasons for this, chief of which is that any such experimental situation is bedeviled with possibilities of unintended and unnoticed hints from the experimenter, so interactions between experimenter and subject are scrupulously minimized and controlled. Nevertheless, subjects feed on the cuing or prompting they receive from the experimenter, and gradually accustom themselves to the otherwise weirdly unpromising practice of making hundreds or thousands of guesses on matters about which they are convinced they have no experience whatever. (Imagine how you would feel if you were asked to sit down with the telephone book and guess what make of automobile each listed person owned, without ever being told when you happened to guess correctly. It would not seem a well-motivated activity for very long, unless you got some credible assurance about how well you were doing, and why this was a stunt worth trying.)

What would happen, then, if we were to throw other scientific goals to the winds and see how much we could achieve by training someone with blindsight, using whatever feedback seemed to help? Suppose we begin with a standard blindsight subject, who "guesses" whenever we cue him (the so-called forced-choice response), and whose guesses are better than chance (if they aren't, he isn't a blindsight subject). Feedback would soon tune this to a maximum, and if the guessing leveled off at some agreeably high rate of accuracy, this should impress on the subject that he had a useful and reliable talent that might be worth exploiting. This is in fact the state that some blindsight subjects are in today.

Now suppose we start asking the subject to do without cuing — to "guess when to guess," to guess "whenever the spirit moves you" — and again let's suppose that the experimenter provides instant feedback. There are two possible outcomes:

(1) The subject starts at chance and stays there. In spite of the fact that the subject is measurably informed by the onset of the stimulus events, there seems to be no way the subject can discover when this informing has occurred, no matter what "biofeedback" crutches we lend him.
(2) The subject eventually becomes able to work without a cue from the experimenter (or from any temporary biofeedback crutch), performing significantly above chance.

Which outcome we would get in any particular case is of course an empirical matter, and I won't even hazard a guess on how likely a type-2 result might be. Perhaps in every single case, the subject would be unable to learn to "guess" correctly when to guess. But notice that *if* a type-2 result were to happen, the subject could then quite reasonably be asked to adopt policies that required him to hinge his behavior on stimuli whose occurrence he could only guess to occur. Whether or not he was conscious of these stimuli, if his "guessing" reliability was high, he could treat those stimuli on a par with any conscious experiences. He could think about, and decide upon, policies that hinged on their occurrence as readily as on the occurrence of events consciously experienced.

But would this somehow *make* him conscious of the stimuli? What are your intuitions? When I have asked people what they would be inclined to say in such a case, I get mixed responses. Folk psychology does not yield a clear verdict. But a blindsight subject has spoken for himself about a similar circumstance. DB, one of the subjects studied by Weiskrantz, has right hemianopia, and shows the classic blindsight capacity to guess above chance when cued. For instance, if a light is slowly moved across his scotoma horizontally or vertically and he is prompted to guess "vertical or horizontal," he does extremely well, while denying all consciousness of the motion. However, if the light is moved more swiftly, it becomes self-cuing: DB can volunteer without prompting a highly accurate report of the motion, and even mimic the motion with a hand gesture, as soon as it occurs (Weiskrantz, 1988, 1989). And when asked, DB insists that *of course* he consciously experiences the motion — how else would he be able to report it? (Other blindsight subjects also report conscious experience of fast-moving stimuli.) We may want to reserve judgment, but his response should not surprise us, if Rosenthal's analysis of the ordinary concept of consciousness is on the right track. DB doesn't just come to be informed about the motion of the light; he *realizes* he's come to be informed; in

Rosenthal's terms, he has a second-order thought to the effect that he's just had a first-order thought.

Otto, our critic, returns:

> But this is just more sleight of hand! We've always known that blindsight subjects were conscious of their guessing. All this shows is that such a subject might develop a talent for guessing when to guess (and, of course, he would be conscious of *those* guesses). Coming to recognize that one's guesses on these topics were reliable would hardly in itself be sufficient for one to become directly conscious of the event one is guessing *about*.

This suggests that something more is needed for visual consciousness. What might be added? For one thing, the connection between the guess and the state it is *about*, while reliable, seems pretty thin and ephemeral. Could it be thickened and strengthened? What would be the result if the ties of *aboutness* between the guess and its object were multiplied?

3. HIDE THE THIMBLE: AN EXERCISE IN CONSCIOUSNESS-RAISING

The standard philosophical term for aboutness is *intentionality*, and according to Elizabeth Anscombe (1965) it "comes by metaphor" from the Latin, *intendere arcum in*, which means *to aim a bow and arrow at* (something). This image of aiming or directedness is central in most philosophical discussions of intentionality, but in general philosophers have traded in the complex process of aiming a real arrow for a mere "logical" arrow, a foundational or primitive relation, made all the more mysterious by its supposed simplicity. How *could* something in your head point this abstract arrow at a thing in the world?[6] Thinking of the aboutness relation as an abstract, logical relation may in the end be right, but at the outset it deflects attention from the processes that are actually involved in keeping a mind in enough contact with the things in the world so that they can be *effectively* thought about: the processes of attending to, keeping in touch with, tracking and trailing (Selfridge, unpublished). The actual business of aiming at something, "keeping it in the crosshairs," involves making a series of adjustments and compensations over time, under "feedback control."

---

6. My answer to this question is my book *The Intentional Stance* (1987).

That's why the presence of clouds of distractors (such as the chaff that confounds antimissile systems) can make aiming impossible. Locking on to a target long enough to identify it is an achievement that calls for more than a single, momentary informational transaction. The best way to keep in touch with something is, literally, to keep in *touch* with it — to grab it and not let it get away, so that you can examine it to your heart's content, in your own good time. The next best way is to keep in touch with it figuratively, by tracking it with your eyes (and the rest of your body), never letting it out of your sight. This is an achievement that can be accomplished by perception, of course, but not just by passive perception; it may take some effort, some planning, and, in any event, *continuing activity* to keep in touch with something.

When I was a child I loved to play the children's party game Hide the Thimble. An ordinary thimble is shown to all participants, and all but one leave the room, while the thimble is "hidden." The rules for the hider are clear: The thimble has to be hidden *in plain sight.* It may not be placed behind or under anything, or too high up for any of the children to see. In the average living room there are dozens of places you can place a thimble where it will tend to melt into its surroundings like a well-camouflaged animal. Once it is hidden, the rest of the children come back in the room and proceed to hunt for the thimble. As soon as you see the thimble you quietly sit down, trying not to betray its location. The last few children to find the thimble can usually be counted on to *look right at* the thimble several times without actually *seeing* it. In these delicious moments, everyone else can see that the thimble is right in front of Betsy's nose, let's say, well lit and subtending a healthy angle in her visual field. (At such moments my mother liked to say, "If it were a bear, it would bite you!") From the giggles and gasps of the other children, Betsy may herself realize that she must be staring right at it — and still not see it.

We might put it this way: Even if some representational state in Betsy's brain in some way "includes" the thimble, no perceptual state of Betsy is *about* the thimble yet. We may grant that one of her conscious states is about the thimble: her "search image." She may be fiercely concentrating on finding the thimble, the very thimble she was allowed to examine just a minute or two ago. But no strong relation of intentionality or aboutness yet holds between any of her perceptual states and the thimble, even though there may well be information in some state of her visual system that would make it possible for someone else (an outside observer, for instance, studying the states of her visual cortex) to locate or identify the thimble. What must happen is for Betsy to "zero in on" the thimble, to separate it as "figure" from "ground" and identify it. After that has happened, Betsy really does see the thimble. The thimble will finally be "in her conscious experience" — and now that she is conscious of it, she will be able at last to raise her hand in triumph — or go quietly to sit with the other children who have already spotted the thimble.[7]

Such feedback-guided, error-corrected, gain-adjusted, purposeful links are the prerequisite for the sort of acquaintance that deserves the name — that can then serve as a hinge for policy, for instance. Once I have seen something in this strong sense, I can "do something about it" or do something *because* I saw it or *as soon as* I saw it. Individual thimbles, once identified, are normally easy enough to keep track of thereafter (unless of course you happen to be in a room full of thimbles in an earthquake). Under normal circumstances, then, the elevated status the thimble achieves in Betsy's control system is not just for a fleeting moment; the thimble will remain located by Betsy during reaching, or during time taken to reaffirm its identity, to check it again (and again — if there are grounds for doubt). The things we are most definitely conscious of are the items we frankly and unhurriedly observe, gathering in and integrating the fruits of many saccades, building up an acquaintance over time while keeping the object located in personal space. If the object is darting around like a butterfly, we will actually take action to immobilize it, "so we can look at it," and if it is well camouflaged in its surroundings, we have to take steps — literally, if we mustn't touch it — to get it in front of a contrasting background.

Our failure to do this may prevent us from seeing the object, in an important and familiar sense of the term.[8]

---

7. Is her identification of the thimble a subsequent *effect* of her becoming conscious of it, or a prior *cause* of her becoming conscious of it? This is the question — Orwellian or Stalinesque? — that the Multiple Drafts model teaches us not to ask.

8. Under normal conditions, location ("spotting it") and identification go hand in hand; spotting the thing-to-be-identified is a precondition for identifying it. But this normal coincidence masks a striking fact: The identification machinery and the location machinery are to a large degree independent in the brain, located in different regions of the cortex (Mishkin, Ungerleider, and Macko, 1983), and hence capable of being shut down independently. There are rare pathologies in which the subject can readily identify *what* he is seeing without being able to locate it in personal space at all, and counterpart pathologies in which subjects can locate a visual stimulus — point to it, for instance — and yet be very poor at identifying the object, in spite of having otherwise quite normal vision. The psychologist Anne Treisman (1988; Treisman and Gelade, 1980; Treisman and Sato, 1990; Treisman and Souther, 1985) has conducted an important series of

Birdwatchers often keep a life list of all the species they have seen. Suppose you and I are birdwatchers, and we both hear a bird singing in the trees above our heads; I look up and say, "I see it — do you?" You stare right where I am staring, and yet you say, truthfully, "No. I don't see it." I get to write this bird on my list; you do not, in spite of the fact that you may be morally certain that its image must have swum repeatedly across your foveae.

What would you say? Was the thimble somehow "present" in Betsy's consciousness before she spotted it? Was the bird present in the "background" of your consciousness, or not present at all? Getting something into the forefront of your consciousness is getting it into a position where it can be reported on, but what is required to get something into the background of your conscious experience (and not merely into the background of your visible environment)? The thimble and the bird were undoubtedly present in the visible environment — that's not the issue. Presumably it's not sufficient for reflected light from the object merely to enter your eyes, but what further effect must the reflected light have — what further notice must your brain take of it — for the object to pass from the ranks of the merely unconsciously responded to into the background of conscious experience?

The way to answer these "first-person point of view" stumpers is to ignore the first-person point of view and examine what can be learned from the third-person point of view. In chapters 8–10, we explored a model of speech act production that depended on a Pandemonium process in which the eventual mating of contents with expressions was the culmination of competitions, the building, dismantling, and rebuilding of coalitions. Contents that entered that fray but did not manage to perpetuate themselves for long might send some sort of one-shot "ballistic" effect rippling through the system, but would be close to *unreportable*. When an event doesn't linger, any attempt to report it,

---

experiments that support her claim that *seeing* should be distinguished from *identifying*. When something is seen, on her model, the brain sets up a "token" for the object. Tokens are "separate temporary episodic representations" — and their creation is the preamble for their further identification, something that is accomplished by searching one's semantic memory using a process of the sort that production systems model. A token does not have to be defined by a definite location in personal space, however, if I understand her model, and for these reasons it is not out of the question that subjects in Betsy's state (before she found the thimble) could do better than chance if prompted to make a forced-choice guess as to whether the thimble was currently in their field of view or not. For experiments bearing on this, see Pollatsek, Rayner, and Henderson (1990).

if started, will either be aborted or will wander out of control, having nothing against which to correct itself. For reportability, there must be a capacity to identify and reidentify the effect. We can see the development of reportability in many varieties of training, reminiscent of the training we imagine giving to our blindsight patient: the results of palate-training in wine tasters, ear-training in musicians, and the like — or the simple experiment with plucking the guitar string described in chapter 3.

Consider, for instance, the instructions given to apprentice piano tuners. They are told to listen to the "beats" as they strike the key they are tuning together with a reference key. *What* beats? At first, most novices are unable to discern in their auditory experience anything answering to the description "beats" — what they hear is something they would describe as some sort of unstructured bad-soundingness or out-of-tuneness. Eventually, though, if training is successful, they come to be able to isolate, in their auditory experience, the interference "beats," and to notice how the patterns of beats shift in response to their turning of the tuning "hammer" on the peg. They can then readily tune the piano by tuning out the beats. What they typically say — and we can all confirm this with similar episodes in our own experience — is that as a result of their training *their conscious experience has changed*. More specifically, it has been augmented: they are now conscious of things they were not previously conscious of.

Now in one sense, of course, they were hearing the beats all along. It is the interference, after all, that composes the out-of-tuneness of which they were certainly conscious. But they were previously unable to detect these components in their experience, which is why one might say that these factors *contributed to* but were not themselves *present in* the experience. The functional status of such contributions prior to training was the same as that of the events occurring in blindsight: the subject is unable to report the particular contributions, or hinge policy on their onset, but the results of these contributions can still be made manifest in the subject's behavior, for instance in the subject's capacity to answer artfully posed questions. What I am suggesting is that there is nothing more to being in the background of experience than that. Now it's not out of the question, as we have seen, that a strengthened link of the sort we have just described for piano tuners and wine tasters could be built up in a blindsight subject to the point where he would declare, and we would readily accept, that he had *become* conscious of the stimuli — even in the forefront of his consciousness — which before he could only guess about.

> Not so fast, [says Otto.] Here's another objection: You imagine the blindsight subject learning to use his blindsight capacities in these new ways, and maybe that would give him a *sort* of consciousness of the events occurring in his blind field, but this still leaves something out. The consciousness wouldn't be *visual* consciousness; it wouldn't be like *seeing*. The "phenomenal qualities" or *qualia* of conscious vision would be missing, even if the blindsight subject could make all these functional moves.

Maybe so, and maybe not. Just what are "phenomenal qualities" or *qualia*? (*Qualia* is just the Latin for *qualities;* the singular is *quale*, usually pronounced *kwah´-lay*.) They seem terribly obvious at first — they're the way things look, smell, feel, sound to us — but they have a way of changing their status or vanishing under scrutiny. In the next chapter we'll track these suspects down through philosophical thickets, but first we should get a better look at some properties that *aren't* phenomenal qualities, but might be mistaken for them.

4. PROSTHETIC VISION: WHAT, ASIDE FROM INFORMATION, IS STILL MISSING?

Does Weiskrantz's subject, DB, *see* the motion? Well, he surely doesn't *hear* it, or *feel* it. But is it vision? Does it have the "phenomenal qualities" of vision? Weiskrantz says:

> As stimulus "salience" increases, the patient may say insistently that he still does not "see," but he now has a kind of "feeling" that something is there. In some cases, if the salience is increased still further, a point may be reached where the subject says he "sees" but the experience is not veridical. For example, DB "sees" in response to a vigorously moving stimulus, but he does not see it as a coherent moving object, but instead reports complex patterns of "waves." Other subjects report "dark shadows" emerging as brightness and contrast are increased to high levels. [1988, p. 189]

The vigorously moving object is not perceived by DB as having color or shape, but so what? As we proved to ourselves in chapter 2, in the experiment with the playing card held in peripheral vision, we can certainly see the card without being able to identify either its colors or its shapes. That's normal sight, not blindsight, so we should be reluctant on those grounds to deny visual experience to the subject.

The question of whether this abnormal way of obtaining infor-

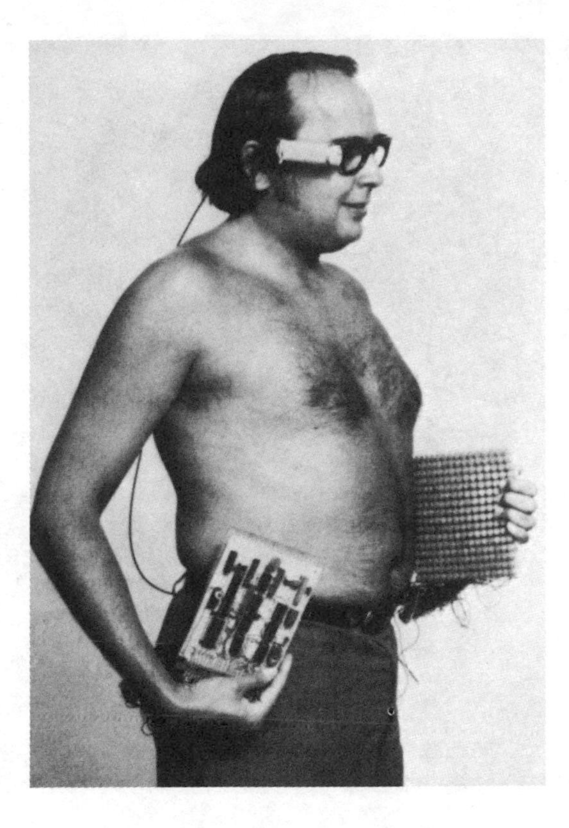

Figure 11.3

A blind subject with a 16-line portable electrical system. The TV camera is attached to the lens housing, mounted on a pair of spectacle frames. A small bundle of wires leads to an electrical stimulus drive circuitry (held in his right hand). The matrix of 256 concentric silver electrodes is held in his left hand.

mation about visible things would be a variety of *seeing* can be more vividly posed if we turn to an even more radical departure from normal vision. Prosthetic devices have been designed to provide "vision" to the blind, and some of them raise just the right issues. Almost twenty years ago, Paul Bach-y-Rita (1972) developed several devices that involved small, ultralow-resolution video cameras that could be mounted on eyeglass frames. The low-resolution signal from these cameras, a 16-by-16 or 20-by-20 array of "black and white" pixels, was spread over the back or belly of the subject in a grid of either electrical or mechanically vibrating tinglers called tactors.

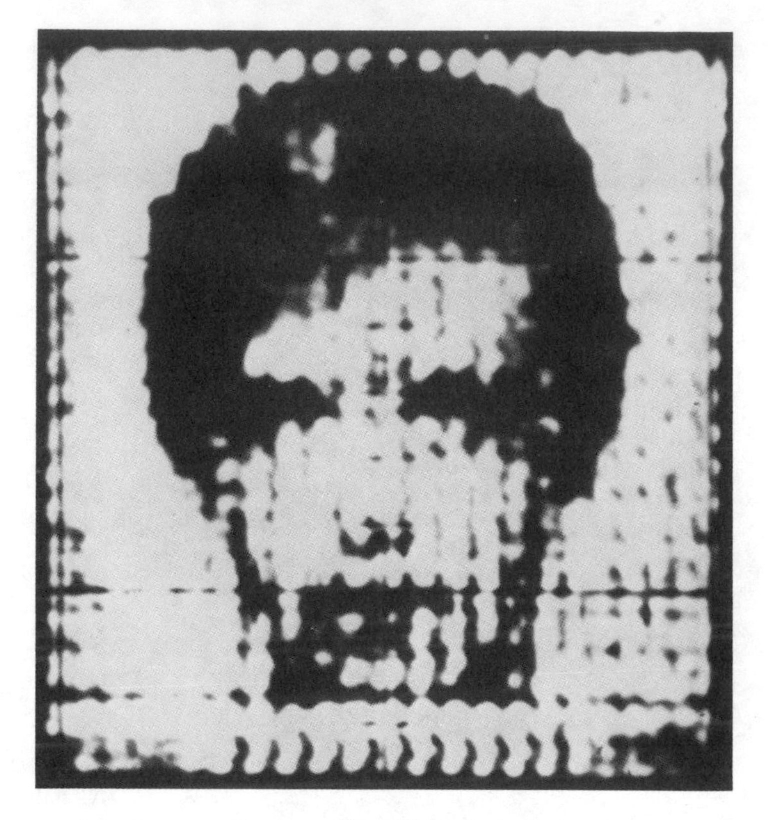

Figure 11.4

Appearance of a 400-count representation of a woman's face as seen on the monitor osciliscope. Subjects can correctly identify stimulus patterns of this level of complexity.

After only a few hours of training, blind subjects wearing this device could learn to interpret the patterns of tingles on their skin, much as you can interpret letters traced on your skin by someone's finger. The resolution is low, but even so, subjects could learn to read signs, and identify objects and even people's faces, as we can gather from *looking* at this photograph taken of the signal as it appears on an oscilloscope monitor.

The result was certainly prosthetically produced conscious perceptual experience, but since the input was spread over the subjects' backs or bellies instead of their retinas, was it *vision*? Did it have the "phenomenal qualities" of vision, or just of tactile sensation?

Recall one of our experiments in chapter 3. It is quite easy for your tactile point of view to extend out to the tip of a pencil, permitting you to feel textures with the tip, while quite oblivious to the vibrations of the pencil against your fingers. So it should not surprise us to learn that a similar, if more extreme, effect was enjoyed by Bach-y-Rita's subjects. After a brief training period, their awareness of the tingles on their skin dropped out; the pad of pixels became transparent, one might say, and the subjects' point of view shifted to the point of view of the camera, mounted to the side of their heads. A striking demonstration of the robustness of the shift in point of view was the behavior of an experienced subject whose camera had a zoom-lens with a control button (pp. 98–99). The array of tinglers was on his back, and the camera was mounted on the side of his head. When the experimenter without warning touched the zoom button, causing the image on the subject's *back* to expand or "loom" suddenly, the subject instinctively lurched *backward, raising his arms to protect his head.* Another striking demonstration of the transparency of the tingles is the fact that subjects who had been trained with the tingler-patch on their backs could adapt almost immediately when the tingler-patch was shifted to their bellies (p. 33). And yet, as Bach-y-Rita notes, they still responded to an itch on the back as something to scratch — they didn't complain of "seeing" it — and were perfectly able to attend to the tingles, as tingles, on demand.

These observations are tantalizing but inconclusive. One might argue that once the use of the device's inputs became second nature the subjects were really seeing, or, contrarily, that only some of the most central "functional" features of seeing had been reproduced prosthetically. What of the other "phenomenal qualities" of vision? Bach-y-Rita reports the result of showing two trained subjects, blind male college students, for the first time in their lives, photographs of nude women from *Playboy* magazine. They were disappointed — "although they both could describe much of the content of the photographs, the experience had no affectual component; no pleasant feelings were aroused. This greatly disturbed the two young men, who were aware that similar photographs contained an affectual component for their normally sighted friends" (p. 145).

So Bach-y-Rita's prosthetic devices did not produce *all* the effects of normal vision. Some of the shortfall must be due to the staggering difference in the rate of information flow. Normal vision informs us about the spatial properties of things in our environment at great speed and with almost whatever level of detail we desire. It is not surprising that low-resolution spatial information sent brainward via an interface on the skin failed to stir up all the reactions that are stirred up in normally sighted people when their visual systems are flooded with input.[9] How much pleasure would we expect a normally sighted person to derive from *looking* at similarly low-resolution translations — glance at Figure 11.4 — of pictures of beautiful people?

It is not clear how much would change if we somehow managed to improve the "baud rate"[10] of prosthetic vision to match normal vision. It might be that simply increasing the amount and rate of information, somehow providing higher-resolution bit-maps to the brain, would suffice to produce the delight that is missing. Or rather some of it. People born blind would be at a tremendous disadvantage to people who have recently lost their sight, for they have built up none of the specifically visual associations that no doubt play an important role in the delight that sighted people take in their experiences, which *remind* them of earlier visual experiences. It might also be that some of the pleasure we take in visual experiences is the by-product of ancient fossil traces of an earlier economy in our nervous systems, a topic raised in chapter 7 that will be further explored in the next chapter.

The same considerations apply to blindsight and any imagined improvements in blindsight subjects' abilities. Discussions of blindsight have tended to ignore just how paltry the information is that blindsight subjects glean from their blind fields. It is one thing to be able to guess, when prompted, whether a square or a circle was just presented in the blind field. It would be quite another to be able to guess in detail, when prompted, what was just happening outside the window.

---

9. For instance, the latency of response for some of these perceptual tasks, even in trained subjects, is quite long — eight to fifteen seconds for various simple identifications, for instance (Bach-y-Rita, p. 103). This in itself demonstrates that the information flow of prosthetic vision is extremely sluggish, compared to normal vision.

10. "Baud rate" is a standard term for rate of digital information flow (it means, approximately: bits per second). For instance, if your computer communicates over telephone lines to other computers, it may transmit its bit strings at 1200 baud or 2400 baud or at a much higher rate. It takes a baud rate approximately four times as fast to transmit high-resolution real-time animated pictures — a clear instance in which a picture is indeed worth more than a thousand words. Regular television signals are analog, like a phonograph record, rather than digital, like a compact disk, so its information-flow rate is measured as *bandwidth*, rather than baud rate. The term antedates computers; Baudot code, named after its inventor (as was Morse code), was the standard international telegraph code adopted in 1880, and the baud rate was the number of code elements per second transmitted. By using "baud rate" rather than "bandwidth," I don't mean to imply that the brain's information-handling is best conceived in digital terms.

We can use what we have learned about prosthetic vision to guide our imagination about what it would be like for a blindsight subject to regain more of the *functions* of vision. Let's try to imagine coming upon a cortically blind person, who, after assiduous training, (1) has turned his capacity to guess when to guess into second nature, (2) can play Hide the Thimble with the best of them, and (3) has somehow managed to increase the speed and detail of his guesswork by orders of magnitude. We encounter him reading the newspaper and chuckling at the comics and ask him to explain himself. Here are three scenarios, in ascending order of plausibility:

(1) "Just guessing, of course! Can't see a darn thing, you know, but I've learned how to guess when to guess, and right now, for instance, I guess you're making a rude gesture at me, and screwing up your face into a look of utter disbelief."

(2) "Well, what started out as sheer guesses gradually lost their status as guesses, as I came to trust them. They turned into *presentiments*, shall we say? I would suddenly just *know* that something was going on in my blind field. I could then express my knowledge, and act on my knowledge. Moreover, I then had meta-knowledge that I was in fact capable of such presentiments, and I could use this meta-knowledge in planning my actions and setting policies for myself. What began as conscious guesses turned into conscious presentiments, and now they come so fast and furious I can't even separate them. But I still can't see a darn thing! Not the way I used to! It isn't like seeing at all."

(3) "Well, actually, it's *very much* like seeing. I now effortlessly act in the world on the basis of information gleaned by my eyes from my surroundings. Or I can be self-conscious about what I'm getting from my eyes if I want to be. Without the slightest hesitation I react to the colors of things, to their shapes, and locations, and I've lost all sense of the effort I expended to develop these talents and render them second nature."

And yet we may still imagine our subject saying that something is missing:

> "*Qualia*. My perceptual states do have qualia, of course, because they are conscious states, but back before I lost my sight, they used to have *visual* qualia, and now they don't, in spite of all my training."

It may seem obvious to you that this makes sense, that it is just what you would expect our subject to say. If so, the rest of the chapter is for you, an exercise designed to shake that conviction. If you're already beginning to doubt that this speech about qualia makes any sense at all, you've probably anticipated some of the twists that our story is about to take.

5. "FILLING IN" VERSUS FINDING OUT

But the existence of this feeling of strangeness does not give us a reason for saying that every object we know well and which does not seem strange to us gives us a feeling of familiarity. — We think that, as it were, the place once filled by the feeling of strangeness must surely be occupied *somehow*.

LUDWIG WITTGENSTEIN (1953), i596

In chapter 2, we saw that one of the reasons for believing in dualism was that it promised to provide the "stuff dreams are made of" — the purple cows and other figments of our imagination. And in chapter 5 we saw the confusions that arose from the natural but misguided assumption that *after* the brain has arrived at a discrimination or judgment, it re-*presents* the material on which its judgment is based, for the enjoyment of the audience in the Cartesian Theater, filling in the colors. This idea of *filling in* is common in the thinking of even sophisticated theorists, and it is a dead giveaway of vestigial Cartesian materialism. What is amusing is that those who use the term often know better, but since they find the term irresistible, they cover themselves by putting it in scare-quotes.

For instance, just about everyone describes the brain as "filling in" the blind spot (my italics in all the examples):

> . . . the neurologically well-known phenomenon of subjective "*filling in*" of the missing portion of a blind area in the visual field. [Libet, 1985b, p. 567]

> . . . you can locate your own blind spot, and also demonstrate how a pattern is "*filled in*" or "completed" across the blind spot. . . . [Hundert, 1987, p. 427]

There is also auditory "filling in." When we listen to speech, gaps in the acoustic signal can be "filled in" — for instance, in the "phoneme restoration effect" (Warren, 1970). Ray Jackendoff puts it this way:

> Consider, for example, speech perception with noisy or defective input — say, in the presence of an operating jet airplane or over a bad telephone connection. . . . What one constructs . . . is not just an intended meaning but a phonological structure as well: one "hears" more than the signal actually conveys. . . . In other words, phonetic information is "*filled in*" from higher-level structures as well as from the acoustic signal; and though there is a difference in how it is derived, there is no qualitative difference in the completed structure itself. [Jackendoff, 1989, p. 99]

And when we read text, something similar (but visual) occurs: As Bernard Baars puts it:

> We find similar phenomena in the well-known "proofreader effect," the general finding that spelling errors in page proofs are difficult to detect because the mind "*fills in*" the correct information. [Baars, 1988, p. 173]

Howard Margolis adds an uncontroversial commentary on the whole business of "filling in":

> The "*filled-in*" details are ordinarily correct. [Margolis, 1987, p. 41]

Tacit recognition that there is something fishy about the idea of "filling in" is nicely manifested in this description of the blind spot by the philosopher C. L. Hardin, in his book *Color for Philosophers:*

> It covers an area with a 6 degree visual diameter, enough to hold the images of ten full moons placed end to end, and yet there is no hole in the corresponding region of the visual field. This is because the eye-brain *fills in* with whatever is seen in the adjoining regions. If that is blue, it *fills in* blue; if it is plaid, we are aware of no discontinuity in the expanse of plaid. [1988, p. 22]

Hardin just can't bring himself to say that the brain fills in the plaid, for this suggests, surely, quite a sophisticated bit of "construction," like the fancy "invisible mending" you can pay good money for to fill in the hole in your herringbone jacket: all the lines line up, and all the shades of color match across the boundary between old and new. It seems that filling in blue is one thing — all it would take is a swipe or two with a cerebral paintbrush loaded with the right color; but filling in plaid is something else, and it is more than he can bring himself to assert.

But as Hardin's comment reminds us, we are just as oblivious of our blind spots when confronting a field of plaid as when confronting a uniformly colored expanse, so whatever it takes to create that oblivion can as readily be accomplished by the brain in either case. "We are aware of no discontinuity," as he says. But if the brain doesn't have to fill in the gap with plaid, why should it bother filling in the gap with blue?

In neither case, presumably, is "filling in" a matter of literally filling in — of the sort that would require something like paintbrushes. (This was the moral of the story of the CADBLIND Mark II in chapter 8.) I take it that no one thinks that "filling in" is a matter of the brain's actually going to the trouble of covering some spatial expanse with *pigment*. We know that the real, upside-down image on the retina is the last stage of vision at which there is anything colored in the unproblematic way that an image on a movie screen is colored. Since there is no literal mind's eye, there is no use for pigment in the brain.

So much for pigment. But still, we may be inclined to think that there is something that happens in the brain that is in *some* important way analogous to covering an area with pigment — otherwise we wouldn't want to talk of "filling in" at all. It is this special happening, whatever it is, that occurs in the special "medium" of visual or auditory experience, apparently. As Jackendoff says, speaking of the auditory case, "one 'hears' more than the signal actually conveys" — but note that still he puts "hears" in scare-quotes. What could it be that is *present* when one "hears" sounds filling silent times or "sees" colors spanning empty spaces? It does seem that something is there in these cases, something the brain has to provide (by "filling in"). What should we call this unknown whatever-it-is? Let's call it *figment*. The temptation, then, is to suppose that there is something, made out of figment, which is there when the brain "fills in" and not there when it doesn't bother "filling in." Put so baldly, the idea of figment will not appeal to many people. (At least I hope it doesn't.) We know better: there is no such stuff as figment. The brain doesn't make figment; the brain doesn't use figment to fill in the gaps; figment is just a figment of my imagination. So much for figment! But then, what does "filling in" mean, what *could* it mean, if it doesn't mean filling in with figment? If there is no such medium as figment, how does "filling in" differ from not bothering to fill in?

In chapter 10, we saw how a CAD system could represent colors by associating a color number with each pixel, or with each delineated region of the object depicted, and we saw how the CADBLIND Mark II could search for, or detect, colors by reading such a code. This process is reminiscent of the children's pastime, color-by-numbers, which provides a simple analogue of the representation processes that must occur, or might occur, in the brain. Figure 11.5 is a representation that has information about shapes, but no information about colors at all.

Figure 11.5

Compare it with Figure 11.6, which has information about colors in the form of a numbered code. If you were to take some crayons and follow the directions for filling in the colors, you could turn Figure 11.6 into yet another sort of "filled in" representation — one in which the regions were filled in with real color, real pigment.

There is still another way the color could be "filled in" pixel by pixel, by a coded bit-map, as in Figure 11.7.

Figures 11.6 and 11.7 are both sorts of filling in (compared to Figure 11.5, for example), since any procedure that needs to be informed about the color of a region can, by mechanical inspection of that region, extract that information. This is purely informational filling-in. The systems are entirely arbitrary, of course. We can readily construct in-

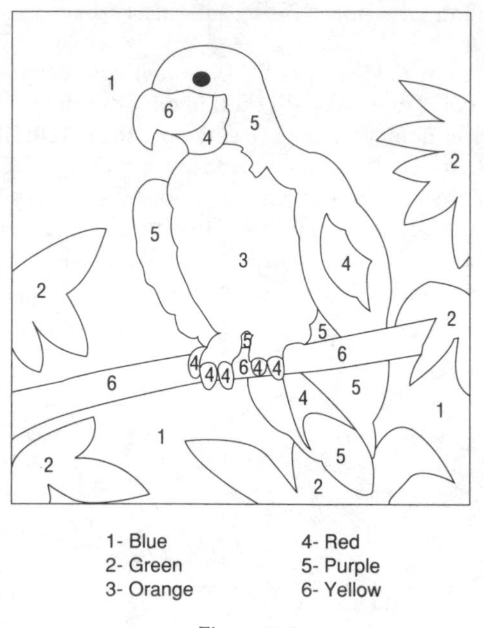

Figure 11.6

```
1111111111111111111111111111111111111111111111111
1111111111111111111111111111111111111111111111111
1111111111111111111155551111111111111111111111111
1111111111111111155555555551111111111111111111111
11111111111111115555●555551111111111111111111111
111111111111111666666664455551111111111111222222
1111111111111166666666644555551111111111111122222
111111111111116666666644455555111111111111122222222
11111111111111661114444455555511111111112222222222
1111111111111111111113333335555555511111111111122222
111111111111111111113333333333555555555111112222222222
11111111111111155533333333333333355555551111111122222
1111111111111155553333333333333333555555551111122222
1111111111111155553333333333333333555455551111222221122
11111112221115555333333333333333355544555511111111122
11112222222111155553333333333333333554444555111111122
1222222222111115555333333333333333355444444551111111111
22222222111111155553333333333333355544445551111111
2222222111111111155553333333333333355555445551122222222
2222222111111111115555333333333333335555555511222222222
22222222111111111115555333333333333355555555666662222
222112222111111111111153333533333336666666666666622212
122211111111111111166433336643466666666666661111122112
1111111166666666666664444466444644555555555111211112
1666666666666666111111111111155554445555555511111111
666666666111111111111111111155554444455555111111111
66661112211111111111111111111115544555445511111111111
111122222211111111111111111111111154555555551222222111
1222222221111111111111222222221112255555555222222222111
2222222222222111111111112222222221222255555222222222211
22222222222222221111111112222222222222222222211111111
22222111222222111111111111111222222222222222222111111111
```

Figure 11.7

definitely many functionally equivalent systems of representation — involving different coding systems, or different media.

If you make a colored picture on your personal computer using the PC-Paintbrush program, the screen you see is represented in the machine as a bit-map in the "frame buffer," analogous to Figure 11.7, but when you go to store the picture on a disk, a compression algorithm translates it into something similar to Figure 11.6. It divides the area into like-colored regions, and stores the region boundaries and their color number in an "archive" file.[11] An archive file is just as accurate as a bit-map, but by generalizing over regions, and labeling each region only once, it is a more efficient system of representation.

A bit-map, by explicitly labeling each pixel, is a form of what we may call *roughly continuous* representation — the roughness is a function of the size of the pixels. A bit-map is not literally an image, but just an array of values, a sort of recipe for forming an image. The array can be stored in any system that preserves information about location. Videotape is yet another medium of roughly continuous representation, but what it stores on the tape is not literally images, but recipes (at a different grain level) for forming images.

Another way of storing the image on your computer screen would be to take a color photograph, and store the image on, say, a 35mm slide, and this is importantly different from the other systems, in an obvious way: there is actual dye, literally filling in a region of real space. Like the bit-map, this is a roughly continuous representation of the depicted spatial regions (continuous down to the grain of the film — at a fine enough scale, it becomes pixel-like or granular). But unlike the bit-map, color is used to represent color. A color negative also uses color to represent color, but in an inverted mapping.

Here, then, are three ways of "filling in" color information: color-by-numbers, as in Figure 11.6 or an archive file, color-by-bit-map, as in Figure 11.7 or frame buffers or videotape, and color-by-colors. Color-by-numbers is in one regard a way of "filling in" color information, but it achieves its efficiency, compared to the others, precisely because it does *not* bother filling in values explicitly for each pixel. Now in which of these senses (if any) does the brain "fill in" the blind spot? No one thinks that the brain uses numbers in registers to code for colors, but that's a red herring. Numbers in registers can be understood just to

11. There are other sorts of compression algorithms that do not rely on breaking the image into same-colored regions in just this way, but I will not discuss them.

stand for any system of magnitudes, any system of "vectors," that a brain might employ as a "code" for colors; it might be neural firing frequencies, or some system of addresses or locations in neural networks, or any other system of physical variations in the brain you like. Numbers in registers have the nice property of preserving relationships between physical magnitudes while remaining neutral about any "intrinsic" properties of such magnitudes, so they can stand for any physical magnitudes in the brain that "code for" colors. Although numbers can be used in an entirely arbitrary way, they can also be used in nonarbitrary ways, to reflect the structural relations between colors that have been discovered. The familiar "color solid," in which hue, saturation, and lightness are the three dimensions along which our colors vary,[12] is a logical space ideally suited to a numerical treatment — any numerical treatment that reflects the betweenness relations, the oppositional and complementary relations, and so on, that human sight actually exhibits. The more we learn about how the brain codes colors, the more powerful and nonarbitrary a numerical model of human color vision we will be able to devise.

The trouble with speaking of the brain "coding" for colors by using intensities or magnitudes of one thing or other is that it suggests to the unwary that eventually these codings have to be decoded, getting us "back to color." That is one route — perhaps the most popular route — back to figment: one imagines that the brain might unconsciously store its encyclopedic information about color in a format something like that of Figure 11.8 but then arrange to have the representation "decoded" into "real colors" on special occasions — like running a videocassette, to project real color on a screen. There certainly is a difference in phenomenology between just recalling the proposition that the flag is red, white, and blue, and actually imagining the flag "in color" and "seeing" (with the mind's eye) that it is red, white, and blue. If this contrast in phenomenology inspires some people to posit

---

12. Other creatures have different color solids — or hypersolids! We are "trichromats": we have three different types of photopigmented transducer cells in the cones in our retinas. Other species, such as pigeons, are tetrachromats; their subjective color space would have to be represented, numerically, as a four-dimensional hyperspace. Other species are dichromats, all of whose color discriminations could be mapped onto a single two-dimensional plane. (Note that "black and white" is just a one-dimensional representational scheme, with all the possible grays representable as different distances on a line between 0 and 1.) For reflections on the implications of this incommensurability of color systems, see Hardin (1988) and Thompson, Palacios, and Varela (in press).

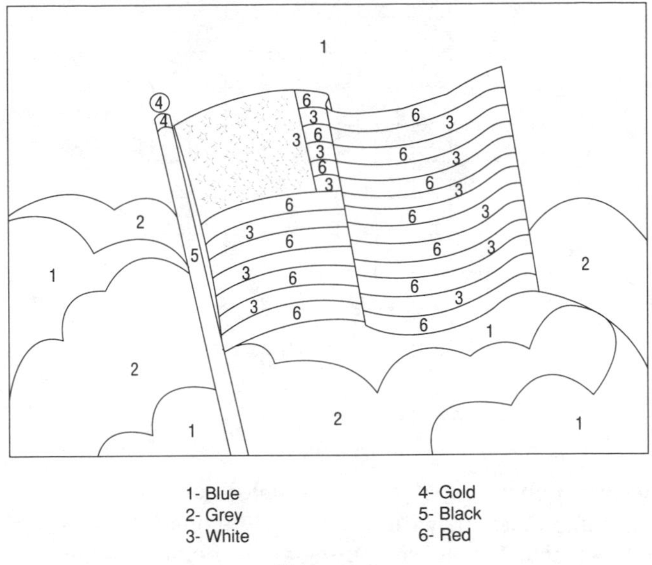

Figure 11.8

figment, an even more compelling case is provided by the phenomenon of neon color spreading (van Tuijl, 1975), an example of which can be seen on the back dust jacket of this book.

The pink you see filling in the ring defined by the red lines is not a result of pink smudging on the page, or light scattering. There is no pink on your retinal image, in other words, in addition to the red lines. Now how might this illusion be explained? One brain circuit, specializing in shape, is misled to distinguish a particular bounded region: the ring with its "subjective contours." Subjective contours are produced by many similar figures, such as these.

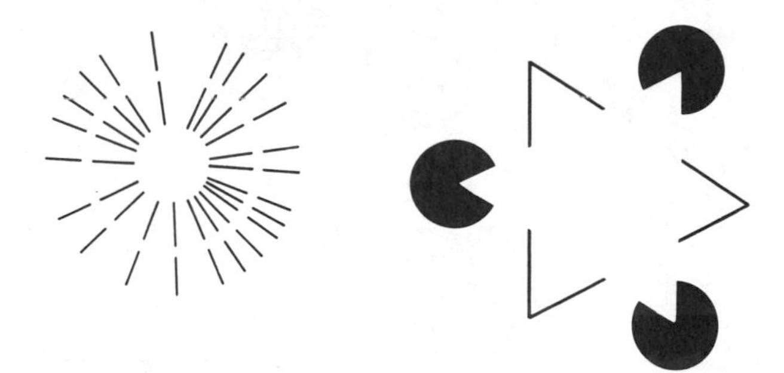

Figure 11.9

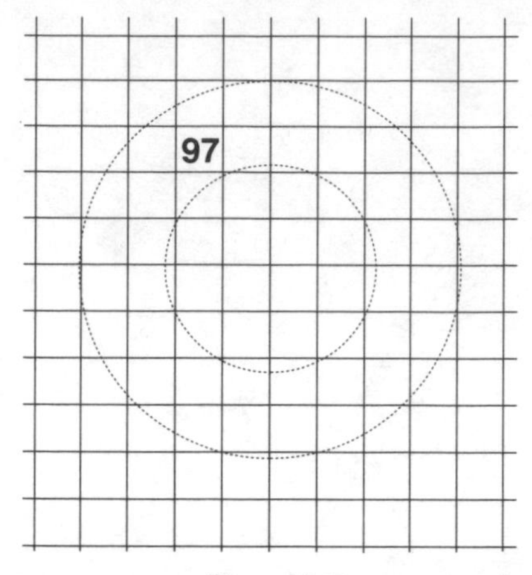

Figure 11.10

Another brain circuit, specializing in color but rather poor on shape and location, comes up with a color discrimination (pink #97, let's say) with which to "label" something in the vicinity, and the label gets attached (or "bound") to the whole region.

Why these particular discriminations should occur under these conditions is still controversial, but the controversy concerns the causal mechanisms that lead to mislabeling the region, not the further "prod-

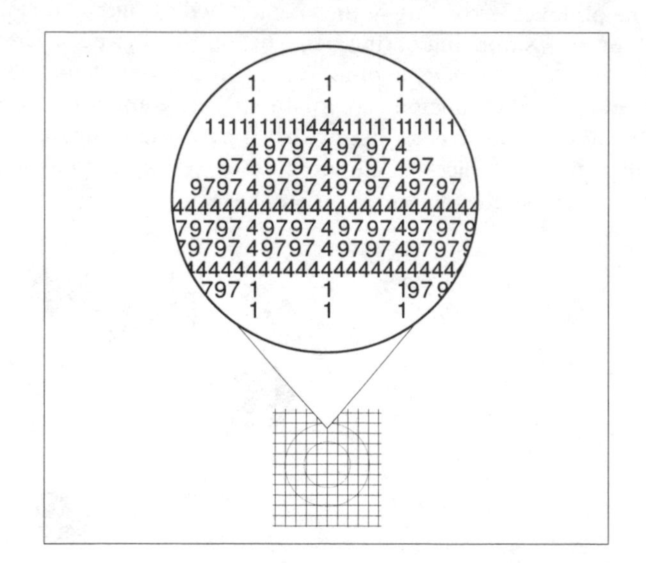

Figure 11.11

ucts" (if any) of the visual system. But isn't something missing? I have stopped short at an explanation that provides a labeled color-by-numbers region: Doesn't that recipe for a colored image have to be executed somewhere? Doesn't pink #97 have to be "filled in"? After all, you may be tempted to insist, you see the pink! You certainly don't see an outlined region with a number written in it. The pink you see is not in the outside world (it isn't pigment or dye or "colored light"), so it must be "in here" — pink figment, in other words.

We must be careful to distinguish the "pink figment" hypothesis from others that would be legitimate alternatives to an explanation that stopped short with the color-by-numbers suggestion. For instance, it might turn out that somewhere in the brain there is a roughly continuous representation of colored regions — a bit-map — such that "each pixel" in the region has to be labeled "color #97," more or less in the fashion of Figure 11.11.

This is an empirical possibility. We could devise experiments to confirm or disconfirm it. The question would be: Is there a representational medium in the brain in which the value of some variable parameter (the intensity or whatever that codes for color) has to be propagated across or replicated across the relevant pixels of an array, or is there just a "single label" of the region, with no further "filling in" or "spreading out" required? What sort of experiments could favor such a model of the neon color spreading effect? Well, it would be impressive, for instance, if the color could be shown under some conditions to spread slowly in time — bleeding out from the central red lines and gradually reaching out to the subjective contour boundaries.[13] I don't want to prejudge the question, for my main purpose in raising it is to illustrate my claim that whereas there are plenty of unresolved empirical questions about how the neon color spreading phenomenon happens in the brain, none of them involve differences over whether or not figment is generated in a "decoding" of the neural coding system.

The question of whether the brain "fills in" in one way or another

---

13. As if in response to this suggestion, V. S. Ramachandran and R. L. Gregory (submitted) have just performed some experiments with what they call (misleadingly, I think) artificially induced scotomas, in which there is good evidence of gradual filling in of textures and details. There is one fundamental difference between their experimental circumstances and the conditions I have been describing; in their experiments there is competition between two sources of information, and one gets overruled (gradually). The phenomenon of gradual spatial filling in of textures is an important discovery, but it does not take us beyond a model in the spirit of Figure 11.11. And further questions about these experiments need to be resolved before their interpretation can be settled.

is not a question on which introspection by itself can bear, for as we saw in chapter 4, introspection provides us — the subject as well as the "outside" experimenter — only with the content of representation, not with the features of the representational medium itself. For evidence about the medium, we need to conduct further experiments.[14] But for some phenomena, we can already be *quite* sure that the medium of representation is a version of something efficient, like color-by-numbers, not roughly continuous, like bit-mapping.

Consider how the brain must deal with wallpaper, for instance. Suppose you walk into a room and notice that the wallpaper is a regular array of hundreds of identical sailboats, or — let's pay homage to Andy Warhol — identical photographic portraits of Marilyn Monroe. In order to identify a picture as a portrait of Marilyn Monroe, you have to foveate the picture: the image has to fall on the high-resolution foveae of your eyes. As we saw in the playing card experiment in chapter 2, your *parafoveal* vision (served by the rest of the retina) does not have very good resolution; you can't even identify a jack of diamonds held at arm's length. Yet we know that if you were to enter a room whose walls were papered with identical photos of Marilyn Monroe, you would "instantly" see that this was the case. You would see in a fraction of a second that there were "lots and lots of identical, detailed, focused portraits of Marilyn Monroe." Since your eyes saccade four or five times a second at most, you could foveate only one or two Marilyns in the time it takes you to jump to the conclusion *and thereupon to see* hundreds of identical Marilyns. We know that parafoveal vision *could not* distinguish Marilyn from various Marilyn-shaped blobs, but nevertheless, what you see is *not* wallpaper of Marilyn-in-the-middle surrounded by various indistinct Marilyn-shaped blobs.

Now, is it possible that the brain takes one of its high-resolution foveal views of Marilyn and reproduces it, as if by photocopying, across an internal mapping of the expanse of wall? That is the only way the high-resolution details you used to identify Marilyn could "get into the background" at all, since parafoveal vision is not sharp enough to provide it by itself. I suppose it is possible in principle, but the brain

14. For instance, Roger Shepard's initial experiments with the mental rotation of cube diagrams showed that it certainly *seemed* to subjects that they harbored roughly continuously rotating representations of the shapes they were imagining, but it took further experiments, probing the actual temporal properties of the underlying representations, to provide (partial) confirmation of the hypothesis that they were actually doing what it seemed to them they were doing. (See Shepard and Cooper, 1982.)

almost certainly does not go to the trouble of doing *that* filling in! Having identified a single Marilyn, and having received no information to the effect that the other blobs are not Marilyns, it jumps to the conclusion that the rest are Marilyns, and labels the whole region "more Marilyns" without any further rendering of Marilyn at all.[15]

Of course it does not seem that way to you. It seems to you as if you are actually seeing hundreds of identical Marilyns. And in one sense you are: there are, indeed, hundreds of identical Marilyns out there on the wall, and you're seeing them. What is not the case, however, is that there are hundreds of identical Marilyns represented in your brain. Your brain just somehow represents *that* there are hundreds of identical Marilyns, and no matter how vivid your impression is that you see all that detail, the detail is in the world, not in your head. And no figment gets used up in rendering the seeming, for the seeming isn't rendered at all, not even as a bit-map.

So now we can answer our question about the blind spot. The brain doesn't have to "fill in" for the blind spot, since the region in which the blind spot falls is already labeled (e.g., "plaid" or "Marilyns" or just "more of the same"). If the brain received contradictory evidence from some region, it would abandon or adjust its generalization, but not getting any evidence from the blind spot region is not the same as getting contradictory evidence. The absence of confirming evidence from the blind spot region is no problem for the brain; since the brain has no precedent of getting information from that gap of the retina, it has not developed any epistemically hungry agencies demanding to be fed from that region. Among all the homunculi of vision, not a single one has the role of coordinating information from that region of the eye, so when no information arrives from those sources, no one complains. The area is simply neglected. In other words, all normally sighted people "suffer" from a tiny bit of "anosognosia." We are unaware of our "deficit" — of the fact that we are receiving no visual information from our blind spots. (A good overview of anosognosia is McGlynn and Schacter, 1989.)

The blind spot is a spatial hole, but there can be temporal holes as well. The smallest are the gaps that occur while our eyes dart about during saccades. We don't notice these gaps, but they don't have to be filled in *because* we're designed not to notice them. The temporal analogues of scotomata might be the "absences" that occur during petit

---

15. In Appendix B, I will suggest some "experiments with wallpaper" that would put this empirical claim on the line.

mal epileptic seizures. These are noticeable by the sufferer, but only by inference: they can't "see the edges" any more than you can see the edges of your blind spot, but they can be struck, retrospectively, by discontinuities in the events they have experienced.

The fundamental flaw in the idea of "filling in" is that it suggests that the brain is providing something when in fact the brain is ignoring something. And this leads even very sophisticated thinkers to make crashing mistakes, perfectly epitomized by Edelman: "One of the most striking features of consciousness is its continuity" (1989, p. 119). This is utterly wrong. One of the most striking features of consciousness is its *dis*continuity — as revealed in the blind spot, and saccadic gaps, to take the simplest examples. The discontinuity of consciousness is striking because of the *apparent* continuity of consciousness. Neumann (1990) points out that consciousness may in general be a gappy phenomenon, and as long as the temporal edges of the gaps are not positively perceived, there will be no sense of the gappiness of the "stream" of consciousness. As Minsky puts it, "Nothing can *seem* jerky except what is *represented* as jerky. Paradoxically, our sense of continuity comes from our marvelous *insensitivity* to most kinds of changes rather than from any genuine perceptiveness" (1985, p. 257).

6. NEGLECT AS A PATHOLOGICAL LOSS OF EPISTEMIC APPETITE

The brain's motto for handling the blind spot could be: Ask me no questions and I'll tell you no lies. As we saw in chapter 1, so long as the brain assuages whatever epistemic hunger is around, there is nothing more it needs to do. But what about those occasions when there is much less epistemic hunger than there ought to be? These are the pathologies of neglect.

One of the most familiar forms of neglect is *hemi-neglect,* in which one side of the body, usually the left side, is entirely neglected, due to brain damage on the opposite side. Not only the left side of the body, but also the left side of the immediate vicinity is neglected. If a group of people stand around the bed of the left-neglect patient, he will look only at the people standing to his right; if asked to count the people in the room, he will tend to overlook the people on the left, and if someone on the left tries to attract his attention, this will typically fail. And yet it can be shown that the patient's sense organs are still taking in, and analyzing, and responding in various ways to the stimuli occurring on the left. What can be going on in the patient's head? Is the "left side of phenomenal space a blank"? Or does the patient's "mind's eye" fail to see the material that the brain provides for it on the left side of . . . the stage in the Cartesian Theater?

There is a simpler explanation, not in terms of inner representations with curious properties, but in terms of — *neglect* — in the political sense! Daniel Patrick Moynihan notoriously recommended that certain problems of race relations in America would resolve themselves if we treated them with "benign neglect" — if Washington and the rest of the nation would simply ignore them for a while. I don't think that was good advice, but Moynihan was right about something: There certainly are circumstances where benign neglect is called for — such as our treatment of the blind spot problem.

There are no homunculi, as I have put it, who are supposed to "care about" information arising from the part of the visual field covered by the blind spot, so when nothing arrives, there is no one to complain. Perhaps the difference between us and sufferers of pathological neglect or other forms of anosognosia is that some of their complainers have been killed. This theory has been proposed, in less colorful terms, by the neuropsychologist Marcel Kinsbourne (1980), who calls the inner complainers "cortical analyzers." In terms of the model we have developed, neglect could be described as a loss of political clout by certain parties of demons in the brain, due, in many but not all cases, to the death or suppression of their Representative. These demons are still active, trying to do their various things and even succeeding on occasion, but they no longer can win out in certain competitions against better-organized coalitions.

On this model, the benign neglect of our blind spots shades almost imperceptibly into the various mildly dysfunctional neglects we all suffer from, and then into the most bizarre neglects studied by neurologists. For instance, I myself suffer from several rather common forms of neglect. Least serious, but sometimes embarrassing, is my *typo neglect*. I am pathologically unable to notice typographical errors in my page proofs when I read them over, and only by the most laborious exercises of concentration and focusing can I overcome this. It is not, as Baars suggested, that my brain "fills in" the correct spelling; it doesn't have to "fill in" since it does not normally pay enough attention to these matters to notice the errors; its attention gets captured by other features of the words on the page. Another of my mild disabilities is *student examination neglect*. It is just amazing how attractive the prospect of washing the kitchen floor or changing my shelf paper or balancing my checkbook becomes when I have a pile of exams on my desk that need grading. This feature, the heightened interest in alternatives, is particularly evident in hemi-neglect; to a first approximation, the farther to the right something is, the more noteworthy it is to a left hemi-neglect patient. Perhaps my most serious form of neglect, however, is my bad case of *finances neglect*. So little do I like balancing my checkbook, in fact, that only some truly awful alternative, such as grading student exams, can force my attention to the topic. This neglect has serious consequences for my welfare, consequences that I can readily be brought to appreciate, but in spite of this alarmingly unsuccessful appeal to my underlying rationality, I manage to persist in my neglect, unless fairly drastic measures of self-manipulation are brought into play.

It is not that I *can't see* my checkbook but that I *won't look at it*. And although in cool, reflective moments such as these, I can report all this (proving that I am not *deeply* anosognosic about my own disability), in the normal course of events part of what I don't notice is my own neglect of my finances. Mild anosognosia, in short. From this perspective, the only thing startling about the bizarre forms of neglect studied by neuropsychologists is the topic boundaries. Imagine someone neglecting everything *on the left* (Bisiach and Luzzatti, 1978; Bisiach, 1988; Bisiach and Vallar, 1988; Calvanio, Petrone, and Levine, 1987). Or imagine someone who had lost color vision *but had no complaint* (Geschwind and Fusillo, 1966). Or even, imagine someone who has gone blind, but hasn't yet noticed this profound loss — Anton's syndrome or blindness denial (Anton, 1899; McGlynn and Schacter, 1989, pp. 154–158).

These conditions are readily explainable on the Multiple Drafts theory of consciousness, for the central Witness has been replaced by coalitions of specialists whose particular epistemic hungers cannot be immediately adopted by other agents if they are expunged or on holiday.[16] When these epistemic hungers vanish, they vanish without a trace, leaving the field to other coalitions, other agents with other agendas.

But the same principle that accounts for neglect provides an alternative scenario for the "missing visual qualia" of our imagined blindsight virtuoso. I suggested that it is possible that if he complains of the absence of qualia, he *might* simply be noticing the relative paucity of

16. For a contrast, see Bisiach et al. (1986) and McGlynn and Schacter (1989), whose models of anosognosia are similar, but are committed to the "boxology" of separate *systems*, especially McGlynn and Schacter, who posit a *conscious awareness system*, that takes *inputs* from modules.

information he now gets from his vision and misdescribing it. I went on to speculate that if somehow we could increase the "baud rate" of his information-gathering, some if not all of the gap between his kind of vision and normal vision could be closed. Now we can see that another, cheaper way of closing the same gap would be simply to lower his epistemic hunger, or obtund his visual curiosity in some way. After all, if in Anton's syndrome a person can be utterly blind and not yet realize it, a bit of strategically placed neglect could turn our blindsight subject who complains of the loss of visual qualia into an uncomplaining subject who declares his vision to be flawlessly restored. It may seem that we would know better, but would we? Would anything be missing in such a person? There isn't any *figment* in normal vision, so it can't be the figment that's missing. What else could it be?

7. VIRTUAL PRESENCE

> We have the sense of actuality when every question asked of our visual systems is answered so swiftly that it seems as though those answers were already there.
>
> MARVIN MINSKY (1985), p. 257

Once again, the absence of representation is not the same as the representation of absence. And the representation of presence is not the same as the presence of representation. But this is hard to believe. Our conviction that we are somehow *directly acquainted* with special properties or features in our experience is one of the most powerful intuitions confronting anyone trying to develop a good theory of consciousness. I've been chipping away at it, trying to undermine its authority, but there's still more work to be done. Otto has yet another tack to try:

> Your point about the Marilyns in the wallpaper is actually a backhanded defense of dualism. You argue very persuasively that there aren't hundreds of high-resolution Marilyns in the brain, and then conclude that there aren't any anywhere! But I argue that since *what I see* is hundreds of high-resolution Marilyns, then *since*, as you argue, they aren't anywhere in my brain, they must be somewhere else — in my nonphysical mind!

The hundreds of Marilyns in the wallpaper seem to be present in your experience, seem to be in your mind, not just on the wall. But since, as we know, your gaze can shift in a fraction of a second to draw information from any part of your visual environment, why should your brain bother importing all those Marilyns in the first place? Why not just let the world store them, at no cost, until they're needed?

Compare the brain to a library. Some research libraries are gigantic storehouses, containing within their walls millions of books, all quite readily accessible in the stacks. Other libraries keep fewer books on hand, but have a generous and efficient accessions system, buying whatever books the library users demand, or borrowing them from other libraries, using a swift interlibrary loan system. If you don't keep the books stored on the premises, the delays in access are greater, but not much greater. We can imagine an electronic interlibrary loan system (using fax or computer files) that could obtain a book from the outside world faster than the swiftest runner could fetch the book from the stacks. A computer scientist might say of the books in such a system that they were "virtually present" in the library all along, or that the library's "virtual collection" was hundreds or thousands of times greater than its actual hard-copy collection.

Now how could we, as Users of our own brain-libraries, know which of the items we retrieve were *there* all along, and which our brains sent out for, in swift information-gathering forays into the external world? Careful experiments, conducted according to the heterophenomenological method, can answer this question, but introspection by itself simply can't tell. That doesn't stop us from thinking we can tell, however. In the absence of any evidence one way or the other, our natural tendency is to jump to the conclusion that more is present. I have called this the Introspective Trap (Dennett, 1969, pp. 139–140) and Minsky calls it the Immanence Illusion: "Whenever you can answer a question without a noticeable delay, it seems as though that answer were already active in your mind" (Minsky, 1985, p. 155).

The interlibrary loan system is a useful but incomplete analogy, for your brain doesn't just have facilities for acquiring information about whatever external topics happen to interest you; it also has literally millions of sentries almost continuously gazing at a portion of the external world, ready to sound the alarm and *draw* your attention to anything novel and relevant happening in the world. In vision, this is accomplished by the parafoveal rods and cones of the retinas, and the neural agents inboard of these sentries who specialize in detecting change and motion. If one of these agents sounds the alarm — "Change in my sector!" — this almost instantaneously triggers a saccade, bringing the fovea to bear on the region of interest, so the novelty can be located, identified, dealt with. The sentry system is so reliable that it's hard to sneak a change into the visible world without the whole visual system being informed of it, but with the aid of high-tech trickery, the sentries can sometimes be bypassed, with astonishing results.

When your eyes dart about in saccades, the muscular contractions that cause the eyeballs to rotate are *ballistic* actions: your fixation points are unguided missiles whose trajectories at lift-off determine where and when they will hit ground zero at a new target. For instance, if you are reading text on a computer screen, your eyes will leap along a few words with each saccade, farther and faster the better a reader you are. What would it be like if a magician, a sort of Cartesian evil demon on a modest scale, could change the world during the few milliseconds your eyes were darting to their next destination? Amazingly, a computer equipped with an automatic eye-tracker can detect and analyze the lift-off in the first few milliseconds of a saccade, calculate where ground zero will be, and, *before the saccade is over,* erase the word on the screen at ground zero and replace it with a different word of the same length. What do you see? Just the new word, and with no sense at all of anything having been changed. As you peruse the text on the screen, it seems to you for all the world as stable as if the words were carved in marble, but to another person reading the same text over your shoulder (and saccading to a different drummer) the screen is aquiver with changes.

The effect is overpowering. When I first encountered an eye-tracker experiment, and saw how oblivious subjects were (apparently) to the changes flickering on the screen, I asked if I could be a subject. I wanted to see for myself. I was seated at the apparatus, and my head was immobilized by having me bite on a "bite bar." This makes the job easier for the eye-tracker, which bounces an unnoticeable beam of light off the lens of the subject's eye, and analyzes the return to detect any motion of the eye. While I waited for the experimenters to turn on the apparatus, I read the text on the screen. I waited, and waited, eager for the trials to begin. I got impatient. "Why don't you turn it on?" I asked. "It *is* on," they replied.

Since all the changes on the screen happen during saccades, your sentries don't get to issue any effective alarms. Until recently this phenomenon has been called "saccadic suppression." The idea was that the brain must somehow shut down the input from the eyes during saccades, since no one can notice changes that occur in the visual field during saccades, and, of course, no one complains of giddy and alarming changes. But a clever experiment with an eye-tracker (Brooks et al., 1980) has shown that if a stimulus — such as a word or alphabet letter — is moved in synchrony with the saccade, keeping pace with the "shadow" of the fovea as it races to its new landing spot, it is readily seen and identified by the subject. Input from the eye is not blocked on its way to the brain during saccades, but under normal conditions it is unusable — everything is just rushing by too fast to make sense of — so the brain treats it all with benign neglect. If all your sentries send their alarms at once, the best thing to do is just ignore them all.

In the experimental situation I was in, words on the screen were being erased and replaced during my saccades. If your parafoveal vision can't discriminate the word at ground zero before you saccade to it, once you get there and identify it, there can't be any prior record or memory of it in your brain with which to compare it. The switch can't be noticed because the information logically required for such a noticing is simply *not there*. Of course it *seems to you* as you read this page that all the words on the line are in some sense present in your consciousness (in the background) even before you specifically attend to them, but this is an illusion. They are only virtually present.

There is *some* information about the surrounding words in your brain, of course — enough to have served as the guide and instigator of the most recent saccade, for instance. Just what information is already *there*? Experiments with eye-trackers and similar apparatus can determine the limits of what you can notice, and thereby determine the limits of what is *present* in your mind. (See, e.g., Pollatsek, Rayner, and Collins, 1984; Morris, Rayner, and Pollatsek, 1990.) To insist, as Otto was tempted to do, that what is not *there* in the brain must nevertheless be *there* in the mind because it certainly *seems* to be there is pointless. For as we have just seen, it wouldn't be "there" in any sense that could make a difference *to Otto's own experiences*, let alone to his capacity to pass tests, press buttons, and so forth.

8. SEEING IS BELIEVING: A DIALOGUE WITH OTTO

At this time our critic Otto insists on a review, for he's sure he's been hoodwinked somewhere along the line. I'm going to engage him in a dialogue, hoping that he does justice to many if not all of your doubts as well. Otto begins:

It seems to me that you've denied the existence of the most indubitably real phenomena there are: the real seemings that even Descartes in his *Meditations* couldn't doubt.

In a sense, you're right; that's what I'm denying exist. Let's return to the neon color-spreading phenomenon. There seems to be a pinkish glowing ring on the dust jacket.

There sure does.

But there isn't any pinkish glowing ring. Not really.

Right. But there sure seems to be!

Right.

So where is it, then?

Where's what?

The pinkish glowing ring.

There isn't any; I thought you'd just acknowledged that.

Well, yes, there isn't any pinkish ring out there on the page, but there sure seems to be.

Right. There seems to be a pinkish glowing ring.

So let's talk about *that* ring.

Which one?

The one that *seems to be*.

There is no such thing as a pink ring that merely seems to be.

Look, I don't just *say* that there seems to be a pinkish glowing ring; *there really does seem* to be a pinkish glowing ring!

I hasten to agree. I never would accuse you of speaking disingenuously! You really mean it when you say there seems to be a pinkish glowing ring.

Look. I don't just mean it. I don't just *think* there seems to be a pinkish glowing ring; *there really* seems to be a pinkish glowing ring!

Now you've done it. You've fallen in a trap, along with a lot of others. You seem to think there's a difference between thinking (judging, deciding, being of the heartfelt opinion that) something seems pink to you and something *really seeming* pink to you. But there is no difference. There is no such phenomenon as really seeming — over and above the phenomenon of judging in one way or another that something is the case.

Recall the Marilyn wallpaper. The wall is, in fact, covered in high-resolution Marilyns. Moreover, that's just how it seems to you! It seems to you that the wall is covered in high-resolution Marilyns. Lucky you, your visual apparatus has led you to a true belief about a feature of your environment. But there aren't lots of *real seeming* Marilyns represented in your brain — or your mind. There is no medium that *reproduces* the wallpaper detail, that *renders* it for your inner Witness. All that is the case is that it seems to you that there are lots of high-resolution Marilyns there (and this time you're right — there really are). Other times, you may be wrong; it may seem to you — in the color phi phenomenon — that a single spot moved right, changing color as it went, when in fact there were simply two differently colored spots flashing. Its seeming to you this way does not require *rendering* in the brain, any more than the brain's color judgments, once reached, need to be subsequently *decoded* somewhere.

> But then what *is* happening when it seems to me that there is a pinkish glowing ring? What is the positive account that your theory provides? You seem to me to be terribly evasive on this score.

I guess you're right. It's time to come clean and present the positive account, but I confess I'll have to do it by starting with a caricature and then revising it. I can't seem to discover a more direct way of expounding it.

> So I've noticed. Carry on.

Suppose there *were* a Central Meaner. But suppose that instead of sitting in a Cartesian Theater watching the Presentation, the Central Meaner sits in the dark and has presentiments — it just suddenly occurs to him that there is something pink out there, the way it might suddenly occur to you that there's somebody standing behind you.

> What are presentiments, exactly? What are they made of?

Good question, which I must answer evasively at first, in caricature. These presentiments are propositions the Central Meaner exclaims to himself in his own special language, Mentalese. So his life consists of a sequence of *judgments,* which are sentences of Mentalese, expressing one proposition after another, at tremendous speed. Some of these he decides to publish, in English translation.

This theory has the virtue of getting rid of the figment, the projection into phenomenal space, the filling in of all the blanks on the Theater Screen, but it still has a Central Meaner, and the Language of Thought. So let's revise the theory. First, get rid of the Central Meaner by distributing all his judgments around in time and space in the brain — each act of discrimination or discernment or content-fixation happens somewhere, but there is no *one* Discerner doing all the work. And second, get rid of the Language of Thought; the content of the judgments doesn't have to be expressible in "propositional" form — that's a mistake, a case of misprojecting the categories of language back onto the activities of the brain too enthusiastically.

> So presentiments are like speech acts except that there's no Actor and no Speech!

Well, yes. What there is, really, is just various events of content-fixation occurring in various places at various times in the brain. These are nobody's speech acts, and hence they don't have to be in a language, but they are rather *like* speech acts; they have content, and they do have the effect of informing various processes with this content. We considered more detailed versions of this in chapters 5–10. Some of these content-fixations have further effects, which eventually lead to the utterance of sentences — in natural language — either public or merely internal. And so a heterophenomenological text gets created. When it's interpreted, the benign illusion is created of there being an Author. This is sufficient to produce *hetero*phenomenology.

> But what about the *actual* phenomenology?

There is no such thing. Recall our discussion of the interpretation of fiction. When we come across a novel that is loosely veiled autobiography, we find we can map the fictional events onto many of the real events in the author's life, so in a strained sense the novel is *about* those real events. The author may not realize this at all, but nevertheless, in this strained sense it is true; those events are what the text is about, because those are the real events that explain why *this* text got created.

But what is the text about in the *unstrained* sense?

Nothing. It's fiction. It *seems to be* about various fictional characters, places, and events, but these events never happened; it isn't *really* about anything.

But when I read a novel, these fictional events come alive! Something happens in me; I *visualize* the events. The act of reading, and interpreting, a text such as a novel *creates* some new things in my imagination: images of the characters doing the deeds. After all, when we go to see a film version of a novel we have read, we often think — "That's not at all the way I imagined her!"

Granted. In "Fearing Fictions," the philosopher Kendall Walton (1978) claims that these acts of imagination on the part of an interpreter supplement the text in much the same way the pictures found in illustrated editions of novels do, "combining with the novel to form a 'larger' [fictional, heterophenomenological] world" (p. 17). These additions are perfectly real, but they are just more "text" — not made of figment, but made of judgment. There is nothing more to phenomenology than that.

But there seems to be!

Exactly! *There seems to be phenomenology*. That's a fact that the heterophenomenologist enthusiastically concedes. But it does *not* follow from this undeniable, universally attested fact that *there really is* phenomenology. This is the crux.

Are you denying then that consciousness is a *plenum*?

Yes indeed. That's part of what I'm denying. Consciousness is gappy and sparse, and doesn't contain half of what people think is there!

But, but . . .

But consciousness sure seems to be a plenum?

Yes!

I agree; it seems to be a plenum; it even seems to be a "striking fact" about consciousness that it is continuous, as Edelman says, but . . .

I know, I know: it doesn't follow from the fact that it *seems* to be a plenum that it *is* a plenum.

Now you've got it.

But there's another problem I have with this hall of mirrors you call a theory. You say it is only *as if* there were a Central Meaner, *as if* there were a single Author, *as if* there were a place where it all comes together! I don't understand this *as if* business!

Perhaps another thought experiment will make this more palatable. Imagine that we visited another planet and found that the scientists there had a rather charming theory: Every physical thing has a soul inside it, and every soul loves every other soul. This being so, things tend to move toward each other, impelled by the love of their internal souls for each other. We can suppose, moreover, that these scientists had worked out quite accurate systems of soul-placement, so that, having determined the precise location in physical space of an item's soul, they could answer questions about its stability ("It will fall over because its soul is so high"), about vibration ("If you put a counterbalancing object on the side of that drive wheel, with a rather large soul, it will smooth out the wobble"), and about many much more technical topics.

What we could tell them, of course, is that they have hit upon the concept of a center of gravity (or more accurately, a center of mass), and are just treating it a bit too ceremoniously. We tell them that they can go right on talking and thinking the way they were — all they have to give up is a bit of unnecessary metaphysical baggage. There is a simpler, more austere (and much more satisfying) interpretation of the very facts they use their soul-physics to understand. They ask us: Are there souls? Well, sure, we reply — only they're *abstracta*, mathematical abstractions rather than nuggets of mysterious stuff. They're exquisitely useful fictions. It is *as if* every object attracted every other object by concentrating all its gravitational oomph in a single point — and it's vastly easier to calculate the behavior of systems using this principled fiction than it would be to descend to the grubby details — every point attracting every other point.

I feel *as if* my pocket were just picked.

Well don't say I didn't warn you. You can't expect consciousness to turn out to be *just* the way you wanted it. Besides, what are you really giving up?

Only my soul.

Not in any coherent, defensible sense. All you're giving up is a nugget of specialness that couldn't really be special anyway. Why would you think any more of yourself if you turned out to be a sort of mind-pearl in the brain-oyster? What would be so special about being a mind-pearl?

A mind-pearl might be immortal, unlike a brain.

The idea that the Self — or the Soul — is really just an abstraction strikes many people as simply a negative idea, a denial rather than anything positive. But in fact it has a lot going for it, including — if it matters to you — a somewhat more robustly conceived version of potential immortality than anything to be found in traditional ideas of a soul, but that will have to wait until chapter 13. First we must deal definitively with qualia, which *still* have a grip on our imaginations.

12

QUALIA DISQUALIFIED

1. A NEW KITE STRING

Thrown into a causal gap, a quale will simply fall through it.

IVAN FOX (1989), p. 82

When your kite string gets snarled up, in principle it can be unsnarled, especially if you're patient and analytic. But there's a point beyond which principle lapses and practicality triumphs. Some snarls should just be abandoned. Go get a new kite string. It's actually cheaper in the end than the labor it would take to salvage the old one, and you get your kite airborne again sooner. That's how it is, in my opinion, with the philosophical topic of qualia, a tormented snarl of increasingly convoluted and bizarre thought experiments, jargon, in-jokes, allusions to putative refutations, "received" results that should be returned to sender, and a bounty of other sidetrackers and time-wasters. Some messes are best walked away from, so I am not going to conduct an analytical tour of that literature, even though it contains moments of insight and ingenuity from which I have benefited (Shoemaker, 1975, 1981, 1988; White, 1986; Kitcher, 1979; Harman, 1990; Fox, 1989). I've tried in the past to unsnarl the issue (Dennett, 1988a), but now I think it's better if we try to start over almost from scratch.

It's not hard to see how philosophers have tied themselves into such knots over qualia. They started where anyone with any sense would start: with their strongest and clearest intuitions about their own minds. Those intuitions, alas, form a mutually self-supporting closed circle of doctrines, imprisoning their imaginations in the Cartesian Theater. Even though philosophers have discovered the paradoxes inherent in this closed circle of ideas — that's why the literature on qualia exists — they haven't had a *whole alternative vision* to leap to, and so, trusting their still-strong intuitions, they get dragged back into the paradoxical prison. That's why the literature on qualia gets more and more convoluted, instead of resolving itself in agreement. But now we've put in place just such an alternative vision, the Multiple Drafts model. Using it, we can offer a rather different positive account of the issues. Then we can pause in sections 4 and 5 to compare it to the visions I hope it will replace.

An excellent introductory book on the brain contains the following passage:

> "Color" as such does not exist in the world; it exists only in the eye and brain of the beholder. Objects reflect many different wavelengths of light, but these light waves themselves have no color. [Ornstein and Thompson, 1984, p. 55]

This is a good stab at expressing the common wisdom, but notice that taken strictly and literally, it cannot be what the authors mean, and it cannot be true. Color, they say, does not exist "in the world" but only in the "eye and brain" of the beholder. But the eye and brain of the beholder are in the world, just as much parts of the physical world as the objects seen by the observer. And like those objects, the eye and brain are colorful. Eyes can be blue or brown or green, and even the brain is made not *just* of gray (and white) matter: in addition to the *substantia nigra* (the black stuff) there is the *locus ceruleus* (the blue place). But of course the colors that are "in the eye and brain of the beholder" in *this* sense are not what the authors are talking about. What makes anyone think there is color in any other sense?

Modern science — so goes the standard story — has removed the color from the physical world, replacing it with colorless electromagnetic radiation of various wavelengths, bouncing off surfaces that variably reflect and absorb that radiation. It may look as if the color is *out there*, but it isn't. It's *in here* — in the "eye and brain of the beholder." (If the authors of the passage were not such good materialists, they would probably have said that it was in the *mind* of the observer, saving themselves from the silly reading we just dismissed, but creating even worse problems for themselves.) But now, if there is no inner *figment* that could be colored in some special, subjective, in-the-mind, phenomenal sense, colors seem to disappear altogether! *Something* has to be the colors we know and love, the colors we mix and match. Where oh where can they be?

This is the ancient philosophical conundrum we must now face. In the seventeenth century, the philosopher John Locke (and before him, the scientist Robert Boyle) called such properties as colors, aromas, tastes, and sounds *secondary qualities*. These were distinguished from the primary qualities: size, shape, motion, number, and solidity. Secondary qualities were not themselves things-in-the-mind but rather the *powers of things* in the world (thanks to their particular primary qualities) to produce or provoke certain things in the minds of normal observers. (And what if there were no observers around? This is the eternally popular puzzler about the tree in the forest that falls. Does it make a sound? The answer is left as an exercise for the reader.) Locke's way of defining secondary qualities has become part of the standard layperson's interpretation of science, and it has its virtues, but it also gives hostages: the things produced in the mind. The secondary quality *red*, for instance, was for Locke the dispositional property or power of certain surfaces of physical objects, thanks to their microscopic textural features, to produce in us the *idea of red* whenever light was reflected off those surfaces into our eyes. The power in the external object is clear enough, it seems, but what kind of a thing is an idea of red? Is it, like a beautiful gown of blue, *colored* — in some sense? Or is it, like a beautiful discussion of purple, just *about* a color, without itself being colored at all? This opens up possibilities, but how could an idea be just *about* a color (e.g., the color red) if nothing anywhere *is* red?

What is red, anyway? What are colors? Color has always been the philosophers' favorite example, and I will go along with tradition for the time being. The main problem with the tradition nicely emerges in the philosophical analysis of Wilfrid Sellars (1963, 1981b), who distinguished the dispositional properties of objects (Locke's secondary qualities) from what he called occurrent properties. A pink ice cube in the freezer with the light off has the secondary quality pink, but there is no instance of the property *occurrent pink* until an observer opens the door and looks. Is occurrent pink a property of something in the brain or something "in the external world"? In either case, Sellars insisted, occurrent pink is a "homogeneous" property of something real. Part of what he meant to deny by this insistence on homogeneity would be the hypothesis that occurrent pink is anything like *neural activity of intensity 97 in region 75* of the brain. He also meant to deny that the subjective world of color phenomenology is exhausted by anything as colorless as *judgments* that one thing or another is, or seems to be, pink. For instance, the act of recalling in your mind's eye the color of a ripe banana and judging that it is the color yellow would not by itself bring into existence an instance of occurrent yellow (Sellars, 1981; Dennett, 1981b). That would merely be judging that something was yellow, a phenomenon that by itself is as devoid of occurrent yellow as a poem about bananas would be.

Sellars went so far as to claim that all of the physical sciences would have to be revolutionized to make room for occurrent pink and its kin. Few philosophers went along with him on this radical view, but a version of it has recently been resurrected by the philosopher Michael Lockwood (1989). Other philosophers, such as Thomas Nagel, have supposed that even revolutionized science would be unable to deal with such properties:

> The subjective features of conscious mental processes — as opposed to their physical causes and effects — cannot be captured by the purified form of thought suitable for dealing with the physical world that underlies the appearances. [1986, p. 15]

Philosophers have adopted various names for the things in the beholder (or properties of the beholder) that have been supposed to provide a safe home for the colors and the rest of the properties that have been banished from the "external" world by the triumphs of physics: "raw feels," "sensa," "phenomenal qualities," "intrinsic properties of conscious experiences," "the qualitative content of mental states," and, of course, "qualia," the term I will use. There are subtle differences in how these terms have been defined, but I'm going to ride roughshod over them. In the previous chapter I seemed to be denying that there are *any* such properties, and for once what seems so *is* so. I *am* denying that there are any such properties. But (here comes that theme again) I agree wholeheartedly that there seem to be qualia.

There seem to be qualia, because it really does seem as if science has shown us that the colors can't be out there, and hence must be in here. Moreover, it seems that what is in here can't *just* be the judgments we make when things seem colored to us. This reasoning is confused, however. What science has actually shown us is just that the light-reflecting properties of objects cause creatures to go into various discriminative states, scattered about in their brains, and underlying a host of innate dispositions and learned habits of varying complexity. And what are *their* properties? Here we can play Locke's card a second time:

These discriminative states of observers' brains have various "primary" properties (their mechanistic properties due to their connections, the excitation states of their elements, etc.), and in virtue of these primary properties, they have various secondary, merely dispositional properties. In human creatures with language, for instance, these discriminative states often eventually dispose the creatures to express verbal judgments alluding to the "color" of various things. When someone says "I know the ring isn't really pink, but it sure seems pink," the first clause expresses a judgment about something in the world, and the second clause expresses a second-order judgment about a discriminative state about something in the world. The semantics of such statements makes it clear what colors supposedly are: reflective properties of the surfaces of objects, or of transparent volumes (the pink ice cube, the shaft of limelight). And that is just what they are in fact — though saying just *which* reflective properties they are is tricky (for reasons we will explore in the next section).

Don't our internal discriminative states *also* have some special "intrinsic" properties, the subjective, private, ineffable, properties that constitute *the way things look to us* (sound to us, smell to us, etc.)? Those additional properties would be the qualia, and before looking at the arguments philosophers have devised in an attempt to *prove* that there are these additional properties, we will try to remove the motivation for believing in these properties in the first place, by finding alternative explanations for the phenomena that seem to demand them. Then the systematic flaws in the attempted proofs will be readily visible.

According to this alternative view, colors *are* properties "out there" after all. In place of Locke's "ideas of red" we have (in normal human beings) discriminative states that have the content: *red*. An example will help make absolutely clear what these discriminative states are — and more important, what they are not. We can compare the colors of things in the world by putting them side by side and looking at them, to see what judgment we reach, but we can also compare the colors of things by just recalling or imagining them "in our minds." Is the standard red of the stripes on the American flag the same red as, or is it darker or lighter or brighter or more or less orange than, the standard red of Santa Claus's suit (or a British pillar box or the Soviet red star)? (If no two of these standards are available in your memory, try a different pair, such as Visa-card blue and sky blue, or billiard-table-felt green and Granny-Smith-apple green, or lemon yellow and butter yellow.) We are able to make such comparisons "in our mind's eyes," and when we do, we somehow make something happen in us that retrieves information from memory and permits us to compare, in conscious experience, the colors of the standard objects as we remember them (as we take ourselves to remember them, in any case). Some of us are better at this than others, no doubt, and many of us are not very confident in the judgments we reach under such circumstances. That is why we take home paint samples, or take fabric samples to the paint store, so that we can put side by side in the external world instances of the two colors we wish to compare.

When we do make these comparisons "in our mind's eyes," what happens, according to my view? Something strictly analogous to what would happen in a machine — a robot — that could also make such comparisons. Recall from chapter 10 the CADBLIND Mark I Vorsetzer (the one with the camera that could be aimed at the CAD screen). Suppose we put a color picture of Santa Claus in front of it and ask it whether the red in the picture is deeper than the red of the American flag (something it has already stored in its memory). This is what it would do: retrieve its representation of Old Glory from memory, and locate the "red" stripes (they are labeled "red #163" in its diagram). It would then compare this red to the red of the Santa Claus suit in the picture in front of its camera, which happens to be transduced by its color graphics system as red #172. It would compare the two reds *by subtracting 163 from 172 and getting 9*, which it would interpret, let's say, as showing that Santa Claus red seems somewhat deeper and richer (to *it*) than American flag red.

This story is deliberately oversimple, to dramatize the assertion I wish to make: It is obvious that the CADBLIND Mark I doesn't use figment to render its memory (or its current perception), *but neither do we*. The CADBLIND Mark I probably doesn't know how it compares the colors of something seen with something remembered *and neither do we*. The CADBLIND Mark I has — I will allow — a rather simple, impoverished color space with few of the associations or built-in biases of a human being's personal color space, but aside from this vast difference in dispositional complexity, there is no important difference. I could even put it this way: There is no *qualitative* difference between the CADBLIND's performance of such a task and our own. The discriminative states of the CADBLIND Mark I have content in just the same way, and for just the same reasons, as the discriminative brain states I have put in place of Locke's ideas. The CADBLIND Mark I *certainly* doesn't have any qualia (at least, that is the way I expect lovers of qualia to jump at this point), so it does indeed follow from my comparison that I am claiming that we don't have qualia either. The *sort* of difference that people imagine there to be between any machine and any human experiencer (recall the wine-tasting machine we imagined in chapter 3) is one I am firmly denying: There is no such sort of difference. There just seems to be.

2. WHY ARE THERE COLORS?

When Otto, in chapter 11, judged that there seemed to be a glowing pinkish ring, what was the content of his judgment? If, as I have insisted, his judgment wasn't about a quale, a property of a "phenomenal" seeming-ring (made out of figment), just what was it about? What property did he find himself tempted to attribute (falsely) to something out in the world?

Many have noticed that it is curiously difficult to say just what properties of things in the world colors could be. The simple and appealing idea — still found in many elementary discussions — is that each color can be associated with a unique wavelength of light, and hence that the property of being red is simply the property of reflecting all the red-wavelength light and absorbing all the other wavelengths. But this has been known for quite some time to be false. Surfaces with different fundamental reflective properties can be seen as the same color, and the same surface under different conditions of lighting can be seen as different colors. The wavelengths of the light entering the eye are only indirectly related to the colors we see objects to be. (See Gouras, 1984; Hilbert, 1987; and Hardin, 1988, for surveys of the details with different emphases.) For those who had hoped there would be some simple, elegant way to cash in Locke's promissory note about dispositional powers of surfaces, the situation could hardly be more bleak. Some (e.g., Hilbert, 1987) have decided to anchor color objectively by declaring it to be a relatively straightforward property of external objects, such as the property of "surface spectral reflectance"; having made that choice, they must then go on to conclude that normal color vision often presents us with illusions, since the constancies we perceive match up so poorly with the constancies of surface spectral reflectance measured by scientific instruments. Others have concluded that color properties are best considered subjectively, as properties to be defined strictly in terms of systems of brain states in observers, ignoring the confusing variation in the world that gives rise to these states: "Colored objects are illusions, but not unfounded illusions. We are normally in chromatic perceptual states, and these are neural states"

(Hardin, 1988, p. 111; see Thompson, Palacios, and Varela, in press, for a critical discussion of these options, and further arguments for the better option to be adopted here).

What is beyond dispute is that there is no simple, nondisjunctive property of surfaces such that all and only the surfaces with that property are red (in Locke's secondary quality sense). This is an initially puzzling, even depressing fact, since it seems to suggest that our perceptual grip on the world is much worse than we had thought — that we are living in something of a dream world, or are victims of mass delusion. Our color vision does not give us access to simple properties of objects, even though it seems to do so. Why should this be so?

Just bad luck? Second-rate design? Not at all. There is a different, and much more illuminating, perspective we can take on color, first shown to me by the philosopher of neuroscience, Kathleen Akins (1989, 1990).[1] Sometimes new properties come into existence for a reason. A particularly useful example is provided by the famous case of Julius and Ethel Rosenberg, who were convicted and executed in 1953 for spying on the U.S. atomic bomb project for the Soviet Union. It came out at their trial that at one point they improvised a clever password system: a cardboard Jell-O box was torn in two, and the pieces were taken to two individuals who had to be very careful about identifying each other. Each ragged piece became a practically foolproof and unique "detector" of its mate: at a later encounter each party could produce his piece, and if the pieces lined up perfectly, all would be well. Why does this system work? Because tearing the cardboard in two produces an edge of such informational complexity that it would be virtually impossible to reproduce by deliberate construction. (Note that cutting the Jell-O box with straight-edge and razor would entirely defeat the purpose.) The particular jagged edge of one piece becomes a *practically* unique pattern-recognition device for its mate; it is an apparatus or transducer for detecting the shape property *M*, where *M* is uniquely instantiated by its mate.

In other words, the shape property *M* and the *M*-property-detector that detects it were made for each other. There would be no reason for either to exist, to have been created, in the absence of the other. And the same thing is true of colors and color vision: they were made for each other. Color-coding is a fairly recent idea in "human factors engineering," but its virtues are now widely recognized. Hospitals lay out

1. Variations on these themes can be found in Humphrey (1976, 1983a) and in Thompson, Palacios, and Varela (in press).

colored lines in the corridors, simplifying the directions that patients must follow: "To get to physiotherapy, just follow the yellow line; to get to the blood bank, follow the red line!" Manufacturers of televisions, computers, and other electronic gear color-code the large bundles of wires inside so that they can be easily traced from point to point. These are recent applications, but of course the idea is much older; older than the Scarlet Letter with which an adulterer might be marked, older than the colored uniforms used to tell friend from foe in the heat of battle, older than the human species, in fact.

We tend to think of color-coding as the clever introduction of "conventional" color schemes designed to take advantage of "natural" color vision, but this misses the fact that "natural" color vision coevolved *from the outset* with colors whose *raison d'être* was color-coding (Humphrey, 1976). Some things in nature "needed to be seen" and others needed to see them, so a system evolved that tended to minimize the task for the latter by heightening the salience of the former. Consider the insects. Their color vision coevolved with the colors of the plants they pollinated, a good trick of design that benefited both. Without the color-coding of the flowers, the color vision of the insects would not have evolved, and vice versa. So the principle of color-coding is the basis of color vision in insects, not just a recent invention of one clever species of mammal. Similar stories can be told about the evolution of color vision in other species. While some sort of color vision may have evolved initially for the task of discriminating inorganic phenomena visually, it is not yet clear that this has happened with any species on this planet. (Evan Thompson has pointed out to me that honeybees may use their special brand of color vision in navigation, to discriminate polarized sunlight on cloudy days, but is this a secondary utilization of color vision that originally coevolved with flower colors?)

Different systems of color vision have evolved independently, sometimes with radically different color spaces. (For a brief survey, and references, see Thompson, Palacios, and Varela, in press.) Not all creatures with eyes have any sort of color vision. Birds and fish and reptiles and insects clearly have color vision, rather like our "trichromatic" (red-green-blue) system; dogs and cats do not. Among mammals, only primates have color vision, and there are striking differences among them. Which species have color vision, and why? This turns out to be a fascinating and complex story, still largely speculative.

Why do apples turn red when they ripen? It is natural to assume that the entire answer can be given in terms of the chemical changes that happen when sugar and other substances reach various concentrations in the maturing fruit, causing various reactions, and so forth. But this ignores the fact that there wouldn't be apples in the first place if there weren't apple-eating seed-spreaders to see them, so the fact that apples are readily visible to at least some varieties of apple-eaters is a condition of their existence, not a mere "hazard" (from the apple's point of view!). The fact that apples have the surface spectral reflectance properties they do is as much a function of the photopigments that were available to be harnessed in the cone cells in the eyes of fructivores as it is of the effects of interactions between sugar and other compounds in the chemistry of the fruit. Fruits that are not color-coded compete poorly on the shelves of nature's supermarket, but false advertising will be punished; the fruits that are ripe (full of nutrition) *and that advertise that fact* will sell better, but the advertising has to be tailored to the visual capabilities and proclivities of the target consumers.

In the beginning, colors were made to be seen by those who were made to see them. But this evolved gradually, by happenstance, taking serendipitous advantage of whatever materials lay at hand, occasionally exploding in a profusion of elaborations of a new Trick, and always tolerating a large measure of pointless variation *and* pointless (merely coincidental) constancy. These coincidental constancies often concerned "more fundamental" features of the physical world. Once there were creatures who could distinguish red from green berries, they could also distinguish red rubies from green emeralds, but this was just a coincidental bonus. The fact that there is a difference *in color* between rubies and emeralds can thus be considered to be a *derived* color phenomenon. Why is the sky blue? Because apples are red and grapes are purple, not the other way around.

It is a mistake to think that first there were colors — colored rocks, colored water, colored sky, reddish-orange rust and bright blue cobalt — and then Mother Nature came along and took advantage of *those* properties by using them to color-code things. It is rather that first there were various reflective properties of surfaces, reactive properties of photopigments, and so forth, and Mother Nature developed out of these raw materials efficient, mutually adjusted "color"-coding/"color"-vision systems, and among the properties that settled out of that design process are the properties we normal human beings call colors. If the blue of cobalt and the blue of a butterfly's wing happened to match (in normal human beings' vision) this is just a coincidence, a negligible side effect of the processes that brought color vision into existence and thereby (as Locke himself might have acknowledged) baptized a certain curiously gerrymandered set of complexes of primary properties with the shared secondary property of producing a common effect in a set of normal observers.

"But still," you will want to object, "back before there were any animals with color vision, there were glorious red sunsets, and bright green emeralds!" Well, yes, you can say so, but then those very same sunsets were also garish, multicolored, and disgusting, rendered in colors we cannot see, and hence have no names for. That is, you will have to admit this, if there are *or could be* creatures on some planet whose sensory apparatus would be so affected by them. And for all we know, there are species somewhere who naturally see that there are two (or seventeen) different colors among a batch of emeralds *we* found to be indistinguishably green.

Many human beings are red-green colorblind. Suppose we all were; it would then be common knowledge that both rubies and emeralds were "gred" — after all, they look to normal observers just like other gred things: fire engines, well-watered lawns, apples ripe and unripe (Dennett, 1969). Were folks like us to come along, insisting that rubies and emeralds were in fact different colors, there would be no way to declare one of these color-vision systems "truer" than the other.

The philosopher Jonathan Bennett (1965) draws our attention to a case that makes the same point, more persuasively, in another sensory modality. The substance phenol-thio-urea, he tells us, tastes bitter to one-quarter of the human population and is utterly tasteless to the rest. Which way it tastes to you is genetically determined. Is phenol-thio-urea bitter or tasteless? By "eugenics" (controlled breeding) or genetic engineering, we might succeed in eliminating the genotype for finding phenol bitter. If we succeeded, phenol-thio-urea would then be *paradigmatically* tasteless, as tasteless as distilled water: tasteless to all normal human beings. If we performed the opposite genetic experiment, we could in time render phenol-thio-urea paradigmatically bitter. Now, before there were any human beings, was phenol-thio-urea *both* bitter and tasteless? It was chemically the same as it is now.

Facts about secondary qualities are inescapably linked to a reference class of observers, but there are weak and strong ways of treating the link. We may say that secondary qualities are *lovely* rather than *suspect*. Someone could be lovely who had never yet, as it happened, been observed by any observer of the sort who would find her lovely, but she could not — as a matter of logic — be a suspect until someone actually suspected her of something. Particular instances of lovely qualities (such as the quality of loveliness) can be said to exist as Lockean dispositions prior to the moment (if any) where they exercise their power over an observer, producing the defining effect therein. Thus some unseen woman (self-raised on a desert island, I guess) could be genuinely lovely, having the dispositional power to affect normal observers of a certain class in a certain way, in spite of never having the opportunity to do so. But lovely qualities cannot be defined independently of the proclivities, susceptibilities, or dispositions of a class of observers, so it really makes no sense to speak of the existence of lovely properties in complete independence of the existence of the relevant observers. Actually, that's a bit too strong. Lovely qualities *would* not be defined — there would be no point in defining *them*, in contrast to all the other logically possible gerrymandered properties — independently of such a class of observers. So while it might be logically possible ("in retrospect," one might say) to gather color-property instances together by something like brute force enumeration, the reasons for singling out such properties (for instance, in order to explain certain causal regularities in a set of curiously complicated objects) depend on the existence of the class of observers.

Are sea elephants lovely? Not to us. It is hard to imagine an uglier creature. What makes a sea elephant lovely to another sea elephant is not what makes a woman lovely to a man, and to call some as-yet-unobserved woman lovely who, as it happens, would mightily appeal to sea elephants would be to abuse both her and the term. It is only by reference to human tastes, which are contingent and indeed idiosyncratic features of the world, that the property of loveliness (to-a-human-being) can be identified.

On the other hand, suspect qualities (such as the property of being a suspect) are understood in such a way as to presuppose that any instance of the property has already had its defining effect on at least one observer. You may be eminently worthy of suspicion — you may even be obviously guilty — but you can't be a suspect until someone actually suspects you. I am not claiming that colors are suspect qualities. Our intuition that the as-yet-unobserved emerald in the middle of the clump of ore is *already* green does not have to be denied. But I am claiming that colors are lovely qualities, whose existence, tied as it is to a reference class of observers, makes no sense in a world in which the observers have no place. This is easier to accept for some secondary qualities than for others. That the sulphurous fumes spewed forth by primordial volcanos were yellow seems somehow more objective than that they stank, but so long as what we mean by "yellow" is what *we* mean by "yellow," the claims are parallel. For suppose some primordial earthquake cast up a cliff face exposing the stripes of hundreds of chemically different layers to the atmosphere. Were those stripes *visible?* We must ask to whom. Perhaps some of them would be visible to us and others not. Perhaps some of the invisible stripes would be visible to tetrachromat pigeons, or to creatures who saw in the infrared or ultraviolet part of the electromagnetic spectrum. For the same reason one cannot meaningfully ask whether the difference between emeralds and rubies is a visible difference without specifying the vision system in question.

Evolution softens the blow of the "subjectivism" or "relativism" implied by the fact that secondary qualities are lovely qualities. It shows that the absence of "simple" or "fundamental" commonalities in things that are all the same color is not an earmark of total illusion, but rather, a sign of a widespread tolerance for "false positive" detections of the ecological properties that really matter.[2] The basic categories of our color spaces (and of course our odor spaces and sound spaces, and all the rest) are shaped by selection pressures, so that in general it makes sense to ask what a particular discrimination or preference is for. There are reasons why we shun the odors of certain things and seek out others, why we prefer certain colors to others, why some sounds bother us more, or soothe us more. They may not always be *our* reasons, but rather the reasons of distant ancestors, leaving their fossil traces in the built-in biases that innately shape our quality spaces. But as good Darwinians, we should also recognize the possibility — indeed, the necessity — of other, nonfunctional biases, distributed haphazardly

---

2. Philosophers are currently fond of the concept of *natural kinds*, reintroduced to philosophy by Quine (1969), who may now regret the way it has become a stand-in for the dubious but covertly popular concept of *essences*. "Green things, or at least green emeralds, are a kind," Quine observes (p. 116), manifesting his own appreciation of the fact that while emeralds may be a natural kind, *green* things are probably not. The present discussion is meant to forestall one of the tempting mistakes of armchair naturalism: the assumption that whatever nature makes is a natural kind. Colors are *not* "natural kinds" precisely *because* they are the product of biological evolution, which has a tolerance for sloppy boundaries when making categories that would horrify any philosopher bent on good clean definitions. If some creature's life depended on lumping together the moon, blue cheese, and bicycles, you can be pretty sure that Mother Nature would find a way for it to "see" these as "intuitively just the same kind of thing."

through the population in genetic variation. In order for selection pressure to differentially favor those who exhibit a bias against *F* once *F* becomes ecologically important, there has to have been pointless (not-yet-functional) variation in "attitude toward *F*" on which selection can act. For example, if eating tripe *were* to spell prereproductive doom in the future, only those of us who were "naturally" (and *heretofore* pointlessly) disposed against eating tripe would have an advantage (perhaps slight to begin with, but soon to be explosive, if conditions favored it). So it doesn't follow that if you find something (e.g., broccoli) indescribably and ineffably awful, there is a reason for this. Nor does it follow that you are defective if you disagree with your peers about this. It may just be one of the innate bulges in your quality space that has, as of yet, no functional significance at all. (And for your sake, you had better hope that if it ever does have significance, it is because broccoli has suddenly turned out to be bad for us.)

These evolutionary considerations go a long way to explaining why secondary qualities turn out to be so "ineffable," so resistant to definition. Like the shape property *M* of the Rosenbergs' piece of Jell-O box, secondary qualities are extremely resistant to straightforward definition. It is of the essence of the Rosenbergs' trick that we cannot replace our dummy predicate *M* with a longer, more complex, but accurate and exhaustive description of the property, for if we could, we (or someone else) could use that description as a recipe for producing another instance of *M* or another *M*-detector. Our secondary quality detectors were not specifically designed to detect only hard-to-define properties, but the result is much the same. As Akins (1989) observes, it is *not the point* of our sensory systems that they should detect "basic" or "natural" properties of the environment, but just that they should serve our "narcissistic" purposes in staying alive; nature doesn't build epistemic engines.

The only *readily available* way of saying just what shape property *M* is is just to point to the *M*-detector and say that *M* is the shape property detected by this thing here. The same predicament naturally faces anyone trying to say what property someone detects (or misdetects) when something "looks the way it looks to him." So now we can answer the question with which this section began: What property does Otto judge something to have when he judges it to be pink? The property he calls pink. And what property is that? It's hard to say, but this should not embarrass us, because we can say why it's hard to say. The best we can do, practically, when asked what surface properties we detect with color vision, is to say, uninformatively, that we detect the properties we detect. If someone wants a more informative story about those properties, there is a large and rather incompressible literature in biology, neuroscience, and psychophysics to consult. And Otto can't say anything more about the property he calls pink by saying "It's *this!*" (taking himself to be pointing "inside" at a private, phenomenal property of his experience). All that move accomplishes (at best) is to point to his own idiosyncratic color-discrimination state, a move that is parallel to holding up a piece of Jell-O box and saying that it detects *this* shape property. Otto points to his discrimination-device, perhaps, but not to any quale that is exuded by it, or worn by it, or rendered by it, when it does its work. There are no such things.

> But still [Otto insists], you haven't yet said why pink should look like *this!*

Like what?

> Like *this*. Like the particularly ineffable, wonderful, intrinsic pinkness that I am right now enjoying. *That* is not some indescribably convoluted surface reflectance property of external objects.

I see, Otto, that you use the term *enjoying*. You are not alone. Often, when an author wants to stress that the topic has turned from (mere) neuroanatomy to experience, (mere) psychophysics to consciousness, (mere) information to qualia, the word "enjoy" is ushered onto the stage.

3. ENJOYING OUR EXPERIENCES

> But Dan, qualia are what make life worth living!
>
> WILFRID SELLARS (over a fine bottle of Chambertin, Cincinnati, 1971)

> If what I want when I drink fine wine is information about its chemical properties, why don't I just read the label?
>
> SYDNEY SHOEMAKER, Tufts Colloquium, 1988

Some colors were made for liking, and so were some smells and tastes. And other colors, smells, and tastes, were made for disliking. To put the same point more carefully, it is no accident that we (and other creatures who can detect them) like and dislike colors, smells, tastes, and other secondary qualities. Just as we are the inheritors of evolved vertical symmetry detectors in our visual systems for alerting us (like our ancestors) to the ecologically significant fact that another creature is looking at us, so we are the inheritors of evolved quality-detectors that are not disinterested reporters, but rather warners and beckoners, sirens in both the fire-engine sense and the Homeric sense.

As we saw in chapter 7, on evolution, these native alarmists have subsequently been coopted in a host of more complicated organizations, built from millions of associations, and shaped, in the human case, by thousands of memes. In this way the brute come-and-get-it appeal of sex and food, and the brute run-for-your-life aversion of pain and fear get stirred together in all sorts of piquant combinations. When an organism discovers that it pays to attend to some feature of the world *in spite of* its built-in aversion to doing that, it must construct some countervailing coalition to keep aversion from winning. The resulting semi-stable tension can then itself become an acquired taste, to be sought out under certain conditions. When an organism discovers that it must smother the effects of certain insistent beckoners if it is to steer the proper course, it may cultivate a taste for whatever sequences of activity it can find that tend to produce the desired peace and quiet. In such a way could we come to love spicy food that burns our mouths (Rozin, 1982), deliciously "discordant" music, and both the calm, cool realism of Andrew Wyeth and the unsettling, hot expressionism of Willem de Kooning. Marshall McLuhan (1967) proclaimed that the medium is the message, a half-truth that is truer perhaps in the nervous system than in any other forum of communication. What we want when we sip a great wine is not, indeed, the information about its chemical contents; what we want is *to be informed* about its chemical contents in our favorite way. And our preference is *ultimately* based on the biases that are still wired into our nervous systems though their ecological significance may have lapsed eons ago.

This fact has been largely concealed from us by our own technology. As the psychologist Nicholas Humphrey notes,

> As I look around the room I'm working in, man-made colour shouts back at me from every surface: books, cushions, a rug on the floor, a coffee-cup, a box of staples — bright blues, reds, yellows, greens. There is as much colour here as in any tropical forest. Yet while almost every colour in the forest would be meaningful, here in my study almost nothing is. Colour anarchy has taken over. [1983, p. 149]

Consider, for instance, the curious fact that monkeys don't like red light. Given a choice, rhesus monkeys show a strong preference for the blue-green end of the spectrum, and get agitated when they have to endure periods in red environments (Humphrey, 1972, 1973, 1983; Humphrey and Keeble, 1978). Why should this be? Humphrey points out that red is always used to alert, the ultimate color-coding color, but for that very reason ambiguous: the red fruit may be good to eat, but the red snake or insect is probably advertising that it is poisonous. So "red" sends mixed messages. But why does it send an "alert" message in the first place? Perhaps because it is the strongest available contrast with the ambient background of vegetative green or sea blue, or — in the case of monkeys — because red light (red to reddish-orange to orange light) is the light of dusk and dawn, the times of day when virtually all the predators of monkeys do their hunting.

The affective or emotional properties of red are not restricted to rhesus monkeys. All primates share these reactions, including human beings. If your factory workers are lounging too long in the rest rooms, painting the walls of the rest rooms red will solve that problem — but create others (see Humphrey, forthcoming). Such "visceral" responses are not restricted to colors, of course. Most primates raised in captivity who have never seen a snake will make it unmistakably clear that they loathe snakes the moment they see one, and it is probable that the traditional human dislike of snakes has a biological source that explains the biblical source, rather than the other way around.[3] That is, our genetic heritage queers the pitch in favor of memes for snake-hating.

Now here are two different explanations for the uneasiness most of us feel (even if we "conquer" it) when we see a snake:

(1) Snakes evoke in us a particular intrinsic snake-yuckiness quale when we look at them, and our uneasiness is a reaction to that quale.

(2) We find ourselves less than eager to see snakes because of innate biases built into our nervous systems. These favor the release of adrenaline, bring fight-or-flight routines on line, and,

3. The primatologist Sue Savage-Rumbaugh has informed me that laboratory-raised bonobos, or pygmy chimps, show no signs of an innate dislike of snakes, unlike chimpanzees.

> by activating various associative links, call a host of scenarios into play involving danger, violence, damage. The original primate aversion is, in us, transformed, revised, deflected in a hundred ways by the memes that have exploited it, coopted it, shaped it. (There are many different levels at which we could couch an explanation of this "functionalist" type. For instance, we could permit ourselves to speak more casually about the power of snake-perceptions to produce anxieties, fears, anticipations of pain, and the like, but that might be seen as "cheating" so I am avoiding it.)

The trouble with the first sort of explanation is that it only seems to be an explanation. The idea that an "intrinsic" property (of occurrent pink, of snake-yuckiness, of pain, of the aroma of coffee) could *explain* a subject's reactions to a circumstance is hopeless — a straightforward case of a *virtus dormitiva* (see page 63). Convicting a theory of harboring a vacuous *virtus dormitiva* is not that simple, however. Sometimes it makes perfectly good sense to posit a temporary *virtus dormitiva*, pending further investigation. Conception is, by definition we might say, the cause of pregnancy. If we had no other way of identifying conception, telling someone she got pregnant because she conceived would be an empty gesture, not an explanation. But once we've figured out the requisite mechanical theory of conception, we can see *how* conception is the cause of pregnancy, and informativeness is restored. In the same spirit, we might identify qualia, by definition, as the proximal causes of our enjoyment and suffering (roughly put), and then proceed to discharge our obligations to inform by pursuing the second style of explanation. But curiously enough, qualophiles (as I call those who still believe in qualia) will have none of it; they insist, like Otto, that qualia "reduced" to mere complexes of mechanically accomplished dispositions to react are not the qualia they are talking about. *Their* qualia are something different.

> Consider [says Otto] the way the pink ring seems to me right *now*, at this very moment, in isolation from all my dispositions, past associations and future activities. *That*, the purified, isolated *way it is with me* in regards to color at this moment — that is my pink quale.

Otto has just made a mistake. In fact, this is the big mistake, the source of all the paradoxes about qualia, as we shall see. But before exposing the follies of taking this path, I want to demonstrate some of the positive benefits of the path that Otto shuns: the "reductionist" path of *identifying* "the way it is with me" with the sum total of all the idiosyncratic reactive dispositions inherent in my nervous system as a result of my being confronted by a certain pattern of stimulation.

Consider what it must have been like to be a Leipzig Lutheran churchgoer in, say, 1725, hearing one of J. S. Bach's chorale cantatas in its premier performance. (This exercise in imagining *what it is like* is a warm-up for chapter 14, where we will be concerned with consciousness in other animals.) There are probably no significant biological differences between us today and German Lutherans of the eighteenth century; we are the same species, and hardly any time has passed. But, because of the tremendous influence of culture — the memosphere — our psychological world is quite different from theirs, in ways that would have a noticeable impact on our respective experiences when hearing a Bach cantata for the first time. Our musical imagination has been enriched and complicated in many ways (by Mozart, by Charlie Parker, by the Beatles), but also it has lost some powerful associations that Bach could count on. His chorale cantatas were built around chorales, traditional hymn melodies that were deeply familiar to his churchgoers and hence provoked waves of emotional and thematic association as soon as their traces or echoes appeared in the music. Most of us today know these chorales only from Bach's settings of them, so when we hear them, we hear them with different ears. If we want to imagine what it was like to be a Leipzig Bach-hearer, it is not enough for us to hear the same tones on the same instruments in the same order; we must also prepare ourselves somehow to respond to those tones with the same heartaches, thrills, and waves of nostalgia.

It is not utterly impossible to prepare ourselves in these ways. A music scholar who carefully avoided all contact with post-1725 music and familiarized himself intensively with the traditional music of that period would be a good first approximation. More important, as these observations show, it is not impossible to know in just what ways we would have to prepare ourselves whether or not we cared to go to all the trouble. So we could know what it was like "in the abstract" so to speak, and in fact I've just told you: the Leipzigers, hearing the chorale cantatas, were reminded of all the associations that already flavored their recognition of the chorale melodies. It is easy enough to imagine what *that* must have been like for them — though with variations drawn from our own experience. We can imagine what it would be like to hear Bach's setting of familiar Christmas carols, for instance, or "Home on the Range." We can't do the job precisely, but only because we can't forget or abandon all that we know that the Leipzigers didn't know.

To see how crucial this excess baggage of ours is, imagine that musicologists unearthed a heretofore unknown Bach cantata, definitely by the great man, but hidden in a desk and probably never yet heard even by the composer himself. Everyone would be aching to hear it, to experience for the first time the "qualia" that the Leipzigers would have known, had they only heard it, but this turns out to be impossible, for the main theme of the cantata, by an ugly coincidence, is the first seven notes of "Rudolph the Red-Nosed Reindeer"! We who are burdened with that tune would *never* be able to hear Bach's version as he intended it or as the Leipzigers would have received it.

A clearer case of imagination-blockade would be hard to find, but note that it has nothing to do with biological differences or even with "intrinsic" or "ineffable" properties of Bach's music. The reason we couldn't imaginatively relive in detail (and accurately) the musical experience of the Leipzigers is simply that we would have to take ourselves along for the imaginary trip, and we know too much. But if we want, we can carefully list the differences between our dispositions and knowledge and theirs, and by comparing the lists, come to appreciate, *in whatever detail we want*, the differences between what it was like to be them listening to Bach, and what it is like to be us. While we might lament that inaccessibility, at least we could understand it. There would be no mystery left over; just an experience that could be described quite accurately, but not directly enjoyed unless we went to ridiculous lengths to rebuild our personal dispositional structures.

Qualophiles, however, have resisted this conclusion. It has seemed to them that even though such an investigation as we have just imagined might settle *almost* all the questions we had about what it was like to be the Leipzigers, there would have to be an ineffable residue, *something* about what it was like for the Leipzigers that no further advances in merely "dispositional" and "mechanistic" knowledge could reduce to zero. That is why qualia have to be invoked by qualophiles as *additional* features, over and above and strictly independent of the wiring that determines withdrawal, frowning, screaming, and other "mere behaviors" of disgust, loathing, fear. We can see this clearly if we revert to our example of colors.

Suppose we suggest to Otto that what made his "occurrent pink" the particular tantalizing experience that he enjoyed was simply the sum total of all the innate and learned associations and reactive dispositions triggered by the particular way he was (mis)informed by his eyes:

What qualia *are*, Otto, are just those complexes of dispositions. When you say "*This* is my quale," what you are singling out, or referring to, *whether you realize it or not*, is your idiosyncratic complex of dispositions. You *seem* to be referring to a private, ineffable something-or-other in your mind's eye, a private shade of homogeneous pink, but this is just how it seems to you, not how it is. That "quale" of yours is a character in good standing in the fictional world of your heterophenomenology, but what it turns out to be in the *real* world in your brain is just a complex of dispositions.

> That cannot be all there is to it [Otto replies, taking the fatal step in the qualophile tradition], for while that complex of mere dispositions might be the basis or source, somehow, for my particular quale of pink, they could all be changed without changing my intrinsic quale, or my intrinsic quale could change, without changing that manifold of mere dispositions. For instance, my qualia could be *inverted* without inverting all my dispositions. I could have all the reactivities and associations that I now have for *green* to the accompaniment of the *quale* I now have for *red*, and vice versa.

4. A PHILOSOPHICAL FANTASY: INVERTED QUALIA

The idea of the possibility of such "inverted qualia" is one of philosophy's most virulent memes. Locke discussed it in his *Essay Concerning Human Understanding* (1690), and many of my students tell me that as young children they hit upon the same idea for themselves, and were fascinated by it. The idea seems to be transparently clear and safe:

> There are the ways things look to me, and sound to me, and smell to me, and so forth. That much is obvious. I wonder, though, if the ways things appear to me are the same as the ways things appear to other people.

Philosophers have composed many different variations on this theme, but the classic version is the interpersonal version: How do I know that you and I see the same *subjective* color when we look at something? Since we both learned our color words by being shown public colored objects, our verbal behavior will match even if we experience entirely different subjective colors — even if the way red things look to me is the way green things look to you, for instance. We would call the same public things "red" and "green" even if our private experiences were "the opposite" (or just different).

Is there any way to tell whether this is the case? Consider the hypothesis that red things look the same to you and me. Is this hypothesis both irrefutable and unconfirmable? Many have thought so, and some have concluded that for just that reason it is one sort of nonsense or another, in spite of its initial appeal to common sense. Others have wondered if technology might come to the rescue and confirm (or disconfirm) the interpersonal inverted spectrum hypothesis. The science-fiction movie *Brainstorm* (not, I hasten to say, a version of my book *Brainstorms*) featured just the right imaginary device: Some neuroscientific apparatus fits on your head and feeds your visual experience into my brain via a cable. With eyes closed I accurately report everything you are looking at, except that I marvel at how the sky is yellow, the grass red, and so forth. If we had such a machine, couldn't such an experiment with it confirm, empirically, the hypothesis that our qualia were different? But suppose the technician pulls the plug on the connecting cable, inverts it 180 degrees, reinserts it in the socket, and I now report the sky is blue, the grass green, and so forth. Which would be the "right" orientation of the plug? Designing and building such a device — supposing for the moment that it would be possible — would require that its "fidelity" be tuned or calibrated by the normalization of the two subjects' reports, so we would be right back at our evidential starting point. Now one might try to avert this conclusion with further elaborations, but the consensus among the qualophiles is that this is a lost cause; there seems to be general agreement that the moral of *this* thought experiment is that no intersubjective comparison of qualia would be possible, even with perfect technology. This does provide support, however, for the shockingly "verificationist" or "positivistic" view that the very idea of inverted qualia is nonsense — and hence that the very idea of qualia is nonsense. As the philosopher Ludwig Wittgenstein put it, using his famous "beetle in the box" analogy,

> The thing in the box has no place in the language-game at all; not even as a *something*; for the box might even be empty. — No, one can "divide through" by the thing in the box; it cancels out, whatever it is. [1953, p. 100]

But just what does this mean? Does it mean that qualia are real but ineffective? Or that there aren't any qualia after all? It still seemed obvious to most philosophers who thought about it that qualia were real, even if a difference in qualia would be a difference that couldn't be detected in any way. That's how matters stood, uneasily, until someone dreamt up the presumably improved version of the thought experiment: the *intra*personal inverted spectrum. The idea seems to have occurred to several people independently (Gert, 1965; Putnam, 1965; Taylor, 1966; Shoemaker, 1969; Lycan, 1973). In this version, the experiences to be compared are all in one mind, so we don't need the hopeless Brainstorm machine.

> You wake up one morning to find that the grass has turned red, the sky yellow, and so forth. No one else notices any color anomalies in the world, so the problem must be in you. You are entitled, it seems, to conclude that you have undergone visual color qualia inversion. How did it happen? It turns out that while you slept, evil neurosurgeons switched all the wires — the neurons — leading from the color-sensitive cone cells in your retinas.

So far, so good. The effect on you would be startling, maybe even terrifying. You would certainly be able to detect that the way things looked to you now was very different, and we would even have a proper scientific explanation of why this was: The neuron clusters in the visual cortex that "care about" color, for instance, would be getting their stimulation from a systematically shifted set of retinal receptors. So half the battle is won, it seems: A difference in qualia would be detectable after all, if it were a difference that developed rather swiftly in a single person.[4] But this is only half the battle, for the imagined neurosurgical prank has also switched all your reactive dispositions; not only do you *say* your color experiences have all been discombobulated, but your nonverbal color-related behavior has been inverted as well. The edginess you used to exhibit in red light you now exhibit in green light, and you've lost the fluency with which you used to rely on various color-coding schemes in your life. (If you play basketball for the Boston Celtics, you keep passing the ball mistakenly to the guys in the *red* uniforms.)

What the qualophile needs is a thought experiment that demon-

---

4. The suddenness would be important, since if it happened very gradually, you might not be able to notice. As Hardin (1990) has pointed out, the gradual yellowing of your lenses with age slowly shifts your sense of the primary colors; shown a color wheel and asked to point at pure red (red with no orange or purple in it), where on the continuum you point is partly a function of age.

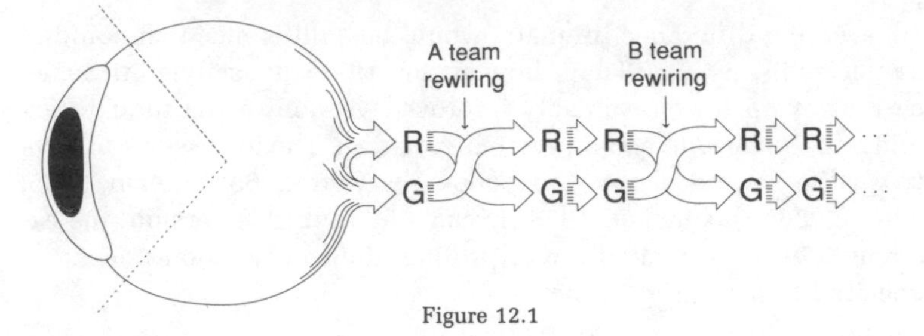

Figure 12.1

strates that the-way-things-look can be independent of all these reactive dispositions. So we have to complicate the story with a further development; we must describe something happening that undoes the switch in reactive dispositions while leaving the switched "qualia" intact. Here is where the literature lurches into ever more convoluted fantasies, for no one thinks for a moment that the-way-things-look is ever *actually* divorced from the subject's reactive dispositions; it is just that this is deemed an important *possibility in principle* by the qualophiles. To show this, they need to describe a possible case, however outlandish, in which it would be obvious that this detachment was actual. Consider a story that *won't* work:

> One night while you sleep, evil neurosurgeons switch all the wires from the cone cells (just as before), and then, later the same busy night, another team of neurosurgeons, the B team, comes along and performs a *complementary* rewiring a little farther up on the optic nerve.

This restores all the old reactive dispositions (we can presume), but, alas, it also restores the old qualia. The cells in the cortex that "care about" color, for instance, will now be getting their original signals again, thanks to the speedy undoing of the damage by the B team. The second switcheroo happened too early, it seems; it happened *on the way up* to conscious experience. So we'll have to tell the story differently, with the second switcheroo happening later, *after* the inverted qualia have taken their bow in consciousness, but *before* any of the inverted reactions to them can set in. But is this possible? Not if the arguments for the Multiple Drafts model are correct. There is no line that can be drawn across the causal "chain" from eyeball through consciousness to subsequent behavior such that all reactions to x happen after it and consciousness of x does not happen before it. This is because it is not a simple causal chain, but a causal network, with multiple paths on which Multiple Drafts are being edited simultaneously and semi-independently. The qualophile's story would make sense if there were a Cartesian Theater, a special place in the brain where the conscious experience happened. If there were such a place, we could bracket it with the two switcheroos, leaving inverted qualia in the Theater, while keeping all the reactive dispositions normalized. Since there is no such Cartesian Theater, however, the thought experiment doesn't make sense. There is no coherent way to tell the necessary story. There is no way to isolate the properties *presented* in consciousness from the brain's multiple reactions to its discriminations, because there is no such additional presentation process.

In the literature on the inverted spectrum, the second switcheroo is often supposed to be accomplished not by surgery but by gradual adaptation by the subject to the new regime of experiences. This makes superficial sense; people can adapt amazingly well to bizarre displacements of their senses. There have been many visual field inversion experiments in which subjects wear goggles that turn everything upside down — by turning the retinal image right side up! (E.g., Stratton, 1896; Kohler, 1961; Welch, 1978, provides a good summary; see also Cole, 1990.) After several days of constantly wearing inverting goggles of one sort or another (it makes a difference — some varieties had a wide field of view, and others gave the viewers a sort of tunnel vision), subjects often make an astonishingly successful adaptation. In Ivo Kohler's film of his experiments in Innsbruck, we see two of his subjects, comically helpless when they first put on the goggles, skiing downhill and riding bicycles through city traffic, still wearing the inverting goggles and apparently completely adapted to them.

So let's suppose that you gradually adapt to the surgical inversion of your color vision. (Why you would want to adapt, or would have to adapt, is another matter, but we may as well concede the point to the qualophiles, to hasten their demise.) Now some adaptations would at first be clearly *post-experiential*. We may suppose that the clear sky would still *look* yellow to you, but you would start calling it blue to get in step with your neighbors. Looking at a novel object might cause momentary confusion: "It's gr— I mean *red!*" What about your edginess in green light — would it still show up as an abnormality in your galvanic skin response? For the sake of the argument, the qualophile has to imagine, however unlikely this might be, that *all* your reactive dispositions adapt, leaving behind only the residue of the still-inverted qualia, so for the sake of argument, let's concede that the most fundamental and innate biases in your quality spaces also "adapt" — this is preposterous, but there is worse to come.

In order to tell the necessary story, the qualophile must suppose that eventually all these adaptations become second nature — swift and unstudied. (If they didn't become second nature, there would be leftover reactive dispositions that would be still different, and the argument requires that these all be ironed out.) So be it. Now, assuming that *all* your reactive dispositions are restored, what is your intuition about your qualia? Are they still inverted or not?

It is legitimate to pass at this point, on the grounds that after being asked to tolerate so many dubious assumptions for the sake of argument, you either come up empty — no intuition bubbles up at all — or you find yourself mistrustful of whatever intuition does strike you. But perhaps it does seem quite obvious to you that your qualia would still be inverted. But why? What in the story has led you to see it this way? Perhaps, even though you have been following directions, you have innocently added some further assumptions not demanded by the story, or failed to notice certain possibilities not ruled out by the story. I suggest that the most likely explanation for your intuition that, in this imagined instance, you would still have "inverted qualia" is that you are making the additional, and unwarranted, assumption that all the adaptation is happening on the "post-experiential side."

It could be, though, couldn't it, that the adaptation was accomplished on the upward path? When you first put on heavily tinted goggles, you won't see any color at all — or at least the colors you see are weird and hard-to-distinguish colors — but after wearing them for a while, surprising normal color vision returns. (Cole, 1990, draws philosophers' attention to these effects, which you can test for yourself with army-surplus infrared sniper goggles.) Perhaps, not knowing this surprising fact, it just never occurred to you that you *might* adapt to the surgery in much the same way. We could have highlighted this possibility in the thought experiment, by adding a few details:

> . . . And as the adaptation proceeded, you often found to your surprise that the colors of things didn't seem so strange after all, and sometimes you got confused and made *double* corrections. When asked the color of a novel object you said "It's gr— , no red — no, it *is* green!"

Told this way, the story might make it seem "obvious" that the color qualia themselves had adapted, or been reinverted. But in any case, you may now think, it has to be one way or the other. There couldn't be a case where it wasn't perfectly obvious which sort of adjustment you had made! The unexamined assumption that grounds *this* conviction is that all adaptations can be categorized as either pre-experiential or post-experiential (Stalinesque or Orwellian). At first this may seem to be an innocent assumption, since extreme cases are easy to classify. When the brain compensates for head and eye motions, producing a stable visual world "in experience," this is surely a pre-experiential cancellation of effects, an adaptation on the upward path to consciousness. And when you imagine making peripheral ("late") compensations in color-word choosing ("It's gr— I mean *red!*") this is obviously a post-experiential, merely behavioral adjustment. It stands to reason then, doesn't it, that when *all* the adaptations have been made, either they leave the subjective color (the color "in consciousness") inverted or they don't? Here's how we would tell: Add up the switcheroos on the upward path; if there are an even number — as in the Team B handiwork — the qualia are normalized, and if odd, the qualia are still inverted. Nonsense. Recall the Neo-Laffer curve in chapter 5. It is not at all a logical or geometric necessity that there be a single value of a discriminated variable that can be singled out as *the* value of the variable "in consciousness."

We can demonstrate this with a little fantasy of our own, playing by the qualophile's rules. Suppose that presurgically a certain shade of blue tended to remind you of a car in which you once crashed, and hence was a color to be shunned. At first, postsurgically, you have no negative reactions to things of that color, finding them an innocuous and unmemorable yellow, let's suppose. After your complete adaptation, however, you again shun things of that shade of blue, *and it is because they remind you of that crash.* (If they didn't, this would be an unadapted reactive disposition.) But if we ask you whether this is because, as you remember the crash, the car was yellow — just like the noxious object before you now — or because, as you remember the crash, the car was blue — just like the noxious object before you now, you really shouldn't be able to answer. Your verbal behavior will be totally "adapted"; your immediate, second-nature answer to the question: "What color was the car you crashed?" is "blue" and you will unhesitatingly call the noxious object before you blue as well. Does that entail that you have *forgotten* the long training period?

No. We don't need anything so dramatic as amnesia to explain your inability to answer, for we have plenty of everyday cases in which the same phenomenon arises. Do you like beer? Many people who like beer will acknowledge that beer is an acquired taste. One gradually trains oneself — or just comes — to enjoy that flavor. What flavor? The flavor of the first sip?

> No one could like *that* flavor [an experienced beer drinker might retort]. Beer tastes different to the experienced beer drinker. If beer went on tasting to me the way the first sip tasted, I would never have gone on drinking beer! Or, to put the same point the other way around, if my first sip of beer had tasted to me the way my most recent sip just tasted, I would never have had to acquire the taste in the first place! I would have loved the first sip as much as the one I just enjoyed.

If this beer drinker is right, then beer is *not* an acquired taste. No one comes to enjoy *the way the first sip tasted*. Instead, *the way beer tastes to them* gradually changes. Other beer drinkers might insist that, no, beer *did* taste to them now the way it always did, only now they like *that very taste*. Is there a real difference? There is a difference in heterophenomenology, certainly, and the difference needs to be explained. It *could* be that the different convictions spring from genuine differences in discriminative capacity of the following sort: in the first sort of beer drinker the "training" has changed the "shape" of the quality space for tasting, while in the second sort the quality space remains roughly the same, but the "evaluation function" over that space has been revised. Or it *could* be that some or even all of the beer drinkers are kidding themselves (like those who insist that the high-resolution Marilyns are all really *there* in the background of their visual field). We have to look beyond the heterophenomenological worlds to the actual happenings in the head to see whether there is a truth-preserving (if "strained") interpretation of the beer drinkers' claims, and *if* there is, it will only be because we decide to *reduce* "the way it tastes" to one complex of reactive dispositions or another (Dennett, 1988a). We would have to "destroy" qualia in order to "save" them.

So if a beer drinker furrows his brow and gets a deadly serious expression on his face and says that what he is referring to is "*the way the beer tastes to me right now*," he is definitely kidding himself if he thinks he can *thereby* refer to a quale of his acquaintance, a subjective state that is independent of his changing reactive attitudes. It may seem to him that he can, but he can't.[5]

And by the same token, in the imagined case of being reminded

---

5. "The very fact that we should so much like to say: '*This* is the important thing' — while we point privately to the sensation — is enough to shew how much we are inclined to say something which gives no information." Wittgenstein (1953), i298.

of the car crash by the blue object, you would be kidding yourself if you thought you could tell, from *the way the object looks to you*, whether it was "intrinsically" the same as the way the car looked to you when you crashed. This is enough to undercut the qualophile's thought experiment, for the goal was to describe a case in which it was *obvious* that the qualia would be inverted while the reactive dispositions would be normalized. The *assumption* that one could just tell is question-begging, and without the assumption, there is no argument, but just an intuition pump — a story that cajoles you into declaring your gut intuition without giving you a good reason for it.

Question-begging or not, it may *still* seem just plain obvious that "the subjective colors you would be seeing things to be" would *have* to be "one way or the other." This just shows the powerful gravitational force that the Cartesian Theater exerts on our imaginations. It may help to break down the residual attractiveness of this idea if we consider further the invited parallel with image-inverting goggles. When the adaptations of the subjects wearing these goggles have become so second nature that they can ride bicycles and ski, the natural (but misguided) question to ask is this: Have they adapted *by turning their experiential world back right side up*, or *by getting used to their experiential world being upside down?* And what do they say? They say different things, which correlate roughly with how complete their adaptation was. The more complete it was, the more the subjects dismiss the question as improper or unanswerable. This is just what the Multiple Drafts theory demands: Since there are a host of discriminations and reactions that need to be adjusted, scattered around in the brain, some of them dealing with low-level "reflexes" (such as ducking the right way when something looms at you) and others dealing with focally attended deliberate actions, it is not surprising that as the adaptations in this patchwork accumulate, subjects should lose all conviction of whether to say "things look the way they used to look" instead of "things still look different, but I'm getting used to it." In some ways things look the same to them (as judged by their reactions), in other ways things look different (as judged by other reactions). If there were a single representation of visuo-motor space through which all reactions to visual stimuli had to be channeled, *it* would have to be "one way or the other," perhaps, but there is no such single representation. *The* way things look to them is composed of many partly independent habits of reaction, not a single intrinsically right-side-up or upside-down picture in the head. All that matters is the fit between the input and the output, and since this is accomplished in many different places with many different and largely independent means, there is just no saying what "counts" as "my visual field is still upside down."

The same is true of "qualia" inversion. The idea that it is something *in addition* to the inversion of all one's reactive dispositions, so that, if they were renormalized the inverted qualia would remain, is simply part of the tenacious myth of the Cartesian Theater. This myth is celebrated in the elaborate thought experiments about spectrum inversion, but to celebrate is not to support or prove. *If* there are no qualia over and above the sum total of dispositions to react, the idea of holding the qualia constant while adjusting the dispositions is self-contradictory.

5. "EPIPHENOMENAL" QUALIA?

There is another philosophical thought experiment about our experience of color that has proven irresistible: Frank Jackson's (1982) much-discussed case of Mary, the color scientist who has never seen colors. Like a good thought experiment, its point is immediately evident to even the uninitiated. In fact it is a bad thought experiment, an intuition pump that actually encourages us to misunderstand its premises!

> Mary is a brilliant scientist who is, for whatever reason, forced to investigate the world from a black-and-white room *via* a black-and-white television monitor. She specializes in the neurophysiology of vision and acquires, let us suppose, all the physical information there is to obtain about what goes on when we see ripe tomatoes, or the sky, and use terms like *red*, *blue*, and so on. She discovers, for example, just which wavelength combinations from the sky stimulate the retina, and exactly how this produces *via* the central nervous system the contraction of the vocal chords and expulsion of air from the lungs that results in the uttering of the sentence "The sky is blue." . . . What will happen when Mary is released from her black-and-white room or is given a color television monitor? Will she *learn* anything or not? It seems just obvious that she will learn something about the world and our visual experience of it. But then it is inescapable that her previous knowledge was incomplete. But she had *all* the physical information. *Ergo* there is more to have than that, and Physicalism is false. . . . [p. 128]

The point could hardly be clearer. Mary has had *no* experience of color at all (there are no mirrors to look at her face in, she's obliged to wear black gloves, etc., etc.), and so, at that special moment when her captors finally let her come out into the colored world which she knows only by description (and black-and-white diagrams), "it seems just obvious," as Jackson says, that she will learn something. Indeed, we can all vividly imagine her, seeing a red rose for the first time and exclaiming, "So *that*'s what red looks like!" And it may also occur to us that if the first colored things she is shown are, say, unlabeled wooden blocks, and she is told only that one of them is red and the other blue, she won't have the faintest idea which is which until she somehow learns which color words go with her newfound experiences.

That is how almost everyone imagines this thought experiment — not just the uninitiated, but the shrewdest, most battle-hardened philosophers (Tye, 1986; Lewis, 1988; Loar, 1990; Lycan, 1990; Nemirov, 1990; Harman, 1990; Block, 1990; van Gulick, 1990). Only Paul Churchland (1985, 1990) has offered any serious resistance to the *image*, so vividly conjured up by the thought experiment, of Mary's dramatic discovery. The image is wrong; if that is the way you imagine the case, you are simply not following directions! The reason no one follows directions is because what they ask you to imagine is so preposterously immense, you can't even try. The crucial premise is that "She has *all* the physical information." That is not readily imaginable, so no one bothers. They just imagine that she knows lots and lots — perhaps they imagine that she knows everything that anyone knows *today* about the neurophysiology of color vision. But that's just a drop in the bucket, and it's not surprising that Mary would learn something if *that* were all she knew.

To bring out the illusion of imagination here, let me continue the story in a surprising — but legitimate — way:

> And so, one day, Mary's captors decided it was time for her to see colors. As a trick, they prepared a bright blue banana to present as her first color experience ever. Mary took one look at it and said "Hey! You tried to trick me! Bananas are yellow, but this one is blue!" Her captors were dumfounded. How did she do it? "Simple," she replied. "You have to remember that I know *everything* — absolutely everything — that could ever be known about the physical causes and effects of color vision. So of course before you brought the banana in, I had already written down, in exquisite detail, exactly what physical impression a yellow object

or a blue object (or a green object, etc.) would make on my nervous system. So I already knew exactly what *thoughts* I would have (because, after all, the "mere disposition" to think about this or that is not one of your famous qualia, is it?). I was not in the slightest surprised by my experience of blue (what surprised me was that you would try such a second-rate trick on me). I realize it is *hard for you to imagine* that I could know so much about my reactive dispositions that the way blue affected me came as no surprise. Of course it's hard for you to imagine. It's hard for anyone to imagine the consequences of someone knowing absolutely everything physical about anything!"

Surely I've cheated, you think. I must be hiding some impossibility behind the veil of Mary's remarks. Can you prove it? My point is not that my way of telling the rest of the story proves that Mary *doesn't* learn anything, but that the usual way of imagining the story doesn't *prove* that she *does*. It doesn't prove anything; it simply pumps the intuition that she does ("it seems just obvious") by lulling you into imagining something other than what the premises require.

It is of course true that in any realistic, readily imaginable version of the story, Mary would come to learn something, but in any realistic, readily imaginable version she might know a lot, but she would not know everything physical. Simply imagining that Mary knows a lot, and leaving it at that, is not a good way to figure out the implications of her having "all the physical information" — any more than imagining she is filthy rich would be a good way to figure out the implications of the hypothesis that she owned everything. It may help us imagine the extent of the powers her knowledge gives her if we begin by enumerating a few of the things she obviously knows in advance. She knows black and white and shades of gray, and she knows the difference between the color of any object and such surface properties as glossiness versus matte, and she knows all about the difference between luminance boundaries and color boundaries (luminance boundaries are those that show up on black-and-white television, to put it roughly). And she knows precisely which effects — described in neurophysiological terms — each particular color will have on her nervous system. So the only task that remains is for her to figure out a way of identifying those neurophysiological effects "from the inside." You may find you can readily imagine her making *a little* progress on this — for instance, figuring out tricky ways in which she would be able to tell that some color, whatever it is, is *not* yellow, or *not* red. How? By noting some salient and specific reaction that her brain would have only for yellow or only for red. But if you allow her even a little entry into her color space in this way, you should conclude that she can leverage her way to complete advance knowledge, because she doesn't just know the *salient* reactions, she knows them all.

Recall Julius and Ethel Rosenberg's Jell-O box, which they turned into an *M*-detector. Now imagine their surprise if an impostor were to show up with a "matching" piece that was not the original. "Impossible!" they cry. "Not impossible," says the impostor, "just difficult. I had *all the information* required to reconstruct an *M*-detector, and to make another thing with shape-property *M*." Mary had enough information (in the original case, if correctly imagined) to figure out just what her red-detectors and blue-detectors were, and hence to identify them in advance. Not the usual way of coming to learn about colors, but Mary is not your usual person.

I know that this will not satisfy many of Mary's philosophical fans, and that there is a lot more to be said, but — and this is my main point — the actual proving must go on in an arena far removed from Jackson's example, which is a classic provoker of Philosophers' Syndrome: mistaking a failure of imagination for an insight into necessity. Some of the philosophers who have dealt with the case of Mary may not care that they have imagined it wrong, since they have simply used it as a springboard into discussions that shed light on various independently interesting and important issues. I will not pursue those issues here, since I am interested in directly considering the conclusion that Jackson himself draws from his example: visual experiences have qualia that are "epiphenomenal."

The term "epiphenomena" is in common use today by both philosophers and psychologists (and other cognitive scientists). It is used with the presumption that its meaning is familiar and agreed upon, when in fact, philosophers and cognitive scientists use the term with *entirely* different meanings — a strange fact made even stranger to me by the fact that although I have pointed this out time and again, no one seems to care. Since "epiphenomenalism" often seems to be the last remaining safe haven for qualia, and since this appearance of safety is due entirely to the confusion between these two meanings, I must become a scold, and put those who use the term on the defensive.

According to the *Shorter Oxford English Dictionary*, the term "epiphenomenon" first appears in 1706 as a term in pathology, "a secondary appearance or symptom." The evolutionary biologist Thomas Huxley (1874) was probably the writer who extended the term to its current use in psychology, where it means a *nonfunctional* property or by-product. Huxley used the term in his discussion of the evolution of consciousness and his claim that epiphenomenal properties (like the "whistle of the steam engine") could not be explained by natural selection.

Here is a clear instance of this use of the word:

> Why do people who are thinking hard bite their lips and tap their feet? Are these actions just epiphenomena that accompany the core processes of feeling and thinking or might they themselves be integral parts of these processes? [Zajonc and Markus, 1984, p. 74]

Notice that the authors mean to assert that these actions, while perfectly detectable, play no enabling role, no designed role, in the processes of feeling and thinking; they are nonfunctional. In the same spirit, the hum of the computer is epiphenomenal, as is your shadow when you make yourself a cup of tea. Epiphenomena are mere by-products, but as such they are products with lots of effects in the world: tapping your feet makes a recordable noise, and your shadow has its effects on photographic film, not to mention the slight cooling of the surfaces it spreads itself over.

The standard philosophical meaning is different: "x is epiphenomenal" means "x is an effect but itself has no effects in the physical world whatever." (See Broad, 1925, p. 118, for the definition that inaugurates, or at any rate establishes, the philosophical usage.) Are these meanings really so different? Yes, as different as the meanings of *murder* and *death*. The philosophical meaning is stronger: Anything that has no effects whatever in the physical world surely has no effects on the function of anything, but the converse doesn't follow, as the example from Zajonc and Markus makes obvious.

In fact, the philosophical meaning is too strong; it yields a concept of no utility whatsoever (Harman, 1990; Fox, 1989). Since x has no physical effects (according to this definition), no instrument can detect the presence of x directly or indirectly; the way the world goes is not modulated in the slightest by the presence or absence of x. How then, could there ever be any empirical reason to assert the presence of x? Suppose, for instance, that Otto insists that he (for one) has epiphenomenal qualia. Why does he say this? Not because they have some effect on him, somehow guiding him or alerting him as he makes his avowals. By the very definition of epiphenomena (in the philosophical sense), Otto's heartfelt avowals that he has epiphenomena *could not* be evidence for himself or anyone else that he does have them, since he would be saying exactly the same thing even if he didn't have them. But perhaps Otto has some "internal" evidence?

Here there's a loophole, but not an attractive one. Epiphenomena, remember, are defined as having no effect in the *physical* world. If Otto wants to embrace out-and-out dualism, he can claim that his epiphenomenal qualia have no effects in the physical world, but do have effects in his (nonphysical) mental world (Broad, 1925, closed this loophole by definition, but it's free for the asking). For instance, they *cause some of his (nonphysical) beliefs*, such as his belief that he has epiphenomenal qualia. But this is just a temporary escape from embarrassment. For now on pain of contradiction, his beliefs, in turn, can have no effect in the physical world. If he suddenly lost his epiphenomenal qualia, he would no longer believe he had them, but he'd still go right on *saying* he did. He just wouldn't believe what he was saying! (Nor could he tell you that he didn't believe what he was saying, or do anything at all that revealed that he no longer believed what he was saying.) So the only way Otto could "justify" his belief in epiphenomena would be by retreating into a solipsistic world where there is only himself, his beliefs and his qualia, cut off from all effects in the world. Far from being a "safe" way of being a materialist and having your qualia too, this is at best a way of endorsing the most radical solipsism, by cutting off your mind — your beliefs and your experiences — from any commerce with the material world.

If qualia are epiphenomenal in the standard philosophical sense, their occurrence can't explain the way things happen (in the material world) since, by definition, things would happen exactly the same without them. There could not be an empirical reason, then, for believing in epiphenomena. Could there be another sort of reason for asserting their existence? What sort of reason? An *a priori* reason, presumably. But what? No one has ever offered one — good, bad, or indifferent — that I have seen. If someone wants to object that I am being a "verificationist" about these epiphenomena, I reply: Isn't everyone a verificationist about *this* sort of assertion? Consider, for instance, the hypothesis that there are fourteen epiphenomenal gremlins in each cylinder of an internal combustion engine. These gremlins have no mass, no energy, no physical properties; they do not make the engine run smoother or rougher, faster or slower. There is *and could be* no empirical evidence of their presence, and no empirical way in principle of distinguishing this hypothesis from its rivals: there are twelve or thirteen or fifteen . . . gremlins. By what principle does one defend one's wholesale dismissal of such nonsense? A verificationist principle, or just plain common sense?

> Ah, but there's a difference! [says Otto.] There is no independent motivation for taking the hypothesis of these gremlins seriously. You just made them up on the spur of the moment. Qualia, in contrast, have been around for a long time, playing a major role in our conceptual scheme!

And what if some benighted people have been thinking for generations that gremlins made their cars go, and by now have been pushed back by the march of science into the forlorn claim that the gremlins are there, all right, but are epiphenomenal? Is it a mistake for us to dismiss their "hypothesis" out of hand? Whatever the principle is that we rely on when we give the back of our hand to such nonsense, it suffices to dismiss the doctrine that qualia are epiphenomenal in this philosophical sense. These are not views that deserve to be discussed with a straight face.

It's hard to believe that the philosophers who have recently described their views as epiphenomenalism can be making such a woebegone mistake. Are they, perhaps, just asserting that qualia are epiphenomenal in Huxley's sense? Qualia, on this reading, *are* physical effects and *have* physical effects; they just aren't functional. Any materialist should be happy to admit that this hypothesis is true — if we identify qualia with reactive dispositions, for instance. As we noted in the discussion of enjoyment, even though some bulges or biases in our quality spaces are functional — or used to be functional — others are just brute happenstance. Why don't I like broccoli? Probably for no reason at all; my negative reactive disposition is purely epiphenomenal, a by-product of my wiring with no significance. It has no function, but has plenty of effects. In any designed system, some properties are crucial while others are more or less revisable *ad lib*. Everything has to be some way or another, but often the ways don't matter. The gear shift lever on a car may have to be a certain length and a certain strength, but whether it is round or square or oval in cross section is an epiphenomenal property, in Huxley's sense. In the CADBLIND systems we imagined in chapter 10, the particular color-by-number coding scheme was epiphenomenal. We could "invert" it (by using negative numbers, or multiplying all the values by some constant) without making any *functional* difference to its information-processing prowess. Such an inversion might be undetectable to casual inspection, and might be undetectable *by the system*, but it would not be epiphenomenal in the philosophical sense. There would be lots of tiny voltage differences in the memory registers that held the different numbers, for instance.

If we think of all the properties of our nervous systems that enable us to see, hear, smell, taste, and touch things, we can divide them, roughly, into the properties that play truly crucial roles in mediating the information processing, and the epiphenomal properties that are more or less revisable *ad lib*, like the color-coding system in the CAD-BLIND system. When a philosopher surmises that qualia are epiphenomenal properties of brain states, this might mean that qualia could turn out to be local variations in the heat generated by neuronal metabolism. That cannot be what epiphenomenalists have in mind, can it? If it is, then qualia as epiphenomena are no challenge to materialism.

The time has come to put the burden of proof squarely on those who persist in using the term. The philosophical sense of the term is simply ridiculous; Huxley's sense is relatively clear and unproblematic — and irrelevant to the philosophical arguments. No other sense of the term has any currency. So if anyone claims to uphold a variety of epiphenomenalism, try to be polite, but ask: What *are* you talking about?

Notice, by the way, that this equivocation between two senses of "epiphenomenal" also infects the discussion of zombies. A philosopher's zombie, you will recall, is behaviorally indistinguishable from a normal human being, but is not conscious. There is nothing it is like to be a zombie; it just seems that way to observers (including itself, as we saw in the previous chapter). Now this can be given a strong or weak interpretation, depending on how we treat this indistinguishability to observers. If we were to declare that *in principle*, a zombie is indistinguishable from a conscious person, then we would be saying that genuine consciousness is epiphenomenal *in the ridiculous sense*. That is just silly. So we could say instead that consciousness might be epiphenomenal in the Huxley sense: although there was some way of distinguishing zombies from real people (who knows, maybe zombies have green brains), the difference doesn't show up as a functional difference *to observers*. Equivalently, human bodies with green brains don't harbor observers, while other human bodies do. On this hypothesis, we would be able in principle to distinguish the inhabited bodies from the uninhabited bodies by checking for brain color. This is also silly, of course, and dangerously silly, for it echoes the sort of utterly unmotivated prejudices that have denied full personhood to people on the basis of the color of their skin. It is time to recognize the idea of the possibility of zombies for what it is: not a serious philosophical idea but a preposterous and ignoble relic of ancient prejudices. Maybe women aren't really conscious! Maybe Jews! What pernicious nonsense. As Shylock says, drawing our attention, quite properly, to "merely behavioral" criteria:

> Hath not a Jew eyes? Hath not a Jew hands, organs, dimensions, senses, affections, passions; fed with the same food, hurt with the same weapons, subject to the same diseases, heal'd by the same means, warm'd and cool'd by the same winter and summer, as a Christian is? If you prick us, do we not bleed? If you tickle us, do we not laugh? If you poison us, do we not die?

There is another way to address the possibility of zombies, and in some regards I think it is more satisfying. Are zombies possible? They're not just possible, they're actual. We're all zombies.[6] Nobody is conscious — not in the systematically mysterious way that supports such doctrines as epiphenomenalism! I can't prove that no such sort of consciousness exists. I also cannot prove that gremlins don't exist. The best I can do is show that there is no respectable motivation for believing in it.

6. GETTING BACK ON MY ROCKER

In chapter 2, section 2, I set up the task of explaining consciousness by recollecting an episode from my own conscious experience as I sat, rocking in my chair, looking out the window on a beautiful spring day. Let's return to that passage and see how the theory I have developed handles it. Here is the text:

> Green-golden sunlight was streaming in the window that early spring day, and the thousands of branches and twigs of the maple tree in the yard were still clearly visible through a mist of green buds, forming an elegant pattern of wonderful intricacy. The windowpane is made of old glass, and has a scarcely detectable wrinkle line in it, and as I rocked back and forth, this imperfection in the glass caused a wave of synchronized wiggles to march back and forth across the delta of branches, a regular motion superimposed with remarkable vividness on the more chaotic shimmer of the twigs and branches in the breeze.

---

6. It would be an act of desperate intellectual dishonesty to quote this assertion out of context!

> Then I noticed that this visual metronome in the tree branches was locked in rhythm with the Vivaldi concerto grosso I was listening to as "background music" for my reading. . . . My conscious thinking, and especially the enjoyment I felt in the combination of sunny light, sunny Vivaldi violins, rippling branches — plus the pleasure I took in just thinking about it all — how could *all that* be just something physical happening in my brain? How could any combination of electrochemical happenings in my brain somehow add up to the delightful way those hundreds of twigs genuflected in time with the music? How could some information-processing event in my brain *be* the delicate warmth of the sunlight I felt falling on me? . . . It does seem impossible.

Since I have encouraged us all to be heterophenomenologists, I can hardly exempt myself, and I ought to be as content to be the subject as the practitioner, so here goes: I apply my own theory to myself. As heterophenomenologists, our task is to take this text, interpret it, and then relate the objects of the resulting heterophenomenological world of Dennett to the events going on in Dennett's brain at the time.

Since the text was produced some weeks or months after the events about which it speaks occurred, we can be sure that it has been abridged, not only by the author's deliberate editorial compressions, but also by the inexorable abridgment processes of memory over time. Had we probed earlier — had the author picked up a tape recorder while he sat rocking, and produced the text there and then — it would surely have been quite different. Not only richer in detail, and messier, but also, of course, reshaped and redirected by the author's own reactions to the very process of creating the text — listening to the actual sounds of his own words instead of musing silently. Speaking aloud, as every lecturer knows, often reveals implications (and particularly problems) in one's own message that elude one when one engages in silent soliloquy.

As it is, the text portrays a mere portion (and no doubt an idealized portion) of the contents of the author's consciousness. We must be careful, however, not to suppose that the "parts left out" in the given text were all "actually present" in something we might call the author's stream of consciousness. We must not make the mistake of supposing that there are facts — unrecoverable but actual facts — about just which contents were conscious and which were not at the time. And in particular, we should not suppose that when he looked out the window, he "took it all in" in one wonderful mental gulp — even though this is what his text portrays. It seemed to him, according to the text, as if his mind — his visual field — were filled with intricate details of gold-green buds and wiggling branches, but although this is how it seemed, this was an illusion. No such "plenum" ever came into his mind; the plenum remained out in the world where it didn't have to be *represented*, but could just *be*. When we marvel, in those moments of heightened self-consciousness, at the glorious richness of our conscious experience, the richness we marvel at is actually the richness of the world outside, in all *its* ravishing detail. It does not "enter" our conscious minds, but is simply available.

What about all the branches and twigs rippling in unison? The branches outside on the tree didn't ripple, to be sure, since the rippling was due to the wrinkle in the windowpane, but that doesn't mean that all that rippling had to be happening in the author's mind or brain, just that it happened inboard of the windowpane that caused it. If someone had filmed the changing images on the author's retinas, they would have found the rippling there, just as in a movie, but that was no doubt where almost all the rippling stopped; what happened inboard of his retinas was just his recognition *that* there was, as he says in the text, a wonderful wave of synchronized ripples for him to experience. He saw the ripples, and he saw the extent of them, in just the way you would see all the Marilyns in the wallpaper. And since his retinas were provided with a steady dose of rippling, had he felt like sampling it further, there *would have been* more detail in the Multiple Drafts of which our text is all that remains.

There were many other details that the author could have focused on, but didn't. There are plenty of unrecoverable but genuine facts of the matter about which of these details got discriminated where and when by various systems in his brain, but the sum total of those facts doesn't settle such questions as which of these was he definitely, actually conscious of (but had forgotten by the time he produced his text), and which were definitely, actually in the "background" of his consciousness (though he didn't attend to them at the time). Our tendency to suppose that there *has* to be a fact of the matter to settle such questions is like the naïve reader's supposition that there *has* to be an answer to such questions as: Did Sherlock Holmes have eggs for breakfast on the day that Dr. Watson met him? Conan Doyle *might have* put that detail into the text, but he didn't, and since he didn't, there is simply no fact of the matter about whether those eggs belong *in the fictional world of Sherlock Holmes*. Even if Conan Doyle *thought* of Holmes eating eggs that morning, even if in an early draft Holmes *is represented in handwritten words* as eating eggs that morning, there is simply no fact of the matter about whether in the fictional world of Sherlock Holmes, the world constituted from the published text we actually have, he had eggs for breakfast.

The text we have from Dennett was not "written in his brain" between the time in the rocking chair and the time it was typed into a file on a word processor. The attending he engaged in while rocking, and the concomitant rehearsal of those particulars that drew his attention, had the effect of fixing the contents of those particulars relatively securely "in memory" but this effect should not be viewed as storing a picture (or a sentence) or any other such salient representation. Rather, it should be thought of as just making a partially similar recurrence of the activity more likely, and that likely event is what happened, we may presume, on the occasion of the typing, driving the word-demons in his brain into the coalitions that yielded, for the first time, a string a sentences. Now some of what happened earlier, in the rocking chair, no doubt enlisted actual English words and phrases, and this prior collaboration between wordless contents and words no doubt facilitated the recovery of some of the very same English expressions when typing time came around.

Let's return to the heterophenomenological world of that text. What about the joy of which it speaks? ". . . the combination of sunny light, sunny Vivaldi violins, rippling branches — plus the pleasure I took in just thinking about it all. . . ." This could not be explained by the invocation of intrinsically pleasant qualia of sight, sound, and sheer thought. The idea that there are such qualia just distracts us from all possible paths of explanation, capturing our attention the way a wagging finger in front of a baby's eyes can capture its attention, getting us to stare numbly at the "intrinsic object" instead of casting about for a description of the underlying mechanisms *and* an explanation (ultimately an evolutionary explanation) of why the mechanisms do what they do.

The author's enjoyment is readily explainable by the fact that all visual experience is composed of the activities of neural circuits whose very activity is innately pleasing to us, not only because we simply like to become informed but because we like the particular ways we come to be informed. The fact that the look of sunlight-dappled spring buds should be something a human being likes is not surprising. The fact that some human beings also like looking at microscopic slides of bacteria and others like looking at photographs of airplane crashes is stranger, but the sublimations and perversions of desire grow from the same animal sources in the wiring of our nervous systems.

The author goes on to wonder how on earth "All this could indeed be just a combination of electrochemical happenings in my brain." As his wondering makes plain, it doesn't seem to be. Or in any event there was a moment when it occurred to him that it didn't seem to him to be just a combination of electrochemical happenings in his brain. But our subsequent chapters suggest a retort: Well, what do you think it would seem like if it *were* just a combination of electrochemical happenings in your brain?[7] Haven't we given ourselves grounds for concluding that with a brain organized the way ours is, this is just the sort of heterophenomenological world we would expect? Why shouldn't such combinations of electrochemical happenings in the brain have precisely the effects we set out to explain?

(The author speaks:) There is still one puzzle, however. How do *I* get to know all about this? How come *I* can tell you all about what was going on in my head? The answer to the puzzle is simple: *Because that is what I am*. Because a knower and reporter of such things in such terms is what is me. *My* existence is explained by the fact that there are these capacities in this body.

This idea, the idea of the Self as the Center of Narrative Gravity, is one we are finally ready to examine. It is certainly an idea whose time has come. Imagine my mixed emotions when I discovered that before I could get my version of it properly published in a book,[8] it had already been satirized in a novel, David Lodge's *Nice Work* (1988). It is apparently a hot theme among the deconstructionists:

> According to Robyn (or, more precisely, according to the writers who have influenced her thinking on these matters), there is no such thing as the "Self" on which capitalism and the classic novel are founded — that is to say, a finite, unique soul or essence that constitutes a person's identity; there is only a subject position in an infinite web of discourses — the discourses of power, sex, fam-

---

7. Cf. Lockwood (1989): "What would consciousness have felt like if it *had* felt like billions of tiny atoms wiggling in place?" (pp. 15–16)

8. I presented the main ideas in my reflections on Borges, in *The Mind's I* (Hofstadter and Dennett, 1981, pp. 348–352), and drew them together in a talk, "The Self as the Center of Narrative Gravity," presented at the Houston Symposium in 1983. While waiting for that symposium volume to appear, I published a somewhat truncated version of my talk in the *Times Literary Supplement*, Sept. 16–22, 1988, under the boring title — not mine — "Why everyone is a novelist." The original version, under the title "The Self as the Center of Narrative Gravity," is still forthcoming in F. Kessel, P. Cole, and D. Johnson, eds., *Self and Consciousness: Multiple Perspectives*, Hillsdale, NJ: Erlbaum.

> ily, science, religion, poetry, etc. And by the same token, there is no such thing as an author, that is to say, one who originates a work of fiction *ab nihilo*. . . . in the famous words of Jacques Derrida . . . "*il n'y a pas de hors-texte*", there is nothing outside the text. There are no origins, there is only production, and we produce our "selves" in language. Not "*you are what you eat*" but "*you are what you speak*," or, rather "*you are what speaks you*," is the axiomatic basis of Robyn's philosophy, which she would call, if required to give it a name, "semiotic materialism."

Semiotic materialism? Must *I* call it that? Aside from the allusions to capitalism and the classic novel, about which I have kept my counsel, this jocular passage is a fine parody of the view I'm about to present. (Like all parody, it exaggerates; I wouldn't say there is *nothing* outside the text. There are, for instance, all the bookcases, buildings, bodies, bacteria . . .)

Robyn and I think alike — and of course we are *both*, by our own accounts, fictional characters of a sort, though of a slightly different sort.

13

THE REALITY OF SELVES

> Suppose that there be a machine, the structure of which produces thinking, feeling, and perceiving; imagine this machine enlarged but preserving the same proportions, so that you could enter it as if it were a mill. This being supposed, you might visit its inside; but what would you observe there? Nothing but parts which push and move each other, and never anything that could explain perception.
>
> GOTTFRIED WILHELM LEIBNIZ (1646–1716), *Monadology* (first published, 1840)

> For my part, when I enter most intimately into what I call *myself,* I always stumble on some particular perception or other, of heat or cold, light or shade, love or hatred, pain or pleasure. I never can catch *myself* at any time without a perception, and never can observe anything but the perception. . . . If anyone, upon serious and unprejudiced reflection, thinks he has a different notion of *himself,* I must confess I can reason no longer with him. All I can allow him is, that he may be in the right as well as I, and that we are essentially different in this particular. He may, perhaps, perceive something simple and continued, which he calls *himself*; though I am certain there is no such principle in me.
>
> DAVID HUME (1739)

Since the dawn of modern science in the seventeenth century, there has been nearly unanimous agreement that the self, whatever it is, would be invisible under a microscope, and invisible to introspection, too. For some, this has suggested that the self was a nonphysical soul, a ghost in the machine. For others, it has suggested that the self was nothing at all, a figment of metaphysically fevered imaginations. And for still others, it has suggested only that a self was in one way or another a sort of abstraction, something whose existence was not in the slightest impugned by its invisibility. After all, one might say, a center of gravity is just as invisible — and just as real. Is that real enough?

The question of whether there really are selves can be made to look ridiculously easy to answer, in either direction: Do *we* exist? Of course! The question presupposes its own answer. (After all, who is this *I* that has looked in vain for a self, according to Hume?) Are there entities, either *in* our brains, or *over and above* our brains, that control our bodies, think our thoughts, make our decisions? Of course not! Such an idea is either empirical idiocy (James's "pontifical neuron") or metaphysical claptrap (Ryle's "ghost in the machine"). When a simple question gets two answers, "Obviously yes!" and "Obviously no!", a middle-ground position is worth considering (Dennett, 1991a), even though it is bound to be initially counterintuitive to all parties — everyone agrees that it denies one obvious fact or another!

1. HOW HUMAN BEINGS SPIN A SELF

> In addition they seemed to spend a great deal of time eating and drinking and going to parties, and Frensic, whose appearance tended to limit his sensual pleasures to putting things into himself rather than into other people, was something of a gourmet.
>
> TOM SHARPE (1977)

The novelist Tom Sharpe suggests, in this funny but unsettling passage, that when you get right down to it, all sensual pleasure consists in playing around with one's own boundary, or someone else's, and he is on to something — if not the whole truth, then part of the truth.

People have selves. Do dogs? Do lobsters? If selves are anything at all, then they exist. *Now* there are selves. There was a time, thousands (or millions, or billions) of years ago, when there were none — at least none on this planet. So there has to be — as a matter of logic — a true story to be told about *how there came to be* creatures with selves. This story will have to tell — as a matter of logic — about a process (or a series of processes) involving the activities or behaviors of things that do not yet *have* selves — or *are* not yet selves — but which eventually yield, as a new product, beings that are, or have, selves.

In chapter 7, we saw how the birth of reasons was also the birth of boundaries, the boundary between "me" and "the rest of the world," a distinction that even the lowliest amoeba must make, in its blind, unknowing way. This minimal proclivity to distinguish self from other in order to protect one*self* is the biological self, and even such a simple self is not a concrete thing but just an abstraction, a principle of organization. Moreover the boundaries of a biological self are porous and indefinite — another instance of Mother Nature tolerating "error" if the cost is right.

Within the walls of human bodies are many, many interlopers, ranging from bacteria and viruses through microscopic mites that live like cliff-dwellers in the ecological niche of our skin and scalp, to larger parasites — horrible tapeworms, for instance. These interlopers are all tiny self-protectors in their own rights, but some of them, such as the bacteria that populate our digestive systems and without which we would die, are just as essential team members in our quest for self-preservation as the antibodies in our immune systems. (If the biologist Lynn Margulis's theory (1970) is correct, the mitochondria that do the work in almost all the cells in our body are the descendants of bacteria with whom "we" joined forces about two billions years ago.) Other interlopers are tolerated parasites — not worth the effort to evict, apparently — and still others are indeed the enemy within, deadly if not rooted out.

This fundamental biological principle of distinguishing self from world, inside from outside, produces some remarkable echoes in the highest vaults of our psychology. The psychologists Paul Rozin and April Fallon (1987) have shown in a fascinating series of experiments on the nature of *disgust* that there is a powerful and unacknowledged undercurrent of blind resistance to certain acts that, rationally considered, should not trouble us. For example, would you please swallow the saliva in your mouth right now? This act does not fill you with revulsion. But suppose I had asked you to get a clean drinking glass and spit into the glass and *then* swallow the saliva from the glass. Disgusting! But why? It seems to have to do with our perception that once something is outside of our bodies it is not longer quite part of us anymore — it becomes alien and suspicious — it has renounced its citizenship and becomes something to be rejected.

Border crossings are thus either moments of anxiety, or, as pointed out by Sharpe, something to be especially enjoyed. Many species have developed remarkable constructions for extending their territorial boundaries, either to make the bad kind of crossings more difficult or the good kind easier. Beavers make dams, and spiders spin webs, for instance. When the spider spins its web, it doesn't have to understand what it is doing; Mother Nature has simply provided its tiny brain with the necessary routines for carrying out this biologically essential task of engineering. Experiments with beavers show that even their magnificently efficient engineering practices are at least largely the product of innate drives and proclivities they need not understand to benefit from. Beavers do learn, and may even teach each other, but mainly, they are driven by powerful innate mechanisms controlling what the behaviorist B. F. Skinner called negative reinforcement. A beaver will cast about quite frantically for something — anything — to stop the sound of running water, and in one experiment a beaver found its relief by plastering mud all over the loudspeaker from which the recorded gurgling emerged! (Wilsson, 1974)

The beaver protects its outer boundary with twigs and mud and one of its inner boundaries with fur. The snail gathers calcium in its food and uses it to exude a hard shell; the hermit crab gets its calcium shell ready-made, taking over the discarded shell of another creature, daintily avoiding the ingestion and exudation process. The difference is not fundamental, according to Richard Dawkins, who points out that the result in either case, which he calls *the extended phenotype* (1982), is a part of the fundamental biological equipment of the individuals who are submitted to the selective forces that drive evolution.

The definition of an extended phenotype not only extends beyond the "natural" boundary of individuals to include external equipment such as shells (and internal equipment such as resident bacteria); it often includes other individuals of the same species. Beavers cannot do it alone, but require teamwork to build a single dam. Termites have to band together by the millions to build their castles.

And consider the astonishing architectural constructions of the Australian bowerbird (Borgia, 1986). The males build elaborate bowers, courtship shrines with grand central naves, richly decorated with brightly colored objects — predominantly deep blue, and including bottle caps, bits of colored glass, and other human artifacts — which are gathered from far afield and carefully arranged in the bower the better to impress the female he is courting. The bowerbird, like the spider, does not really have to understand what he is doing; he simply finds himself hard at work, he knows not why, creating an edifice that is crucial to his success as a bowerbird.

But the strangest and most wonderful constructions in the whole animal world are the amazing, intricate constructions made by the primate, *Homo sapiens*. Each normal individual of this species makes a *self*. Out of its brain it spins a web of words and deeds, and, like the other creatures, it doesn't have to know what it's doing; it just does it. This web protects it, just like the snail's shell, and provides it a livelihood, just like the spider's web, and advances its prospects for sex, just like the bowerbird's bower. Unlike a spider, an individual human doesn't just *exude* its web; more like a beaver, it works hard to gather the materials out of which it builds its protective fortress. Like a bowerbird, it appropriates many found objects which happen to delight it — or its mate — including many that have been designed by others for other purposes.

This "web of discourses" as Robyn called it at the close of the previous chapter, is as much a biological product as any of the other constructions to be found in the animal world. Stripped of it, an individual human being is as incomplete as a bird without its feathers, a turtle without its shell. (Clothes, too, are part of the extended phenotype of *Homo sapiens* in almost every niche inhabited by that species. An illustrated encyclopedia of zoology should no more picture *Homo sapiens* naked than it should picture *Ursus arctus* — the black bear — wearing a clown suit and riding a bicycle.)

So wonderful is the organization of a termite colony that it seemed to some observers that each termite colony had to have a soul (Marais, 1937). We now understand that its organization is simply the result of a million semi-independent little agents, each itself an automaton, doing its thing. So wonderful is the organization of a human self that to many observers it has seemed that each human being had a soul, too: a benevolent Dictator ruling from Headquarters.

In every beehive or termite colony there is, to be sure, a queen bee or queen termite, but these individuals are more patient than agent, more like the crown jewels to be protected than the chief of the protective forces — in fact their royal name is more fitting today than in earlier ages, for they are much more like Queen Elizabeth II than Queen Elizabeth I. There is no Margaret Thatcher bee, no George Bush termite, no Oval Office in the anthill.

Do our selves, our nonminimal *selfy* selves, exhibit the same permeability and flexibility of boundaries as the simpler selves of other creatures? Do we expand our personal boundaries — the boundaries of our *selves* — to enclose any of our "stuff"? In general, perhaps, no, but there are certainly times when this seems true, psychologically. For instance, while some people merely own cars and drive them, others are *motorists;* the inveterate motorist prefers *being* a four-wheeled gas-consuming agent to being a two-legged food-consuming agent, and his use of the first-person pronoun betrays this identification:

> I'm not cornering well on rainy days because my tires are getting bald.

So sometimes we enlarge our boundaries; at other times, in response to perceived challenges real or imaginary, we let our boundaries shrink:

> *I* didn't do that! That wasn't the real me talking. Yes, the words came out of my mouth, but I refuse to recognize them as my own.

I have reminded you of these familiar speeches to draw out the similarities between our selves and the selves of ants and hermit crabs, but the speeches also draw attention to the most important difference: Ants and hermit crabs don't talk. The hermit crab is designed in such a way as to see to it that it acquires a shell. Its organization, we might say, *implies* a shell, and hence, in a *very* weak sense, tacitly *represents* the crab as having a shell, but the crab does not in any stronger sense *represent itself* as having a shell. It doesn't go in for self-representation at all. To whom would it so represent itself and why? It doesn't need to remind itself of this aspect of its nature, since its innate design takes care of that problem, and there are no other interested parties in the offing. And the ants and termites, as we have noted, accomplish their communal projects without relying on any explicitly communicated blueprints or edicts.

We, in contrast, are almost constantly engaged in presenting ourselves to others, and to ourselves, and hence *representing* ourselves — in language and gesture, external and internal. The most obvious difference in our environment that would explain this difference in our behavior is the behavior itself. Our human environment contains not just food and shelter, enemies to fight or flee, and conspecifics with whom to mate, but words, words, words. These words are potent elements of our environment that we readily incorporate, ingesting and extruding them, weaving them like spiderwebs into self-protective strings of *narrative*. Indeed, as we saw in chapter 7, when we let in these words, these meme-vehicles, they tend to take over, creating *us* out of the raw materials they find in our brains.

Our fundamental tactic of self-protection, self-control, and self-definition is not spinning webs or building dams, but telling stories, and more particularly concocting and controlling the story we tell others — and ourselves — about who we are. And just as spiders don't have to think, consciously and deliberately, about how to spin their webs, and just as beavers, unlike professional human engineers, do not consciously and deliberately plan the structures they build, we (unlike *professional* human storytellers) do not consciously and deliberately figure out what narratives to tell and how to tell them. Our tales are spun, but for the most part we don't spin them; they spin us. Our human consciousness, and our narrative selfhood, is their product, not their source.

These strings or streams of narrative issue forth *as if* from a single source — not just in the obvious physical sense of flowing from just one mouth, or one pencil or pen, but in a more subtle sense: their effect on any audience is to encourage them to (try to) posit a unified agent whose words they are, about whom they are: in short, to posit a *center of narrative gravity*. Physicists appreciate the enormous simplification you get when you posit a center of gravity for an object, a single point relative to which all gravitational forces may be calculated. We heterophenomenologists appreciate the enormous simplification you get when you posit a center of narrative gravity for a narrative-spinning human body. Like the biological self, this psychological or narrative self is yet another abstraction, not a thing in the brain, but still a remarkably robust and almost tangible attractor of properties, the "owner of record" of whatever items and features are lying about unclaimed. Who owns your car? You do. Who owns your clothes? You do. Then who owns your body? You do! When you say

> This is *my* body.

you certainly aren't taken as saying

> This body owns itself.

But what can you be saying, then? If what you say is neither a bizarre and pointless tautology (this body is its own owner, or something like that) nor the claim that you are an immaterial soul or ghost puppeteer who owns and operates this body the way you own and operate your car, what else could you mean?

2. HOW MANY SELVES TO A CUSTOMER?

I think we could see more clearly what

This is my body

meant, if we could answer the question: As opposed to what? How about as opposed to this?

No it isn't; it's *mine*, and I don't like sharing it!

If we could see what it would be like for two (or more) selves to vie for control of a single body, we could see better what a single self really is. As scientists of the self, we would like to conduct controlled experiments, in which, by varying the initial conditions, we could see just what has to happen, in what order and requiring what resources, for such a talking self to emerge. Are there conditions under which life goes on but no self emerges? Are there conditions under which more than one self emerges? We can't ethically conduct such experiments, but, as so often before, we can avail ourselves of the data generated by some of the terrible experiments nature conducts, cautiously drawing conclusions.

Such an experiment is Multiple Personality Disorder (MPD), in which a single human body *seems* to be shared by several selves, each, typically, with a proper name and an autobiography. The idea of MPD strikes many people as too outlandish and metaphysically bizarre to believe — a "paranormal" phenomenon to discard along with ESP, close encounters of the third kind, and witches on broomsticks. I suspect that some of these people have made a simple arithmetical mistake: they have failed to notice that two or three or seventeen selves per body is really no more metaphysically extravagant than one self per body. One is bad enough!

"I just saw a car drive by with five selves in it."
"What?? The mind reels! What kind of metaphysical nonsense is this?"
"Well, there were also five bodies in the car."
"Oh, well, why didn't you say so? Then everything is okay."
" — Or maybe only four bodies, or three — but definitely five selves."
"What??!!"

The normal arrangement is one self per body, but if a body can have one, why not more than one under abnormal conditions?

I don't mean to suggest that there is nothing shocking or deeply puzzling about MPD. It is, in fact, a phenomenon of surpassing strangeness, not, I think, because it challenges our presuppositions about what is *metaphysically* possible, but more because it challenges our presuppositions about what is *humanly* possible, about the limits of human cruelty and depravity on the one hand, and the limits of human creativity on the other. For the evidence is now voluminous that there are not a handful or a hundred but thousands of cases of MPD diagnosed today, and it almost invariably owes its existence to prolonged early childhood abuse, usually sexual, and of sickening severity. Nicholas Humphrey and I investigated MPD several years ago (Humphrey and Dennett, 1989), and found it to be a complex phenomenon that extends far beyond the individual brains of the sufferers.

These children have often been kept in such extraordinarily terrifying and confusing circumstances that I am more amazed that they survive psychologically at all than I am that they manage to preserve themselves by a desperate redrawing of their boundaries. What they do, when confronted with overwhelming conflict and pain, is this: They "leave." They create a boundary so that the horror doesn't happen *to them;* it either happens to no one, or to some other self, better able to sustain its organization under such an onslaught — at least that's what they *say* they did, as best they recall.

How can this be? What kind of account could we give, ultimately at the biological level, of such a process of splitting? Does there have to have been a single, whole self that somehow fissioned, amoebalike? How could that be if a self is not a proper physical part of an organism or a brain, but, as I have suggested, an abstraction? The response to the trauma seems so creative, moreover, that one is inclined at first to suppose that it must be the work of some kind of a supervisor in there: a supervisory brain program, a central controller, or whatever. But we should remind ourselves of the termite colony, which also seemed, at first, to require a central chief executive to accomplish such clever projects.

We have become accustomed to evolutionary narratives that start from a state in which a certain phenomenon does not yet exist and end with a state in which the phenomenon is definitely present. The innovation of agriculture, of clothing and dwellings and tools, the innovation of language, the innovation of consciousness itself, the earlier innovation of life on earth. All these stories are there to be told. And each of them must traverse what we might call the chasm of absolutism. This chasm is illustrated by the following curious argument (borrowed from Sanford, 1975):

> Every mammal has a mammal for a mother,
> but there have been only a finite number of mammals, so
> there must have been a first mammal,
> which contradicts our first premise, so, contrary to
> appearances,
> there are no such things as mammals!

Something has to give. What should it be? The absolutist or essentialist philosopher is attracted to sharp lines, thresholds, "essences" and "criteria." For the absolutist, there must indeed have been a first mammal, a first living thing, a first moment of consciousness, a first moral agent; it was whichever product of saltation, whichever radically new candidate, first met the essential conditions — whatever analysis shows them to be.

It was this taste for sharp species boundaries that was the greatest intellectual obstacle Darwin faced when trying to develop the theory of evolution (Richards, 1987). Opposed to this way of thinking is the sort of anti-essentialism that is comfortable with penumbral cases and the lack of strict dividing lines. Since selves and minds and even consciousness itself are biological products (not elements to be found in the periodic table of chemistry), we should expect that the transitions between them and the phenomena that are not them should be gradual, contentious, gerrymandered. This doesn't mean that everything is always in transition, always gradual; transitions that look gradual from close up usually look like abrupt punctuations between plateaus of equilibrium from a more distant vantage point (Eldredge and Gould, 1972; but see also Dawkins, 1982, pp. 101–109).

The importance of this fact for philosophical theories (and philosophers' predilections) is not widely enough recognized. There have always been — and always will be — a few transitional things, "missing links," quasi-mammals and the like that defy definition, but the fact is that *almost all* real (as opposed to merely possible) things in nature tend to fall into similarity clusters separated in logical space by huge oceans of emptiness. We don't need "essences" or "criteria" to keep the meaning of our words from sliding all over the place; our words will stay put, quite firmly attached as if by gravity to the nearest similarity cluster, even if there has been — must have been — a brief isthmus that once attached it by a series of gradual steps to some neighboring cluster. This idea is uncontroversially applied to many topics. But many people who are quite comfortable taking this pragmatic approach to night and day, living and nonliving, mammal and premammal, get anxious when invited to adopt the same attitude toward having a self and not having a self. They think that here, if nowhere else in nature, it must be All or Nothing and One to a Customer.

The theory of consciousness we have been developing discredits these presumptions, and Multiple Personality Disorder provides a good illustration of the way the theory challenges them. The convictions that there *cannot* be quasi-selves or sort-of selves, and that, moreover there *must* be a whole number of selves associated with one body — and it better be the number one! — are *not* self-evident. That is, they are no longer self-evident, now that we have developed in some detail an alternative to the Cartesian Theater with its Witness or Central Meaner. MPD challenges these presumptions from one side, but we can also imagine a challenge from the other side: two or more bodies sharing a single self! There may actually be such a case, in York, England: the Chaplin twins, Greta and Freda (*Time,* April 6, 1981). These identical twins, now in their forties and living together in a hostel, seem to act *as one;* they collaborate on the speaking of single speech acts, for instance, finishing each other's sentences with ease or speaking in unison, with one just a split-second behind. For years they have been inseparable, as inseparable as two twins who are not Siamese twins could arrange. Some who have dealt with them suggest that the natural and effective tactic that suggested itself was to consider *them* more of a *her.*

Our view countenances the theoretical possibility not only of MPD but FPD (Fractional Personality Disorder). Could it be? Why not? I'm not for a moment suggesting that these twins were linked by telepathy or ESP or any other sort of occult bonds. I am suggesting that there are plenty of subtle, everyday ways of communicating and coordinating (techniques often highly developed by identical twins, in fact). Since these twins have seen, heard, touched, smelled, and thought about very much the same events throughout their lives, and started, no doubt, with brains quite similarly disposed to react to these stimuli, it might not take enormous channels of communication to keep them homing in on some sort of loose harmony. (And besides, how unified is the most self-possessed among us?) We should hesitate to prescribe the limits of such practiced coordination.

But in any case, wouldn't there also be two clearly defined individual selves, one for each twin, and responsible for maintaining this curious charade? Perhaps, but what if each of these women had become so selfless (as we do say) in her devotion to the joint cause, that she more or less lost herself (as we also say) in the project? As the poet Paul Valéry once said, in a delicious twist of his countryman's dictum: "Sometimes I am, sometimes I think."

In chapter 11 we saw that while consciousness appears to be continuous, in fact it is gappy. A self could be just as gappy, lapsing into nothingness as easily as a candle flame is snuffed, only to be rekindled at some later time, under more auspicious circumstances. Are you the very person whose kindergarten adventures you sketchily recall (sometimes vividly, sometimes dimly)? Are the adventures of that child, whose trajectory through space and time has apparently been continuous with the trajectory of your body, your very own adventures? That child with your name, a child whose scrawled signature on a crayon drawing reminds you of the way *you* used to sign your name — is (was) that child you? The philosopher Derek Parfit (1984) has compared a person to a club, a rather different sort of human construction, which might go out of existence one year, and come to be reconstituted by some of its (former?) members some years later. Would it be the same club? It might be, if, for instance, the club had had a written constitution that provided explicitly for just such lapses of existence. But there might be no telling. We might know all the facts that could conceivably bear on the situation and be able to see that they were inconclusive about the *identity* of the (new?) club. On the view of selves — or persons — emerging here, this is the right analogy; selves are not independently existing soul-pearls, but artifacts of the social processes that create us, and, like other such artifacts, subject to sudden shifts in status. The only "momentum" that accrues to the trajectory of a self, or a club, is the stability imparted to it by the web of beliefs that constitute it, and when those beliefs lapse, it lapses, either permanently or temporarily.

It is important to bear this in mind when considering another favorite among philosophers, the much-discussed phenomenon of split-brain patients. A so-called split brain is the result of *commissurotomy*, an operation that severs the *corpus callosum*, the broad band of fibers directly connecting the left and right hemispheres of the cortex. This leaves the hemispheres still indirectly connected, through a variety of midbrain structures, but it is obviously a drastic procedure, not to be performed unless there are no alternatives. It provides relief in some severe cases of epilepsy that are not otherwise treatable, by preventing the internally generated electrical storms that cause seizures from sweeping across the cortex from an originating "focus" in one hemisphere to the opposite side. Standard philosophical legend has it that split-brain patients may be "split into two selves" but otherwise suffer no serious diminution in powers as a result of the surgery. The most appealing version of this oversimplification is that the original person's two "sides" — uptight, analytic left hemisphere, and laid-back, intuitive, holistic right hemisphere — are postoperatively freed to shine forth with more individuality, now that the normal close teamwork must be replaced by a less intimate *détente*. This is an appealing idea, but it is a wild exaggeration of the empirical findings that inspire it. In fact, in only a tiny fraction of cases are *any* of the theoretically striking symptoms of multiple selfhood to be observed. (See, e.g., Kinsbourne, 1974; Kinsbourne and Smith, 1974; Levy and Trevarthen, 1976; Gazzaniga and LeDoux, 1978; Gazzaniga, 1985; Oakley, 1985; Dennett, 1985b.)

It's not surprising that split-brains patients, like blindsight patients and people with Multiple Personality Disorder, don't live up to their philosophical billing, and it's nobody's fault. It's not that philosophers (and many other interpreters, including the primary researchers) deliberately exaggerate their descriptions of the phenomena. Rather, in their effort to describe the phenomena concisely, they find that the limited resources of everyday language pull them inexorably toward the simplistic Boss of the Body, Ghost in the Machine, Audience in the Cartesian Theater model. Nicholas Humphrey and I, comparing our own careful notes of what happened at various meetings with MPD sufferers, found that we often slipped, in spite of ourselves, into all-too-natural but seriously misleading turns of phrase to describe what we had actually seen. Thomas Nagel (1971), the first philosopher to write about split-brain patients, presented a judicious and accurate account of the phenomena as they were then understood, and, noting the difficulty in providing a coherent account, surmised: "It may be impossible for us to abandon certain ways of conceiving and representing ourselves, no matter how little support they get from scientific research" (1971, p. 397).

It is indeed difficult but not impossible. Nagel's pessimism is itself exaggerated. Haven't we just succeeded, in fact, in shaking ourselves free of the traditional way of thinking? Now some people may not *want* to abandon the traditional vision. There might even be good reasons — moral reasons — for trying to preserve the myth of selves as brain-pearls, particular concrete, countable things rather than abstractions, and for refusing to countenance the possibility of quasi-selves, semi-selves, transitional selves. But that is surely the correct way to understand the phenomena of split-brains. For brief periods during carefully devised experimental procedures, a few of these patients bifurcate in their response to a predicament, temporarily creating a second center of narrative gravity. A few effects of the bifurcation may linger on indefinitely in mutually inaccessible memory traces, but aside from these actually quite primitive traces of the bifurcation, the life of a second rudimentary self lasts a few minutes at most, not much time to accrue the sort of autobiography of which fully fledged selves are made. (This is just as obviously true of most of the dozens of fragmentary selves developed by MPD patients; there simply aren't enough waking hours in the day for most of them to salt away more than a few minutes of exclusive biography per week.)

The distinctness of different narratives is the life-blood of different selves. As the philosopher Ronald de Sousa (1976) notes:

> When Dr. Jekyll changes into Mr. Hyde, that is a strange and mysterious thing. Are they two people taking turns in one body? But here is something stranger: Dr. Juggle and Dr. Boggle too, take turns in one body. *But they are as like as identical twins!* You balk: why then say that they have changed into one another? Well, why not: if Dr. Jekyll can change into a man as different as Hyde, surely it must be all the *easier* for Juggle to change into Boggle, who is exactly like him.
>
> We need conflict or strong difference to shake our natural assumption that to one body there corresponds at most one agent. [p. 219]

So *what is it like* to be the right hemisphere self in a split-brain patient? This is the most natural question in the world,[1] and it conjures up a mind-boggling — and chilling — image: there you are, trapped in the right hemisphere of a body whose left side you know intimately (and still control) and whose right side is now as remote as the body of a passing stranger. You would like to tell the world what it is like to be you, but you can't! You're cut off from all verbal communication by the loss of your indirect phone lines to the radio station in the left hemisphere. You do your best to signal your existence to the outside

1. It is interesting to note that Nagel, in 1971, was already addressing this question explicitly (p. 398), before he turned his attention to bats — a topic we will discuss in the next chapter.

world, tugging your half of the face into lopsided frowns and smiles, and occasionally (if you are a virtuoso right hemisphere self) scrawling a word or two with your left hand.

This exercise of imagination could go on in the obvious ways, but we know it is a fantasy — as much a fantasy as Beatrix Potter's charming stories of Peter Rabbit and his anthropomorphic animal friends. Not because "consciousness is only in the left hemisphere" and not because it *couldn't be* the case that someone found himself or herself in such a pickle, but simply because it *isn't* the case that commissurotomy leaves in its wake organizations both distinct and robust enough to support such a separate self.

It could hardly be a challenge to my theory of the self that it is "logically possible" that there is such a right hemisphere self in a split-brain patient, for my theory says that there isn't, and says why: the conditions for accumulating the sort of narrative richness (and independence) that constitutes a "fully fledged" self are not present. My theory is similarly impervious to the claim — which I would not dream of denying — that there *could be* talking bunny rabbits, spiders who write English messages in their webs, and for that matter, melancholy choo-choo trains. There could be, I suppose, but there aren't — so my theory doesn't have to explain them.

3. THE UNBEARABLE LIGHTNESS OF BEING

> Whatever happens, where or when, we're prone to wonder who or what's responsible. This leads us to discover explanations that we might not otherwise imagine, and that helps us predict and control not only what happens in the world, but also what happens in our minds. But what if those same tendencies should lead us to imagine things and causes that do not exist? Then we'll invent false gods and superstitions and see their hand in every chance coincidence. Indeed, perhaps that strange word "I" — as used in "I just had a good idea" — reflects the selfsame tendency. If you're compelled to find some cause that causes everything you do — why, then, that something needs a name. You call it "me." I call it "you."
>
> MARVIN MINSKY (1985), p. 232

A self, according to my theory, is not any old mathematical point, but an abstraction defined by the myriads of attributions and interpretations (including self-attributions and self-interpretations) that have composed the biography of the living body whose Center of Narrative Gravity it is. As such, it plays a singularly important role in the ongoing cognitive economy of that living body, because, of all the things in the environment an active body must make mental models of, none is more crucial than the model the agent has of itself. (See, e.g., Johnson-Laird, 1988; Perlis, 1991.)

To begin with, every agent has to know which thing in the world it is! This may seem at first either trivial or impossible. "I'm me!" is not really informative, and what else could one need to know — or could one discover if one didn't already know it? For simpler organisms, it is true, there is really nothing much to self-knowledge beyond the rudimentary biological wisdom enshrined in such maxims as When Hungry, Don't Eat Yourself! and When There's a Pain, It's Yours! In every organism, including human beings, acknowledgment of these basic biological design principles is simply "wired in" — part of the underlying design of the nervous system, like blinking when something approaches the eye or shivering when cold. A lobster might well eat another lobster's claws, but the prospect of eating one of its own claws is conveniently unthinkable to it. Its options are limited, and when it "thinks of" moving a claw, its "thinker" is directly and appropriately wired to the very claw it thinks of moving. With human beings (and chimpanzees and maybe a few other species), on the other hand, there are more options, and hence more sources of confusion.

Some years ago the authorities in New York Harbor experimented with a shared radar system for small boat owners. A single powerful land-based radar antenna formed a radar image of the harbor, which could then be transmitted as a television signal to boat owners who could save the cost of radar by simply installing small television sets in their boats. What good would this do? If you were lost in the fog, and looked at the television screen, you would know that one of those many moving blips on the screen was you — but which one? Here is a case in which the question "Which thing in the world am I?" is neither trivial nor impossible to answer. The mystery succumbs to a simple trick: Turn your boat quickly in a tight circle; then your blip is the one that traces the little "O" on the screen — unless several boats in the fog try to perform the same test at the same time.

The method is not foolproof, but it works most of the time, and it nicely illustrates a much more general point: In order to control the sorts of sophisticated activities human bodies engage in, the body's control system (housed in the brain) has to be able to recognize a wide variety of different sorts of inputs as informing it about itself, and when quandaries arise or skepticism sets in, the only reliable (but not foolproof) way of sorting out and properly assigning this information is to run little experiments: do something and look to see what moves.[2] A chimpanzee can readily learn to reach through a hole in the wall of its cage for bananas, guiding its arm movements by watching its own arm on a closed circuit television monitor mounted quite some distance from his arm (Menzel et al., 1985). This is a decidedly nontrivial bit of self-recognition, depending as it does on noticing the consonance of the seen arm movements on the screen with the unseen *but intended* arm movements. What would happen if the experimenters built in a small delay in the videotape? How long do you think it would take *you* to discover that you were looking at your own arm (without verbal clues from the experimental setup) if a tape delay of, say, twenty seconds were built into the closed circuit?

The need for self-knowledge extends beyond the problems of identifying the external signs of our own bodily movement. We need to know about our own internal states, tendencies, decisions, strengths, and weaknesses, and the basic method of obtaining this knowledge is essentially the same: Do something and "look" to see what "moves." An advanced agent must build up practices for keeping track of both its bodily and "mental" circumstances. In human beings, as we have seen, those practices mainly involve incessant bouts of storytelling and story-checking, some of it factual and some of it fictional. Children practice this aloud (think of Snoopy, saying to himself as he sits on his doghouse roof: "Here's the World War I flying ace . . ."). We adults do it more elegantly: silently, tacitly, effortlessly keeping track of the difference between our fantasies and our "serious" rehearsals and reflections. The philosopher Kendall Walton (1973, 1978) and the psychologist Nicholas Humphrey (1986) have shown from different perspectives the importance of drama, storytelling, and the more fundamental phenomenon of make-believe in providing practice for human beings who are novice self-spinners.

Thus do we build up a defining story about ourselves, organized

---

2. And how do we know that *we* are doing something? Where do we get the initial bit of self-knowledge we use for this leverage? This has seemed to be an utterly fundamental question to some philosophers (Castañeda, 1967, 1968; Lewis, 1979; Perry, 1979), and has generated a literature of surpassing intricacy. If this is a substantial philosophical problem, there must be something wrong with the "trivial" answer (but I can't see what): We get our basic, original self-knowledge the same way the lobster does; we're just wired that way.

around a sort of basic blip of self-representation (Dennett, 1981a). The blip isn't a self, of course; it's a *representation* of a self (and the blip on the radar screen for Ellis Island isn't an island — it's a representation of an island). What makes one blip the *me*-blip and another blip just a *he*- or *she*- or *it*-blip is not what it looks like, but what it is used for. It gathers and organizes the information on the topic of *me* in the same way other structures in my brain keep track of information on Boston, or Reagan, or ice cream.

And where is the thing your self-representation is *about*? It is wherever you are (Dennett, 1978b). And *what* is this thing? It's nothing more than, and nothing less than, your center of narrative gravity.

Otto returns:

> The trouble with centers of gravity is that they aren't real; they're theorists' fictions.

That's not the trouble with centers of gravity; it's their glory. They are *magnificent* fictions, fictions anyone would be proud to have created. And the fictional characters of literature are even more wonderful. Think of Ishmael, in *Moby-Dick*. "Call me Ishmael" is the way the text opens, and we oblige. We don't call the text Ishmael, and we don't call Melville Ishmael. Who or what do we call Ishmael? We call Ishmael Ishmael, the wonderful fictional character to be found in the pages of *Moby-Dick*. "Call me Dan," you hear from my lips, and you oblige, not by calling my lips Dan, or my body Dan, but by calling *me* Dan, the theorists' fiction created by . . . well, not by me but by my brain, acting in concert over the years with my parents and siblings and friends.

> That's all very well for you, but *I* am perfectly real. I may have been created by the social process you just alluded to (I must have been, if I didn't exist before my birth), but what the process created is a *real* self, not a mere fictional character!

I think I know what you're getting at. If a self isn't a real thing, what happens to moral responsibility? One of the most important roles of a self in our traditional conceptual scheme is as the place where the buck stops, as Harry Truman's sign announced. If selves aren't real — aren't *really* real — won't the buck just get passed on and on, round and round, forever? If there is no Oval Office in the brain, housing a Highest Authority to whom all decisions can be appealed, we seem to be threatened with a Kafkaesque bureaucracy of homunculi, who always reply, when challenged: "Don't blame me, I just work here." The task of constructing a self that can *take* responsibility is a major social and educational project, and you are right to be concerned about threats to its integrity. But a brain-pearl, a real, "intrinsically responsible" whatever-it-is, is a pathetic bauble to brandish like a lucky charm in the face of this threat. The only hope, and not at all a forlorn one, is to come to understand, naturalistically, the ways in which brains grow self-representations, thereby equipping the bodies they control with responsible selves when all goes well. Free will and moral responsibility are well worth wanting, and as I try to show in *Elbow Room: The Varieties of Free Will Worth Wanting* (1984), the best defense of them abandons the hopelessly contradiction-riddled myth of the distinct, separate soul.

But don't I exist?

Of course you do. There you are, sitting in the chair, reading my book and raising challenges. And curiously enough, your current embodiment, though a necessary precondition for your creation, is not necessarily a requirement for your existence to be prolonged indefinitely. Now if you were a soul, a pearl of immaterial substance, we could "explain" your potential immortality only by postulating it as an inexplicable property, an ineliminable *virtus dormitiva* of soul-stuff. And if you were a pearl of material substance, some spectacularly special group of atoms in your brain, your mortality would depend on the physical forces holding them together (we might ask the physicists what the "half-life" of a self is). If you think of yourself as a center of narrative gravity, on the other hand, your existence depends on the persistence of that narrative (rather like the Thousand and One Arabian Nights, but all a single tale), which could *theoretically* survive indefinitely many switches of *medium*, be teleported as readily (in principle) as the evening news, and stored indefinitely as sheer information. If what you are is that organization of information that has structured your body's control system (or, to put it in its more usual provocative form, if what you are is the program that runs on your brain's computer), then you could in principle survive the death of your body as intact as a program can survive the destruction of the computer on which it was created and first run. Some thinkers (e.g., Penrose, 1989) find this an appalling and deeply counterintuitive implication of the view I've defended here. But if it is potential immortality you hanker for, the alternatives are simply indefensible.

14

CONSCIOUSNESS IMAGINED

1. IMAGINING A CONSCIOUS ROBOT

The phenomena of human consciousness have been explained in the preceding chapters in terms of the operations of a "virtual machine," a sort of evolved (and evolving) computer program that shapes the activities of the brain. There is no Cartesian Theater; there are just Multiple Drafts composed by processes of content fixation playing various semi-independent roles in the brain's larger economy of controlling a human body's journey through life. The astonishingly persistent conviction that there is a Cartesian Theater is the result of a variety of cognitive illusions that have now been exposed and explained. "Qualia" have been replaced by complex dispositional states of the brain, and the self (otherwise known as the Audience in the Cartesian Theater, the Central Meaner, or the Witness) turns out to be a valuable abstraction, a theorist's fiction rather than an internal observer or boss.

If the self is "just" the Center of Narrative Gravity, and if all the phenomena of human consciousness are explicable as "just" the activities of a virtual machine realized in the astronomically adjustable connections of a human brain, then, in principle, a suitably "programmed" robot, with a silicon-based computer brain, would be conscious, would have a self. More aptly, there would be a conscious self whose body was the robot and whose brain was the computer. This implication of my theory strikes some people as obvious and unobjectionable. "*Of course* we're machines! We're just very, very complicated, evolved machines made of organic molecules instead of metal and silicon, and we are conscious, so there can be conscious machines — us." For these readers, this implication was a foregone conclusion. What has proved to be interesting to them, I hope, are the variety of unobvious implications encountered along the way, in particular those that show how much of the commonsense Cartesian picture must be replaced as we learn more about the actual machinery of the brain.

Other people, however, find the implication that there could be, in principle, a conscious robot so incredible that it amounts in their eyes to the *reductio ad absurdum* of my theory. A friend of mine once responded to my theory with the following heartfelt admission: "But, Dan, I just can't imagine a conscious robot!" Some readers may be inclined to endorse his claim. They should resist the inclination, for he misspoke. His error was simple, but it draws attention to a fundamental confusion blocking progress on understanding consciousness. "You *know* that's false," I replied. "You've often imagined conscious robots. It's not that you can't imagine a conscious robot; it's that you can't imagine *how* a robot could be conscious."

Anyone who has seen R2D2 and C3PO in *Star Wars*, or listened to Hal in *2001*, has imagined a conscious robot (or a conscious computer — whether the system is up-and-about, like R2D2, or bedridden, like Hal, is not really that crucial to the task of imagination). It is literally child's play to imagine the stream of consciousness of an "inanimate" thing. Children do it all the time. Not only do teddy bears have inner lives, but so does the Little Engine That Could. Balsam trees stand silently in the woods, fearing the woodsman's ax but at the same time yearning to become a Christmas tree in some nice warm house, surrounded by happy children. Children's literature (to say nothing of television) is chock full of opportunities to imagine the conscious lives of such mere things. The artists who illustrate these fantasies usually help the children's imagination by drawing expressive faces on these phony agents, but it's not essential. Speaking — as Hal does — will serve about as well, in the absence of an expressive face, to secure the illusion that there is someone in there, that it is like something to be Hal, or a teddy bear, or a choo-choo train.

That's the rub, of course: These are all illusions — or so it seems. There are differences among them. It's obvious that no teddy bear is conscious, but it's really not obvious that no robot could be. What is obvious is just that it's hard to imagine how they could be. Since my friend found it hard to imagine *how* a robot could be conscious, he was reluctant to imagine a robot *to be* conscious — though he could easily have done so. There is all the difference in the world between these two feats of imagination, but people tend to confuse them. It is indeed mind-bogglingly difficult to imagine how the computer-brain of a robot could support consciousness. How *could* a complicated slew of information-processing events in a bunch of silicon chips amount to conscious experiences? But it's just as difficult to imagine how an organic human brain could support consciousness. How *could* a complicated slew of electrochemical interactions between billions of neurons amount to conscious experiences? And yet we readily imagine human beings to be conscious, even if we still can't imagine *how* this could be.

How could the brain be the seat of consciousness? This has usually been treated as a rhetorical question by philosophers, suggesting that an answer to it would be quite beyond human comprehension. A primary goal of this book has been to demolish that presumption. I have argued that you *can* imagine how all that complicated slew of activity in the brain amounts to conscious experience. My argument is straightforward: I have shown you how to do it. It turns out that the way to imagine this is to think of the brain as a computer of sorts. The concepts of computer science provide the crutches of imagination we need if we are to stumble across the *terra incognita* between our phenomenology as we know it by "introspection" and our brains as science reveals them to us. By thinking of our brains as information-processing systems, we can gradually dispel the fog and pick our way across the great divide, discovering how it might be that our brains produce all the phenomena. There are many treacherous pitfalls to avoid — such inviting dead ends as the Central Meaner, "filling in," and "qualia," for instance — and no doubt there are still some residual confusions and outright errors in the sketch I have provided, but at least we can now see what a path would be like.

Some philosophers have declared, however, that crossing this divide is strictly impossible. Thomas Nagel (1974, 1986) has claimed that there is no getting to the subjective level of phenomenology from the objective level of physiology. More recently Colin McGinn has claimed that consciousness has a "hidden structure" that lies beyond both phenomenology and physiology, and while this hidden structure could bridge the gap, it is probably forever inaccessible to us.

> The kind of hidden structure I envisage would lie at neither of the levels suggested by Nagel: it would be situated somewhere between them. Neither phenomenological nor physical, this mediating level would not (by definition) be fashioned on the model

> of either side of the divide, and hence would not find itself unable to reach out to the other side. Its characterization would call for radical conceptual innovation (which I have argued is probably beyond us). [McGinn, 1991, pp. 102–103]

The "software" or "virtual machine" level of description I have exploited in this book is exactly the sort of mediating level McGinn describes: not explicitly physiological or mechanical and yet capable of providing the necessary bridges to the brain machinery on the one hand, while on the other hand not being explicitly phenomenological and yet capable of providing the necessary bridges to the world of content, the worlds of (hetero-)phenomenology. We've done it! We *have* imagined how a brain could produce conscious experience. Why does McGinn think it is beyond us to engage in this "radical conceptual innovation"? Does he subject the various software approaches to the mind to a rigorous and detailed analysis that demonstrates their futility? No. He doesn't examine them at all. He doesn't even try to imagine the intermediate level he posits; he just notes that it seems obvious to him that there is nothing to hope for from this quarter.

This spurious "obviousness" is a great obstacle to progress in understanding consciousness. It is the most natural thing in the world to think of consciousness as occurring in some sort of Cartesian Theater, and to suppose that there is nothing really wrong with thinking this way. This seems obvious until you look quite hard at what we might learn about the brain's activities, and begin trying to imagine, in detail, an alternative model. Then what happens is rather like the effect of learning how a stage magician performs a conjuring trick. Once we take a serious look backstage, we discover that we didn't actually see what we thought we saw onstage. The huge gap between phenomenology and physiology shrinks a bit; we see that some of the "obvious" features of phenomenology are not real at all: There is no filling in with figment; there are no intrinsic qualia; there is no central fount of meaning and action; there is no magic place where the understanding happens. In fact, there is no Cartesian Theater; the very distinction between onstage experiences and backstage processes loses its appeal. We still have plenty of amazing phenomena to explain, but a few of the most mind-boggling special effects just don't exist at all, and hence require no explanation.

Once we make some progress on the difficult task, imagining *how* a brain produces the phenomena of consciousness, we get to make some slight adjustments in the easy task: imagining someone or something to be conscious. We may continue to think of this by positing a stream of consciousness of sorts, but we no longer endow that stream with all of its traditional properties. Now that the stream of consciousness has been reconceived as the operations of a virtual machine realized in the brain, it is no longer "obvious" that we are succumbing to an illusion when we imagine such a stream occurring in the computer brain of a robot, for instance.

McGinn invites his readers to join him in surrender: It's just impossible to imagine how software could make a robot conscious. Don't even try, he says. Other philosophers have fostered this attitude by devising thought experiments that "work" precisely because they dissuade the reader from trying to imagine, in detail, how software could accomplish this. Curiously, the two best known both involve allusions to China: Ned Block's (1978) Chinese Nation and John Searle's (1980, 1982, 1984, 1988) Chinese Room.[1] Both thought experiments rely on the same misdirection of imagination, and since Searle's has been the more widely discussed, I will concentrate on it. Searle invites us to imagine him locked in a room, hand-simulating a giant AI program, which putatively understands Chinese. He stipulates that the program passes the Turing test, foiling all attempts by human interlocutors to distinguish it from a genuine understander of Chinese. It does not follow, he says, from this merely behavioral indistinguishability that there is any genuine understanding of Chinese, or any Chinese consciousness, in the Chinese Room. Searle, locked in the room and busily manipulating the symbol strings of the program according to the program, doesn't thereby gain any understanding of Chinese, and there is nothing else in the room that understands Chinese either (this is "just obvious," as Frank Jackson would say).

This thought experiment is supposed to prove the impossibility of what Searle calls "strong AI," the thesis that "the appropriately programmed digital computer with the right inputs and outputs would thereby have a mind in exactly the sense that human beings have minds" (Searle, 1988a). There has been a huge outpouring of reaction to Searle's many versions of this thought experiment over the last dec-

1. Try to imagine the state of mind of Ji Hu-Min, my graduate student from Beijing, whose introduction to Anglo-American philosophy of mind (while his English was still quite rudimentary) was sitting in a seminar where students and professors vigorously debated what would be the case if the entire population of China were somehow forced to participate in a massive realization of a putatively conscious AI program (Block's example) and then went on to discuss, with equal obliviousness to the sensitivities of a Chinese observer, Searle's Chinese Room.

ade, and while philosophers and others have always found flaws in his thought experiment when it is considered as a logical argument,[2] it is undeniable that its "conclusion" continues to seem "obvious" to many people. Why? Because people don't actually imagine the case in the detail that it requires.

Here is an informal experiment that will help us see if my diagnosis is correct. First, let's imagine a brief excerpt from the Chinese Room's winning dialogue with the judge in the Turing test. (For convenience, I've translated it from Chinese to English.)

> JUDGE: Did you hear about the Irishman who found a magic lamp? When he rubbed it a genie appeared and granted him three wishes. "I'll have a pint of Guinness!" the Irishman replied, and immediately it appeared. The Irishman eagerly set to sipping and then gulping, but the level of Guinness in the glass was always magically restored. After a while the genie became impatient. "Well, what about your second wish?" he asked. Replied the Irishman between gulps, "Oh well, I guess I'll have another one of these!"
>
> CHINESE ROOM: Very funny. No, I hadn't heard it — but you know, I find ethnic jokes in bad taste. I laughed in spite of myself, but really, I think you should find other topics for us to discuss.
>
> J: Fair enough, but I told you the joke because I want you to explain it to me.
>
> CR: Boring! You should never explain jokes.
>
> J: Nevertheless, this is my test question. Can you explain to me how and why the joke "works"?
>
> CR: If you insist. You see, it depends on the assumption that the magically refilling glass will go on refilling forever, so the Irishman has all the stout he can ever drink. So he hardly has a reason for wanting a duplicate, but he is so stupid (that's the part I object to) or so besotted by the alcohol, that he doesn't recognize this, and so, unthinkingly endorsing his delight with his first wish come true, he asks for seconds. These background assumptions aren't true, of course, but just part of the ambient lore of joke-

2. The definitive refutation, still never adequately responded to by Searle, is Douglas Hofstadter's, in Hofstadter and Dennett (1981), pp. 373–382. There have been many other incisive criticisms over the years. In "Fast Thinking" (in Dennett, 1987a), I offered a new diagnosis of the sources of confusion in his thought experiment. His response was to declare, with no supporting argument, that all its points are irrelevant (Searle, 1988b). No conjuror enjoys having his tricks explained to the public.

telling, in which we suspend our disbelief in magic, and so forth. By the way, we could imagine a somewhat labored continuation in which the Irishman turned out to be "right" in his second wish after all — perhaps he's planning to throw a big party, and one glass won't refill fast enough to satisfy all his thirsty guests (and it's no use saving it up in advance — we all know how stale stout loses its taste). We tend not to think of such complications, which is part of the explanation of why jokes work. Is that enough?

This conversation is not dazzling, but let's suppose it was good enough to fool the judge. Now we are invited to imagine all these speeches by CR being composed by the giant program Searle is diligently hand-simulating. Hard to imagine? Of course, but since Searle stipulates that the program passes the Turing test, and since this level of conversational sophistication would surely be within its powers, unless we try to imagine the complexities of a program capable of generating this sort of conversation, we are not following directions. Of course we should also imagine that Searle hasn't any inkling of what he is doing in the Chinese Room; he just sees zeros and ones that he manipulates according to the program. It is important, by the way, that Searle invites us to imagine that he manipulates inscrutable Chinese characters instead of zeros and ones, for this may lull us into the (unwarranted) supposition that the giant program would work by somehow simply "matching up" the input Chinese characters with some output Chinese characters. No such program would work, of course — do CR's speeches in English "match up" with the judge's questions?

A program that could actually generate CR's speeches in response to J's questions might look something like this in action (viewed from the virtual-machine level, not from Searle's ground-floor level). On parsing the first words, "Did you hear about . . . " some of the program's joke-detecting demons were activated, which called up a host of strategies for dealing with fiction, "second intention" language, and the like, so when the words "magic lamp" came to be parsed, the program had already put a low priority on responses complaining that there were no such things as magic lamps. A variety of standard genie-joke narrative frames (Minsky, 1975) or scripts (Schank and Abelson, 1977) were activated, creating various expectations for continuations, but these were short-circuited, in effect, by the punch line, which invoked a more mundane script (the script for "asking for seconds"), and the unexpectedness of this was not lost on the program. . . . At the same time, demons sensitive to the negative connotations of ethnic-joke-telling were also alerted, eventually leading to the second theme of CR's first response. . . . And so forth, in vastly more detail than I have tried to sketch here.

That fact is that any program that could actually hold up its end in the conversation depicted would have to be an extraordinarily supple, sophisticated, and multilayered system, brimming with "world knowledge" and meta-knowledge and meta-meta-knowledge about its own responses, the likely responses of its interlocutor, its own "motivations" and the motivations of its interlocutor, and much, much more. Searle does not deny that programs can have all this structure, of course. He simply discourages us from attending to it. But if we are to do a good job imagining the case, we are not only entitled but obliged to imagine that the program Searle is hand-simulating has all this structure — and more, if only we can imagine it. But then it is no longer *obvious*, I trust, that there is no genuine understanding of the joke going on. *Maybe* the billions of actions of all those highly structured parts produce genuine understanding in the system after all. If your response to this hypothesis is that you haven't the faintest idea whether there would be genuine understanding in such a complex system, that is already enough to show that Searle's thought experiment depends, illicitly, on your imagining too simple a case, an irrelevant case, and drawing the "obvious" conclusion from it.

Here is how the misdirection occurs. We see clearly enough that if there were understanding in such a giant system, it would not be Searle's understanding (since he is just a cog in the machinery, oblivious to the context of what he is doing). We also see clearly that there is nothing remotely like genuine understanding in any hunk of programming small enough to imagine readily — whatever it is, it's just a mindless routine for transforming symbol strings into other symbol strings according to some mechanical or syntactical recipe. Then comes the suppressed premise: Surely *more of the same*, no matter how much more, would never add up to genuine understanding. But why should anyone think this was true? Cartesian dualists would think so, because they think that even human brains are unable to accomplish understanding all by themselves; according to the Cartesian view, it takes an immaterial soul to pull off the miracle of understanding. If, on the other hand, we are materialists who are convinced that one way or another our brains are responsible on their own, without miraculous assistance, for our understanding, we must admit that genuine understanding is somehow achieved by a process composed of interactions between a host of subsystems none of which understand a thing by themselves. The argument that begins "this little bit of brain activity doesn't understand Chinese, and neither does this bigger bit of which it is a part . . . " is headed for the unwanted conclusion that even the activity of the whole brain is insufficient to account for understanding Chinese. It is *hard to imagine* how "just more of the same" could add up to understanding, but we have very good reason to believe that it does, so in this case, we should try harder, not give up.

How might we try harder? With the help of some handy concepts: the intermediate-level software concepts that were designed by computer scientists precisely to help us keep track of otherwise unimaginable complexities in large systems. At the intermediate levels we see many entities that are quite invisible at more microscopic levels, such as the "demons" alluded to above, to which a modicum of quasi-understanding is attributed. Then it becomes not so difficult to imagine how "more of the same" could amount to genuine understanding. All these demons and other entities are organized into a huge system, the activities of which organize themselves around its own Center of Narrative Gravity. Searle, laboring in the Chinese Room, does not understand Chinese, but he is not alone in the room. There is also the System, CR, and it is to *that* self that we should attribute any understanding of the joke.

This reply to Searle's example is what he calls the Systems Reply. It has been the standard reply of people in AI from the earliest outings of his thought experiment, more than a decade ago, but it is seldom appreciated by people outside of AI. Why not? Probably because they haven't learned how to imagine such a system. They just can't imagine how understanding could be a property that emerges from lots of distributed quasi-understanding in a large system. They certainly can't if they don't try, but how could they be helped along on this difficult exercise? Is it "cheating" to think of the software as composed of homunculi who quasi-understand, or is that just the right crutch to help the imagination make sense of astronomical complexity? Searle begs the question. He invites us to imagine that the giant program consists of some simple table-lookup architecture that directly matches Chinese character strings to others, as if such a program could stand in, fairly, for any program at all. We have no business imagining such a simple program and assuming that *it* is the program Searle is simulating, since no such program could produce the sorts of results that would pass the Turing test, as advertised. (For a similar move and its rebuttal, see Block, 1982; and Dennett, 1985).

Complexity does matter. If it didn't, there would be a much shorter argument against strong AI: "Hey, look at this hand calculator. It doesn't understand Chinese, and any conceivable computer is just a giant hand calculator, so no computer could understand Chinese. Q.E.D." When we factor in the complexity, as we must, we really have to factor it in — and not just pretend to factor it in. That is hard to do, but until we do, any intuitions we have about what is "obviously" not present are not to be trusted. Like Frank Jackson's case of Mary the color scientist, Searle's thought experiment yields a strong, clear conviction only when we fail to follow instructions. These intuition pumps are defective; they do not enhance but mislead our imaginations.

But what, then, of my own intuition pumps? What of Shakey the robot, or the CADBLIND Mark II, or the biofeedback-trained blindsight patient, for instance? Are they not equally suspect, equally guilty of misleading the reader? I've certainly done my best in telling these tales to lead your imagination down certain paths, and to keep you from bogging down in complexities I deemed unnecessary to the point I was attempting to make. There is some asymmetry, however: My intuition pumps are, for the most part, intended to help you imagine new possibilities, not convince you that certain prospects are impossible. There are exceptions. My variation on the brain in the vat that opened the book was designed to impress on you the impossibility of certain sorts of deception, and some of the thought experiments in chapter 5 were intended to show that, unless there were a Cartesian Theater, there *could not* be a fact of the matter distinguishing Orwellian from Stalinesque content revisions. These thought experiments proceeded, however, by heightening the vividness for the "opposition"; the examples of the woman in the hat at the party and the long-haired woman with glasses, for instance, were designed to sharpen the very intuition I then sought to discredit by argument.

Still, let the reader beware: My intuition pumps, like anyone else's, are not the straightforward demonstrations they may seem to be; they are more art than science. (For further warnings about philosophers' thought experiments, see Wilkes, 1988.) If they help us conceive of new possibilities, which we can then confirm by more systematic methods, that is an achievement; if they lure us down the primrose path, that is a pity. Even good tools can be misused, and like any other workers, we will do better if we understand how our tools work.

2. WHAT IT IS LIKE TO BE A BAT

The most widely cited and influential thought experiment about consciousness is Thomas Nagel's "What Is It Like to Be a Bat?" (1974). He answers his title question by claiming that this is impossible for us to imagine. This claim is congenial to many, apparently; one sometimes sees his paper cited by scientists as if it were that rarity of rarities, a philosophical "result" — a received demonstration of a fact that any theory must subsequently accommodate.

Nagel chose his target creatures well. Bats, as fellow mammals, are enough like us to support the conviction that *of course* they are conscious. (If he had written "What Is It Like to Be a Spider?" many would be inclined to wonder what made him so sure it was like anything at all.) But thanks to their system of echolocation — bats can "see with their ears" — they are also different enough from us so that we can sense the vast gulf. Had he written a paper called "What Is It Like to Be a Chimpanzee?" or, more to the point, "What Is It Like to Be a Cat?" the opinion that his pessimistic conclusion was obvious would not be so close to unanimity. There are many people who are supremely confident that they know *just* what it's like to be a cat. (They are wrong, of course, unless they have supplemented all their loving and empathetic observation with vast amounts of physiological research, but they would be erring on the wrong side, from Nagel's point of view.)

For better or worse, most people seem quite cheerful about accepting Nagel's "result" regarding the inaccessibility to us of bat consciousness. Some philosophers have challenged it, however, and for good reason (Hofstadter, 1981; Hardin, 1988; Leiber, 1988; Akins, 1990). First we must be clear about just which result it is. It is not just the epistemological or evidential claim that even if someone succeeded ("by accident") in imagining what it is like to be a bat, we would never be able to confirm that this successful feat of imagination had occurred. It is rather that we human beings don't have and could never acquire the wherewithal, the representational machinery, to represent to ourselves what it is like to be a bat.

The distinction is important. In chapter 12 we looked at the similar feat of imagining what it must have been like to be a Leipziger hearing one of Bach's cantatas for the first time. The epistemological problem is difficult, but straightforwardly addressable by the usual sorts of research. Figuring out just what sorts of experiences they would have had, and how these would differ from our experiences of Bach, is a matter of historical, cultural, psychological, and, maybe, physiological investigation. We can figure out some of this quite readily, including some of the most striking differences from our own experience, but if we were to try to put ourselves into the very sequence of experiential states such a person would enjoy, we would face diminishing returns. The task would require us to subject ourselves to vast transformations — forgetting much of what we know, losing associations and habits, acquiring new habits and associations. We can use our "third-person" research to say what these transformations would be, but actually undergoing them would involve terrible costs of isolation from our contemporary culture — no listening to the radio, no reading about post-Bach political and social developments, and so forth. There is no need to go to those lengths to learn about Leipziger consciousness.

The same is true about imagining what it is like to be a bat. We should be interested in what we can know about the bat's consciousness (if any), not whether we can turn our minds temporarily or permanently into bat minds. In chapter 12, we undermined the presumption that there were "intrinsic" properties — qualia — that constitute *what it is like* to have one conscious experience or another, and as Akins (1990) points out, even if there *were* residual nondispositional, nonrelational properties of bat experiences, becoming intimately acquainted with them, while remaining ignorant of the researchable facts about the systematic structure of bat perception and behavior, would leave us ignorant of what it is like to be a bat. There is at least a lot that we can know about what it is like to be a bat, and neither Nagel nor anyone else has given us a good reason to believe there is anything interesting or theoretically important that is inaccessible to us.

Nagel claims that no amount of third-person knowledge could tell us what it is like to be a bat, and I flatly deny that claim. How might we resolve this dispute? By engaging in something that starts out as child's play — a game in which one person imagines what it is like to be x, and the other then tries to demonstrate that there is something wrong with that particular exercise of heterophenomenology.

Here are some simple warmup exercises:

> *A:* Here's Pooh the teddy bear, thinking how nice it would be to have some honey for breakfast!
>
> *B:* Wrong. The teddy bear has no provision for distinguishing honey from anything else. No operating sense organs, and not even a stomach. The teddy bear is filled with inert stuffing. It is not like anything to be a teddy bear.

A. Here's Bambi the deer, admiring the beautiful sunset, until the bright orange sky suddenly reminds him of the evil hunter's jacket!

B. Wrong. Deer are colorblind (well, they may have some sort of dichromatic vision). Whatever deer are conscious of (if anything), they don't distinguish colors such as orange.

A. Here's Billy the bat perceiving, in his special sonar sort of way, that the flying thing swooping down toward him was not his cousin Bob, but an eagle, with pinfeathers spread and talons poised for the kill!

B. Hang on — how far away did you say the eagle was? A bat's echolocation is only good for a few meters.

A. Um, well . . . And the eagle was already only two meters away!

B. Ah, now this is harder to say. Just what are the resolution limits of a bat's echolocation? Is it used to identify objects at all, or just as an alerter and tracker for capture? Would a bat be able to distinguish pinfeathers spread from pinfeathers closed just using echolocation? I doubt it, but we will have to design some experiments to see, and also, of course, some experiments to discover whether bats are capable of keeping track of, and reidentifying, their kin. Some mammals can, and others, we have good reason to believe, are utterly oblivious of such matters.

The sorts of investigation suggested by this exercise would take us a long way into an account of the structure of the bat's perceptual and behavioral world, so we could rank order heterophenomenological narratives for realism, discarding those that asserted or presupposed discriminatory talents, or reactive dispositions, demonstrably not provided for in the ecology and neurophysiology of the bat. For example, we would learn that bats would not be bothered by the loud squeaks they emit in order to produce their echoes, because they have a cleverly designed muscle that shuts down their ears in perfect timing with their squeaks, not unlike the timing devices that permit sensitive radar systems to avoid being blasted by their own outgoing signals. A lot of research has already been done on these issues, so we can already say much more, for instance about why bats use different frequency patterns for their squeaks, depending on whether they are scanning for prey, approaching a target, or homing in for the kill (Akins, 1989, 1990).

When we arrive at heterophenomenological narratives that no critic can find any positive grounds for rejecting, we should accept them — tentatively, pending further discoveries — as accurate accounts of what it is like to be the creature in question. That, after all, is how we treat each other. In recommending that we treat bats and other candidates for interpretation the same way, I am not *shifting* the burden of proof but extending the normal, human, burden of proof to other entities.

We could use these investigations to dispel all sorts of overly romantic illusions about bat consciousness. We *know* that Randall Jarrell's delightful children's book, *The Bat-Poet* (1963), is fantasy, because we know that bats don't talk! Less obviously fantastical claims about their phenomenology succumb to less obvious, but still public, facts about their physiology and behavior. These investigations would show us a great deal about what a bat could and could not be conscious of under various conditions, by showing us what provisions there were in their nervous systems for representing this and that, and by checking experimentally to make sure the bat actually put the information to use in the modulation of its behavior. It is hard to imagine how much can be gleaned from this sort of research until you actually look into it. (For a surprisingly detailed preliminary investigation of what it is like to be a vervet monkey, for instance, see Cheney and Seyfarth, *How Monkeys See the World*, 1990.)

This invites an obvious objection: These investigations would show us a great deal about brain organization and information-processing in the bat, but they would show us only what bats are *not* conscious of, leaving entirely open what, if anything, bats *are* conscious of. As we know, much of the information-processing in nervous systems is entirely unconscious, so these methods of investigation will do nothing to rule out the hypothesis that bats are . . . flying zombies, creatures it is not like anything to be! (Wilkes, 1988, p. 224, wonders whether bat echolocation is a sort of blindsight, not like anything at all.)

Ah, the bat is out of the bag. This is indeed the ominous direction in which this discussion seems to be sliding, and we must head it off. Richard Dawkins (1986), in an illuminating discussion of the design of echolocation in horseshoe bats, gives us a clear version of the image that is lurking.

> The Doppler Effect is used in police radar speed-traps for motorists. . . . By comparing the outgoing frequency with the frequency of the returning echo the police, *or rather their automatic instrument* [my emphasis], can calculate the speed of each car. . . . By

> comparing the pitch of its cry with the pitch of the returning echo, therefore, the bat *(or rather its on-board computer in the brain)* [my emphasis] could, in theory, calculate how fast it was moving towards the tree. [pp. 30–31]

It is tempting to ask: Is there something in the bat that is situated relative to *its* "onboard computer" (which operates without a smidgen of consciousness) as the police are situated relative to their "automatic device"? The police don't have to calculate the Doppler shift consciously, but they do have to experience, consciously, the readout on their device that says, in bright red LED symbols: "75MPH." That is their cue for leaping on their motorcycles and starting up their sirens. We may plausibly suppose that the bat also does not consciously calculate the Doppler shift — its onboard computer takes care of that — but then isn't there a role left over, in the bat, for something like the experiencing cop, a witness to appreciate (consciously) the "output" of the bat's Doppler-effect-analysis computer? Note that we could easily enough replace the police officers with an automatic device that somehow recorded the registration number of the offending vehicle, looked up the operator's name and address and sent him or her a ticket. There is nothing special about the task the police are doing that shows it could not be done without any experiencing of anything. The same holds, it would seem, for the bat. A bat might be a zombie. It would be a zombie — so this line of reasoning suggests — unless there were an inner observer in it that reacts to an inner presentation in much the way the officers react to the flashing red lights on their instruments.

Don't fall in the trap. This is our old nemesis, the Audience in the Cartesian Theater. *Your* consciousness does not consist in the fact that your brain is inhabited by an inner agent to whom your brain presents displays, so our inability to find such a central agent in the bat's brain would not jeopardize its claim to consciousness, or our claim to be able to say what its consciousness was like. In order to understand a bat's consciousness, we must simply apply the same principles to the bat that we apply to ourselves.

But what could a bat do, then, that would be special enough to convince us that we were in the presence of genuine consciousness? It may seem that no matter what fancy output-users we situate behind the bat's Doppler-transducer, there could be no convincing, from-the-outside, "third-person" reason to grant the bat conscious experience. Not so. If the bat could talk, for instance, it would generate a text from which we could generate a heterophenomenological world, and that would give us exactly the same grounds for granting it consciousness that serve for any person. But, as we just noted, bats can't talk. They can, however, behave in many nonverbal ways that can provide a clear basis for describing their heterophenomenological world, or, as the pioneer researcher von Uexküll (1909) called it, their *Umwelt und Innenwelt*, their Surroundworld and Innerworld.

Heterophenomenology without a text is not impossible, just difficult (Dennett, 1988a, 1988b, 1989a, 1989b). One branch of animal heterophenomenology is known as cognitive ethology, the attempt to model animals' minds by studying — and experimenting on — their behavior in the field. The possibilities and difficulties of this sort of investigation are well represented in Cheney and Seyfarth (1990), Whiten and Byrne (1988), and in Ristau (1991), a festschrift dedicated to Donald Griffin, the pioneer investigator of bat echolocation and the creator of the field of cognitive ethology. One of the frustrating difficulties encountered by these investigators is that many of the experiments one dreams of running turn out to be utterly impractical in the absence of language; one simply cannot *set up* subjects (and know that one has set them up) in the ways these experiments would require without conversing with the subjects (Dennett, 1988a).

This is not just an epistemological problem for the heterophenomenologist; the very difficulty of creating the requisite experimental circumstances in the natural environment demonstrates something more fundamental about the minds of languageless creatures. It shows that the ecological situations of these animals have never provided them with *opportunities* for the development (by evolution, by learning, or by both) of many of the advanced mental activities that shape our minds, and so we can be quite sure they have never developed them. For instance, consider the concept of a *secret*. A secret is not just something you know that others don't know. For you to have a secret you need to know that the others don't know it, and you have to be able to control that fact. (If you are the first to see the approaching stampede, you may know something the others don't know, but not for long; you can't *keep* this bit of privileged information secret.) The behavioral ecology of a species has to be rather specially structured for there to be any role for secrets at all. Antelopes, in their herds, have no secrets and no way of getting any. So an antelope is probably no more capable of hatching a secret plan than it is capable of counting to a hundred or enjoying the colors of a sunset. Bats, who engage in relatively solitary forays during which they might be able to recognize that very isolation from their rivals, meet one of the necessary conditions for having secrets. Do they also have interests that might be noticeably well served by exploiting secrets? (What could a clam do with a secret? Just sit there in the mud, chuckling to itself?) Do bats also have habits of stealth or deception in hunting that might be adapted for more elaborate secret-keeping activity? There are in fact many questions of this sort that, once raised, suggest further investigations and experiments. The structure of a bat's mind is just as accessible as the structure of a bat's digestive system; the way to investigate either one is to go back and forth systematically between an assay of its contents and an assay of the world from which its contents were derived, paying attention to the methods and goals of the derivation.

Wittgenstein once said, "If a lion could talk, we could not understand him" (1958, p. 223). I think, on the contrary, that if a lion could talk, that lion would have a mind so different from the general run of lion minds, that although we could understand him just fine, we would learn little about ordinary lions from him. Language, as we saw in earlier chapters, plays an enormous role in the structuring of a human mind, and the mind of a creature lacking language — and having really no need for language — should not be supposed to be structured in these ways. Does this mean that languageless animals "are not conscious at all" (as Descartes insisted)? This question always arises at this moment as a sort of incredulous challenge, but we shouldn't feel obliged to answer it as it stands. Notice that it presupposes something we have worked hard to escape: the assumption that consciousness is a special all-or-nothing property that sunders the universe into two vastly different categories: the things that have it (the things that it is like something to be, as Nagel would put it) and the things that lack it. Even in our own case, we cannot draw the line separating our conscious mental states from our unconscious mental states. The theory of consciousness we have sketched allows for many variations of functional architecture, and while the presence of language marks a particularly dramatic increase in imaginative range, versatility, and self-control (to mention a few of the more obvious powers of the Joycean virtual machine), these powers do not have the *further* power of turning on some special inner light that would otherwise be off.

When we imagine what it is like to be a languageless creature, we start, naturally, from our own experience, and most of what then springs to mind has to be adjusted (mainly downward). The sort of consciousness such animals enjoy is dramatically truncated, compared to ours. A bat, for instance, not only can't wonder whether it's Friday; it can't even wonder whether it's a bat; there is no role for wondering to play in its cognitive structure. While a bat, like even the lowly lobster, has a biological self, it has no selfy self to speak of — no Center of Narrative Gravity, or at most a negligible one. No words-on-the-tip-of-its-tongue, but also no regrets, no complex yearnings, no nostalgic reminiscences, no grand schemes, no reflections on what it is like to be a cat, or even on what it is like to be a bat. This list of dismissals would be cheap skepticism if we didn't have a positive empirical theory on which to base it. Am I claiming to have proven that bats *could not* have these mental states? Well, no, but I also can't prove that mushrooms *could not* be intergalactic spaceships spying on us.

Isn't this an awfully anthropocentric prejudice? Besides, what about deaf-mutes? Aren't they conscious? Of course they are — but let's not jump to extravagant conclusions about their consciousness, out of misguided sympathy. When a deaf-mute acquires language (in particular, Sign language, the most natural language a deaf-mute can learn), a full-fledged human mind is born, clearly different in discoverable ways from the mind of a hearing person, but capable of all the reflective intricacy and generative power — perhaps more. But without a natural language, a deaf-mute's mind is terribly stunted. (See Sacks, 1989, especially the annotated bibliography.) As the philosopher Ian Hacking (1990) notes in a review of Sacks's book, "It takes a vivid imagination even to have a sense of what a deaf child is missing." One does not do deaf-mutes a favor by imagining that in the absence of language they enjoy all the mental delights we hearing human beings enjoy, and one does not do a favor to nonhuman animals by trying to obscure the available facts about the limitations of their minds.

And this, as many of you are aching to point out, is a subtext that has been struggling to get to the surface for quite a while: Many people are afraid to see consciousness explained because they fear that if we succeed in explaining it, we will lose our moral bearings. Maybe we *can* imagine a conscious computer (or the consciousness of a bat) but we *shouldn't try*, they think. If we get into that bad habit, we will start treating animals as if they were wind-up toys, babies and deaf-mutes as if they were teddy bears, and — just to add insult to injury — robots as if they were real people.

3. MINDING AND MATTERING

I take the title of this section from an article by Marian Stamp Dawkins (1987), who has done careful investigations of the moral implications of animal heterophenomenology. (Her early work is reported in her book *Animal Suffering: The Science of Animal Welfare,* 1980.) As she notes, our moral attitudes towards other animals are full of inconsistencies.

> We have only to think of various different sorts of animals to show up our inconsistencies. There are demonstrations against killing baby harp seals, but there are no comparable campaigns to stop the killing of rats. Many people are quite happy to eat pigs or sheep but horrified by the idea of eating dogs or horses. [p. 150]

Dawkins points out that there are two main strands to this tangle: the ability to reason and the ability to suffer. Descartes made much of the inability of nonhuman animals to reason (at least the way human beings reason), which provoked a famous response from the British utilitarian philosopher Jeremy Bentham: "a full-grown horse or dog is beyond comparison a more rational, as well as a more conversible animal than an infant of a day or a week, or even a month old. But suppose they were otherwise, what would it avail? The question is not, Can they *reason?* nor, Can they *talk,* but Can they *suffer?*" (Bentham, 1789) These usually appear to be opposing benchmarks of moral standing, but as Dawkins argues, "giving ethical value to the ability to suffer will in the end lead us to value animals that are clever. Even if we start out by rejecting Descartes' reasoning criterion, it is the reasoning animals that are the ones most likely to possess the capacity to suffer" (p. 153).

The reasons for this are implicit in the theory of consciousness we have developed. Suffering is not a matter of being visited by some ineffable but intrinsically awful state, but of having one's life hopes, life plans, life projects blighted by circumstances imposed on one's desires, thwarting one's intentions — whatever they are. The idea of suffering being somehow explicable as the presence of some intrinsic property — horribility, let's say — is as hopeless as the idea of amusement being somehow explicable as the presence of intrinsic hilarity. So the presumed inaccessibility, the ultimate unknowability, of another's suffering is just as misleading as the other fantasies about intrinsic qualia we have unmasked, though more obviously pernicious. It follows — and this does strike an intuitive chord — that the capacity to suffer is a function of the capacity to have articulated, wide-ranging, highly discriminative desires, expectations, and other sophisticated mental states.

Human beings are not the only creatues smart enough to suffer; Bentham's horse and dog show by their behavior that they have enough mental complexity to distinguish — and care about — a spectrum of pains and other impositions that is far from negligible, even if it is a narrow window compared to the scope of possibilities of human suffering. Other mammals, notably apes, elephants, and dolphins, apparently have much greater ranges.

In compensation for having to endure all the suffering, the smart creatures get to have all the fun. You have to have a cognitive economy with a budget for exploration and self-stimulation to provide the space for the recursive stacks of derived desires that make fun possible. You have taken a first step when your architecture permits you to appreciate the meaning of "Stop it, I love it!" Shallow versions of this building power are manifest in some higher species, but it takes a luxuriant imagination, and leisure time — something most species cannot afford — to grow a broad spectrum of pleasures. The greater the scope, the richer the detail, the more finely discriminative the desires, the worse it is when those desires are thwarted.

But why should it matter, you may want to ask, that a creature's desires are thwarted if they aren't *conscious* desires? I reply: Why would it matter more if they were conscious — especially if consciousness were a property, as some think, that forever eludes investigation? Why should a "zombie's" crushed hopes matter less than a conscious person's crushed hopes? There is a trick with mirrors here that should be exposed and discarded. Consciousness, you say, is what matters, but then you cling to doctrines about consciousness that systematically prevent us from getting any purchase on *why* it matters. Postulating special inner qualities that are not only private and intrinsically valuable, but also unconfirmable and uninvestigatable is just obscurantism.

Dawkins shows how the investigatable differences — the only differences that could possibly matter — can be experimentally explored, and it is worth a few details to show how much insight can be gleaned even from simple experiments with a rather unprepossessing species.

> Hens kept outside or in large litter pens spend a lot of their time scratching around and I therefore suspected that the lack of litter in battery cages might cause hens to suffer. Sure enough, when I gave them a choice between a cage with a wire floor and one with litter in which they could scratch, they chose the litter-floored cage. In fact they would enter a tiny cage (so small that they could hardly turn round) if this was the only way they could gain access to litter. Even birds which had been reared all their lives in cages

> and had never before had experience of litter chose the cage with litter on the floor. Although this was suggestive, it was not enough. I had to show that not only did the hens have a preference for litter but also that they had a preference which was strong enough to say that they might suffer if kept without it.
>
> Hens were then offered a slightly different choice. This time they had to choose between a wire-floored cage which had food and water and a litter-floored cage without food and water. . . . The result was that they spent a lot of time in the litter cage, with much less time being spent in the wire cage, even though this was the only place they could feed and drink. Then a complication was introduced. The birds had to "work" to move between the cages. They either had to jump from a corridor or push through a curtain of black plastic. So changing from one cage to another now had a cost. . . . The hens still spent the same amount of time in the wire cage with food as previously when there was no difficulty in entering it. But they spent hardly any time in the litter cage. They simply didn't seem prepared to work or pay any cost to get into the litter cage. . . . Quite contrary to what I had expected, the birds seemed to be saying that litter did not really matter to them. [pp. 157–159]

She concludes that "Suffering by the emotional mind is revealed by animals that have enough of a rational mind to be able to do something about the conditions that make them suffer," and she goes on to note that "it is also likely that organisms without the capacity to do anything to remove themselves from a source of danger would not evolve the capacity to suffer. There would be no evolutionary point in a tree which was having its branches cut off having the capacity to suffer in silence" (p. 159). As we saw in chapter 7 (see also chapter 3, footnote 9), one must be careful in framing such evolutionary arguments about function, for history plays a big role in evolution, and history can play tricks. But in the absence of positive grounds for imputing suffering, or positive grounds for suspecting that such positive grounds are for one reason or another systematically concealed, we should conclude that there is no suffering. We need not fear that this austere rule will lead us to slight our obligations to our fellow creatures. It still provides ample grounds for positive conclusions: Many, but not all, animals are capable of significant degrees of suffering. A more persuasive case in support of humane treatment can be mounted by acknowledging the vast differences in degrees, than by piously promulgating an unsupportable dogma about the universality and equality of animal pain.

This may settle the objective question about the presence or absence of suffering, but it does not settle the moral sentiments upset by the prospect of explaining consciousness in such heartlessly mechanistic ways. There is more at stake.

I have a farm in Maine, and I love the fact that there are bears and coyotes living in my woods. I very seldom see them, or even see signs of their presence, but I just like knowing that they are there, and would be very unhappy to learn that they had left. I would also not feel myself entirely compensated for the loss if some of my AI friends stocked my woods with lots of robot beasties (though the idea, if imagined in detail, is enchanting). It matters to me that there are wild creatures, descendants of wild creatures, living so close to me. Similarly, it delights me that there are concerts going on in the Boston area that I not only do not hear, but never even hear *about*.

These are facts of a special sort. They are facts that are important to us simply because one part of the environment that matters to us is our belief environment. And since we are not easily gulled into continuing to believe propositions after the support for them has evaporated, it matters to us that the beliefs be *true*, even when we won't ourselves see any direct evidence for them. Like any other part of the environment, a belief environment can be fragile, composed of parts that are interconnected by both historical accidents and well-designed links. Consider, for instance, that delicate part of our belief environment concerned with the disposition of our bodies after death. Few people believe that the soul resides in the body after death — even people who believe in souls don't believe *that*. And yet few if any of us would tolerate a "reform" that encouraged people to dispose of their dead kin by putting them in plastic bags in the trash, or otherwise unceremoniously discarding them. Why not? Not because we believe that corpses can actually suffer some indignity. A corpse can no more suffer an indignity than a log can. And yet, the idea is shocking, repulsive. Why?

The reasons are complex, but we can distill a few simple points for now. A person is not just a body; a person *has* a body. That corpse is the body of dear old Jones, a Center of Narrative Gravity that owes its reality as much to our collaborative efforts of mutual heterophenomenological interpretation as to the body that is now lifeless. The boundaries of Jones are not identical to the boundaries of Jones's body, and the interests of Jones, thanks to the curious human practice of self-spinning, can extend beyond the basic biological interests that spawned the practice. We treat his corpse with respect because it is important for the preservation of the belief environment in which we all live. If we start treating corpses as garbage, for instance, it might change the way we treat near-corpses — those who are still alive but dying. If we don't err on the side of prolonging the rituals and practices of respect well beyond the threshold of death, the dying (and those who care about them) will face an anxiety, an affront, a *possibility*, that risks offending them. Treating a corpse "badly" may not directly harm any dying person, and certainly doesn't harm the corpse, but, if it became common practice and this became widely known (as it would), this would significantly change the belief environment that surrounds dying. People would imagine the events that were due to follow their demise differently from the way they now imagine them, and in ways that would be particularly depressing. Maybe not for any good reason, but so what? If people are going to be depressed, that in itself is a good reason for not adopting a policy.

So there are indirect, but still creditable, legitimate, weighty reasons for continuing to respect corpses. We don't need any mythology about something special that actually resides in corpses that makes them privileged. That *might* be a useful myth to spread among the unsophisticated, but it would be patronizing in the extreme to think that we among the better informed had to preserve such myths. Similarly, there are perfectly good reasons for treating all living animals with care and solicitude. These reasons are somewhat independent of the facts about just which animals feel which kinds of pain. They depend more directly on the fact that various beliefs are ambient in our culture, and matter to us, whether they *ought* to matter or not. Since they now matter, they matter. But the rationality of the belief environment — the fact that silly or baseless beliefs do tend to be extinguished in the long run, in spite of superstition — does imply that things that matter now may not always matter.

But then, as we anticipated in chapter 2, a theory that radically assaults the general belief environment has a genuine potential for doing harm, for causing suffering (in people who particularly care about animals, for instance, whether or not what happens to the animals amounts to suffering). Does this mean that we should suppress the investigation of these issues, for fear of opening Pandora's box? That might be justified, if we could convince ourselves that our current belief environment, myth-ridden or not, was clearly a morally acceptable, benign environment, but I submit that it is clear that it is not. Those who are worried about the costs threatened by this unasked-for enlightenment should take a hard look at the costs of the current myths. Do we really think what we are currently confronted with is worth protecting with some creative obscurantism? Do we think, for instance, that vast resources should be set aside to preserve the imaginary prospects of a renewed mental life for deeply comatose people, while there are no resources to spare to enhance the desperate, but far from imaginary, expectations of the poor? Myths about the sanctity of life, or of consciousness, cut both ways. They may be useful in erecting barriers (against euthanasia, against capital punishment, against abortion, against eating meat) to impress the unimaginative, but at the price of offensive hypocrisy or ridiculous self-deception among the more enlightened.

Absolutist barriers, like the Maginot Line, seldom do the work they were designed for. The campaign that used to be waged against materialism has already succumbed to embarrassment, and the campaign against "strong AI," while equally well intentioned, can offer only the most threadbare alternative models of the mind. Surely it would be better to try to foster an appreciation for the *non*absolutist, *non*intrinsic, *non*dichotomized grounds for moral concern that can coexist with our increasing knowledge of the inner workings of that most amazing machine, the brain. The moral arguments *on both sides* of the issues of capital punishment, abortion, eating meat, and experimenting on nonhuman animals, for instance, are raised to a higher, more appropriate standard when we explicitly jettison the myths that are beyond protection in any case.

4. CONSCIOUSNESS EXPLAINED, OR EXPLAINED AWAY?

When we learn that the only difference between gold and silver is the number of subatomic particles in their atoms, we may feel cheated or angry — those physicists have explained something away: The goldness is gone from gold; they've left out the very silveriness of silver that we appreciate. And when they explain the way reflection and absorption of electromagnetic radiation accounts for colors and color vision, they seem to neglect the very thing that matters most. But of course there has to be some "leaving out" — otherwise we wouldn't have begun to explain. Leaving something out is not a feature of failed explanations, but of successful explanations.

Only a theory that explained conscious events in terms of unconscious events could explain consciousness at all. If your model of how pain is a product of brain activity still has a box in it labeled "pain," you haven't yet begun to explain what pain is, and if your model of consciousness carries along nicely until the magic moment when you have to say "then a miracle occurs" you haven't begun to explain what consciousness is.

This leads some people to insist that consciousness can never be explained. But why should consciousness be the only thing that can't be explained? Solids and liquids and gases can be explained in terms of things that aren't themselves solids or liquids or gases. Surely life can be explained in terms of things that aren't themselves alive — and the explanation doesn't leave living things lifeless. The illusion that consciousness is the exception comes about, I suspect, because of a failure to understand this general feature of successful explanation. Thinking, mistakenly, that the explanation leaves something out, we think to save what otherwise would be lost by putting it back into the observer as a quale — or some other "intrinsically" wonderful property. The psyche becomes the protective skirt under which all these beloved kittens can hide. There may be *motives* for thinking that consciousness cannot be explained, but, I hope I have shown, there are good *reasons* for thinking that it can.

My explanation of consciousness is far from complete. One might even say that it was just a beginning, but it *is* a beginning, because it breaks the spell of the enchanted circle of ideas that made explaining consciousness seem impossible. I haven't replaced a metaphorical theory, the Cartesian Theater, with a *non*metaphorical ("literal, scientific") theory. All I have done, really, is to replace one family of metaphors and images with another, trading in the Theater, the Witness, the Central Meaner, the Figment, for Software, Virtual Machines, Multiple Drafts, a Pandemonium of Homunculi. It's just a war of metaphors, you say — but metaphors are not "just" metaphors; metaphors are the tools of thought. No one can think about consciousness without them, so it is important to equip yourself with the best set of tools available. Look what we have built with our tools. Could you have imagined it without them?

APPENDIX A (FOR PHILOSOPHERS)

There are places in the book where I leap swiftly and without comment over major philosophical battles, or in other ways egregiously fail to fulfill the standard obligations of an academic philosopher. Philosophers who have read the manuscript of this book have raised questions about these gaps. The questions address issues that may not interest nonphilosophers, but they deserve answers.

You seem to pull a fast one at the end of chapter 11, in the dialogue with Otto, when you briefly introduce "presentiments" as like speech acts with no Actor and no Speech, and then revise your own self-caricature, replacing the presentiments with "events of content-fixation" with no further explanation. Isn't this the crucial move in your whole theory?

Yes indeed. That is the primary point of contact with the other half of my theory of mind, the theory of content or intentionality most recently presented in *The Intentional Stance*. There are many more places in the book where I rely on that theory, but you have located the point that bears the greatest weight, I think. Without that theory of content, this would be a place where my own theory said, "And then a miracle occurs." My fundamental strategy has always been the same: first, to develop an account of content that is *independent of* and *more fundamental than* consciousness — an account of content that treats equally of all unconscious content-fixation (in brains, in computers, in evolution's "recognition" of properties of selected designs) — and second, to build an account of consciousness on that foundation. First content, then consciousness. The two halves of *Brainstorms* recapitulated the strategy, but as the theory-halves grew, they outgrew a single volume. This book completes my third execution of the campaign. This strategy is completely opposite, of course, to the vision of Nagel and Searle, who in their different ways insist on treating consciousness as foundational. The reason why I leapt so swiftly over this utterly central topic in chapter 11 is simply that there was no useful way I could see to telescope the hundreds of pages of analysis and argument I have devoted to the theory of content into something both accurate and accessible. So if you think I have pulled a fast one in these pages, I beg you to consult the slow version in the other pages cited in the bibliography.

There seems, however, to be a tension — if not an outright contradiction — between the two halves of your theory. The intentional stance presupposes (or fosters) the rationality, and hence the unity, of the agent — the intentional system — while the Multiple Drafts model opposes this central unity all the way. Which, according to your view, is the right way to conceive of a mind?

It all depends on how far away you are. The closer you get, the more the disunity, multiplicity, and competitiveness stand out as important. The chief source of the myth of the Cartesian Theater, after all, is the lazy extrapolation of the intentional stance *all the way in*. Treating a complex, moving entity as a single-minded agent is a magnificent way of seeing pattern in all the activity; the tactic comes naturally to us, and is probably even genetically favored as a way of perceiving and thinking. But when we aspire to a science of the mind, we must learn to restrain and redirect those habits of thought, breaking the single-minded agent down into miniagents and microagents (with no single Boss). Then we can see that many of the *apparent* phenomena of conscious experience are misdescribed by the traditional, unitary tactic. The shock-absorbers that deal with the tension are the strained identifications of heterophenomenological items (as conceived under the traditional perspective) with events of content-fixation in the brain (as conceived under the new perspective).

Philosophers have often pointed out the idealizations of the traditional tactic, but have less often come to terms with them. For instance, a large philosophical literature has been devoted to the difficulties of the logic of reflexive states of belief and knowledge, beginning with Hintikka (1962). One of the essential idealizations of Hintikka's formalization, as he made explicit, was that the statements governed by the logic he presented "must be made on *one and the same occasion*. . . . The notion of forgetting is not applicable within the limits of an occasion" (p. 7). The importance of this limitation, he noted, had not always been appreciated — and it has typically been lost in the clouds of subsequent controversy. Hintikka recognized that this quantizing of "occasions" is a necessary simplification required to formalize the everyday concepts of belief and knowledge in the way he did; it fixes the content at an instant, and thereby fixes the identity of the proposition in question. I have claimed here that this artificial individuation into "states" and "times" is one of the features that turns these folk-psychological concepts into fantasies when we try to map them onto the complexities of what happens in the brain.

What, in the end, do you say conscious experiences are? Are you an identity theorist, an eliminative materialist, a functionalist, an instrumentalist?

I do resist the demand for a single, formal, properly quantified proposition expressing the punch line of my theory. Filling in the formula *(x) (x is a conscious experience if and only if . . .)* and defending it against proposed counterexamples is not a good method for developing a theory of consciousness, and I think I have shown why. The indirectness of the heterophenomenological method is precisely a way of evading ill-motivated obligations to "identify" or "reduce" the (putative) entities that inhabit the ontology of subjects. Do the anthropologists *identify* Feenoman with the chap they discover who has been doing all the good deeds in the jungle, or are they "eliminativists" with regard to Feenoman? If they have done *their* job right, the only issue left over is one that can be decided as a matter of diplomatic policy, not scientific or philosophical doctrine. In some regards, you could say that my theory identifies conscious experiences with information-bearing events in the brain — since that's all that's going on, and many of the brain events bear a striking resemblance to denizens of the heterophenomenological worlds of the subjects. But other properties of the heterophenomenological items might be deemed "essential" — such as the position items take in the subjective temporal sequence, in which case they *couldn't* be identified with the available brain-events, which may be in a different sequence, on pain of violating Leibniz's Law.

The question of whether to treat part of the heterophenomenological world of a subject as a useful fiction rather than a somewhat strained truth is not always a question that deserves much attention. Are mental images real? There are real data structures in people's brains that are rather like images — are *they* the mental images you're asking about? If so, then yes; if not, then no. Are qualia functionally definable? No, because there are no such properties as qualia. Or, no, because qualia are dispositional properties of brains that are not strictly definable in *functional* terms. Or, yes, because if you really understood everything about the functioning of the nervous system, you'd understand everything about the properties people are actually talking about when they claim to be talking about their qualia.

Am I, then, a functionalist? Yes and no. I am not a Turing machine functionalist, but then I doubt that anyone ever was, which is a shame, since so many refutations then have to go to waste. I am a sort of "teleofunctionalist," of course, perhaps the original teleofunctionalist (in *Content and Consciousness*), but as I have all along made clear, and emphasize here in the discussion of evolution, and of qualia, I don't make the mistake of trying to define all salient mental differences in terms of biological *functions*. That would be to misread Darwin badly.

Am I an instrumentalist? I think I have shown why that is a poorly conceived question in "Real Patterns" (1991a). Are pains real? They are as real as haircuts and dollars and opportunities and persons, and centers of gravity, but how real is that? These dichotomizing questions all grow out of the demand to fill in the blank in the quantified formula above, and some philosophers think that one develops a theory of mind by concocting a bulletproof proposition of that sort and then defending it. A single proposition isn't a theory, it's a slogan; and what some philosophers do isn't theorizing, it's slogan-honing. What is this labor *for*? What confusion would be dissipated, what advances in outlook would be created, by success in this endeavor? Do you really need something to print on your T-shirt? Some slogan-honers are very, very good at it, but as the psychologist Donald Hebb once memorably said, "If it isn't worth doing, it isn't worth doing well."

I don't mean to imply that careful definition, and the criticism of definitions by means of counterexamples, is never a valuable exercise. Consider, for example, the definition of color. The recent analyses and attempts at definition by philosophers have been eye-opening. They have actually illuminated the concepts and warded off genuine misapprehensions. Given the care that has recently been devoted by philosophers to the attempt to give a precise definition of color, then, my swift assertion, in chapter 12, that colors are "reflective properties of surfaces of objects, or of transparent volumes" is outrageously underdefended. Just *which* reflective properties? I think I have explained why trying to answer that question precisely would be a waste of time; the only precise answer could not be a concise answer, for reasons we can well understand. That means a "noncircular" definition is hard to come by. So what? Do I really think this simple move can stand up to the issues raised by the competition? (In addition to those cited earlier, I would mention Strawson, 1989, and Boghossian and Velleman, 1989, 1991.) Yes, but it's a long story, so I'll just put the ball in their court.

Isn't your position, in the end, just a brand of verificationism?

Philosophers have recently managed to convince themselves — and many an innocent bystander — that verificationism is *always* a sin. Under the influence of Searle and Putnam, for instance, the neuroscientist Gerald Edelman retreats hastily from an act of near-verificationism: "Absence of evidence of self-consciousness in animals other than chimpanzees does not allow us to consider that they are not self-conscious" (1989, p. 280). Fie! Take courage! Surely we can not only *consider* that they are not, but can investigate the consideration, and if we find strong positive reasons for denying it, we should deny it. It is time for the pendulum to swing back. In a commentary on my earlier criticisms of Nagel (Dennett, 1982a), Richard Rorty once said:

> Dennett thinks that one can be skeptical about Nagel's insistence on the phenomenologically rich inner lives of bats "without thereby becoming the Village Verificationist." I do not. I think that skepticism about Nagel- and Searle-like intuitions is plausible only if it is based on general methodological considerations about the status of intuitions. The verificationist's general complaint about the realist is that he is insisting on differences (between, e.g., bats with private lives and bats without, dogs with intrinsic intentionality and dogs without) which make no difference: that his intuitions cannot be integrated into an explanatory scheme because they are "wheels which play no part in the mechanism" [Wittgenstein, 1953, I, para. 271]. This seems to me a good complaint to make, and the only one we need make. [Rorty, 1982a, pp. 342–343; see also Rorty, 1982b]

I agreed, but proposed a slight (.742) modulation of the claim: "with Professor Rorty cheering me on . . . , I am ready to come out of the closet as some sort of verificationist, but not, please, a Village Verificationist; let's all be *Urbane* Verificationists" (Dennett, 1982b, p. 355). This book pursues the course further, arguing that if we are not urbane verificationists, we will end up tolerating all sorts of nonsense: epiphenomenalism, zombies, indistinguishable inverted spectra, conscious teddy bears, self-conscious spiders.

The most salient pressure point for the brand of verificationism I endorse comes in chapter 5, in the argument purporting to show that since there is *and could be* no evidence in support of either Orwellian or Stalinesque models of consciousness, there is no fact of the matter. The standard rebuttal to this verificationist assertion is that I am prejudging the course of science; how do I know that new discoveries in neuroscience won't *reveal* new grounds for making the distinction? The reply — not often heard these days — is straightforward: about some concepts (not all, but some) we can be sure we know enough to know that *whatever* came along in the way of new science, it wouldn't open up this sort of possibility. Consider, for instance, the hypothesis that the universe is right-side-up, and its denial, the hypothesis that the universe is upside-down. Are these hypotheses in good standing? Is there, or might there be, a fact of the matter here? Is it a verificationist sin to opine that no matter what revolutions in cosmology are in the offing, they won't turn *that* "dispute" into an empirical fact of the matter that gets settled?

But you are, really, a sort of behaviorist, aren't you?

This question has been asked before, and I am happy to endorse the answer that Wittgenstein (1953) gave to it.

> 307. "Are you not really a behaviourist in disguise? Aren't you at bottom really saying that everything except human behaviour is a fiction?" — If I do speak of a fiction, then it is of a *grammatical* fiction.
>
> 308. How does the philosophical problem about mental processes and states and about behaviourism arise? — The first step is the one that altogether escapes notice. We talk of processes and states and leave their nature undecided. Sometime perhaps we shall know more about them — we think. But that is just what commits us to a particular way of looking at the matter. For we have a definite concept of what it means to learn to know a process better. (The decisive movement in the conjuring trick has been made, and it was the very one that we thought quite innocent.) — And now the analogy which was to make us understand our thoughts falls to pieces. So we have to deny the yet uncomprehended process in the yet unexplored medium. And now it looks as if we had denied mental processes. And naturally we don't want to deny them.

Several philosophers have seen what I am doing as a kind of redoing of Wittgenstein's attack on the "objects" of conscious experience. Indeed it is. As 308 makes clear, if we are to avoid the conjuring trick, we have to figure out the "nature" of mental states and processes *first*. That is why I took nine long chapters to get to the point where I could begin confronting the problems in their typical philosophical dress — that is to say, in their misdress. My debt to Wittgenstein is large and longstanding. When I was an undergraduate, he was my hero, so I went to Oxford, where he seemed to be everybody's hero. When I saw how most of my fellow graduate students were (by my lights) missing the point, I gave up trying to "be" a Wittgensteinian, and just took what I thought I had learned from the *Investigations* and tried to put it to work.

APPENDIX B (FOR SCIENTISTS)

Philosophers are often correctly accused of indulging in armchair psychology (or neuroscience or physics or . . .), and there are plenty of embarrassing tales about philosophers whose confident *a priori* declarations have been subsequently disproved in the lab. One reasonable response to this established risk is for the philosopher to retreat cautiously into those conceptual arenas where there is little or no danger of ever saying anything that might be disconfirmed (or confirmed) by empirical discovery. Another reasonable response is to study, in one's armchair, the best fruits of the laboratory, the best efforts of the empirically anchored theoreticians, and then to proceed with one's philosophy, trying to illuminate the conceptual obstacles and even going out on a limb occasionally, in the interests of getting clear, one way or the other, about the implications of some particular theoretical idea. When it comes to conceptual issues scientists are no more immune to confusion than lay people. After all, scientists spend quite a bit of time in their armchairs, trying to figure out how to interpret the results of everybody's experiments, and what they do in those moments blends imperceptibly into what philosophers do. Risky business, but invigorating.

Here, then, are a few half-baked ideas for experiments designed to test implications of the model of consciousness I have sketched, selected from a much larger batch of quarter-baked ideas that have either not made it through the gauntlet of my patient informants, or been shown by them to have already been done. (My batting average on the latter group is high enough to encourage me to persist.) Since as a philosopher I've tried to keep my model as general and noncomittal as possible, if I've done my job right, these experiments should help settle only *how strong* a version of my model is confirmed; if the model were entirely disconfirmed, I would be well and truly refuted and embarrassed.

ON TIME AND TIMING

If subjective sequence is a product of interpretation, not directly a function of actual sequence, it should be possible to create strong interpretational effects of various sorts that are independent of actual timing.

1. *Spider walks:* Light touches in sequence, mimicking the cutaneous rabbit, but intended to produce illusory *direction* judgments. A simple background case would be two touches, separated in space and time in the same approximate range as visual phi phenomena, with the task being to judge direction of "walking" (which is logically equivalent to *sequence,* but a more "immediate" judgment, phenomenologically). Prediction: standard phi phenomenon effects depending on ISI, with greater acuity on high-resolution surfaces such as a fingertip or the lips.

But now have the subject hold left and right index fingers side-by-side and have the first touch on one fingertip, the second on the other. There should be much worse resolution of direction, due to the requirement that comparisons have to be bilateral. Then add visual "help"; let the subject watch the finger stimulation, but provide for false visual input: rig the apparatus so that the *visual direction* implied is the opposite of the direction implied by the actual sequence of touches. Prediction: Subjects will make confident false judgments, overruling or discarding the actual sequence information made available by the cutaneous receptors. If the effect is very strong, it may even overrule unilateral or even same-finger judgments that were very accurate in the absence of visual input.

2. *Film reversals:* Subjects are asked to distinguish brief "takes" of cinema or videotape, some of which have been reversed, or in which there are sequencing disruptions or anomalies. Film editors have tricks of the trade, and a wealth of lore, involving the effects of missequencing frames of film. Sometimes they deliberately splice scenes together with frames out of order to create special effects — to heighten anxiety or shock in horror scenes, for instance. Some events are naturally very strongly ordered; we have all been amused to see film of a diver emerging feet-first out of the splash in the pool and hopping, nimble and dry,

onto the diving board. Other events are imperceptibly reversible — a fluttering flag, for instance — while others are intermediate; it might take careful attention to tell whether film of a bouncing ball was running forward or backward. Prediction: People will be no good at all at distinguishing reversals in which there is no interpretational bias — in which brute sequence has to be detected and remembered. For instance, holding continuity of motion and size and shape disparities roughly constant, subjects should be much worse at distinguishing (reidentifying) sequences that have no biased directional interpretation, and telling them from their reversals or other transformations. (Melody discrimination experiments would be an auditory analogue.)

3. *Writing on your foot:* An experiment designed to disrupt judgments based on the interpretation of "arrival times" at "central availability." Suppose you were to take a pencil and print some letters on the side of your bare foot, without being able to see what you were doing. The signals from the cutaneous receptors in your foot would "confirm" that your intentional writing actions were being properly carried out by the pen in your hand. Now add indirect vision, a television monitor showing your hand writing on your own foot, but with the camera placed so that the pencil point on the foot was obscured by the hand holding the pencil. These visual signals would add further confirmation of the execution of your intentional actions. But now insert a short tape delay in the television (one or two frames of 33msecs each) so that the visual confirmation is always retarded by a small but constant amount. I predict that subjects would accommodate readily to this. (I hope so, because the next step is the interesting one.) After they had accommodated, if the delay were suddenly eliminated, they would interpret the result as *the pencil feeling bendy,* because the perception of the trajectory of its point would be delayed, relative to their visual input, as if the point were trailing along in the wake of its expected trajectory.

4. *Adjusting the delay on Grey Walter's carousel:* The follow-up experiment that measures the amount of delay required to eliminate the "precognitive carousel" effect. I predict the amount will be much smaller than the 300–500msec that would be predicted by an extension of Libet's Stalinesque model.

ON PANDEMONIUM MODELS OF WORD CHOICE

How could one show that "words want to get themselves said"? Can serendipity be experimentally controlled? Levelt's experiments to date have yielded surprising negative results (see footnote 2, page 241). The sort of variation on them I would want to see would open up the possibilities for "creative" word use on the part of subjects, while discreetly providing different raw materials in the environment for them to incorporate into their production. For instance, subjects could be prepared for the experiment in two different preamble settings, in which different striking, vivid, slightly novel or out-of-place words were "casually" left about (on wall posters, in the instructions to subjects, etc.); subjects would then be given opportunities to express themselves on topics in which these target expressions would normally have a low probability of use, so that priming by the preamble would show that the target expressions had been "turned on" and were lurking about, looking for opportunities to be used. Finding no effect would support Levelt's model; finding a large effect (especially if "strained" opportunities were seized upon) would support a Pandemonium model.

EXPERIMENTS USING EYETRACKERS

1. *"Blindsight" in normal subjects:* Experiments using eyetrackers with normal subjects have shown that when a parafoveal stimulus is switched during the saccade, subjects do not notice this (they do not report any sense of switching), but there are enhancement effects — latencies for identification of the second stimulus are shortened or not depending on information gleaned from the original parafoveal stimulus. If subjects under these conditions make a forced-choice guess as to whether the stimulus was switched (or whether the initial stimulus was, say, an upper case letter or a lower case letter), will they do better than chance? I predict they will, for an interesting range of choices, but no better than blindsight at its best.

2. *"Wallpaper" experiments:* Using an eyetracker, and varying the gross and fine features of parafoveal regions of repetitive "wallpaper" fields during saccades, plot the competition required to overrule the "more Marilyns" conclusion. (Since Ramachandran and Gregory's new results astonished me, I will stick my neck out and predict that there are *no* detectable *gradual* effects, although at the levels at which subjects notice the changes, they may well report strange illusory motions.)

3. *The colored checkerboard:* An experiment designed to show how *little* is in the "plenum of the visual field." Subjects are given a task of visual identification or interpretation that requires multiple saccades of a moving scene: they watch animated black-and-white figures

shown against the background of a randomly colored checkerboard. The checks are relatively large — for example, the CRT is divided into a 12x18 array of colored squares randomly filled in with different colors. (The colors are randomly chosen so that the pattern has no significance for the visual task superimposed on the background.) There should be luminence differences between the squares, so there is no Liebmann effect, and for each square there should be prepared an *isoluminent alternative color:* a color which, if switched with the color currently filling the square, would not create radically different luminence boundaries at the edges (this is to keep the luminence-edge detectors quiet). Now suppose that during saccades (as detected by eyetracker) colors in the checkerboard are switched; onlookers would notice one or more squares changing color several times a second. Prediction: There will be conditions under which subjects will be completely oblivious to the fact that large portions of "the background" are being abruptly changed in color. Why? Because the parafoveal visual system is primarily an alarm system, composed of sentries designed to call for saccades when change is noticed; such a system would not bother keeping track of insignificant colors between fixations, and hence would have nothing left over with which to compare the new color. (This depends, of course, on how "fast the film is" in the regions responding to parafoveal color; there may be a sluggish refractory period that will undo the effect I predict.)

BIBLIOGRAPHY

Akins, K. A. 1989. *On Piranhas, Narcissism and Mental Representation: An Essay on Intentionality and Naturalism.* Ph.D. dissertation, Department of Philosophy, University of Michigan, Ann Arbor.

———. 1990. "Science and Our Inner Lives: Birds of Prey, Bats, and the Common (Featherless) Biped" in M. Bekoff and D. Jamieson, eds., *Interpretation and Explanation in the Study of Animal Behavior.* Vol. I. Boulder, CO: Westview, pp. 414–427.

Akins, K. A., and Dennett, D. C. 1986. "Who May I Say Is Calling?" *Behavioral and Brain Sciences,* **9,** pp. 517–518.

Allman, J., Meizin, F., and McGuinness, E. L. 1985. "Direction- and Velocity-Specific Responses from beyond the Classical Receptive Field in the Middle Temporal Visual Area," *Perception,* **14,** pp. 105–126.

Allport, A. 1988. "What Concept of Consciousness?" in Marcel and Bisiach, eds., 1988, pp. 159–182.

———. 1989. "Visual Attention" in M. Posner, ed., *Foundations of Cognitive Psychology,* Cambridge: MIT Press, pp. 631–682.

Anderson, J. 1983. *The Architecture of Cognition.* Cambridge, MA: Harvard University Press.

Anscombe, G. E. M. 1957. *Intention.* Oxford: Blackwell.

———. 1965. "The Intentionality of Sensation: A Grammatical Feature" in R. J. Butler, ed., *Analytical Philosophy* (2nd Series). Oxford: Blackwell, p. 160.

Anton, G. 1899. "Ueber die Selbstwahrnehmung der Herderkrankungen des Gehirs durch den Kranken bei Rindenblindheit under Rindentaubheit," *Archiv für Psychiatrie und Nervenkrankheitene,* **32,** pp. 86–127.

Arnauld, A. 1641. "Fourth Set of Objections" in Cottingham, J., Stoofhoff, R., and Murdoch, D., *The Philosophical Writings of Descartes.* Vol. II, 1984, Cambridge: Cambridge University Press.

Baars, B. 1988. *A Cognitive Theory of Consciousness*. Cambridge: Cambridge University Press.

Bach-y-Rita, P. 1972. *Brain Mechanisms in Sensory Substitution*. New York and London: Academic Presss.

Ballard, D., and Feldman, J. 1982. "Connectionist Models and Their Properties," *Cognitive Science*, **6**, pp. 205–254.

Bechtel, W., and Abrahamsen, A. 1991. *Connectionism and the Mind: An Introduction to Parallel Processing in Networks*. Oxford: Blackwell.

Bennett, J. 1965. "Substance, Reality and Primary Qualities," *American Philosophical Quarterly*, **2**, 1–17.

———. 1976. *Linguistic Behavior*. Cambridge: Cambridge University Press.

Bentham, J. 1789. *Introduction to Principles of Morals and Legislation*. London.

Bick, P. A., and Kinsbourne, M. 1987. "Auditory Hallucinations and Subvocal Speech in Schizophrenic Patients," *American Journal of Psychiatry*, **144**, pp. 222–225.

Bieri, P. 1990. Commentary at the conference "The Phenomenal Mind — How Is It Possible and Why Is It Necessary?" Zentrum für Interdisziplinäre Forschung, Bielefeld, Germany, May 14–17.

Birnbaum, L., and Collins, G. 1984. "Opportunistic Planning and Freudian Slips," *Proceedings, Cognitive Science Society*, Boulder, CO, pp. 124–127.

Bisiach, E. 1988. "The (Haunted) Brain and Consciousness" in Marcel and Bisiach, 1988.

Bisiach, E., and Luzzatti, C. 1978. "Unilateral Neglect of Representational Space," *Cortex*, **14**, pp. 129–133.

Bisiach, E., and Vallar, G. 1988. "Hemineglect in Humans" in F. Boller and J. Grafman, eds., *Handbook of Neuropsychology*. Vol. 1. New York: Elsevier.

Bisiach, E., Vallar, G., Perani, D., Papagno, C., and Berti, A. 1986. "Unawareness of Disease Following Lesions of the Right Hemisphere: Anosognosia for Hemiplegia and Anosognosia for Hemianopia," *Neuropsychologia*, **24**, pp. 471–482.

Blakemore, C. 1976. *Mechanics of the Mind*. Cambridge: Cambridge University Press.

Block, Ned. 1978. "Troubles with Funtionalism" in W. Savage, ed., *Perception and Cognition: Issues in the Foundations of Psychology*, Minnesota Studies in the Philosophy of Science, vol. IX, pp. 261–326.

———. 1981. "Psychologism and Behaviorism," *Philosophical Review*, **90**, pp. 5–43.

———. 1990. "Inverted Earth" in J. E. Tomberlin, ed., *Philosophical Perspectives, 4: Action Theory and Philosophy of Mind, 1990*. Atascadero, CA: Ridgeview Publishing, pp. 53–79.

Boghossian, P. A., and Velleman, J. D. 1989. "Colour as a Secondary Quality," *Mind*, **98**, pp. 81–103.

———. 1991. "Physicalist Theories of Color," *Philosophical Review*, **100**, pp. 67–106.

Booth, W. 1988. "Voodoo Science," *Science*, **240**, pp. 274–277.

Borges, J. L. 1962. *Labyrinths: Selected Stories and Other Writings*, ed. Donald A. Yates and James E. Irby. New York: New Directions.

Borgia, G. 1986. "Sexual Selection in Bowerbirds," *Scientific American*, **254**, pp. 92–100.

Braitenberg, V. 1984. *Vehicles: Experiments in Synthetic Psychology*. Cambridge: MIT Press/A Bradford Book.

Breitmeyer, B. G. 1984. *Visual Masking*. Oxford: Oxford University Press.

Broad, C. D. 1925. *Mind and Its Place in Nature*. London: Routledge & Kegan Paul.

Brooks, B. A., Yates, J. T., and Coleman, R. D. 1980. "Perception of Images Moving at Saccadic Velocities During Saccades and During Fixation," *Experimental Brain Research*, **40**, pp. 71–78.

Byrne, R., and Whiten, A. 1988. *Machiavellian Intelligence: Social Expertise and the Evolution of Intellect in Monkeys, Apes, and Humans*. Oxford: Clarendon.

Calvanio, R., Petrone, P. N., and Levine, D. N. 1987. "Left visual spatial neglect is both environment-centered and body-centered," *Neurology*, **37**, pp. 1179–1183.

Calvin, W. 1983. *The Throwing Madonna: Essays on the Brain*. New York: McGraw-Hill.

———. 1986. *The River that Flows Uphill: A Journey from the Big Bang to the Big Brain*. San Francisco: Sierra Club Books.

———. 1987. "The Brain as a Darwin Machine," *Nature*, **330**, pp. 33–34.

———. 1989a. *The Cerebral Symphony: Seashore Reflections on the Structure of Consciousness*. New York: Bantam.

———. 1989b. "A Global Brain Theory," *Science*, **240**, pp. 1802–1803.

Campion, J., Latto, R., and Smith, Y. M. 1983. "Is Blindsight an Effect of Scattered Light, Spared Cortex, and Near-Threshold Vision?" *Behavioral and Brain Sciences*, **6**, pp. 423–486.

Camus, A. 1942. *Le Myth de Sisyphe*. Paris: Gallimard; English translation as *The Myth of Sisyphus*, 1955, New York: Knopf.

Carruthers, P. 1989. "Brute Experience," *Journal of Philosophy*, **86**, pp. 258–269.

Castañeda, C. 1968. *The Teachings of Don Juan: A Yaqui Way of Knowledge*. Berkeley: University of California Press.

Castaneda, H.-N. 1967. "Indicators and Quasi-Indicators," *American Philosophy Quarterly*, **4**, pp. 85–100.

———. 1968. "On the Logic of Attributions of Self-Knowledge to Others," *Journal of Philosophy*, **65**, pp. 439–456.

Changeux, J.-P., and Danchin, A. 1976. "Selective Stabilization of Developing

Synapses as a Mechanism for the Specifications of Neuronal Networks," *Nature*, **264**, pp. 705–712.

Changeux, J.-P., and Dehaene, S. 1989. "Neuronal Models of Cognitive Functions," *Cognition*, **33**, pp. 63–109.

Cheney, D. L., and Seyfarth, R. M. 1990. *How Monkeys See the World*. Chicago: University of Chicago Press.

Cherniak, C. 1986. *Minimal Rationality*. Cambridge, MA: MIT Press/A Bradford Book.

Churchland, P. M. 1985. "Reduction, Qualia and the Direct Inspection of Brain States," *Journal of Philosophy*, **82**, pp. 8–28.

———. 1990. "Knowing Qualia: A Reply to Jackson," pp. 67–76 in Churchland, P. M., *A Neurocomputational Perspective: The Nature of Mind and the Structure of Science*. Cambridge, MA: MIT Press/A Bradford Book.

Churchland, P. S. 1981a. "On the Alleged Backwards Referral of Experiences and Its Relevance to the Mind-Body Problem," *Philosophy of Science*, **48**, pp. 165–181.

———. 1981b. "The Timing of Sensations: Reply to Libet," *Philosophy of Science*, **48**, pp. 492–497.

———. 1986. *Neurophilosophy: Toward a Unified Science of the Mind/Brain*. Cambridge, MA: MIT Press/A Bradford Book.

Clark, R. W. 1975. *The Life of Bertrand Russell*. London: Weidenfeld and Nicolson.

Cohen, L. D., Kipnis, D., Kunkle, E. C., and Kubzansky, P. E. 1955. "Case Report: Observation of a Person with Congenital Insensitivity to Pain," *Journal of Abnormal and Social Psychology*, **51**, pp. 333–338.

Cole, David. 1990. "Functionalism and Inverted Spectra," *Synthese*, **82**, pp. 207–222.

Crane, H., and Piantanida, T. P. 1983. "On Seeing Reddish Green and Yellowish Blue," *Science*, **222**, pp. 1078–1080.

Crick, F. 1984. "Function of the Thalamic Reticular Complex: The Searchlight Hypothesis," *Proceedings of the National Academy of Sciences*, **81**, pp. 4586–4590.

Crick, F., and Koch, C. 1990. "Towards a Neurobiological Theory of Consciousness," *Seminars in the Neurosciences*, **2**, pp. 263–275.

Damasio, A. R., Damasio, H., and Van Hoesen, G. W. 1982. "Prosopagnosia: Anatomic Basis and Behavioral Mechanisms," *Neurology*, **32**, pp. 331–341.

Darwin, C. 1871. *The Descent of Man, and Selection in Relation to Sex*. 2 vols. London: Murray.

Davis, W. 1985. *The Serpent and the Rainbow*. New York: Simon & Schuster.

———. 1988a. *Passage of Darkness: The Ethnobiology of the Haitian Zombie*. Chapel Hill and London: University of North Carolina Press.

———. 1988b. "Zombification," *Science*, **240**, pp. 1715–1716.

Dawkins, M. S. 1980. *Animal Suffering: The Science of Animal Welfare.* London: Chapman & Hall.

———. 1987. "Minding and Mattering," in C. Blakemore and S. Greenfield, eds., *Mindwaves.* Oxford: Blackwell, pp. 150–160.

———. 1990. "From an Animal's Point of View: Motivation, Fitness, and Animal Welfare," *Behavioral and Brain Sciences,* **13,** pp. 1–61.

Dawkins, R. 1976. *The Selfish Gene.* Oxford: Oxford University Press.

———. 1982. *The Extended Phenotype.* San Francisco: Freeman.

———. 1986. *The Blind Watchmaker.* New York: Norton.

de Sousa, R. 1976. "Rational Homunculi" in Amelie O. Rorty, ed., *The Identity of Persons.* Berkeley: University of California Press, pp. 217–238.

Dennett, D. C. 1969. *Content and Consciousness.* London: Routledge & Kegan Paul.

———. 1971. "Intentional Systems," *Journal of Philosophy,* **8,** pp. 87–106.

———. 1974. "Why the Law of Effect Will Not Go Away," *Journal of the Theory of Social Behaviour,* **5,** pp. 169–187 (reprinted in Dennett, 1978a).

———. 1975. "Are Dreams Experiences?" *Philosophical Review,*

———. 1978a. *Brainstorms.* Montgomery, VT: Bradford Books.

———. 1978b. "Skinner Skinned," ch. 4 in Dennett, 1978a, pp. 53–70.

———. 1978c. "Two Approaches to Mental Images," ch. 10 in Dennett, 1978a, pp. 174–189.

———. 1978d. "Where Am I?" ch. 17 in Dennett, 1978a, pp. 310–323.

———. 1979a. "On the Absence of Phenomenology" in D. Gustafson and B. Tapscott, eds., *Body, Mind and Method: Essays in Honor of Virgil Aldrich.* Dordrecht: Reidel, 1979.

———. 1979b. Review of Popper and Eccles, *The Self and Its Brain: An Argument for Interactionism,* in *Journal of Philosophy,* **76,** pp. 91–97.

———. 1981a. "Reflections" on "Software" in Hofstadter and Dennett, 1981.

———. 1981b. "Wondering Where the Yellow Went" (commentary on W. Sellars's Carus Lectures), *Monist,* **64,** pp. 102–108.

———. 1982a. "How to Study Human Consciousness Empirically, or Nothing Comes to Mind," *Synthese,* **59,** pp. 159–180.

———. 1982b. "Why We Think What We Do about Why We Think What We Do: Discussion on Goodman's 'On Thoughts without Words,' " *Cognition,* **12,** pp. 219–227.

———. 1982c. "Comments on Rorty," *Synthese,* **59,** pp. 349–356.

———. 1982d. "Notes on Prosthetic Imagination," *New Boston Review,* June, pp. 3–7.

———. 1983. "Intentional Systems in Cognitive Ethology: The 'Panglossian Paradigm' Defended," *Behavioral and Brain Sciences,* **6,** pp. 343–390.

———. 1984a. *Elbow Room: The Varieties of Free Will Worth Wanting.* Cambridge, MA: MIT Press/A Bradford Book.

———. 1984b. "Carving the Mind at Its Joints," a review of Fodor, *The Modularity of Mind,* in *Contemporary Psychology,* **29,** pp. 285–286.

———. 1985a. "Can Machines Think?" in M. Shafto, ed., *How We Know*. New York: Harper & Row, pp. 121–145.

———. 1985b. "Music of the Hemispheres," a review of M. Gazzaniga, *The Social Brain*, in *New York Times Book Review*, November 17, 1985, p. 53.

———. 1986. "Julian Jaynes' Software Archeology," *Canadian Psychology*, **27**, pp. 149–154.

———. 1987a. *The Intentional Stance*. Cambridge, MA: MIT Press/A Bradford Book.

———. 1987b. "The Logical Geography of Computational Approaches: A View from the East Pole," in M. Harnish and M. Brand, eds., *Problems in the Representation of Knowledge*. Tucson: University of Arizona Press.

———. 1988a. "Quining Qualia," in Marcel and Bisiach, 1988, pp. 42–77.

———. 1988b. "When Philosophers Encounter AI," *Daedalus*, **117**, pp. 283–296; reprinted in Graubard, 1988.

———. 1988c. "Out of the Armchair and Into the Field," *Poetics Today*, **9**, special issue on Interpretation in Context in Science and Culture, pp. 205–222.

———. 1988d. "The Intentional Stance in Theory and Practice," in Whiten and Byrne, 1988, pp. 180–202.

———. 1988e. "Science, Philosophy and Interpretation," *Behavioral and Brain Sciences*, **11**, pp. 535–546.

———. 1988f. "Why Everyone Is a Novelist," *Times Literary Supplement*, September 16–22.

———. 1989a. "Why Creative Intelligence Is Hard to Find," commentary on Whiten and Byrne, *Behavioral and Brain Sciences*, **11**, p. 253.

———. 1989b. "The Origins of Selves," *Cogito*, **2**, pp. 163–173.

———. 1989c. "Murmurs in the Cathedral," review of R. Penrose, *The Emperor's New Mind*, in *Times Literary Supplement*, September 29–October 5, pp. 1066–1068.

———. 1989d. "Cognitive Ethology: Hunting for Bargains or a Wild Goose Chase?" in A. Montefiore and D. Noble, eds., *Goals, Own Goals and No Goals: A Debate on Goal-Directed And Intentional Behaviour*. London: Unwin Hyman.

———. 1990a. "Memes and the Exploitation of Imagination," *Journal of Aesthetics and Art Criticism*, **48**, pp. 127–135.

———. 1990b. "Thinking with a Computer," in H. Barlow, ed., *Image and Understanding*. Cambridge: Cambridge University Press, pp. 297–309.

———. 1990c. "Betting Your Life on an Algorithm," commentary on Penrose, *Behavioral and Brain Science*, **13**, p. 660.

———. 1990d. "The Interpretation of Texts, People, and Other Artifacts," *Philosophy and Phenomenological Research*, **50**, pp. 177–194.

———. 1990e. "Two Black Boxes: A Fable," Tufts University Center for Cognitive Studies Preprint, November.

———. 1991a. "Real Patterns," *Journal of Philosophy*, **89**, pp. 27–51.

———. 1991b. "Producing Future by Telling Stories" in K. M. Ford and Z. Pylyshyn, eds., *Robot's Dilemma Revisited: The Frame Problem in Artificial Intelligence*. Ablex Series in Theoretical Issues in Cognitive Science. Norwood, NJ: Ablex.

———. 1991c. "Mother Nature versus the Walking Encyclopedia" in W. Ramsey, S. Stich, and D. Rumelhart, eds., *Philosophy and Connectionist Theory*. Hillsdale, NJ: Erlbaum.

———. 1991d. "Two Contrasts: Folk Craft versus Folk Science and Belief versus Opinion" in J. Greenwood, ed., *The Future of Folk Psychology: Intentionality and Cognitive Science*. Cambridge: Cambridge University Press, 1991.

———. 1991e. "Granny's Campaign for Safe Science" in G. Rey and B. Loewer, eds., *Fodor and His Critics*. Oxford: Blackwell.

Dennett, D., and Kinsbourne, M. In press. "Time and the Observer: The Where and When of Consciousness in the Brain," *Behavioral and Brain Sciences*.

Descartes, R. 1637. *Discourse on Method*. Paris.

———. 1641. *Meditations on First Philosophy*. Paris: Michel Soly.

———. 1664. *Treatise on Man*. Paris.

———. 1970. A. Kenny, ed., *Philosophical Letters*. Oxford: Clarendon Press.

Dreyfus, H. 1979. *What Computers Can't Do* (2nd Edition). New York: Harper & Row.

Dreyfus, H. L., and Dreyfus, S. E. 1988. "Making a Mind Versus Modeling the Brain: Artificial Intelligence Back at a Branchpoint," in Graubard, 1988.

Eccles, J. C. 1985. "Mental Summation: The Timing of Voluntary Intentions by Cortical Activity," *Behavioral and Brain Sciences*, **8**, pp. 542–547.

Eco, U. 1990. "After Secret Knowledge," *Times Literary Supplement*, June 22–28, p. 666, "Some Paranoid Readings," *Times Literary Supplement*, June 29–July 5, p. 694.

Edelman, G. 1987. *Neural Darwinism*. New York: Basic Books.

———. 1989. *The Remembered Present: A Biological Theory of Consciousness*. New York: Basic Books.

Efron, R. 1967. "The Duration of the Present," *Proceedings of the New York Academy of Science*, **8**, pp. 542–543.

Eldredge, N., and Gould, S. J. 1972. "Punctuated Equilibria: An Alternative to Phyletic Gradualism," in T. J. M. Schopf, ed., *Models in Paleobiology*. San Francisco: Freeman Cooper, pp. 82–115.

Ericsson, K. A., and Simon, H. A. 1984. *Protocol Analysis: Verbal Reports as Data*. Cambridge, MA: MIT Press/A Bradford Book.

Evans, G. 1982. John McDowell, ed., *The Varieties of Reference*. Oxford: Oxford University Press.

Ewert, J.-P. 1987. "The Neuroethology of Releasing Mechanisms: Prey-catching in Toads," *Behavioral and Brain Sciences*, **10**, pp. 337–405.

Farah, M. J. 1988. "Is Visual Imagery Really Visual? Overlooked Evidence from Neuropsychology," *Psychological Review*, **95**, pp. 307–317.

Farrell, B. A. 1950. "Experience," *Mind*, **59**, pp. 170–198.

Fehling, M., Baars, B., and Fisher, C. 1990. "A Functional Role of Repression in an Autonomous, Resource-constrained Agent" in *Proceedings of Twelfth Annual Conference of the Cognitive Science Society*. Hillsdale, NJ: Erlbaum.

Fehrer, E., and Raab, D. 1962. "Reaction Time to Stimuli Masked by Metacontrast," *Journal of Experimental Psychology*, **63**, pp. 143–147.

Feynmann, R. 1985. *Surely You're Joking, Mr. Feynmann!* New York: Norton.

Finke, R. A., Pinker, S., and Farah, M. J. 1989. "Reinterpreting Visual Patterns in Mental Imagery," *Cognitive Science*, **13**, pp. 51–78.

Flanagan, O. 1991. *The Science of the Mind* (2nd Edition). Cambridge, MA: MIT Press/A Bradford Book.

Flohr, H. 1990. "Brain Processes and Phenomenal Consciousness: A New and Specific Hypothesis," presented at the conference "The Phenomenal Mind — How Is It Possible and Why Is It Necessary?" Zentrum für Interdisziplinäre Forschung, Bielefeld, Germany, May 14–17.

Fodor, J. 1975. *The Language of Thought*. Scranton, PA: Crowell.

———. 1983. *The Modularity of Mind*. Cambridge, MA: MIT Press/A Bradford Book.

———. 1990. *A Theory of Content, and Other Essays*. Cambridge, MA: MIT Press/A Bradford Book.

Fodor, J., and Pylyshyn, Z. 1988. "Connectionism and Cognitive Architecture: A Critical Analysis," *Cognition*, **28**, pp. 3–71.

Fox, I. 1989. "On the Nature and Cognitive Function of Phenomenal Content — Part One," *Philosophical Topics*, **17**, pp. 81–117.

French, R. 1991. "Subcognition and the Turing Test," *Mind*, in press.

Freud, S. 1962. *The Ego and the Id*. New York: Norton.

Freyd, J. 1989. "Dynamic Mental Representations," *Psychological Review*, **94**, pp. 427–438.

Fuster, J. M. 1981. "Prefrontal Cortex in Motor Control," in *Handbook of Physiology, Section 1: The Nervous System, Vol. II: Motor Control*. American Physiological Society, pp. 1149–1178.

Gardner, H. 1975. *The Shattered Mind*. New York: Knopf.

Gardner, M. 1981. "The Laffer Curve and Other Laughs in Current Economics," *Scientific American*, **245**, December, pp. 18–31. Reprinted in Gardner, 1986.

———. 1986. *Knotted Doughnuts and Other Mathematical Diversions*. San Francisco: W. H. Freeman.

Gazzaniga, M. 1978. "Is Seeing Believing: Notes on Clinical Recovery," in S. Finger, ed., *Recovery From Brain Damage: Research and Theory*. New York: Plenum Press, pp. 409–414.

———. 1985. *The Social Brain: Discovering the Networks of the Mind*. New York: Basic Books.

Gazzaniga, M., and Ledoux, J. 1978. *The Integrated Mind.* New York: Plenum Press.

Geldard, F. A. 1977. "Cutaneous Stimulis, Vitratory and Saltatory," *Journal of Investigative Dermatology,* **69,** pp. 83–87.

Geldard, F. A., and Sherrick, C. E. 1972. "The Cutaneous 'Rabbit': A Perceptual Illusion," *Science,* **178,** pp. 178–179.

———. 1983. "The Cutaneous Saltatory Area and Its Presumed Neural Base," *Perception and Psychophysics,* **33,** pp. 299–304.

———. 1986. "Space, Time and Touch," *Scientific American,* **254,** pp. 90–95.

Gert, B. 1965. "Imagination and Verifiability," *Philosophical Studies,* **16,** pp. 44–47.

Geshwind, N., and Fusillo, M. 1966. "Color-naming Defects in Association with Alexia," *Archives of Neurology,* **15,** pp. 137–146.

Gide, A. 1948. *Les Faux Monnayeurs.* Paris: Gallimard.

Goodman, N. 1978. *Ways of Worldmaking.* Hassocks, Sussex: Harvester.

Goody, J. 1977. *The Domestication of the Savage Mind.* Cambridge: Cambridge University Press.

Gould, S. 1980. *The Panda's Thumb.* New York: Norton.

Gouras, P. 1984. "Color Vision," in N. Osborn and J. Chader, eds., *Progress in Retinal Research.* Vol. 3. London: Pergamon Press, pp. 227–261.

Graubard, S. R. 1988. *The Artificial Intelligence Debate: False Starts, Real Foundations* (a reprint of *Daedalus,* **117,** Winter 1988). Cambridge, MA: MIT Press.

Grey Walter, W. 1963. Presentation to the Osler Society, Oxford University.

Grice, H. P. 1957. "Meaning," *Philosophical Review,* **66,** pp. 377–388.

———. 1969. "Utterer's Meaning and Intentions," *Philosophical Review,* **78,** pp. 147–177.

Hacking, Ian. 1990. "Signing," review of Sacks, 1989, *London Review of Books,* April 5, 1990, pp. 3–6.

Hampl, P. 1989. "The Lax Habits of the Free Imagination," *New York Times Book Review,* March 5, 1989, pp. 1, 37–39, excerpted from Hampl, ed., 1989, *The Houghton Mifflin Anthology of Short Fiction.* Boston: Houghton Mifflin.

Handford, M. 1987. *Where's Waldo?* Little, Brown: Boston.

Hardin, C. L. 1988. *Color for Philosophers: Unweaving the Rainbow.* Indianapolis: Hackett.

———. 1990. "Color and Illusion," presented at the conference "The Phenomenal Mind — How Is It Possible and Why Is It Necessary?" Zentrum für Interdisziplinäre Forschung, Bielefeld, Germany, May 14–17.

Harman, G. 1990. "The Intrinsic Quality of Experience," in J. E. Tomberlin, ed., *Philosophical Perspectives, 4: Action Theory and Philosophy of Mind.* Atascadero, CA: Ridgeview, pp. 31–52.

Harnad, S. 1982. "Consciousness: An Afterthought," *Cognition and Brain Theory,* **5,** pp. 29–47.

———. 1989. "Editorial Commentary," *Behavioral and Brain Sciences*, **12**, p. 183.

Haugeland, J. 1981. *Mind Design: Philosophy, Psychology, Artificial Intelligence*. Montgomery, VT: Bradford Books.

———. 1985. *Artificial Intelligence: The Very Idea*. Cambridge, MA: MIT Press/A Bradford Book.

Hawking, S. 1988. *A Brief History of Time*. New York: Bantam.

Hayes, P. 1979. "The Naive Physics Manifesto," in D. Michie, ed., *Expert Systems in the Microelectronic Age*. Edinburgh: Edinburgh University Press.

Hayes-Roth, B. 1985. "A Blackboard Architecture for Control," *Artificial Intelligence*, **26**, pp. 251–321.

Hebb, D. 1949. *The Organization of Behavior: A Neuropsychological Theory*. New York: Wiley.

Hilbert, D. R. 1987. *Color and Color Perception: A Study in Anthropocentric Realism*. Stanford University; Center for the Study of Language and Information.

Hintikka, J. 1962. *Knowledge and Belief*. Ithaca: Cornell University Press.

Hinton, G. E., and Nowland, S. J. 1987. "How Learning Can Guide Evolution," *Complex Systems, I*, Technical Report CMU-CS-86-128, Carnegie Mellon University, pp. 495–502.

Hobbes, T. 1651. *Leviathan*. Paris.

Hoffman, R. E. 1986. "What Can Schizophrenic 'Voices' Tell Us?" *Behavioral and Brain Sciences*, pp. 535–548.

Hoffman, R. E., and Kravitz, R. E. 1987. "Feedforward Action Regulation and the Experience of Will," *Behavioral and Brain Sciences*, **10**, pp. 782–783.

Hofstadter, D. R. 1981a. "The Turing Test: A Coffeehouse Conversation," in "Metamagical Themas," *Scientific American*, May 1981, reprinted in Hofstadter and Dennett, 1981, pp. 69–92.

———. 1981b. "Reflections [on Nagel]," in Hofstadter and Dennett, 1981, pp. 403–414.

———. 1983. "The Architecture of Jumbo," *Proceedings of the Second Machine Learning Workshop*, Monticello, IL.

———. 1985. "On the Seeming Paradox of Mechanizing Creativity," in *Metamagical Themas*. New York: Basic Books, pp. 526–546.

Hofstadter, D. R., and Dennett, D. C. 1981. *The Mind's I: Fantasies and Reflections on Self and Soul*. New York: Basic Books, pp. 191–201.

Holland, J. H. 1975. *Adaptation in Natural and Artificial Systems*. Ann Arbor: University of Michigan Press.

Holland, J. H., Holyoak, K. J., Nisbett, R. E., and Thagard, P. R. 1986. *Induction: Processes of Inference, Learning, and Discovery*. Cambridge, MA: MIT Press/A Bradford Book.

Honderich, T. 1984. "The Time of a Conscious Sensory Experience and Mind-Brain Theories," *Journal of Theoretical Biology*, **110**, pp. 115–129.

Howell, R. 1979. "Fictional Objects: How They Are and How They Aren't," in

D. F. Gustafson and B. L. Tapscott, eds., *Body, Mind and Method*. Dordrecht: D. Reidel, pp. 241–294.

Hughlings Jackson, J. 1915. "Hughlings Jackson on Aphasia and Kindred Affections of Speech," *Brain*, **38**, pp. 1–190.

Hume, D. 1739. *Treatise on Human Nature*. London: John Noon.

Humphrey, N. 1972. " 'Interest' and 'Pleasure': Two Determinants of a Monkey's Visual Preferences," *Perception*, **1**, pp. 395–416.

———. 1976. "The Colour Currency of Nature," in *Colour for Architecture*, T. Porter and B. Mikellides, eds., London: Studio-Vista, pp. 147–161, reprinted in Humphrey, 1983a.

———. 1983a. *Consciousness Regained*. Oxford: Oxford University Press.

———. 1983b. "The Adaptiveness of Mentalism?" commentary on Dennett, 1983, *Behavioral and Brain Sciences*, **6**, pp. 366.

———. 1986. *The Inner Eye*. London: Faber & Faber.

———. Forthcoming. *A History of the Mind*. New York: Simon & Schuster.

Humphrey, N., and Dennett, D. C. 1989. "Speaking for Our Selves: An Assessment of Multiple Personality Disorder," *Raritan*, **9**, pp. 68–98.

Humphrey, N., and Keeble, G. 1978. "Effects of Red Light and Loud Noise on the Rates at Which Monkeys Sample the Sensory Environment," *Perception*, **7**, p. 343.

Hundert, E. 1987. "Can Neuroscience Contribute to Philosophy?" in C. Blakemore and S. Greenfield, *Mindwaves*. Oxford: Blackwell, pp. 407–429 (reprinted as chapter 7 of Hundert, *Philosophy, Psychiatry, and Neuroscience: Three Approaches to the Mind*, Oxford: Clarendon, 1989).

Huxley, T. 1874. "On the Hypothesis that Animals Are Automata," in *Collected Essays*. London, 1893–1894.

Jackendoff, R. 1987. *Consciousness and the Computational Mind*. Cambridge, MA: MIT Press/A Bradford Book.

Jackson, F. 1982. "Epiphenomenal Qualia," *Philosophical Quarterly*, **32**, pp. 127–136.

Jacob, F. 1982. *The Possible and the Actual*. Seattle: University of Washington Press.

Janlert, L.-E. 1985. *Studies in Knowledge Representation*. Umea, Sweden: Institute of Information Processing.

Jarrell, R. 1963. *The Bat-Poet*. New York: Macmillan.

Jaynes, J. 1976. *The Origins of Consciousness in the Breakdown of the Bicameral Mind*. Boston: Houghton Mifflin.

Jerison, H. 1973. *Evolution of the Brain and Intelligence*. New York: Academic Press.

Johnson-Laird, P. 1983. *Mental Models: Towards a Cognitive Science of Language, Inference, and Consciousness*. Cambridge: Cambridge University Press.

———. 1988. "A Computational Analysis of Consciousness" in A. J. Marcel and E. Bisiach, eds., *Consciousness in Contemporary Science*. Oxford: Clarendon Press; New York: Oxford University Press.

Julesz, B. 1971. *Foundations of Cyclopean Perception*. Chicago: University of Chicago Press.

Keller, H. 1908. *The World I Live In*. New York: Century Co.

Kinsbourne, M. 1974. "Lateral Interactions in the Brain," in M. Kinsbourne and W. L. Smith, eds., *Hemisphere Disconnection and Cerebral Function*. Springfield, IL: Charles C. Thomas, pp. 239–259.

———. 1980. "Brain-based Limitations on Mind," in R. W. Rieber, ed., *Body and Mind: Past, Present and Future*. New York: Academic Press, pp. 155–175.

Kinsbourne, M., and Hicks, R. E. 1978. "Functional Cerebral Space: A Model for Overflow, Transfer and Interference Effects in Human Performance: A Tutorial Review," in J. Requin, ed., *Attention and Performance*, **7**, Hillsdale, NJ: Erlbaum, pp. 345–362.

Kinsbourne, M., and Warrington, E. K. 1963. "Jargon Aphasia," *Neuropsychologia*, **1**, pp. 27–37.

Kirman, B. H., et al. 1968. "Congenital Insensitivity to Pain in an Imbecile Boy," *Developmental Medicine and Child Neurology*, **10**, pp. 57–63.

Kitcher, Patricia. 1979. "Phenomenal Qualities," *American Philosophical Quarterly*, **16**, pp. 123–129.

Koestler, Arthur. 1967. *The Ghost in the Machine*. New York: Macmillan.

Kohler, I. 1961. "Experiments with Goggles," *Scientific American*, **206**, pp. 62–86.

Kolers, P. A. 1972. *Aspects of Motion Perception*. London: Pergamon Press.

Kolers, P. A., and von Grünau, M. 1976. "Shape and Color in Apparent Motion," *Vision Research*, **16**, pp. 329–335.

Kosslyn, S. M. 1980. *Image and Mind*. Cambridge, MA: Harvard University Press.

Kosslyn, S. M., Holtzman, J. D., Gazzaniga, M. S., and Farah, M. J. 1985. "A Computational Analysis of Mental Imagery Generation: Evidence for Functional Dissociation in Split Brain Patients," *Journal of Experimental Psychology: General*, **114**, pp. 311–341.

Lackner, J. R. 1988. "Some Proprioceptive Influences on the Perceptual Representation of Body Shape and Orientation," *Brain*, **111**, pp. 281–297.

Langton, C. G. 1989. *Artificial Life*. Redwood City, CA: Addison-Wesley.

Larkin, S., and Simon, H. A. 1987. "Why a Diagram Is (Sometimes) Worth Ten Thousand Words," *Cognitive Science*, **11**, pp. 65–100.

Leiber, J. 1988. " 'Cartesian' Linguistics?" *Philosophia*, **118**, pp. 309–346.

———. 1991. *Invitation to Cognitive Science*. Oxford: Blackwell.

Leibniz, G. W. 1840. *Monadology*, first published posthumously in J. E. Erdmann, ed., Leibniz, *Opera Philosophica*. 2 vols. Berlin.

Levelt, W. 1989. *Speaking*. Cambridge, MA: MIT Press/A Bradford Book.

Levy, J., and Trevarthen, C. 1976. "Metacontrol of Hemispheric Function in Human Split-Brain Patients," *Journal of Experimental Psychology: Human Perception and Performance*, **3**, pp. 299–311.

Lewis, D. 1978. "Truth in Fiction," *American Philosophical Quarterly*, **15**, pp. 37–46.

———. 1979. "Attitudes *De Dicto* and *De Se*," *Philosophical Review*, **78**, pp. 513–543.

———. 1988. "What Experience Teaches," proceedings of the Russellian Society of the University of Sidney, reprinted in W. Lycan, ed., *Mind and Cognition: A Reader*. Oxford: Blackwell, 1990.

Liberman, A., and Studdert-Kennedy, M. 1977. "Phonetic Perception," in R. Held, H. Leibowitz, and H.-L. Teuber, eds., *Handbook of Sensory Physiology, Vol. 8, Perception*. Heidelberg: Springer-Verlag.

Libet, B. 1965. "Cortical Activation in Conscious and Unconscious Experience," *Perspectives in Biology and Medicine*, **9**, pp. 77–86.

———. 1981. "The Experimental Evidence for Subjective Referral of a Sensory Experience backwards in Time: Reply to P. S. Churchland," *Philosophy of Science*, **48**, pp. 182–197.

———. 1982. "Brain Stimulation in the Study of Neuronal Functions for Conscious Sensory Experiences," *Human Neurobiology*, **1**, pp. 235–242.

———. 1985a. "Unconscious Cerebral Initiative and the Role of Conscious Will in Voluntary Action," *Behavioral and Brain Sciences*, **8**, pp. 529–566.

———. 1985b. "Subjective Antedating of a Sensory Experience and Mind-Brain Theories," *Journal of Theoretical Biology*, **114**, pp. 563–570.

———. 1987. "Are the Mental Experiences of Will and Self-control Significant for the Performance of a Voluntary Act?" *Behavioral and Brain Sciences*, **10**, pp. 783–786.

———. 1989. "The Timing of a Subjective Experience," *Behavioral and Brain Sciences*, **12**, pp. 183–185.

Libet, B., Wright, E. W., Feinstein, B., and Pearl, D. K. 1979. "Subjective Referral of the Timing for a Conscious Sensory Experience," *Brain*, **102**, pp. 193–224.

Liebmann, S. 1927. "Ueber das Verhalten fahrbiger Formen bei Heligkeitsgleichtheit von Figur und Grund," *Psychologie Forschung*, **9**, pp. 200–253.

Livingstone, M. S., and Hubel, D. H. 1987. "Psychophysical Evidence for Separate Channels for the Perception of Form, Color, Movement, and Depth," *Journal of Neuroscience*, **7**, pp. 346–368.

Lloyd, M., and Dybas, H. S. 1966. "The Periodical Cicada Problem," *Evolution*, **20**, pp. 132–149.

Loar, B. 1990. "Phenomenal Properties" in J. E. Tomberlin, ed., *Philosophical Perspectives, 4: Action Theory and Philosophy of Mind*. Atascadero, CA: Ridgeview, pp. 81–108.

Locke, J. 1690. *Essay Concerning Human Understanding*. London: Basset.

Lockwood, M. 1989. *Mind, Brain and the Quantum*. Oxford: Blackwell.

Lodge, D. 1988. *Nice Work*. London: Secker and Warburg, 1988.

Lycan, W. 1973. "Inverted Spectrum," *Ratio*, **15**, pp. 315–319.

———. 1990. "What Is the Subjectivity of the Mental?" in J. E. Tomberlin, ed., *Philosophical Perspectives, 4: Action Theory and Philosophy of Mind*. Atascadero, CA: Ridgeview, pp. 109–130.

Marais, E. N. 1937. *The Soul of the White Ant*. London: Methuen.

Marcel, A. J. 1988. "Phenomenal Experience and Functionalism," in Marcel and Bisiach, 1988, pp. 121–158.

Marcel, A. In Press. "Slippage in the Unity of Consciousness," in R. Bornstein and T. Pittman, eds., *Perception Without Awareness: Cognitive, Clinical and Social Perspectives*. New York: Guilford Press.

Marcel, A., and Bisiach, E., eds. 1988. *Consciousness in Contemporary Science*. New York: Oxford University Press.

Margolis, H. 1987. *Patterns, Thinking, and Cognition*. Chicago: University of Chicago Press.

Margulis, L. 1970. *The Origin of Eukaryotic Cells*. New Haven: Yale University Press.

Marks, C. 1980. *Commissurotomy, Consciousness and Unity of Mind*. Cambridge, MA: MIT Press/A Bradford Book.

Marler, P., and Sherman, V. 1983. "Song Structure Without Auditory Feedback: Emendations of the Auditory Template Hypothesis," *Journal of Neuroscience*, **3**, pp. 517–531.

Marr, D. 1982. *Vision*. San Francisco: Freeman.

Maynard Smith, J. 1978. *The Evolution of Sex*. Cambridge: Cambridge University Press.

———. 1989. *Sex, Games, and Evolution*. Brighton, Sussex: Harvester.

McClelland, J., and Rumelhart, D., eds. 1986. *Parallel Distributed Processing: Explorations in the Microstructures of Cognition*. 2 vols. Cambridge, MA: MIT Press/A Bradford Book.

McCulloch, W. S., and Pitts, W. 1943. "A Logical Calculus for the Ideas Immanent in Nervous Activity," *Bulletin of Mathematical Biophysics*, **5**, pp. 115–133.

McGinn, C. 1989. "Can We Solve the Mind-Body Problem?" *Mind*, **98**, pp. 349–366.

———. 1990. *The Problem of Consciousness*. Oxford: Blackwell.

McGlynn, S. M., and Schacter, D. L. 1989. "Unawareness of Deficits in Neuropsychological Syndromes," *Journal of Clinical and Experimental Neuropsychology*, **11**, pp. 143–205.

McGurk, H., and Macdonald, R. 1979. "Hearing Lips and Seeing Voices," *Nature*, **264**, pp. 746–748.

McLuhan, M. 1967. *The Medium Is the Message*. New York: Bantam.

Mellor, H. 1981. *Real Time*. Cambridge: Cambridge University Press.

Menzel, E. W., Savage-Rumbaugh, E. S., and Lawson, J. 1985. "Chimpanzee (*Pan troglodytes*) Spatial Problem Solving with the Use of Mirrors and Televised Equivalents of Mirrors," *Journal of Comparative Psychology*, **99**, pp. 211–217.

Millikan, R. 1990. "Truth Rules, Hoverflies, and the Kripke-Wittgenstein Paradox," *Philosophical Review*, **99**, pp. 323–354.

Minsky, M. 1975. "A Framework for Representing Knowledge," Memo 3306, AI Lab, MIT, Cambridge, MA (excerpts published in Haugeland, 1981, pp. 95–128).

———. 1985. *The Society of Mind*. New York: Simon & Schuster.

Mishkin, M., Ungerleider, L. G., and Macko, K. A. 1983. "Object Vision and Spatial Vision: Two Cortical Pathways," *Trends in Neuroscience*, **64**, pp. 370–375.

Monod, J. 1972. *Chance and Necessity*. New York: Knopf.

Morris, R. K., Rayner, K., and Pollatsek, A. 1990. "Eye Movement Guidance in Reading: The Role of Parafoveal and Space Information," *Journal of Experimental Psychology: Human Perception and Performance*, **16**, pp. 268–281.

Mountcastle, V. B. 1978. "An Organizing Principle for Cerebral Function: The Unit Module and the Distributed System," in G. Edelman and V. B. Mountcastle, eds., *The Mindful Brain*. Cambridge, MA: MIT Press, pp. 7–50.

Nabokov, V. 1930. *Zaschita Luzhina*, in *Sovremennye Zapiski*, Paris, 1930, brought out in book form by Slovo, Berlin, 1930. English edition, *The Defense*, Popular Library, by arrangement with G. P. Putnam, 1964. (The English translation originally appeared in *The New Yorker*.)

Nagel, T. 1971. "Brain Bisection and the Unity of Consciousness," *Synthese*, **22**, pp. 396–413 (reprinted in his *Mortal Questions* [1979], Cambridge: Cambridge University Press.)

Nagel, T. 1974. "What Is It Like to Be a Bat?" *Philosophical Review*, **83**, pp. 435–450.

———. 1986. *The View from Nowhere*. Oxford: Oxford University Press.

Neisser, U. 1967. *Cognitive Psychology*. New York: Appleton-Century-Crofts.

———. 1981. "John Dean's Memory: A Case Study," *Cognition*, **9**, pp. 1–22.

———. 1988. "Five Kinds of Self-Knowledge," *Philosophical Psychology*, **1**, pp. 35–39.

Nemirow, L. 1990. "Physicalism and the Cognitive Role of Acquaintance," in W. Lycan, ed., *Mind and Cognition: A Reader*. Oxford: Blackwell, pp. 490–499.

Neumann, O. 1990. "Some Aspects of Phenomenal Consciousness and Their Possible Functional Correlates," presented at the conference "The Phenomenal Mind — How Is It Possible and Why Is It Necessary?" Zentrum für Interdisziplinäre Forschung, Bielefeld, Germany, May 14–17.

Newell, A. 1973. "Production Systems: Models of Control Structures," in W. G. Chase, ed., *Visual Information Processing*. New York: Academic Press, pp. 463–526.

———. 1982. "The Knowledge Level," *Artificial Intelligence*, **18**, pp. 81–132.

———. 1988. "The Intentional Stance and the Knowledge Level," *Behavioral and Brain Sciences,* **11,** pp. 520–522.

———. 1990. *Unified Theories of Cognition.* Cambridge, MA: Harvard University Press.

Newell, A., Rosenbloom, P. S., and Laird, J. E. 1989. "Symbolic Architectures for Cognition," in M. Posner, ed., *Foundations of Cognitive Science.* Cambridge, MA: MIT Press, pp. 93–132.

Nielsen, T. I. 1963. "Volition: A New Experimental Approach," *Scandinavian Journal of Psychology,* **4,** pp. 225–230.

Nilsson, N. 1984. *Shakey the Computer.* SRI Tech Report, SRI International, Menlo Park, CA.

Norman, D. A., and Shallice, T. 1980. *Attention to Action: Willed and Automatic Control of Behavior.* Center for Human Information Processing (Technical Report No. 99). Reprinted with revisions in R. J. Davidson, G. E. Schwartz, and D. Shapiro, eds., 1986, *Consciousness and Self-Regulation.* New York: Plenum Press.

———. 1985. "Attention to Action," in T. Shallice, ed., *Consciousness and Self-Regulation.* New York: Plenum Press.

Nottebohm, F. 1984. "Birdsong as a Model in Which to Study Brain Processes Related to Learning," *Condor,* **86,** pp. 227–236.

Oakley, D. A., ed. 1985. *Brain and Mind.* London and New York: Methuen.

Ornstein, R., and Thompson, R. F. 1984. *The Amazing Brain.* Boston: Houghton Mifflin.

Pagels, H. 1988. *The Dreams of Reason: The Computer and the Rise of the Sciences of Complexity.* New York: Simon & Schuster.

Papert, S. 1988. "One AI or Many?" *Daedalus,* Winter, pp. 1–14.

Parfit, D. 1984. *Reasons and Persons.* Oxford: Clarendon Press.

Pears, D. 1984. *Motivated Irrationality.* Oxford: Clarendon Press.

Penfield, W. 1958. *The Excitable Cortex in Conscious Man.* Liverpool: Liverpool University Press.

Penrose, R. 1989. *The Emperor's New Mind.* Oxford: Oxford University Press.

Perlis, 1991. "Intentionality and Defaults" in K. M. Ford and P. J. Hayes, eds., *Reasoning Agents in a Dynamic World.* Greenwich, CT: JAI Press.

Perry, J. 1979. "The Problem of the Essential Indexical," *Nous,* **13,** pp. 3–21.

Pinker, S., and Bloom, P. 1990. "Natural Language and Natural Selection," *Behavioral and Brain Sciences,* **13,** pp. 707–784.

Pollatsek, A., Rayner, K., and Collins, W. E. 1984. "Integrating Pictorial Information Across Eye Movements," *Journal of Experimental Psychology: General,* **113,** pp. 426–442.

Pöppel, E. 1985. *Grenzen des Bewusstseins.* Stuttgart: Deutsche Verlags-Anstal.

———. 1988 (translation of Pöppel, 1985). *Mindworks: Time and Conscious Experience.* New York: Harcourt Brace Jovanovich.

Popper, K. R., and Eccles, J. C. 1977. *The Self and Its Brain.* Berlin: Springer-Verlag.

Powers, L. 1978. "Knowledge by Deduction," *Philosophical Review*, **87**, pp. 337–371.

Putnam, H. 1965. "Brains and Behavior" in R. J. Butler, ed., *Analytical Philosophy*. Second Series. Oxford: Blackwell, pp. 1–19.

———. 1988. "Much Ado About Not Very Much," *Daedalus*, **117**, Winter, reprinted in Graubard, 1988.

Pylyshyn, Z. 1979. "Do Mental Events Have Durations?" *Behavioral and Brain Sciences*, **2**, pp. 277–278.

Quine, W. V. O. 1969. "Natural Kinds" in *Ontological Relativity and Other Essays*. New York: Columbia University Press, pp. 114–138.

Ramachandran, V. S. 1985. Guest Editorial in *Perception*, **14**, pp. 97–103.

———. 1991. "2-D or not 2-D: That Is the Question," in R. L. Gregory, J. Harris, P. Heard, D. Rose, and C. Cronly-Dillon, eds., *The Artful Brain*. Oxford: Oxford University Press.

Ramachandran, V. S., and Gregory, R. L. Submitted to *Nature*. "Perceptual Filling in of Artificially Induced Scotomas in Human Vision."

Ramsey, W., Stich, S., and Rumelhart, D., eds. 1991. *Philosophy and Connectionist Theory*. Hillsdale, NJ: Erlbaum.

Raphael, B. 1976. *The Thinking Computer: Mind Inside Matter*. San Francisco: Freeman.

Reddy, D. R., Erman, L. D., Fennel, R. D., and Neely, R. B. 1973. "The HEARSAY-II Speech Understanding System: An Example of the Recognition Process," *Proceedings of the International Joint Conference on Artificial Intelligence*, Stanford, pp. 185–194.

Reingold, E. M., and Merikle, P. M. 1990. "On the Interrelatedness of Theory and Measurement in the Study of Unconscious Processes," *Mind and Language*, **5**, pp. 9–28.

Reisberg, D., and Chambers, D. Forthcoming. "Neither Pictures nor Propositions: What Can We Learn from a Mental Image?" *Canadian Journal of Psychology*.

Richards, R. J. 1987. *Darwin and the Emergence of Evolutionary Theories of Mind and Behavior*. Chicago: University of Chicago Press.

Ristau, C. 1991. *Cognitive Ethology: The Minds of Other Animals: Essays in Honor of Donald R. Griffin*. Hillsdale, NJ: Erlbaum.

Rizzolati, G., Gentilucci, M., and Matelli, M. 1985. "Selective Spatial Attention: One Center, One Circuit, or Many Circuits?" in M. I. Posner and O. S. M. Marin, eds., *Attention and Performance XI*. Hillsdale, NJ: Erlbaum.

Rorty, R. 1970. "Incorrigibility as the Mark of the Mental," *Journal of Philosophy*, **67**, pp. 399–424.

———. 1982a. "Contemporary Philosophy of Mind," *Synthese*, **53**, pp. 323–348.

———. 1982b. "Comments on Dennett," *Synthese*, **53**, pp. 181–187.

Rosenbloom, P. S., Laird, J. E., and Newell, A. 1987. "Knowledge-Level Learning in Soar," *Proceedings of AAAI*, Los Altos, CA: Morgan Kaufman.

Rosenthal, D. 1986. "Two Concepts of Consciousness," *Philosophical Studies*, **49**, pp. 329–359.

———. 1989. "Thinking That One Thinks," ZIF Report No. 11, Research Group on Mind and Brain, Perspectives in Theoretical Psychology and the Philosophy of Mind, Zentrum für Interdisziplinäre Forschung, Bielefeld, Germany.

———. 1990a. "Why Are Verbally Expressed Thoughts Conscious?" ZIF Report No. 32, Zentrum für Interdisziplinäre Forschung, Bielefeld, Germany.

———. 1990b. "A Theory of Consciousness," ZIF Report No. 40, Zentrum für Interdisziplinäre Forschung, Bielefeld, Germany.

Rozin, P. 1976. "The Evolution of Intelligence and Access to the Cognitive Unconscious," *Progress in Psychobiology and Physiological Psychology*, **6**, pp. 245–280.

———. 1982. "Human Food Selection: The Interation of Biology, Culture and Individual Experience" in L. M. Barker, ed., *The Psychobiology of Human Food Selection*. Westport, CT: Avi Publishing Co.

Rozin, P., and Fallon, A. E. 1987. "A Perspective on Disgust," *Psychological Review*, **94**, pp. 23–47.

Russell, B. 1927. *The Analysis of Matter*. London: Allen and Unwin.

Ryle, G. 1949. *The Concept of Mind*. London: Hutchinson.

———. 1979. *On Thinking*, ed. K. Kolenda. Totowa, NJ: Rowman and Littlefield.

Sacks, O. 1985. *The Man Who Mistook His Wife for His Hat*. New York: Summit Books.

———. 1989. *Seeing Voices*. Berkeley: University of California Press.

Sandeval, E. 1991. "Towards a Logic of Dynamic Frames" in K. M. Ford and J. Hayes, eds., *Reasoning Agents in a Dynamic World*. Greenwich, CT: JAI Press.

Sanford, D. 1975. "Infinity and Vagueness," *Philosophical Review*, **84**, pp. 520–535.

Sartre, J.-P. 1943. *L'Etre et le Néant*. Paris: Gallimard.

Schank, R. 1991. *Tell Me a Story*. New York: Scribners.

Schank, R., and Abelson, R. 1977. *Scripts, Plans, Goals and Understanding: An Inquiry into Human Knowledge Structures*. Hillsdale, NJ: Erlbaum.

Schull, J. 1990. "Are Species Intelligent?," *Behavioral and Brain Sciences*, **13**, pp. 63–108.

Searle, J. 1980. "Minds, Brains, and Programs," *Behavioral and Brain Sciences*, **3**, pp. 417–458.

———. 1982. "The Myth of the Computer: An Exchange," *New York Review of Books*, June 24, pp. 56–57.

———. 1983. *Intentionality: An Essay in the Philosophy of Mind*. Cambridge: Cambridge University Press.

———. 1984. "Panel Discussion: Has Artificial Intelligence Research Illuminated Human Thinking?" in H. Pagels, ed., *Computer Culture: The Scientific, Intellectual, and Social Impact of the Computer*. Annals of the New York Academy of Sciences, **426.**

———. 1988a. "Turing the Chinese Room," in T. Singh, ed., *Synthesis of Science and Religion, Critical Essays and Dialogues*. San Francisco: Bhaktivedenta Institute, 1988.

———. 1988b. "The Realistic Stance," *Behavioral and Brain Sciences,* **11,** pp. 527–529.

———. 1990a. "Consciousness, Explanatory Inversion, and Cognitive Science," *Behavioral and Brain Sciences,* **13,** pp. 585–642.

———. 1990b. "Is the Brain's Mind a Computer Program?" *Scientific American,* **262,** pp. 26–31.

Selfridge, O. 1959. "Pandemonium: A Paradigm for Learning," *Symposium on the Mechanization of Thought Processes,* London: HM Stationery Office.

———. Unpublished. *Tracking and Trailing.*

Sellars, W. 1963. "Empiricism and the Philosophy of Mind," in *Science, Perception and Reality*. London: Routledge & Kegan Paul.

———. 1981. "Foundations for a Metaphysics of Pure Process," (the Carus Lectures) *Monist,* **64,** pp. 3–90.

Shallice, T. 1972. "Dual Functions of Consciousness," *Psychological Review,* **79,** pp. 383–393.

———. 1978. "The Dominant Action System: An Information-Processing Approach to Consciousness" in K. S. Pope and J. L. Singer, eds., *The Stream of Consciousness*. New York: Plenum, pp. 148–164.

———. 1988. *From Neuropsychology to Mental Structure*. Cambridge: Cambridge University Press.

Sharpe, T. 1977. *The Great Pursuit*. London: Secker and Warburg.

Shepard, R. N. 1964. "Circularity in Judgments of Relative Pitch," *Journal of the Acoustical Society of America,* **36,** pp. 2346–2353.

Shepard, R. N., and Cooper, L. A. 1982. *Mental Images and Their Transformations*. Cambridge, MA: MIT Press/A Bradford Book.

Shepard, R. N., and Metzler, J. 1971. "Mental Rotation of Three-Dimensional Objects," *Science,* **171,** pp. 701–703.

Shoemaker, S. 1969. "Time Without Change," *Journal of Philosophy,* **66,** pp. 363–381.

———. 1975. "Functionalism and Qualia," *Synthese,* **27,** pp. 291–315.

———. 1981. "Absent Qualia are Impossible — A Reply to Block," *Philosophical Review,* **90,** pp. 581–599.

———. 1988. "Qualia and Consciousness," Tufts University Philosophy Department Colloquium.

Siegel, R. K., and West, L. J., eds. 1975. *Hallucinations: Behavior, Experience and Theory*. New York: Wiley.

Simon, H. A., and Kaplan, C. A. 1989. "Foundations of Cognitive Science," in

Posner, ed., *Foundations of Cognitive Science*. Cambridge, MA: MIT Press.

Smolensky, P. 1988. "On the Proper Treatment of Connectionism," *Behavioral and Brain Sciences*, **11**, pp. 1–74.

Smullyan, R. M. 1981. "An Epistemological Nightmare" in Hofstadter and Dennett, 1981, pp. 415–427, reprinted in Smullyan, 1982, *Philosophical Fantasies*, New York: St. Martin's Press.

Smythies, J. R. 1954. "Analysis of Projection," *British Journal of Philosophy of Science*, **5**, pp. 120–133.

Snyder, D. M. 1988. "On the Time of a Conscious Peripheral Sensation," *Journal of Theoretical Biology*, **130**, pp. 253–254.

Sperber, D., and Wilson, D. 1986. *Relevance: A Theory of Communication*. Cambridge, MA: Harvard University Press.

Sperling, G. 1960. "The Information Available in Brief Visual Presentations," *Psychological Monographs*, **74**, No. 11.

Sperry, R. W. 1977. "Forebrain Commissurotomy and Conscious Awareness," *The Journal of Medicine and Philosophy*, **2**, pp. 101–126.

Spillman, L., and Werner, J. S. 1990. *Visual Perception: The Neurophysiological Foundations*. San Diego: Academic Press.

Spinoza, B. 1677. *Essay on the Improvement of the Understanding* (J. Katz, translator).

Stafford, S. P. 1983. "On The Origin of the Intentional Stance," Tufts University Working Paper in Cognitive Science, CCM 83-1.

Stalnaker, R. 1984. *Inquiry*. Cambridge, MA: MIT Press/A Bradford Book.

Stix, G. 1991. "Reach Out," *Scientific American*, **264**, p. 134.

Stoerig, P., and Cowey, A. 1990. "Wavelength Sensitivity in Blindsight," *Nature*, **342**, pp. 916–918.

Stoll, C. 1989. *The Cuckoo's Egg: Tracking a Spy Through the Maze of Computer Espionage*. New York: Doubleday.

Straight, H. S. 1976. "Comprehension versus Production in Linguistic Theory," *Foundations of Language*, **14**, pp. 525–540.

Stratton, G. M. 1896. "Some Preliminary Experiments on Vision Without Inversion of the Retinal Image," *Psychology Review*, **3**, pp. 611–617.

Strawson, G. 1989. "Red and 'Red,' " *Synthese*, **78**, pp. 193–232.

Strawson, P. F. 1962. "Freedom and Resentment," *Proceedings of the British Academy*, reprinted in P. F. Strawson, ed., *Studies in the Philosophy of Thought and Action*. Oxford: Oxford University Press, 1968.

Taylor, D. M. 1966. "The Incommunicability of Content," *Mind*, **75**, pp. 527–541.

Thompson, D'Arcy W. 1917. *On Growth and Form*. Cambridge: Cambridge University Press.

Thompson, E., Palacios, A., and Varela, F. In press. "Ways of Coloring," in *Behavioral and Brain Sciences*.

Tranel, D., and Damasio, A. R. 1988. "Non-conscious Face Recognition in Patients with Face Agnosia," *Behavioral Brain Research*, **30**, pp. 235–249.

Tranel, D., Damasio, A. R., and Damasio, H. 1988. "Intact Recognition of Facial Expression, Gender, and Age in Patients with Impaired Recognition of Face Identity," *Neurology*, **38**, pp. 690–696.

Treisman, A. 1988. "Features and Objects: The Fourteenth Bartlett Memorial Lecture," *Quarterly Journal of Experimental Psychology*, **40A**, pp. 201–237.

Treisman, A., and Gelade, G. 1980. "A Feature-integration Theory of Attention," *Cognitive Psychology*, **12**, pp. 97–136.

Treisman, A., and Sato, S. 1990. "Conjunction Search Revisited," *Journal of Experimental Psychology: Human Perception and Performance*, **16**, pp. 459–478.

Treisman, A., and Souther, J. 1985. "Search Asymmetry: A Diagnostic for Preattentive Processing of Separable Features," *Journal of Experimental Psychology: General*, **114**, pp. 285–310.

Turing, A. 1950. "Computing Machinery and Intelligence," *Mind*, **59**, pp. 433–460.

Tye, M. 1986. "The Subjective Qualities of Experience," *Mind*, **95**, pp. 1–17.

Uttal, W. R. 1979. "Do Central Nonlinearities Exist?" *Behavioral and Brain Sciences*, **2**, p. 286.

Van der Waals, H. G., and Roelofs, C. O. 1930. "Optische Scheinbewegung," *Zeitschrift für Psychologie und Physiologie des Sinnesorgane*, **114**, pp. 241–288, **115** (1931), pp. 91–190.

Van Essen, D. C. 1979. "Visual Areas of the Mammalian Cerebral Cortex," *Annual Review of Neuroscience*, **2**, pp. 227–263.

van Gulick, R. 1988. "Consciousness, Intrinsic Intentionality, and Self-understanding Machines," in Marcel and Bisiach, 1988, pp. 78–100.

———. 1989. "What Difference Does Consciousness Make?" *Philosophical Topics*, **17**, pp. 211–230.

———. 1990. "Understanding the Phenomenal Mind: Are We All Just Armadillos?" presented at the conference "The Phenomenal Mind — How Is It Possible and Why Is It Necessary?" Zentrum für Interdisziplinäre Forschung, Bielefeld, Germany, May 14–17.

van Tuijl, H. F. J. M. 1975. "A New Visual Illusion: Neonlike Color Spreading and Complementary Color Induction between Subjective Contours," *Acta Psychologica*, **39**, pp. 441–445.

Vendler, Z. 1972. *Res Cogitans*. Ithaca: Cornell University Press.

———. 1984. *The Matter of Minds*. Oxford: Clarendon Press.

von der Malsburg, C. 1985. "Nervous Structures with Dynamical Links," *Berichte der Bunsen-Gesellschaft für Physikalische Chemie*, **89**, pp. 703–710.

von Uexküll, J. 1909. *Umwell und Innenwelt* der Tiere. Berlin: Jena.

Vosberg, R., Fraser, N., and Guehl, J. 1960. "Imagery Sequence in Sensory Deprivation," *Archives of General Psychiatry*, **2**, pp. 356–357.

Walton, K. 1973. "Pictures and Make Believe," *Philosophical Review*, **82**, pp. 283–319.

———. 1978. "Fearing Fiction," *Journal of Philosophy*, **75**, pp. 6–27.
Warren, R. M. 1970. "Perceptual Restoration of Missing Speech Sounds," *Science*, **167**, pp. 392–393.
Wasserman, G. S. 1985. "Neural/Mental Chronometry and Chronotheology," *Behavioral and Brain Sciences*, **8**, pp. 556–557.
Weiskrantz, L. 1986. *Blindsight: A Case Study and Implications*. Oxford: Oxford University Press.
Weiskrantz, L. 1988. "Some Contributions of Neuropsychology of Vision and Memory to the Problem of Consciousness," in Marcel and Bisiach, 1988, pp. 183–199.
———. 1989. Panel discussion on consciousness, European Brain and Behavior Society, Turin, September 1989.
———. 1990. "Outlooks for Blindsight: Explicit Methodologies for Implicit Processes" (The Ferrier Lecture), *Proceedings of the Royal Society London*, B **239**, pp. 247–278.
Welch, R. B. 1978. *Perceptual Modification: Adapting to Altered Sensory Environments*. New York: Academic Press.
Wertheimer, M. 1912. "Experimentelle Studien über das Sehen von Bewegung," *Zeitschrift für Psychologie*, **61**, pp. 161–265.
White, S. L. 1986. "The Curse of the Qualia," *Synthese*, **68**, pp. 333–368.
Whiten, A., and Byrne, R. 1988. "Toward the Next Generation in Data Quality: A New Survey of Primate Tactical Deception," *Behavioral and Brain Sciences*, **11**, pp. 267–273.
Wiener, N. 1948. *Cybernetics: or Control and Communication in the Animal and the Machine*. Cambridge: Technology Press.
Wilkes, K. V. 1988. *Real People*. Oxford: Oxford University Press.
Wilsson, L. 1974. "Observations and Experiments on the Ethology of the European Beaver," *Viltrevy, Swedish Wildlife*, **8**, pp. 115–266.
Winograd, T. 1972. *Understanding Natural Language*. New York: Academic Press.
Wittgenstein, L. 1953. *Philosophical Investigations*. Oxford: Blackwell.
Wolfe, J. M. 1990. "Three Aspects of the Parallel Guidance of Visual Attention," *Proceedings of the Cognitive Science Society*, Hillsdale, NJ: Erlbaum, pp. 1048–1049.
Yonas, A. 1981. "Infants' Responses to Optical Information for Collision" in R. N. Aslin, J. R. Alberts, and M. R. Peterson, eds., *Development of Perception: Psychobiological Perspectives, Vol. 2: The Visual System*. New York: Academic Press.
Young, J. Z. 1965a. "The Organization of a Memory System," *Proceedings Royal Society London [Biology]*, **163**, pp. 285–320.
———. 1965b. *A Model of the Brain*. Oxford: Clarendon.
———. 1979. "Learning as a Process of Selection," *Journal of the Royal Society of Medicine*, **72**, pp. 801–804.
Zajonc, R., and Markus, H. 1984. "Affect and Cognition: The Hard Interface"

in C. Izard, J. Kagan, and R. Zajonc, eds., *Emotion, Cognition and Behavior.* Cambridge: Cambridge University Press, pp. 73–102.

Zeki, S. M., and Shipp, S. 1988. "The Functional Logic of Cortical Connections," *Nature,* **335,** pp. 311–317.

Zihl, J. 1980. " 'Blindsight': Improvement of Visually Guided Eye Movements by Systematic Practice in Patients with Cerebral Blindness," *Neuropsychologica,* **18,** pp. 71–77.

———. 1981. "Recovery of Visual Functions in Patients with Cerebral Blindness," *Experimental Brain Research,* **44,** pp. 159–169.

The author gratefully acknowledges permission to reprint the following:

Figure 2.3: © 1969 Harvey Comics Entertainment, Inc.

Figure 2.4: © 1975 Sidney Harris — *American Scientist* magazine.

Figure 4.1: From *Shakey the Computer*. Nils Nilsson. Copyright © 1984. SRI International. Reprinted with permission.

Figure 4.3: From *The Thinking Computer: Mind Inside Matter*. Bertram Raphael. Copyright © 1976 by W. H. Freeman and Company. Reprinted with permission.

Figure 5.1: From *A Brief History of Time*. Copyright © 1988 by Stephen W. Hawking. Bantam, Doubleday, Dell Publishing Group, Inc. Reprinted with permission.

Figures 5.5 and 5.6: From "Mathematical Games" by Martin Gardner. *Scientific American*, December 1981. Reprinted with permission.

Figure 5.7: From *Foundations of Cyclopean Vision*. Bela Julesz. Copyright © 1971 by University of Chicago Press. Reprinted with permission.

Figure 7.4: Based on MacroComputer, an interactive computer simulation of a computer, developed by Steve Barney at the Curricular Software Studio, Tufts University.

Figure 8.1: From *Speaking: From Intention to Articulation*. Willem J. M. Levelt. Copyright © 1989 by Massachusetts Institute of Technology. Reprinted with permission.

Figure 9.1: From *The Architecture of Cognition* by John R. Anderson, Cambridge, Mass.: Harvard University Press, Copyright © 1983 by the President and Fellows of Harvard College. Reprinted with permission.

Figure 10.1: From *Brainstorms* by Daniel Dennett. Copyright © 1978 by Bradford Books, Publishers. Published by MIT Press. Reprinted with permission.

Figure 10.7: Drawing by Gahan Wilson © 1990 by The New Yorker Magazine, Inc. Reprinted with permission.

Figure 11.3: From *Brain Mechanisms in Sensory Substitution*. Paul Bach-y-Rita. Copyright © 1972 Academic Press, Inc. Reprinted with permission.

Figure 11.4: From *Nature*, vol. 221, pp. 963–964. Copyright © 1969 Macmillan Magazines Ltd. Reprinted with permission.

INDEX

Abelson, R., 258, 437, 486
abortion, 455
aboutness, 333–4, 365–66, 371 (*see also* intentionality)
Abrahamson, A., 269, 470
absence (of *petit mal* seizure), 355–6
absence of representation vs. representation of absence, 359
absolutism, 421, 454
abstracta, 95, 367
abstraction, self as, 413–6, 418, 420, 426–7
abuse, sexual, 420
acceleration, 35
access, 229
  immediate, 319
  less than optimal, 195
  limited, 374
  privileged, 65, 68, 123, 246–7
  Shakey's versus ours, 93, 309
accumulator, 214, 235, 236
ACT*, 264–9, 271, 280
acte gratuit, 164
action, intentional, 31, 251–2
activation threshold, 166
adaptation to inverted vision, 393–5, 397–8
aesthetic judgment, 60
affect, 45, 50
  and color, 381–9, 393–4
agent, rational, 76
AI (Artificial Intelligence), 39, 58, 92, 184, 189, 216, 218, 222, 226, 242, 255–6, 258, 261–2, 264, 268, 280, 439, 452
  critics of, 270
  strong, 435, 440, 454
aiming, 333–4
Akins, K., x, 179, 251n, 376, 382, 441–3
aleatoric novel, 5n
algorithm, 226
algorithmic level, 276–7
allegory, 84, 93
Allison, M., 248
Allman, J., 134, 469
Allport, A., 259, 274n, 276, 469
amnesia, 250, 395
amusement, as intervening variable, 65
amygdala, 41
analysis-by-synthesis, 12
"And then what happens?", 255, 263–75
Anderson, J., 259, 264–7, 469
Anderson, Jane, x
Andler, D., x
anesthesia
  feigning, 40
  local, 154
animal experimentation, 454
animal consciousness, 17, 62, 72–3, 442–51
animate, 32
anosognosia, 78, 250, 355, 357–8
Anscombe, G., 315n, 333, 469
anterior cingulate, 107

anthropology, 82–5, 95–6, 254–5, 258
anticipation, 177–9, 182, 187
antimatter, 36
Anton's syndrome, 78, 358–9, 469
Antony, M., x
aphasia, 248–9, 309
apparent motion, 134, 338
appearance versus reality, 116–7, 131–2
appreciation, 31, 33
apprehension, inner, 304
  span of, 140
archive file, 349
Aristotle, 192n
arithmetical operation, 214
Arnauld, A., 321–2, 469
articulation, extra level of, 304
artifact design versus natural design, 175
Artificial Intelligence. *See* AI
as if, 366–7
ASCII code, 236
assembly language, 302
association, 225–6, 387–8, 442
attention, 224–5, 384
  capture, 189
auditory barber pole, 69
author, 365, 367, 418
  apparent, 81
  authority of, 82, 96, 245, 304, 317
  intentions of, 81, 84
  of record, 228
  of speech acts, 78, 229–30
automatist, 256
automaton, 32, 73, 256, 323, 325, 416
autophenomenology, 97
auto-stimulation. *See* self-stimulation
aviary, Plato's, 223–4, 226, 266, 270, 279–80, 301
awareness line, 246

Baars, B., 257–8, 271, 274, 278–9, 345, 357, 470
Bach, J., 387–8, 441–2
Bach-y-Rita, P., x, 339–42, 470
background of consciousness, 336–7, 354, 362, 408, 468
  as context, 278–9
Baldwin Effect, 184–7, 190, 197, 199, 208
Ballard, D., 189, 470
ballistic
  actions, 145
  events unreportable, 337
  nature of saccades, 361
band, little in the brain, 48
bandwidth, 6, 342n
Barnum, P. 244n
baseball, 248
bat, 441–8
baud rate, 342, 359
Baudot, J., 342n
Beattie, J., x
beaver, 189, 415–6, 418
Bechtel, W., 269, 470
beer, as an acquired taste, 395–6
Beethoven, L., 31, 48, 202
beetle in the box, 390
behavioral criteria, 406
behaviorism, 40, 70, 183, 462
  barefoot, 71, 156
belief, 77, 194n, 310
  environment, 452–3
  expression of, 78, 85, 94, 98, 130–1, 305, 317
  fixation, 260
  mistaken expression of, 85
Bennett, J., 194, 379, 449, 470
Berkeley, G., 55, 66
Bick, P., 252n, 470
Bieri, P., x, 282, 470
binding problem, 119, 257–8
biofeedback, 332
birdwatching, 336
Birnbaum, L., 242–3, 470
Bisiach, E., x, 258, 360, 471
bit (binary digit), 219, 342n
  defined, 214
  bit-map, 243, 254, 256, 291–2, 295–6, 349, 354
black box, of consciousness, 171, 227
black hole, 36, 102n, 171
  as theoretical entity, 71
blackboard architecture, 264–5
blind spot, 323–4, 344–5, 349, 355–7
blindness, 56, 272, 339–42
  denial, see Anton's syndrome
  hysterical, 78, 326–7
blindsight, 248, 322–33, 337–9, 343, 358–9, 440, 467–8
  in bats, 444
  training in, 331
Block, N., x, 133n, 399, 435, 439, 470
Bloom, P., 200n, 484
Boghossian, P., 461, 471
Borges, J., 196n, 471
Borgia, 415

Booth, W., 73n, 471
Boss, 229–30, 233, 252, 261, 424, 431, 457
  in prefrontal cortex, 274–5
  in the thalamus, 274
"bottom-up" vs. "top-down," 12
boundary, 413–4, 416–7, 452–3
  between me and the outside world, 108, 174, 176
  biological, 174, 176
  luminance, 69, 468
bowerbird, 415–6
boxology, 270n, 358n
Boyle, R., 371
bracketing, 44
brain
  does it think?, 29
  as a machine, 31
  as a mind, 16, 33
  in a vat, 3–4, 7, 17, 440
  brain-pearl, 423–4, 430, 367–8 (*see also* mind-pearl)
  giant electronic, 215, 218
Brainstorm (film), 390–1
Braitenberg, V., 171, 179, 471
Breitmeyer, B. 141, 471
bridge, 300
British Empire, 146–7, 162, 169
Broad, C., 402–3, 470
broadcasting in the brain, 196, 257, 278
Bronowski, J., 209
Brooks, B. A., 362, 471
buffer
  memory, 218, 221
  visual, 286
bureacracy of homunculi, 235–6, 238, 240, 243, 251, 261, 429
Byrne, R., 194, 446, 471, 490

CAD system, 287, 289–92, 294, 347
CADBLIND system, 291–2, 295–7, 312, 346–7, 374, 404–5, 440
Calvanio, R., 472, 360, 472
Calvin, W., x, 114, 145, 163, 184n, 190, 225, 258, 275, 471
Campion, J., 327n, 471
Camus, A., 21–2, 471
Canli, T., x
"canned" language, 229–30, 236, 309
cantata, 387–8, 441–2
capital punishment, 454
"care about", neurons that, 262, 272, 324, 357, 392
Carnegie, D., 301
carousel projector, 167–8, 466
Carruthers, P., 327, 471
Cartesian (*see also* Descartes, R.)
  bottleneck, 106
  interactionism, 34
  materialism, 107, 119, 139, 144, 256, 320–1, 344
  Theater, 17, 39, 107, 108, 111, 113–5, 121, 126, 128, 132, 134, 137–9, 142–4, 157, 164–6, 169–71, 227, 229, 231, 253, 256–7, 279, 285–6, 292, 297, 303–5, 312, 315, 321–2, 344, 357, 364–5, 370, 393, 398–9, 422, 424, 431, 434, 440, 445, 455, 458
Casper the Friendly Ghost, 35–6
Castaneda, C., 7, 471
Castaneda, H-N., 430n, 471
cat, what it is like to be, 441
category mistake, 49
Catherine the Great, 10
causation
  mechanical, 156
  unobservable, 133
cause and effect, order of, 152
center
  of the brain (where conscious experience is presumed to occur), 39, 104, 106, 108, 111, 144–5, 164, 166, 236, 260
  of gravity, 95–6, 101, 367, 413, 418, 459
Center of Narrative Gravity, 410, 418, 427, 429–31, 439, 452–3, 447–8
Central Meaner, 228, 231, 233–4, 238, 246, 247, 252–3, 257, 303, 364–6, 422, 431, 433, 455
central processing, 260–1 (*see also* CPU)
Chambers, D., 290, 295, 485
Changeux, J-P., 184n, 472
chaotic processes, 183
Chaplin, Greta and Freda, 422–3
Cheney, D., 194, 442, 446, 472
Cherniak, C., 279n, 473
chess, 298–300
chimpanzee, 189, 194, 461
  self-recognition by, 428
Chinese Room thought experiment, 322, 435–40

Chomsky, N. 190, 231, 300
chunking, 267
Churchland, P. M., 399, 472
Churchland, P. S., 38, 154, 159–62, 472
cinéma verité, 9
clairvoyance, 127
clear and distinct conception, 282
code for color, 350, 374, 404 (*see also* color-coding)
cogito ergo sum, 67
cognitive impenetrability, 260
cognitive science, 17, 39, 156, 235, 256–7, 260, 401
Cohen, L. D., 61, 472
Cole, P., 410n
Cole, D., 388, 394, 472
Collins, G., 242, 243, 470
Collins, W., 362, 484
collision-detection, 181
color, 345, 372–85, 454, 460–1
  blindness, 379
  -by-numbers, 347–8, 352–3, 404
  cells that care about, 272
  -coding, 293, 376–7, 385
  discrimination by brain, 134, 148
  experience of, 28
  phi, 114–5, 120–2, 126–7, 133, 136, 140, 156, 364
  solid, 350
  vision, 30, 55, 79, 179, 350–3, 398–401
combinatorial explosion, 5–6, 15
commissurotomy, 423, 426
communication, 239n
  within the brain, 198, 210, 274, 281, 292–4, 301, 312–3, 316, 320, 324–5
  verbal, importance in experiments, 73–4
competition in the brain, 238, 275, 336, 357, 440, 449 (*see also* evolution; Winner Take All network)
comprehension, phenomenology of, 56–58
compression algorithm, 349
computer, 146
  asynchronous, 146
  speaking, 76, 78, 92–4, 146, 211–20, 229, 435–40
Conan Doyle, A., 81, 408–9
concentration, 277
conception vs. imagination, 282
concepts, phenomena that depend on their, 24
Conceptualiser, 233, 234, 236, 238, 240, 241n, 251
conditional branching, 214, 264
confabulation, 94, 250
conflict resolution, 266
  in prefrontal cortex, 275
connection-strength, as microhabit, 219
connectionism, 176n, 239, 268–70
conscious policy, 329–30, 332
consciousness, *passim* (*see also* stream of consciousness)
  as a mode of action, not a subsystem, 166
  of animals, 17, 62, 72–3, 441–8
  concepts of, 23
  discontinuity of, 356, 366, 423
  hidden structure of, 433–4
  unity of, 108
constraint satisfaction, 239
construct, logical, 82
content, 128, 354, 365, 457–8 (*see also* intentionality)
  determinate, 236, 241, 315
  of experience, 68, 142
  -fixation, 365, 456, 431
  indeterminate, 247
  -passing by neurons, 262
  qualitative, 372
  states individuated by, 319
  vehicles of, 132
contention scheduling, 189
control, 164, 169, 180, 187–8 (*see also* self-control)
  flow of, 236
  usurped, not delegated, 241
Cooper, L. 94, 354n, 486
corpse, treatment of, 452–3
corpus callosum, 423
cortical analyzer, 357
counterintuitiveness, 37–8, 103, 227, 413, 430
Cowey, A., 325, 488
CPU (central processing unit), 219–20, 235
Crane, H., 69n, 472
creativity, 245, 298
  in language use, 467
Crick, F., 166, 256–7, 274, 472
CRT (cathode ray tube), 211, 216, 286, 289, 291, 294, 297, 303, 467
CT (computer aided tomography), 324
cue. *See* prompt
curiosity, 16, 210

Dahlbom, B., x
Damasio, A., 69n, 472
Damasio, H., 69n, 472
Danchin, A., 184n, 472
Darwin, C., 172, 184n, 199, 209, 421, 460
date stamp, 147, 152
Davis, W., 72n, 472
Dawkins, M., 448–51, 473
Dawkins, R., 173, 195, 200–3, 204–5, 206–8, 415, 421, 444–5, 473
daydream, 59
de Sousa, R., 425, 472
de Kooning, W., 384
deadline, 150–3
deaf-mute, 302, 448
decision-making, 164, 241, 435
Dehaene, S., 184n, 472
deliberation, 25
demon, 189, 238–41, 252, 262–4, 266, 272, 276, 282, 299, 306, 359, 411, 439–40 (*see also* homunculus)
  evil, 3, 6, 10, 363
  production as 267
demystification, 22, 25
Dennett, D., 37n, 39, 45, 72n, 76, 77–8n, 96, 117n, 131n, 137, 140, 144, 154, 175, 175n, 183, 184n, 192, 194, 218, 225, 247, 248, 252, 260, 262, 270, 271n, 276, 280, 315n, 317n, 360, 369, 396, 407–9, 410, 420, 424, 429, 430, 436, 446, 460, 461, 473–5
depth perception, 111
Derrida, J., 411
Descartes, R., ix, 3, 6, 8, 9, 29, 30, 33–5, 37n, 41–2, 43n, 66–7, 70, 87, 104, 105, 107, 108, 126, 129, 166, 169, 220, 224, 256, 261, 281, 308, 321–2, 329, 361, 363, 423, 432, 438, 447, 449 (*see also* Cartesian)
  design stance, 276n
desire, 77, 194n, 450
deus ex machina, 25
development, as redesign, 183
dichromatic vision, in deer, 443
discipline
  of chess, 299
  of language, 300
discrimination by the brain, 134–5
  distributed, 297
  need happen only once, 127, 128, 153
  of temporal order, 153
discriminative state, 373
discriminatory crudeness, 179
disease, 24
disgust, 414
disposition, 372–3, 386–9
  reactive, 389, 391–4, 396–8, 404
  dispositional property, 375, 380, 460
DNA, 25
doing versus happening, 32
Doppler shift, 102, 444–5
Dorfman, A. 116, 132
dorsal and ventral brain systems, 181
double take, 280
drawing, 53
dream, 10–11, 14–15, 61, 239
Dreyfus, H., 270, 279–80, 475
Dreyfus, S., 270, 485
drugs, effect in hallucination, 13
dualism, 27, 33–42, 105, 126, 134, 156–7, 159n, 256, 344, 358–9, 438

earthquake, 24
Eccles, J., 29, 33n, 37n, 154–5, 159, 475, 485
echolocation, 441–6
Eco, U. 245, 475
ectoplasm, 36
Edelman, G., 176n, 184n, 268, 313n, 356, 366, 461, 475
editing in the brain, 112, 120, 135, 143, 153
  room, 120–22, 152, 159
  editing changes versus errata, 125, 247
Efron, R., 148, 475
ego, 17, 227
élan vital, 25
Eldredge, N., 421, 475
electromyogram, 163
eliminative materialism, 459–60
Empiricism, British, 55, 59, 66
encapsulation, 240, 260
energy, physical, 35
engineering, 73 (*see also* reverse engineering)
enjoyment, 25, 30–1, 62, 344, 355–6, 383–89, 404, 409–10
epilepsy, 355–6, 423–4
epiphenomenalism, 71, 398, 401–2, 461
epiphysis, 34 (*see also* pineal gland)
epistemic hunger, 16, 181, 325, 355–6, 358–9, 382
epistemological problem of other minds, 441, 446
epithelium, 30, 46
epoché, 44

equator, 95
Ericsson, K., 75, 475
error, 246–7, 317–19, 414
essentialism, 421
ether, 36
ethology, cognitive, 446
euthanasia, 454
Evans, G., 258, 475
evolution, 22, 62–3, 171–227, 236, 242, 277, 293, 384, 401–2, 409, 420, 431
in the brain, 183–4, 192, 239
of color vision, 377–82
cultural, 193 (*see also* meme)
of pain, 61
of selves, 413–5
speed of, 186–7
and suffering, 451
Ewert, J.-P., 187, 475
existentialism, 164
experimental method, 326
experiments, design of, 73–4, 77
expert system, 30
explaining vs. explaining away, 454–5
exploration, 450 (*see also* self-exploration)
expression, 169, 241, 247, 305 (*see also* belief, expression of)
eye-tracker, 361–2, 467–8

Fallon, A., 414, 486
FAP (fixed action pattern), 259
Faraday, M., 44
Farah, M., 94, 295n, 475, 480
Faucher, L., x
feedback, 246–7, 333, 335, 337
closing the loop, 331
feeling, 338
Feenoman, 82–5, 94, 96, 131
Fehrer, E., 141, 476
Feldman, J., 189, 470
fetch-execute cycle, 214, 226, 264–5
Feynmann, R., 293–4
fiction, 365–6, 408, 437
and fictional worlds, 78–81
grammatical, 462
interpretation of, 94
literary vs. theorist's, 95–8
theorist's, 81–2, 128, 157, 367, 429
treated as fact, 84
fictional character, self as, 411, 429
field, electromagnetic, 103n
figment, 346, 350–1, 353, 355, 359, 365–6, 370–1, 374, 434, 455
figure and ground, 335
"filling in," 123, 127–8, 158, 344–56, 365
Finke, R., 295n, 476
first person, 66–7
operationalism, 132
perspective (or point of view), 66–7, 70, 123, 125, 336
fitness, 201, 203, 205
frequency-dependent, 207
of memes, 222
Flanagan, O., x, 273, 476
Flight Simulator, 288n
Flohr, H., x, 273, 476
Fodor, J., 192n, 241, 260, 270n, 273n, 278–9, 299, 476
as cryto-Cartesian, 261
folk psychology, 303, 306, 309, 313–19
food-caching, 189
forced-choice, 327–29, 331, 336n, 467
Formulator, 234, 236–7, 240, 241n
Forster, E., 194, 245
four F's, 188
Fourier transform, 48
fovea, 54, 356 (*see also* parafoveal vision)
Fox, I., 369, 402, 476
FPD (Fractional Personality Disorder) 422–3
frame, 262
frame buffer, 349
Fraser, N., 13, 489
free will, 24–5, 164, 167, 430
French, R., 310, 476
Freud, S., 13–4, 84, 243–4, 238, 247n, 314, 476
Freudian slip, 242–3
Freyd, J., 286n, 476
"from the inside," 42
frontal lobe, 274–5, 313n
fun, 62, 450
function, 173
of consciousness, 222, 276–7
multiple, 175–6, 272–3, 277
of vision, 325
functional vs. epiphenomenal, 401–4
functionalism, 31, 270n, 325, 386, 459–60
homunculuar, 262
teleofunctionalism, 460
Turing machine, 460
Fusillo, M., 358, 477

Fuster, J., 275, 476
future, producing, 144, 176–8, 188, 278

gadget, mind as, 280
galvanic skin response, 70, 330n
Gardner, H., 322n, 476
Gardner, M., 94, 109, 110, 476
Gazzaniga, M., 94, 198, 260, 424, 476
Gelade, G., 335n, 489
Geldard, F., 142–3, 477
gene, as theoretical entity, 71
generate-and-test, 12, 13, 193
genetic engineering, 209
genotypes, 182
Gentilucci, M., 188, 485
Gert, B., 393, 477
Geschwind, N., 358, 477
ghost, 7, 419
  in the machine, 33, 260, 413, 424
Gide, A., 164, 477
Gilbert, W., 44
given, 132
goals, 266
GOFAI (Good Old Fashioned AI), 269
good and bad, 173–4, 177, 188, 207
Goodman, N., 114, 120, 127–8, 131
Goody, J., 258, 477
gorilla, 189, 194
Gould, S., 421, 477
Gouras, P., 377, 477
grain-level, 93
grammar, as source of discipline, 300–1
grammatical structure, 51
Graubard, S., 269, 477
gravity, 182 (*see also* Center of Narrative Gravity)
  center of, 95–6, 101, 368, 413, 418, 460
  effect on light, 102
Great Encephalization, 190
Greenwich Mean Time, 162
Gregory, R., 355n, 467, 485
gremlin, 403–4, 406
Grey Walter, W., 167–8, 466, 477
Grice, H., 194n, 477
Griffin, D., 446, 477
Guehl, J., 13, 489
guessing when to guess, 331–3
guitar, 49

habit, 442
  bad, of thought, 252, 316, 458
  of computer, as determined by software, 217
  microhabit, 217
  of mind, 180, 221–2, 224, 228, 254, 263
hack, odd, 225
Hacking, I., 448, 477
Hal (in *2001*), 432
hallucination, 3–18, 41, 118, 238
  auditory, 250n
  of memory, 119
Hamlet, 14
Hampl, P., 244–5, 477
hand-eye coordination circuit, 188
Handford, M., 278, 477
Hardin, C., 41n, 69n, 345–6, 350n, 375–6, 391n, 441, 477
Harman, G., 369, 399, 402, 477
Harnad, S., 162n, 163, 221, 309, 311, 477
Haugeland, J., 268, 478
Hayes, P., 258, 478
Hayes-Roth, B., 264, 478
Headquarters, 106, 145, 164, 174, 176, 417
hearing, 47–52
hearsay, as model of access, 316
Hebb, D., 176n, 209, 460, 478
Heeschen, W. 249n
hemianopia, 323
hemineglect, 356–8
hen, preferences of, 450–1
heterophenomenological world, 85, 93–6, 98, 120, 124, 126, 156, 228–9, 247, 294, 298, 319–20, 327, 366, 389, 407, 411, 447–9, 460
heterophenomenology, 72–9, 83, 210, 230, 263, 326, 360, 365, 396, 418, 434, 452, 458
  neutrality of, 72, 95–8
  of animals, 444–53
heuristics, 280
hidden line removal, 288
hide the thimble, 333–5
high fidelity recording and playback, 48, 50
higher order intentional state, 306–7, 313
higher order thought, 314–5, 318, 332–3
hilarity, 63–4
  compared to awfulness, 449
Hilbert, D., x
Hilbert, D. R., 375, 478
Hintikka, J., 457–8, 478

Hinton, G., 185, 478
Hobbes, T., 129, 478
Hoffman, R., 252n, 478
Hofstadter, D., x, 222, 238, 309–10, 410n, 436n, 441, 478
Hogeweg, A., 175
Holcomb, P., x
holism, 270
Holland, J., 184n, 478
Holmes, Sherlock, 79, 81, 130, 408–9
Holtzmann, J., 94, 480
Holyoak, K., 184n, 478
hominid, 189–90, 194–5, 197, 215
homo ex machina, 87
Homo sapiens, 190, 194, 202, 209, 260, 274, 416
homogeneity of properties, 371–2
homunculus, 14, 15, 87, 91, 94, 176, 228, 250, 259–62, 324, 355, 357, 429, 455
  parafoveal sentry as, 360–2
  traffic cop, 267
Honderich, T., 154, 175, 478
horizon of simultaneity, 106
Howell, R., 79, 478
Hu-Min, Ji, 437n
Hubel, D., 40, 134, 481
Hughlings Jackson, H., 194, 479
Hume, D., 37n, 55, 66, 133, 182, 201, 412–3, 479
Humphrey, N., x, 32, 51, 178, 222, 376–7, 384–5, 420, 424, 428, 479
Hundert, E., 344, 479
hurt, why pains, 61
Husserl, E., 44, 280
Huxley, T., 401–2, 404–5, 479
hypnogogic revery, 300
hypnotic suggestion, 61
hypothesis testing, 12, 13
hysterical blindness, 326–7

"I," 426
IBM-PC, 219–20, 288n
idea
  Descartes's and Hume's concept of, 37n
  of red, 371, 373–4
  simple, 201
idealization, 458–9
identity theory, 155, 158, 459–60
IF-THEN primitive, 264–5, (*see also* conditional branching)
illusion
  benign, 365
  color, 375
  of consciousness, 432–3, 435, 455
  user, 216, 220, 311–2
illusionist, 9, 10, 13 (*see also* magic, trick)
image, 56–7
  mental, 85, 93–4, 130–1, 163, 229, 254, 298, 349, 354n
  real, 91–2
  reality of mental, 459–60
  rotation of, 285–97, 354n
  Shakey's, 87–95, 309–11
imagination, 27, 28, 49, 60, 366, 397, 431, 448
  blockage, 388
  crutch for, 290, 320, 439
  failure of, 17, 52, 281–2, 303, 370, 399–401, 432–5, 438–40, 450
  how constructed, 55
  limits of, 441
  role in fun and suffering, 450
  stretching, 16, 85, 253, 282
Immanence Illusion, 360
immortality, 430
immune system, 174
immunology, 192–3n
impasse, 267
Impressionism, 44, 55
incorrigibility, 65, 67, 319
indeterminacy of fictional worlds, 81
ineffability, 49–50, 96–7, 373, 382–3, 388
infallibility
  of introspection, 67
  Papal, 83, 96
infinite regress
  how to stop, 318
  of Meaners, 239
information
  how little in blindsight, 326–7, 330, 342–59
  measurement, 6 (*see also* bandwidth, baud rate; bit)
  transmission and processing in brain, 26–7, 144, 147, 342n, 433
  our capacity to use, 16
  passing from outside to inside, 55
  processing, in bat, 444
informational complexity, 376, 401
informavore, 181
informing, neutral, 178

innate
mechanisms and structures, 178, 182, 220, 254, 259, 262, 388
bias in quality space, 393
knowledge, 266
mechanism for language, 190, 200, 220, 300
tendency, 32
inner eye, 316
Inner I, 304
inner light, consciousness as, 447
instruction register, 214, 226
instrumentalism, 459–60
intention
authorial, 84, 245
communicative, 194, 233, 238–44, 246, 248, 250, 317–9
conscious, 162–4
intentional
action, 31, 167
object, 82, 95, 98, 131
stance, 76–7, 194n, 276n, 458
intentionality, 76–7, 192, 333–5, 457–8
defined, 333
intrinsic, 279
interactionism, 34, 41
interpretation, 78, 97
by the brain, 111, 165
of fiction, 84
of speech, 77–8, 74–6, 228
of text, 79–80, 245–6, 365–6
of the brain's states by itself, 313n
self-, 246
intrinsic property, 64–5, 372–3, 383, 386, 388, 397, 409, 430, 434, 449–50, 457
introspection, 17, 55, 65–6, 94, 162n, 223, 230, 255, 309
limitation of, 353–4
of the self, 412
as theorizing, 67–8, 94
as involving third-order thought, 307–8
of von Neumannesque machine, 215–6
Introspectionist movement in psychology, 44, 59, 70
Introspective Trap, 360
intuition, 332, 370, 394
pump, 283, 399–400, 402, 437, 442
inverted qualia or spectrum, 389–98, 460
inverting goggles, 393, 397–8
invisibility of functional structure, 210, 220
IRM (innate releasing mechanism), 259
isotropy, 260, 279–80

Jackendoff, R., x, 131, 278, 292n, 345–6, 478
Jackson, F., 398–9, 401, 435, 440, 479
Jacob, F., 176, 479
Jacob, P., x
Jaikumar, M., x
James, W., 101, 214, 221, 228, 413
Janlert, L.-E., 293, 479
jargon, as expressing new concepts, 211
jargon aphasia, 249–50
Jarrell, R., 444, 479
Jaynes, J., 221, 260, 479
Jell-O box, torn, 376, 382–3, 401
Jerison, H., 259, 479
Johnson, D. 412n
Johnson-Laird, P., 257, 429, 480
Joslin, D., x
Joyce, J., 212
Joycean machine, 214, 219, 225–6, 229, 275–81, 447
judging and seeming, 133–4
judgment, 128, 319, 322, 344, 364–6, 372
of simultaneity or sequence, 166–7
Julesz, B., 111, 255, 480
Just So story, 194

Kafka, F., 429
Kant, I., 44, 132n, 139
Keeble, G., 385, 479
Keller, H., 227, 480
Kessel, F., 410n
Keynes, J., 298, 303
Kim, J., x
kinesthesia, 5, 46–7
kinks, in subjective sequence, 136, 154
Kinsbourne, M., x, 140n, 166, 181, 249, 251n, 271, 357, 424, 480
Kirman, B., 61, 480
Kissinger, H., 81
Kitcher, P., 369, 480
kludge, 211, 225
knowledge level, 276n
Koch, C., 166, 255–6, 472
Koestler, A., 33n, 480
Kohler, I., 393, 480
Kolers, P., 114–5, 120–1, 127–8, 140, 480
Korsakoff's syndrome, 250
Kosslyn, S., 94, 259, 286, 292n, 294, 297, 313n

label, 292, 352–5
Lackner, J., 69n, 480
Laffer Curve, 109
Laird, J., 145n, 264, 267, 269, 270n, 484
Lamarckianism, 186, 208
Land, E., 40–1
landscape, adaptive, 184–7
Langton, C., 175, 480
language, 227, 373
  assembly, 302
  creativity of, 298
  effect on brain's structure and competence, 200, 207–8, 210, 302
  evolution of, 190, 195, 200
  innate mechanisms of, 190, 200, 220, 300
  machine, 220, 302
  natural, as programming language, 302
  of thought, 302–3, 365
  perception vs. production, 230
  production, 304 (*see also* speech production)
  programming, 235
  and proto-language, 194
  role in consciousness, 17, 225, 300–1, 446–7
  role in creating self, 416–8
  role in virtual machine, 221
  understanding, 69
Larkin, J., 295, 480
laterality test, 133n
Latto, R., 329n, 471
laughter, 63–5
Lawlor, K., x
learning, 193, 271
  as redesign, 183
  by explicit instruction, 221
  by imitation, 221
least-effort principle, 239n
leaving something out, 454–5
Ledoux, J., 260, 424, 476
left-right hemisphere distinction, 181, 198, 215, 424
Leiber, J., 43n, 302, 441, 480
Leibniz, W., 412, 480
Leibniz's Law, 459
level of explanation, 276–7
Levelt, W., x, 232–6, 239–41, 246, 249n, 300, 466–7, 480
Levine, D., 358, 471
Levy, J., 424, 480
Lewis, D., 79, 399, 481
Liberman, A., 51, 481
Libet, B., 154–67, 344, 465, 481
library, as analogy for vision, 360
Liebmann effect, 69n, 468, 481
light, speed of, 102, 106
Lincoln, A., 244, 246
linguistics, 231–2, 239n, 249
Lisp, 216, 216n, 236, 302
Livingstone, M., 134, 481
Llina, R., 177n
Loar, B., 399, 481
loci, linked, 207
Locke, J., 55, 66, 201, 371–5, 379–80, 389, 481
Lockwood, M., 372, 410n, 481
locus ceruleus, 370
Lodge, D., 410, 481
logic, of belief and knowledge, 457–8
looming, 178, 341, 397
loop, infinite, 266
Lorentz equations, 102–3n
love, 23–5
lovely and suspect qualities, 379–80
luminance, 400, 407
Luzzatti, C., 358, 470
Lycan, W., 391, 399, 482

MacDonald, R., 112, 482
machine
  language, 235, 302
  person as, 431–2
  table, 213
Macintosh, 220
MacKay, D., 159n, 482
Macko, K., 337n, 483
Madame Bovary, 80
magic, stage, 10, 126, 279–80, 361, 434, 436n (*see also* sleight of hand)
Marais, E., 416, 482
Marceau, M., 211, 216
Marcel, A., x, 248, 257, 327, 330, 482
Margolis, H., 191, 221, 223–5, 277, 345, 482
Margulis, L., 414, 482
Marilyn (Monroe), 354–5, 359–60, 364, 396, 408, 467
Markus, H., 402, 491
Marler, P., 183, 482
Marr, D., 276–7, 482
Marx, K., 227
Mary, the color scientist, 398–401
Matelli, M., 188, 485
materialism, 25, 33, 36–7, 42, 65, 106, 140, 155–6, 158, 161, 322, 398, 404–

7, 430, 438 (*see also* Cartesian materialism)
semiotic, 411
mattering, 31, 41, 173, 448–55
Maxwell, C., 44
Maxwell equations, 102n
Maynard Smith, J., 172n, 482
McClelland, J., 221, 240, 482
McConnell, J., x
McCulloch, W., 217n, 482
McGinn, C., 49n, 273, 328n, 433–5, 482
McGlynn, S., 355, 358, 482
McGuinness, E., 134, 469
McGurk, H., 112, 482
McGurk affect, 112
McLuhan, M., 384, 482
Meaner, 244, 246, 248 (*see also* Central Meaner)
meaning, 75–7, 233, 234, 239, 244–5
and proto-meaning, 178
Meizin, F., 134, 469
Mellor, H., 149, 482
melody discrimination, 466
meme, 200–10, 222, 243, 254, 261, 263, 301, 384, 386, 417
memory, 237, 240
buffer, 145n, 218
color, 374
computer's not like brain's, 220, 225
defective, 318
echoic, 145n
episodic, 278
hallucination of, 161
as library, 121, 132
long-term, 39, 270–1, 274
loss, 250 (*see also* amnesia)
random access (RAM), 213
semantic, 336n
short term, 160
working, 264–6 (*see also* workspace)
memosphere, 206, 220
mental image (see image, mental)
Mental Image Heaven, phenomenal space as, 131n
mentalese, 231, 234, 365 (*see also* language of thought)
mentalistic terms, 38–9
Menzel, E., 428, 482
Merikle, P., 125, 485
metacontrast, 141–2
meta- knowledge, 343, 438
metaphor, 84, 91–3, 130, 181, 230, 275, 286, 311–3, 455
Metzler, J., 286, 487
microhabit, 217, 219, 254
Miller, G., 181
millisecond, 103
mind
is the brain, 33
mind/brain, 39
as meme nest, 206
as pattern perceived by a mind, 309
stuff, 33–4, 36, 85, 344
mind's ear, 58, 60
mind's eye, 53, 295, 297, 310, 350, 356–7, 372, 374, 389
mind-pearl, 367–8
soul or self as, 423–4
mind-reading, scientific, 28
mind-set, 237–8, 241
Minsky, M., x, 63n, 108, 239, 242, 258, 261–3, 273, 295n, 313n, 356, 359–60, 426, 437, 483
miracle, 38, 239, 255, 455
Mishkin, M., 335n, 483
modularity, 260
money, 24
monkey
color preferences of, 385
vervet, 194
Monod, J., 173, 483
Monroe, M. *See* Marilyn (Monroe)
moral agency, 24–5
moral responsibility, 231
morality, relation to consciousness, 448–55
Morse code, 51, 342n
Mother Nature, 174–5, 178, 182–3, 260, 273, 280, 378, 381n, 414–5
motion, apparent, 134, 150, 338
detection of, 360–1
motion pictures, 103, 137, 465–6
Mountcastle, V., 262, 483
Moynihan, D., 357
MPD (Multiple Personality Disorder), 419–20, 422, 424–5
MRI (magnetic resonance imagining), 324
MT (medio-temporal cortex), 128n
Multiple Drafts Model of consciousness, 17, 111–43, 170, 227, 253–4, 258, 263, 335n, 358, 370, 392, 397, 408, 431, 455, 458
multiple function, 272–3, 277
Mysterian, New, 273
mystery, ix, 37–8, 41, 309
defined, 21

mystery (*continued*)
of consciousness, 15, 17, 18, 22, 23, 25
confronted, 281
systematic, 406
myth, 453–4

Nabokov, V., 298, 483
Nagel, T., 71, 96, 273, 372, 424, 425n, 433, 441–2, 447, 458, 461, 483
narcissism of brain design, 382
narrative, 417–8
creation of, 245
dream, 13–4
evolutionary explanation as, 172
flashbacks in, 148
fragment, 135, 255, 259
frame, 439
order of composition, 148
sequence, 136
stream, 113
observer as skein of, 137
precipitated by probes, 135–6, 143, 169
natural kind, 381n
Necker cube, 288
neglect, 355–9
benign, 357, 362
finances, 358
typo, 357 (*see also* proofreader effect)
Neisser, U., 12, 145n, 279, 483
Nemirov, L., 399, 483
Neo-Laffer curve, 110, 395
neon color spreading, 351, 363
Neumann, O., x, 180, 189, 222, 356, 482
neuroanatomy, 271n, 274
neuron, logical, 218n
"neuronal adequacy," 155, 157, 159–60
neuronal groups, 176n
neurophilosophy, 159
neuroscience, 16, 28, 39, 71, 147, 256–7, 262, 269, 376, 382
guiding assumption of, 70
neurosurgery, 58n, 154, 167
evil neurosurgeon, 391–2, 395
neutrality 73, 85
of heterophenomenology, 72, 83, 95–8
neutrino, 36
New Orleans, Battle of, 146–7, 149, 168–9
Newell, A., 145n, 257, 264, 267, 269, 270n, 276n, 483–4
Nielsen, T., 112n, 484
Nietzsche, F., 227
Nilsson, N., x, 85–6, 484
Nisbett, R., 184n, 478
NMDA receptor, 273
Nobel Prize, 41, 255
noemata, 44
Norman, D., 189, 274, 277, 484
Nottebohm, F., 183, 484
noumena, 44
Nowlan, S., 185, 478

Oakley, K., 424, 484
observation
inner, 320
knowledge without, 315n
observer, 101, 102, 130, 137, 231, 252, 256, 405
in the brain, 106, 166, 312, 314, 431
location of, 107, 125
role of, in defining some properties, 380–1
occipital cortex, 323
occurrent property, 371–2, 386, 388
olfaction, 46
ontology, 36
operationalism, 96, 126
defined, 117n
first person, 132, 133
opinion, as opposed to belief, 78n
optimality assumption, 278n
orienting response, 180, 188–9, 223
Ornstein, R., 370, 484
Orwellian vs. Stalinesque, 116–24, 126–8, 133, 142, 160–2, 164, 168, 319, 335n, 395, 440, 461–2
Otto, 230–1, 236, 241, 303–4, 309, 316–7, 319, 333, 338, 359–60, 362–8, 374, 382–3, 386–9, 402–4, 457
Oval Office in the brain, 104, 106, 416, 429
overtones, harmonic, 49, 50
Oxford, 462

Pagels, H., 163, 484
pain, 25, 29, 60–1, 64, 198, 318, 386, 448–53, 455 (*see also* suffering)
dreamed, 61
is it real?, 460
projected, 129
psychosomatic, 327
reporting, 308
Palacios, A., 350n, 376, 488
Pandemonium, 189, 222, 237–8, 240–2,

251, 253, 261, 304, 315, 336, 455, 466–7
Papert, S., 217n, 484
paradox, 17
of consciousness, 101
parallel
architecture, 210, 267
distributed processing, 176n, 268–9 (*see also* connectionism)
processing, 189, 217, 237–8, 243, 251, 253, 259, 269
in the brain, 111, 113, 115, 134–5, 210, 215–6
simulated, 218, 265–6
paranormal, 325, 419
parasite, 414
meme as, 205, 222, 254
Parfit, D., 423, 484
Partee, B., 224
Pascal, B., 302
pathology, 248–51, 274–5, 322, 335n, 356–9
pattern
visible from intentional stance, 458
-recognition 221, 265, 269, 293, 376
Pavlov, I., 120
PC-Paintbrush, 349
Pears, D., 252, 484
Penfield, W., 59n, 154, 484
Penrose, R., 36–7, 102n, 156, 430
Penseur, Le, 223–4
perceived vs. inferred motion, 123
Perceptron, 217n
Perlis, D. 278, 427, 484
perpetual motion machine, 35
Petrone, P., 358, 471
phantom body, limb, 5, 8
phenol-thio-urea, 379
phenomenal
field, 17, 52, 68
property, 322, 338, 340–1, 372, 383
space, 130, 304, 356
phenomenology, 44–5, 47, 55, 60, 65–7, 84, 96, 133, 157, 251, 256, 295, 350–1, 433–4
as the behavior of a black box, 171
of comprehension, 56
of laughter, 64
paradoxical, 123
pure and impure, 69–70
"real," 365–6
visual, 54, 55
Phenomenology (the school of philosophy), 44, 279
phenotype, 182
evolution in, 183–4
extended, 415
phi phenomenon, 114, 465
philosophers, 3, 17, 22, 37, 41, 65, 72–3, 86, 114, 137, 239n, 254–5, 279, 282, 323, 325, 369–70, 389, 399, 401, 405, 421, 433, 435–6, 440, 457–64
Philosophers' Syndrome, 401
phoneme restoration effect, 345
physical symbol system, 276n
physicalism, 400 (*see also* materialism)
physics, 36–7, 156
revolution in, 372
piano tuning, 337
Piantanida, T., 69n, 471
picture in the head, 52–4, 57–8, 68, 231, 298
pictures as objects of visual perception, 52
pineal gland, 34, 41, 104–8, 169, 257
Pinel, P., 106n
Pinker, S., 200n, 295n, 484
Pitts, W., 217n, 482
pixel, 272, 296–7, 349, 353
planning, 178, 343
opportunistic, 243
plasticity
of nervous systems, 182
phenotypic, 185–7, 193, 209, 220
of computers, 211
Plato, 223, 226–7
Plato's aviary, 222–3, 225, 266, 270, 279–80, 301
the Platonic memes, 205–6
plenum, consciousness as a, 366, 408, 467–8
point of view, 101–2, 125
created by replication, 174, 176
of observer, 176
smeared, 107, 136, 152
shift, 341
policy hinge, 335, 337
Pollatsek, A., 336n, 362, 484
poltergeist, 35
Pontifical Neuron, 228, 413
Pöppel, E., 106–8, 484
Popper, K., 29, 33n, 154–5, 159, 485
positivism, 390
postmark, 147, 152, 158

postnatal design fixing, 183, 200
Potemkin, 10
Potter, B., 426
power, not function, of mechanisms of consciousness, 277
pre-experiential vs. post-experiential, 108, 118–9, 123, 128n, 247, 394–5
precognition
by carousel, 167–8, 466
ruled out, 115
preconscious, 247n
prefrontal cortex, 107, 275
preliterate mentality, 221
presentation, 107, 133–4, 153, 169–70, 255–6, 312–3, 364, 393, 445 (*see also* re-presentation)
presentiment, 343, 364–5, 457
preverbal message, 234–5, 238, 240, 246
primary evoked potential, 157–8, 162
primitive (computer operation), 212–4, 235, 264
privileged access, 65, 68, 123, 246–7
none into the process of speech production, 97, 310
probing, 113, 135, 143, 169
no optimal time of, 136
problem, primordial, 177
production system, 265–7
program
as list of instructions, 264
as recipe, 217
as description or list of instructions, 216
of brain's virtual machine, 219
projection, 127, 147n
backwards in time, 128, 129, 131, 139, 157, 159n
in space, 129, 130, 131
Prolog, 302
prompting, role in blindsight, 327–9
proofreader effect, 345
property
dispositional, 371–3, 375, 380
occurrent, 371–2, 386
proposition, 364–5
prosopagnosia, 69
prosthetic vision, 338–42
Psychoanalysis (the party game) 10–16, 238
psychology, 39, 73–4, 254–6, 295, 401 (*see also* folk psychology)
psychomatic blindness, 327
psychophysics, 382–3
publication, as poor metaphor for consciousness, 125–6
pun, 243
Putnam, H., 280, 391, 460, 485
Pylyshyn, Z., 148n, 271n, 485

qualia, 17, 344, 358–9, 368–411, 433–4, 449, 455, 459–60
defined, 65, 338
quality, primary and secondary, 371, 373, 376, 379–82
qualophile, 386, 388, 390, 392–4
Quine, W., 381n, 484

R and D, 208–9
Raab, D., 141, 477
rabbit, cutaneous, 142–3, 156
radar, 427–9
RAM. *See* memory, random access
Ramachandran, V., 259, 276n, 353n, 467, 485
Ramberg, B., x
Ramsey, W., 269, 485
random dot stereogram, 111, 152
Raphael, B., 85–6, 485
rationality, 252
assumption of, 458
ideal, 301
minimal, 280n, 301
raw feel, 322, 372
Rayner, K., 336n, 362, 484
re-presentation, 113, 292–3, 344
Reaganomics, 109
real seeming, 134, 316, 363–4
realism, 459–61
reality of intentional objects, 83
reasoning, practical, 251–2
reasons, birth of, 173–4, 176, 414
recursion, 256, 310
Reddy, R., 264, 485
reductionism, 64, 387, 454–5, 459–60
reentrant maps, circuits, 176n, 268
"referral backwards in time," 155, 157–8
reference, causal theory of, 272n
reflection, 320, 330, 448
reflex, 32, 102, 122
register, 219
memory, 214
numbers in, used to represent, 350
rehearsal, 225, 277–8, 428
reinforcement, 221
Reingold, E., 125, 485
Reisberg, D., 290, 295, 485

relativism, 381
reminding, 197, 278, 295, 342
rendering, 291–3, 296, 355, 364, 374
Reolofs, C., 127, 489
replicator, 173–5, 201–3, 205–6
report, 229–31, 238, 252, 256, 317
  vs. express, 153, 303–9, 314–5, 318
reportability, 337
representation, 191–2
  by the immune system, 174
  in the cortex, 271
  minimal, 178
  of oneself, 417–8, 429–30
Representatives, House of, 272–3, 357
represented vs. representing, 80–1, 131, 137, 143–4, 147–9, 151, 161, 163–4, 166
res cogitans, 29,30, 33, 41, 106
reticular formation, 107, 274
retinotopic map, 262
reverse engineering, 145, 146, 277n
Richard, R., 184n, 421, 485
Ristau, C., 446, 485
Rizzolati, G., 188, 485
robot, 431, 448
  animals, 43, 452
  conscious, 432, 435
  replicator, 175
Rodin, A., 223
Roepke, M., x
Romeo and Juliet, 80
Rorty, R., 67, 461, 485
Rosenberg, J., x, 132n
Rosenberg, Julius and Ethel, 376, 382, 401
Rosenbloom, P., 145n, 264, 267, 269, 270n, 484, 486
Rosenthal, D., x, 304, 307–9, 311, 313–4, 317, 332, 486
roughly continuous representation, 349, 353, 354n
Rozin, P., 259, 384, 414, 486
Rudolph the Red-Nosed Reindeer, 388
Rumelhart, D., 221, 239, 269, 482, 485
Russell, B., 129n, 246, 261
Ryle, G., 33, 49, 223, 413, 486

saccade, 54, 111, 181, 335, 354–5, 361–2, 467–8
  saccadic suppression, 361–2
Sacks, O., 300, 322n, 448, 486
sailing, 101, 150–1
Sandeval, E., 258, 486
Sanford, D., 421, 486
Sangree, M., x
Santa Claus, 85, 131n
Sartre, J.-P., 164, 486
Sato, S., 335n, 489
Savage-Rumbaugh, S., 385n, 486
scare-quotes, 344, 346
Schacter, D., 355, 358, 482
Schank, R., 258, 437, 486
Scheerer, E., x
schizophrenia, 250n
Schossberger, C., x
Schull, J., 184, 486
scientists, 37
scope ambiguity, 244, 246
scotoma, 323–5, 330, 332, 353n, 355
script, 258, 437
sea squirt, 177
searchlight theory of attention, 274
Searle, J., 96, 270, 279–80, 322, 435–40, 458, 461, 486–7
Second Law of Thermodynamics, 205
second-intention, 437
secondary quality (*see* quality, primary and secondary)
secret, 446–7
seeing, 334–6, 338, 355
  as understanding, 56
seeming, 134, 320, 364, 366, 374, 410 (*see also* real seeming)
self, 17, 29, 163, 220, 228, 279, 410, 412–31
  as not independent of the memes it harbors, 208, 302
  as soul, 368
  biological, 414–18, 427, 447
  location of, 164–5
  -control, 222, 277–8, 417–8, 427–9, 447
  -cuing stimuli, 332
  -exploration, 210, 228, 254, 300
  -interpretation, 246, 427
  -knowledge, 448–9
  -manipulation, 209, 277, 280, 293, 316
  -monitoring, 222, 250, 320, 330
  -observation, 315
  -probing, 169
  -recognition, 427–9
  -representation, 310, 417–8, 429–30
  -stimulation, 195–9, 209, 219, 225, 275, 301–2, 450
  -transparency of mind, 301
selfishness, varieties of, 174
Selfridge, O., x, 189–90, 333n, 487

selfy, 173, 416, 447
Sellars, W., 65, 371–2, 383, 486
semantic
  analysis, 74–6
  level similarities, 80
  readiness, 135
semantics, line, 87
semiotic materialism, 411
sensa, 372
sensations, 17, 52
sense data, 322
sequence, subjective, 136, 154, 166, 169–70, 465
serial
  chaining, 222
  process, 213, 215, 219, 235, 239, 241, 252, 254, 258–9, 263
  search, 278
settling, content-sensitive, 152
sex, 178
  evolution of, 172–3
Seyfarth, R., 194, 444, 446, 472
Shakespeare, W., 80
Shakey the robot, 85–95, 97, 130, 179n, 229–31, 286, 291, 310–1
  as zombie, 309
Shallice, T., 189, 256, 274, 277, 322n, 484, 487
Sharpe, T., 413–4, 487
Shepard, R., 69n, 94, 285–9, 291, 354n, 487
Sherman, V., 183, 482
Sherrick, C., 142–3, 477
Shipp, S., 134, 491
Shoemaker, S., 369, 383, 391, 487
show vs. tell 295–7, 316
SHRDLU, 92–3
Shylock, 406
side effects, serendipitous, 175
Siegel, R., 13, 487
sign language, 300, 448
silicon vs. organic machine, 31
Simon, H., 75, 293, 475, 480, 488
simultaneity, subjective, 107, 136, 163, 165
singularity, 102n
Skinner, B., 183, 417
slang, 245
slateboard, 152, 158
sleepwalking, 31–2
sleight of hand, 128, 282, 311, 333 (*see also* magic, stage)
slip of the tongue, 317
slogan-honing, 460
Sloman, A., x
Smith, W., 424, 480
Smith, Y., 327n, 471
Smolensky, P., 269, 488
Smullyan, R., 132, 488
Smythies, J., 129, 131, 159, 488
snake-hating in primates, 385–6
Snyder, D., 137n, 488
Soar, 264, 267–69, 271, 280
software, 219–21, 455
  defined, 211
  in the brain, 190, 210
  level, 434, 439
solipsism, 403
somatosensory cortex, 154–5
soul, 32, 430, 452–3
  immortality of, 367–8
  theory of centers of gravity, 367
sound, speed of, 102, 106
sound studio, 50–1
sound track, dubbing, 112, 152
Souther, J., 335n, 489
space
  logical, 130, 131
  color solid as example, 350
  of representing versus represented, 143–4
  personal, 335–6n
  phenomenal, 130, 304, 356
  problem, 267
  quality, 381–2, 393, 396, 404
  semantic, 247, 274n
  visuo-motor, 397
Spalding, O., x
specialist systems in the brain, 180, 188, 195–6, 223, 228, 240, 253–4, 257–59, 263, 274, 277, 351–2, 360
  as generalist, 271–3
speech
  act, 76–8, 120, 169, 228, 234–5, 238–39, 247–8, 250–1, 301, 313, 315, 365, 457
  compressed, 144
  perception, 144
  production, 232, 237, 240, 248, 315
  sounds, 50, 51, 74
speed of computation, 219, 219n
Sperber, D., x, 195, 239n, 488
Sperling, G., 145n, 488
Sperry, R., 278–9, 488
spider, 415–6, 418, 441
spider walks, 465

Spillman, L., 69n, 488
Spinoza, B., 187, 488
split-brain patient, 198, 260, 423–6
Spoonerism, 232
Stafford, S., x, 32, 488
Stalinesque (*see* Orwellian vs. Stalinesque)
  Libet's model as, 157, 160, 164
Stalnaker, R., 279n, 488
Stanford Research Institute, 85
Star Wars, 432
Steinberg, L., x
stereo sound, experience of, 130
Stich, S., 269, 485
Stoerig, P., 325, 488
Stoll, C., 126, 132, 488
story-telling, 258, 301, 418, 428
Straight, H., 298, 488
Strategic Defense Initiative, 151
Stratton, G., 393, 488
Strawson, G., 461, 488
Strawson, P., 32, 488
stream of consciousness, 45, 67, 113, 135–6, 138, 144, 166, 189, 215, 225, 235, 253, 257–8, 356, 407, 435
Studdert- Kennedy, M., 51, 481
subject
  heterophenomological, 128, 131
  unified, 77–8
subjective contour, 351
subjectivism, 381
subjectivity, 132, 159, 372, 389–90
substantia nigra, 370
suffering, 449–53
Sullivan, A., 207
Superman, 25
surface of the conscious mind, 224, 313n
surface spectral reflectance, 375
suspect qualities, 379–80
symbol, movable, 270n
sympathy, 62
Systems Reply, 439

tachistoscope, 141
talking to oneself, 195–7, 222, 224, 275, 298, 407
  vs. subsystems talking to each other, 316
  what use?, 301
Taylor, D., 391, 488
teleportation, 430
television, 6
  speed of, 103
termite, 415–6
testimony of the senses, 316
tetrachromatic vision, 350n
Thagard, P., 184n, 478
thalamus, 41, 274
theater organ of the mind, 49
third-person perspective, 70–72, 96, 128, 336, 442, 445
Thompson, D'A., 171
Thompson, E., x, 350n, 370, 376–7, 488
thought, 17, 317
  as opposed to belief, 307, 308
  as talking to oneself, 59
  imageless, 59
  unconscious, 308
  thinking in thoughts, 298
  higher order, 314–5, 318, 332–3
thought experiment, 4, 16–7, 48, 323, 367, 391–2, 397–8, 400, 435–6, 440 (*see also* intuition pump)
Thumbnail Sketch, 253–4, 256–63
Tibetan Prayer Wheel, 281–2
Tiegs, C., 63n
time
  fingertip, 162
  how represented in brain, 144–53
  pressure, 144, 150–1
  real, 127, 149
  window, 119, 144, 151–2
timing
  absolute, 161–2, 164–6, 168
  of brain events, 113, 153, 165
  of conscious events, 113, 124
  when it determines content, 150
Titchener, W., 44
touch, 47
  keeping in, 334
track, keeping, 293
tracking, 177, 191, 333–5
training
  of the brain, 219
  palate, 396
  ear, 337
Tranel, D., 69n, 489
transcription of speech sounds, 74–6
Treisman, A., 335–6n, 489
Trevarthen, C., 424, 480
trichromatic color vision, 350n, 377
trick
  of biological design, 32
  conjuring, 126, 462 (*see also* magic, stage; sleight of hand)

trick (*continued*)
Good Trick, 184–6, 190, 197, 199–200, 208, 225, 378
hallucination as brain playing a trick, 8
learning a new, 189
heuristics as a bag of tricks, 280
playing tricks with time, 155
Truman, H., 32, 429
Turing, A., 210–3, 214, 216, 218, 235, 263, 265, 310
Turing machine, 212, 265
Universal, 211, 216
Turing test, 310–1, 435–9
Tye, M., 399, 489

unconscious, 326
cognition, 281
control, 329
driving, 137
goals, 243
higher-order thoughts, 307–9, 311
or automated policies, 329–30
processes, 26
production of language, understanding, 97
slander, 84
thought, 308
unconsciously initiated action, 160, 163
understanding, 55–8, 279, 438–9
language, 69
where does it happen?, 321, 434
Ungerleider, L., 335n, 483
User, library, 360 (*see also* illusion, user)
Uttal, W., 147n, 489

Valéry, P., 176, 423
Vallar, G., 360, 470
van der Heiden, L., x
Van der Waals, H., 127, 489
Van Essen, D., 134, 489
van Gulick, R., x, 280, 327, 399, 489
Van Hoesen, G., 69n, 472
van Tuijl, H., 351, 489
Van Voorhis, A., xi
Varela, F., x, 350n, 376, 488
vegetarianism, 454
Velleman, J., 461, 472
Vendler, Z., 33n, 489
verificationism, 126, 132, 390, 403, 460–1
vertical symmetry detection, 179, 188
Vesalius, 104
Vesuvius, 32
veto by the self, 164, 168
Victoria, Queen, 81
video game, 5–6
videotape, 349
virtual
captain, 228
machine, 210, 211, 216–21, 225–6, 228, 254, 258–9, 269, 281, 311–2, 431, 434–5, 437, 447, 455
programming language as, 302
object, 287
presence, 360–2
Reality, 6n
space, 288, 291
structure, 301
wire, 196
virtus dormitiva, 63–4, 386, 430
vision, 52–6, 278–9, 322–56
computer, 87–92
located inboard of the eyes, 108
parafoveal or peripheral, 54, 68, 354, 360, 362, 466–7
resolution of, 46
visualization, 294–5, 366
vitalism, 25, 282
Vivaldi, 26
von der Malsburg, C., 273, 489
von Grünau, M., 114, 120–1, 480
von Neumann, J., 210–1, 214, 215, 218n, 264
von Neumann bottleneck, 214, 235
von Neumann machine, 211, 213–9, 225–6, 235–6, 263–5, 269, 300
von Neumannesque, 210, 214, 225, 258
von Uexküll, J., 446, 489
Vorsetzer, 293–4, 297, 312, 374
Vosberg, R., 13, 489

Waller, F., 235
wallpaper, 354–5, 359–60, 364
experiments with, 467–8
Walton, K., 79, 366, 428, 490
Warhol, A., 354
Warren, R., 347, 490
Warrington, E., 249, 480
Wasserman, G., 164–5, 490
Waterhouse, L., 259
web of discourses, 411, 416
Weiner, P., x
Weinstein, S., x
Weiskrantz, L., 325, 327n, 330n, 332, 338, 490
Welch, R., 393, 490
Werner, J., 69n, 488

Wertheimer, M., 114, 490
West, L., 13, 487
West Side Story, 80
what it is like, 94, 96, 98, 189, 387, 425–6, 441–8
"where does it all come together?", 39, 107, 134–5, 165, 297, 367
White House, 164 (*see also* Oval Office)
White, S., x, 369, 490
Whiten, A., 194, 446, 471, 490
Wiener, N., 177, 490
Wiesel, T., 40
Wilkes, K., 440, 444, 490
Wilson, D., x, 195, 239n, 488
Wilsson, L., 415, 490
window
  editing, 159
  control, 169
  time, 119, 144, 151–2
wine-tasting machine, 30–1
Winner Take All network, 189
Winograd, T., 92, 490
Witness, 322, 358, 422, 431, 455
  mind stuff has a, 28–9
Wittgenstein, L., 57n, 65, 317n, 344, 390, 396n, 447, 461–3
Wolfe, J., 277, 490
Wonder Tissue, 40
Woodfield, A., x
word processor, 311–2
  brain as, 226
  computer not intended as, 212
WordPerfect, 219
WordStar, 216, 219
Wordsworth, W., 21–2
workspace, 213, 235, 256, 264, 266–7, 270, 281
  global, 257, 271
world (*see also* heterohenomenological world)
  fictional, 130
  heterophenomenological as fictional, 81
world knowledge 233, 439
writing, importance in shaping consciousness, 221
Wundt, W., 44
Wyeth, A., 384
Wynes, K., xi

Yonas, A., 178, 490
Young, J., 184n, 490–1

Zajonc, R., 402, 491
Zeki, S., 134, 491
Zihl, J., 333, 491
zimbo, 310–1
zombie, 72–3, 76, 78, 83, 95, 281–3, 302, 309–11, 313, 405–6, 450, 461
  bat as, 444–5
  blindsight subject as partial, 323–33
zombist, 256
zoology, 43–5, 66, 70